GeoGuide

Series Editors

Wolfgang Eder, Geocentre-Geobiology, University of Göttingen, Göttingen, Niedersachsen, Germany

Peter T. Bobrowsky, Geological Survey of Canada, Natural Resources Canada, Sidney, BC, Canada

Jesús Martínez-Frías, CSIC-Universidad Complutense de Madrid, Instituto de Geociencias, Madrid, Spain

Axel Vollbrecht, Geowissenschaftlichen Zentrum der Universität Göttingen, Göttingen, Germany

W0225793

The GeoGuide series publishes travel guide type short monographs focussed on areas and regions of geo-morphological and geological importance including Geoparks, National Parks, World Heritage areas and Geosites. Volumes in this series are produced with the focus on public outreach and provide an introduction to the geological and environmental context of the region followed by in depth and colourful descriptions of each Geosite and its significance. Each volume is supplemented with ecological, cultural and logistical tips and information to allow these beautiful and fascinating regions of the world to be fully enjoyed.

More information about this series at https://link.springer.com/bookseries/11638

Marc Calvet · Magali Delmas ·
Yanni Gunnell · Bernard Laumonier

Geology
and Landscapes
of the Eastern Pyrenees

A Field Guide with Excursions

 Springer

Marc Calvet
Department of Geography, Laboratoire
Histoire Naturelle de l'Homme
Préhistorique, UMR 7194
Université de Perpignan-Via Domitia
Perpignan, France

Yanni Gunnell
Department of Geography
Environnement, Ville Société, UMR
5600, Université Lumière-Lyon 2
Lyon, France

Magali Delmas
Department of Geography, Laboratoire
Histoire Naturelle de l'Homme
Préhistorique, UMR 7194
Université de Perpignan-Via Domitia
Perpignan, France

Bernard Laumonier
Laboratoire GeoResources, École des
Mines de Nancy
University of Lorraine
Nancy, France

ISSN 2364-6497   ISSN 2364-6500 (electronic)
GeoGuide
ISBN 978-3-030-84265-9   ISBN 978-3-030-84266-6 (eBook)
https://doi.org/10.1007/978-3-030-84266-6

This Springer imprint is published by the registered company Springer Nature Switzerland AG
The registered company address is: Gewerbestrasse 11, 6330 Cham, Switzerland

Foreword

This Geoguide on the eastern Pyrenees is a page-turner for readers interested in the science of scenery, i.e., in understanding how a given landscape has changed over time. While flicking through the book, the reader is instantly impressed by the quality and abundance of the iconography. The carefully crafted illustrations—no fewer than 275 of them—are not merely the perfunctory accompaniment to a narrative: each map and photograph provides the "smoking gun" evidence of what is being highlighted and discussed in the text. From rock fabrics to stratigraphic exposures, geological structures, and landform assemblages, these illustrations showcase the basic building blocks that geoscientists routinely have to contend within the field and on the basis of which Earth–science interpretations are subsequently crafted and interwoven.

Couched in crisp, carefully cross-referenced description, and demonstration, this field guide takes the reader on a journey not just of learning and scholarship, but also of time travel deep into the history of the orogen. Of course, any scientific portrayal of a region will inevitably be a function of the state of the art and of the culture of the authors presenting the evidence. In the present case, even though geological knowledge about the Pyrenees has been accumulating for two and a half centuries, the GeoGuide manages remarkably to generate new outlooks on previously well-known outcrops, exposures, and landscapes while also producing new insights into features that were previously unreported, overlooked, or disregarded as unimportant.

As you work your way through the itineraries, nature is observed first-hand, with mind and hammer—*mente et malleo*—shining through as the basic tools of inquiry. In case anyone needed reminding that observation is the bedrock of scientific interpretation, this book squarely instates the field as the supreme laboratory where everything in geoscience begins but also resolves, above and beyond the physics and chemistry that nonetheless also assist in the chain of analysis and understanding. Whether or not the narrative that grows out of observational skills can lastingly endure scrutiny is ultimately conditional upon the acumen of the grassroots observer. Paul Valéry emphasizes this fundamental truth when he states that *"a carelessly observed fact is more treacherous than faulty reasoning"* (Paul Valéry, *Tel Quel*, 1941). In this book, the standards and finesse of observation are outstanding.

The book is divided in two unequal parts. Part I, intended as a prelude to the field itineraries themselves, explains how mountain building proceeded as a result of plate tectonic processes and provides a general overview of the Pyrenees as background for what comes next. The Pyrenean range is an Alpine collisional orogen with a crustal pro-wedge advancing southward over the Iberian Plate, a retro-wedge thrust northward over the European Plate, and the North-Pyrenean Fault forming the boundary between the two tectonic plates. The elevated core of the mountain range consists of Paleozoic outcrops, which are a distant legacy of Europe's much older Hercynian orogen. However, the position of the Pyrenees in Eurasia's Alpine orogenic belt is overall clear, palaeogeographic reconstructions still struggle to understand the position of those Paleozoic structures in the wider puzzle of the Hercynian mountain belt. The large half-window of pre-Mesozoic rocks, for example, displays widespread occurrences of high-temperature–low-pressure metamorphism which, in many settings, would be characteristic of extensional rather than of collisional tectonics.

Part II contains the nine itineraries. These consistently emphasize how the mountain range was shaped and partly destroyed by long-term denudation. The many keys to this evolution are grounded in geomorphological analysis, and the successive roadside features and trailwalks methodically demonstrate how linking landform assemblages and topography to geological criteria can document landscape evolution in consistent ways. The timescales and successive stages of mountain growth, decay, regrowth, and transformation by a multitude of processes are backed up by multiple methods of rock, sediment, and landform dating. The resulting chronology, tightly and carefully constructed, brings alive the metabolism of the mountain range in all of its four dimensions.

The GeoGuide overall offers well over 2000 km of exploration routes through the eastern Pyrenees, documenting the Hercynian, Alpine, and post-Alpine tectonic cycles. Itineraries 1–4 focus predominantly on Neogene and Quaternary features within a wide radius around Perpignan, whereas itineraries 5–9 take us deeper into the mountains and extend the chronology of events to the more distant geological past. In greater detail, **Itinerary 1** outlines landscape evolution since Oligocene time along the Mediterranean seaboard and across part of its hinterland (Corbières). **Itinerary 2** deals with the Neogene evolution of the Roussillon Basin, with **Itinerary 3** scoping late Holocene fluvial dynamics across the Roussillon coastal plain based on geoarchaeological and sedimentological evidence. **Itinerary 4** makes a loop deep into the Tech River valley between the Spanish border and Mt. Canigou, with a special focus on extreme hydrometeorological events in historical time and their impacts on rivers, hillslopes, and human infrastructure. **Itinerary 5** strikes along and cuts through the Alpine structures of the North-Pyrenean Zone, showing how geomorphological indices of landscape evolution previously outlined in the Corbières remain equally valid for understanding landscape patterns and processes farther west. Advancing westward up the Têt River valley into the Axial Zone, **Itinerary 6** is a cross section through the Paleozoic core of the orogen. Having been only weakly deformed by Alpine collision, the basement outcrops have retained clear legacies of Hercynian rock deformation and tectonics while also displaying sharp overprints of Neogene extensional faulting in the scenery (Conflent Basin). A unique population of vertically distributed, alluvium-filled cave systems in Paleozoic limestone further documents the chronology of Neogene valley incision in response to Neogene and Quaternary regional uplift. Continuing westward, **Itineraries 7** and **8** showcase the late Neogene, intermontane extensional basins of Cerdagne and Capcir and their surrounding basement-cored massifs. The basin fill sequences, multiple generations of range-top and range-flank erosion surfaces, and suites of glacial landforms provide robust constraints on the chronology of denudation in response to crustal uplift as well as clues about the history of climate change. **Itinerary 9** cuts through and strikes along the south-vergent Alpine fold-and-thrust belts of the Iberian pro-wedge (Pedraforca and Cadí nappes). The chronology of tectonic deformation, provenance of conglomerate sequences produced by the eroding mountain range, and extant population of late- to post-orogenic range-top erosion surfaces (similar to those previously encountered in the Axial Zone and retro-wedge) are analyzed and discussed in forensic detail.

The clear, consistent, and persuasive interpretations of structure and landscape at every step of this GeoGuide should provide readers with much material to enrich the sheer aesthetic enjoyment of the scenic landscapes of the Pyrenees and to ultimately ponder the thoughts of the Greek philosopher Epicurus[1] about the essence of naturalist inquiry: *"Remember that you are mortal and that, although*

---

[1] *Vatican sayings,* 10; passage in single quotation marks is an embedded allusion to Homer's *Iliad,* I, 70.

having but a limited life span, you have devoted yourself to discussions on nature for all time, and have seen 'things that are now, are to come, and were before'".

Orléans, France   Philippe Rossi
                                                          President Emeritus, Commission for
                                                               the Geological Map of the World

Preface

This Geoguide is based on the original template of an unpublished, six-day post-conference field excursion guide prepared for the *8th International Conference on Geomorphology, Paris (France) 2013*. The internal structure of the guide, however, has changed substantially as a result of numerous scientific updates, the crafting of additional itineraries, and the inclusion of new original illustrations such as photographs of rock exposures and landscapes.

The initial post-conference excursion in 2013 benefited from a number of contributions, either written or involving oral presentations in situ. Here we wish to thank the collective of co-workers without whom the synthesis presented in this GeoGuide would not be as wide ranging as originally achieved: the late D. Bourlès and R. Braucher (LN2C-CEREGE, Aix-Marseille Université), J. Bordonau, and R. Pallàs (Universitat de Barcelona), J. Corominas (Technical University of Catalonia–BarcelonaTech), J.-M. Carozza (Université de La Rochelle), V. Turu (Igeotest and Fundacio M. Chevalier, Andorra), X. Planas (Ministeri d'Economia i Territori, Departement d'Ordenament Territorial, Govern d'Andorra), and J. Michaux and J.-P. Aguilar (Institut des Sciences de l'Evolution, Université de Montpellier 2). Credit is also due to our karst and cave science colleagues, without whom the sections dealing with Pyrenean cave systems as proxies for quantifying the chronology of valley incision could not have been accomplished: P. Sorriaux (Spéléo-Club du Haut Sabarthès), G. Hez (Spéléo-Club de Villefranche-de-Conflent), S. Jaillet (Université Savoie Mont Blanc), C. Bès and J.-C. Gayet (Comité Départemental de Spéléologie de l'Aude), B. and S. Ournié (Entente Spéléologique du Roussillon), and P. Audra (Université Côte d'Azur).

This GeoGuide deals with four complementary topics: (i) the geology, which involves a brief presentation of the Paleozoic, Mesozoic, and Paleogene rock and structural units that were involved in, or became the products of, mountain building during the Hercynian and Alpine orogenies; (ii) the geodynamics, which also includes the long-term post-orogenic evolution of the Pyrenees since the declining stages of tectonic collision ca. 30 million years ago, and within the wider reference frame of Iberia, the Western Mediterranean, and the Atlantic margin; (iii) the geomorphological processes and landforms that have conspired to shape the eastern part of the French and Spanish Pyrenees in response to base-level and climate-related changes over that same time period; and (iv) the geoheritage dimension, which is addressed through nutshell presentations of distinctive landscape units and type geological sections encountered in the study area. The field guide format is designed to help readers construct modular discovery-based itineraries across the region—with options and variants depending on time and physical ability. This should allow GeoGuide users to discover the key landscape and flagship geoheritage features over the course of 1 week, but ideally much more.

In order to fulfill those multiple goals, the GeoGuide is structured in two parts. Part I is a synthesis of the physiography and geodynamics of the Pyrenean orogen and its foreland basins. This frames the big picture: structure of the orogen and chronology of its formation, overview of its post-orogenic evolution, Quaternary, and recent landscape history. Part II proposes itineraries across the eastern Pyrenees, an area which currently provides the richest, best documented, and most carefully curated database on the post-orogenic evolution of the mountain belt. Nine itineraries are described. Each can be completed in one to three days, but the proposed additional stops and optional loops require longer, particularly the trailwalks.

This GeoGuide is primarily aimed at confirmed geoscientists, at their post-graduate students engaged in field studies, and at curiosity-driven individuals keen to enhance their understanding of spectacular or enigmatic features encountered during their travels or holidays. The proposed itineraries can be undertaken all year round, but the best seasons are summer (expect occasional thunderstorms) and autumn; it often rains in spring. Snow from mid-November to mid-May may impede field observation above 1500 m on the Atlantic side and 2000 m on the Mediterranean side of the mountain range. Hillwalking requires caution until mid-June at higher elevations.

Useful Maps and Online Resources

As a complement to the book and to assist in preparing the itineraries, a series of online resources are recommended.

Géoportail IGN (Institut Géographique National, France)

https://www.geoportail.gouv.fr

This interactive Web site provides free access to displays of up-to-date topographic and geological maps at all commercially available scales, including older maps and mosaics of geometrically corrected aerial photographs. A downloadable archive of aerial photographs dating back to ca. 1920 is also accessible from http://remonterletemps.ign.fr

Folded paper versions of IGN topographic sheets are available from many retail outlets such as bookshops, newsagents, and supermarkets. The most relevant to this field guide are the 1:25,000 scale "Top 25" blue series. About 19 of these sheets cover the study area. Meanwhile, the 1:100,000 scale "Top 100" green series can usefully serve as road maps to travel through the entire Pyrenean range and its northern foreland. Sheets 173 and 174 will specifically cover the study area.

Infoterre BRGM (Bureau de Recherches Géologiques et Minières, France)

http://infoterre.brgm.fr

This interactive Web site provides a mosaic of 1:50,000 scale geological maps, either as an assemblage of scanned paper maps or in the form of a harmonized vector map for the entire country. Only screen captures are possible, but the handbooks (in French) accompanying each map are downloadable in portable document format (PDF). The content of the handbooks depends on the publication date, but the most recent ones, typically up to two hundred pages long, usually provide useful updates on the regional geology (recommended handbooks relevant to the study area are Argelès, Céret, Mont-Louis, Prats-de-Mollo, although some of these may not yet be available from the Web site). The Web site also provides access to a auxiliary information such as the subsoil database (Banque du Sous-Sol), which compiles existing borehole data.

Folded paper versions of 1:50,000 scale BRGM geological sheets (sold with the print version of their respective handbooks) can be ordered from bookshops (rarely stocked) and directly from the BRGM website. The sheet names and code numbers relevant to this GeoGuide are from W to E and N to S: Mirepoix (1058), Limoux (1059), Capendu (1060), Narbonne (1061), Lavelanet (1076), Quillan

(1077), Tuchan (1078), Leucate (1079), Rivesaltes (1090), Perpignan (1091), Fontargente (1093), Mont-Louis (1094), Prades (1095), Céret (1096), Argelès-sur-Mer-Cerbère (1097-1101), Saillagouse (1098), and Prats-de-Mollo (1099). The sheet for Arles-sur-Tech (1100) is displayed on the InfoTerre Web site but remains unpublished at the time of writing. The sheets for Ax-les-Thermes (1088) and Saint-Paul-de-Fenouillet (1089) are still work in progress. Note that production and publication of the 1:50,000 scale sheets have been ongoing for 40 years, with the result that some map and handbook information among the earlier editions is scientifically out of date.

Direction régionale de l'environnement, de l'aménagement et du logement (DREAL) Occitanie (France)

http://www.occitanie.developpement-durable.gouv.fr, then follow the thunbnails for Ecologie —> biodiversité —> patrimoine géologique.

This Web site presents the results of the regional geological heritage inventory, with downloadable site-specific brochures (sites outside the public domain are confidential and thus not documented).

Instituto Geológico y Minero de España (IGME, Spain)

http://info.igme.es/cartografiadigital/geologica
http://www.igme.es/patrimonio/GEOSITES/publication.htm

This Web site provides free access to the 1:50,000 scale "Magna 50" assemblage of geological maps of Spain (Series 2, started in 1972), as well as information about geosites. Individual sheets, their handbooks, and any additional information components are downloadable, but sheets Noarre (150), Tirvia (182), Andorra (183), Sort (214), Seu-de-Urgell (215), Bellver (216), Organa (253), all relevant to the study area, are currently not available. Maps from the older edition (Series 1, 1928–1972) are also accessible, with the option of obtaining Bellver (216).

Some 1:200,000 scale sheets and their handbooks (synthesis of existing maps, 1970) are also available. The relevant sheet names and code numbers are Viella (14), Arties (15), Huesca (23), Berga (24), Figueras (25), Lerida (33), Hospitalet (34), and Barcelona (35).

Instituto Geográfico Nacional (Spain)

http://signa.ign.es/signa/Pege.aspx?
http://www.ign.es/iberpix2/visor/

For any chosen area in Spain, these two Web sites display topographic maps, a digital elevation model, and corresponding orthoimages.

Institut Cartogràfic y Geològic de Catalunya (ICGC, Spain)
http://www.icc.cat
From this Web site, the "Vissir" application displays detailed topographic maps and the corresponding orthoimage. These maps can be downloaded. ICGC 1:25,000 scale geological maps are also useful, free to download, but currently only available for a restricted portion of the southern side of the Pyrenees. Also free to download is a 1:50,000 *Mapa geològic comarcal de Catalunya*, provided as 41 separate sheets. Folded paper editions (one sheet by comarca) are also commercially available at the ICGC and in bookshops, alongside a 1:250,000 scale (and also a 1:300,000 scale) synthesis. The ICGC has also published an *Atles geològic de Catalunya*, which contains maps, cross sections, photographs of geological exposures, and a presentation of remarkable geological sites and landscapes accessible via the Web site.

Other Available Geological Guides (in French)

Guides géologiques régionaux

These field guides are outdated, rather dull by twenty-first century standards, and exclusively in French. They usually ignore the Neogene and Quaternary periods and do not deal with geomorphology, landscape, or geoheritage. They contain black and white line drawings and no photographs, excepting the occasional plate displaying fossil assemblages. Additionally, despite a few exceptions, only the French side of the Pyrenees is covered. Six guides are relevant to the Pyrenees:

- Jaffrezo M., 1977. Pyrénées orientales–Corbières. Masson, Paris, 191 p.
- Mirouse R., 1992. Pyrénées centrales franco-espagnoles. Masson, Paris, 176 p.
- Debourle A., Deloffre R., 1976. Pyrénées occidentales, Béarn, Pays Basque. Masson, Paris, 175 p.
- Gèze B., 1979. Languedoc méditerranéen. Masson, Paris, 191 p.
- Gèze B., Cavaillé A., 1977. Aquitaine orientale. Masson, Paris, 183 p.
- Vigneaux M., 1975. Aquitaine occidentale. Masson, Paris, 223 p.

New Series of Geological Guides (BRGM)

The new series of guide books published by the BRGM (series title: *Curiosités géologiques*) are all in French, richly illustrated, and aimed at a popular audience of hillwalkers and amateur naturalists. They do not deal with geomorphology and are mainly restricted to rock outcrops and sections. Only two volumes have been published for the Pyrenees:

* Mulder T., 2014. Curiosités géologiques de la Côte Basque. BRGM, Orléans, 110 p.
* Le Goff E., Calvet M., Moigne A.M., 2018. Curiosités géologiques des Pyrénées orientales. BRGM, Orléans, 116 p.

The BRGM (subsequently Omniscience éditions) has also started another, more scientific collection (series title: *Guides Géologiques*), still exclusively in French and restricted to general bedrock geology and structure. It is aimed at hillwalkers and offers pedestrian day trips. Volumes tend to focus on an administrative division rather than on geological regions. Two volumes concerning the Pyrenees have been published, but nothing until now on the east-central and eastern Pyrenees:

* Hervouët Y., 2014. Pyrénées Atlantiques. BRGM-Omniscience, Orléans, 256 p.
* Hervouët Y., Péré A., Rossier D., 2016. Hautes Pyrénées. BRGM-Omniscience, Orléans, 256 p.

Further Reading

* Canérot J., 2008. Les Pyrénées. Tome 1: Histoire géologique, 516 p; Tome 2: Itinéraires de découverte, 127 p. Atlantica, Biarritz & BRGM éditions.

Only three out of the 11 *itinéraires de découverte* concern the focus area covered by this GeoGuide. They distinctly emphasize roadside geology.

A Note on Place Names and Site Locations

This GeoGuide straddles an international boundary between France and Spain and encompasses up to four languages: Castillan, French, Catalan, and Occitan. The text and figures have given priority to name spellings as encountered on the topographic and geological sheets of the study area. For a number of sites of interest, place names may also follow the official spelling given on road signs. When several names exist, their variants are often given as well, in brackets, while trying not to clutter the text with too much detail of this nature. On French maps, Catalan and Occitan names are fairly uniform from one source to another, but discrepancies have been found to appear between 1:25,000 scale paper sheets and their digital version on the Géoportail. On Spanish maps, Catalan spelling is usually prevalent, but discrepancies between IGME and Iberpix topographic may appear, and likewise when searching the Iberpix Web site depending on magnification as you zoom in and out. The 1:25,000 scale maps also vary and may turn out to display fewer place names than the ICGC 1:50,000 scale maps. In the absence of place names, some observation points and sites of interest are located in the GeoGuide by their latitude and longitude, each time following the default projection systems used either by the Géoportail (when in France) or by Iberpix (when in Spain).

Marc Calvet
Université de Perpignan-Via Domitia
Perpignan, France

Magali Delmas
Université de Perpignan-Via Domitia
Perpignan, France

Yanni Gunnell
Université Lumière-Lyon 2
Lyon, France

Bernard Laumonier
University of Lorraine
Nancy, France

Introduction

The GeoGuide consists of a general introductory overview of the Pyrenean orogen (Part I), followed by nine field itineraries through the eastern Pyrenees (Part II).

Part I—The Eastern Pyrenees in Their Wider Regional Setting: An Overview

The natural landscapes of the Pyrenees became a focus of scientific research among geologists, geographers, and other scholars (e.g., Palassou 1781, 1815–1823; Ramond 1789; Pasumot 1797) at the end of the eighteenth century, before gaining appeal as a cultural, aesthetic, and tourist attraction in the nineteenth century (see Briffaud 1994 and Besson 2000 for an overview). Like the European Alps, the Pyrenees have been investigated in the hope of elaborating a reference model that would help geologists to understand mountain building processes more generally worldwide. One of the very early scholars of Pyrenean geology was Louis François Élisabeth Ramond, baron of Carbonnières (1755–1827), whose name was given to one of the three peaks of Monte Perdido (Soum de Ramond) as well as to an endemic flowering plant with a Cenozoic evolutionary lineage (*Ramonda pyrenaica*). He wrote about the mountain range in the following terms: *"Je doute qu'il existe une chaîne de montagnes plus propre que celle des Pyrénées, à être observée par le Naturaliste qui veut étudier la structure & la disposition des roches primitives. Simple & régulière dans presque toute son étendue, elle lui rappellera bientôt les idées de l'ordre qui a dû présider à la formation des monts, & des règles auxquelles leur dégradation est soumise ; leurs masses ne seront plus des accumulations informes, leurs intervalles ne seront plus un labyrinthe bizarre; dans la situation, le rapport, l'élévation de ses différentes parties, il reconnaîtra bientôt l'influence de ces lois constantes, dont il avait peine, ailleurs, à démêler l'existence"*[2] (Ramond 1789, p. 1).

The mountain range has attracted vast efforts of international research, whether into the fabric of its Hercynian basement outcrops or into the Mesozoic and Cenozoic cover rocks and their many thrusts, faults, folds, and crumples (e.g., de Margerie 1946; Canerot 2008; Barnolas and Chiron 1996, 2018). A number of seminal scientific papers over the last several decades on these topics would include Mattauer (1968), Souquet et al. (1977), Choukroune (1992), Roure and Choukroune (1998), Vergés et al. (2002), and Teixell et al. (2018). The landscape of the Pyrenees has also been attracting geomorphologists since the late nineteenth century and has involved scholars from Germany (Penck 1883, 1894; Aschauer 1934; Panzer 1926, 1933), the Netherlands (Pannekoek 1935; Boissevain 1934; de Sitter 1952, 1954, 1961), Switzerland (Nussbaum 1930, 1931, 1943), Spain (García Sainz 1940; Solé Sabaris 1951) and France (de Martonne 1910; Blanchard 1914; Sorre 1913; Birot 1937, 1952; Goron 1941a, b; Sermet 1950; Taillefer 1951; Viers 1960, 1962; Barrère 1963; for more detailed reference lists before 1990, see Birot 1937, and Calvet 1996). The orogen at the boundary between the Iberian and European plates is, of course, the structural and geodynamic consequence of continental collision, but the mountain range is also unique by its position as an isthmus between two contrasting maritime domains and a transitional area between two bioclimatic regions: temperate in the north and subtropical in the south. The morphoclimatic distinctiveness of the Pyrenees hinges on the convergence between Atlantic and Mediterranean meteorological influences. Precipitation contrasts between the north- and south-facing range fronts are substantial because of the ~1000-km-long Cantabro-Pyrenean topographic barrier to Atlantic weather systems. These conditions also prevailed throughout the Pleistocene, with a deep imprint left in today's scenery by the glacial and fluvial landforms (Barrère 1954; Taillefer 1982; Calvet 2004). A synthesis of the tectonics, geology, and landscape evolution of the Pyrenees during the last 80–90 million years was crafted by Calvet et al. (2021), and Part I in this GeoGuide offers a simplified overview of the orogen and its evolution, with

---

[2] "I doubt that there exists a mountain range better suited than the Pyrenees to the observation by a naturalist of its structure and rock disposition. Because it is simple and regular throughout most of its extent, the order which presided over the formation of its peaks and the laws which govern their degradation will soon become apparent. The relief will eventually be reduced to shapeless accumulations of debris, and its pattern to a strange labyrinth. Through the location, proportion and elevation of its different parts, the naturalist will readily understand the constants and laws over which he had stumbled in other settings".

emphasis on the eastern and central parts of the mountain range. Part II presents nine detailed itineraries through the eastern Pyrenees.

Part II—Field Itineraries

The Pyrenees in this GeoGuide are presented with a focus on the Mediterranean base level in the east in addition to the more commonly studied southern and northern foreland basins. The basis for this choice is that the eastern third of the mountain range, including the Axial Zone, has been fragmented by a population of Oligocene and Neogene extensional basins, the sedimentary records of which provide unique constraints on the geomorphological and geodynamic evolution of the mountain range. The legacy of Quaternary processes, particularly in the form of alluvial, glacial, and slope deposits, is also conspicuous, and our understanding of landform chronology and the history of topographic growth have recently been enhanced by multi-method dating of rocks, sediments, and landforms. Alluvial deposits and some caves in the Corbières contain evidence of early hominin presence (particularly *Homo erectus tautavelensis*, likely ancestor to the Neanderthals in Europe). The area is also vulnerable to Mediterranean rainstorms and flash flooding. As a result, modern geomorphological processes are intense, and their clear-cut scars in the landscape also help to understand the Holocene and historical evolution of the landscape in a context where climatic change was compounded by human impacts. Records of Neolithic pastoralism and soil erosion, for example, reach back to 7500 yr BP.

Nine detailed itineraries contained within the frames shown in the Figure on page XX are each presented as individual book chapters. The first three focus on the Mediterranean seaboard of the Pyrenees north of the Spanish border, spanning the Corbières and various basins of the Mediterranean rift system in the north (Itinerary 1), to the Roussillon Basin and Albères massif in the south (Itineraries 2 and 3). The following three explore the valleys, basins, and plateaus of the orogen's Axial Zone and flanking thrust wedges extending from the Roussillon Basin southward (Tech valley: Itinerary 4), northwestward (Agly, Fenouillèdes, upper Aude valley: Itinerary 5), and westward (Conflent Basin, Têt valley: Itinerary 6). The next two itineraries investigate the Neogene intermontane basins, elevated massifs, and synorogenic tectonic and chronostratigraphic units of the orogen's Axial Zone: Cerdagne, and Capcir basins (Itineraries 7 and 8, which arguably provide, with Itinerary 6 Conflent basin, the best introduction to tectonic geomorphology anywhere in France). Itinerary 9 mostly focuses on the Cadí and Pedraforca nappes on the south side of the orogen and on legacies in the landscape of the overfilled clastic piedmont environment prevalent during the Paleogene. Landforms and landscape evolution are the main focus throughout.

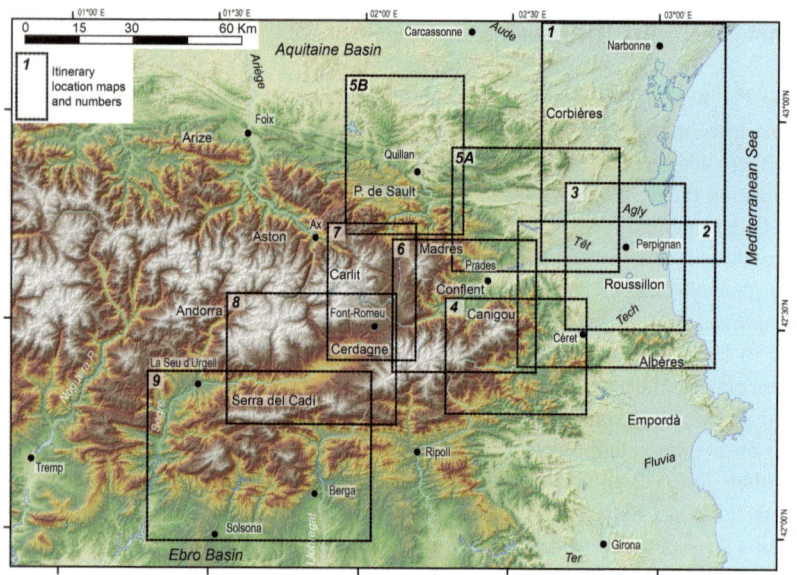

References

Aschauer H (1934) Die östliche Endigung der Pyrenäen. Abh Ges Wiss Göttingen, Math Phys Kl, 10:2–115

Barnolas A, Chiron JC (eds) (1996) Synthèse géologique et géophysique des Pyrénées. – Vol. 1: introduction, géophysique, cycle hercynien. Editions BRGM-ITGE, 729 p

Barnolas A, Chiron JC (eds) (2018) Synthèse géologique et géophysique des Pyrénées. – Vols 2 & 3: Cycle Alpin. AGSO and BRGM

Barrère P (1954) Equilibre glaciaire actuel et quaternaire dans l'Ouest des Pyrénées centrales. Rev Géogr Pyrén Sud-Ouest, XXIV:116–134

Barrère P (1963) La période glaciaire dans l'Ouest des Pyrénées centrales franco-espagnoles. Bull Soc Géol Fr 7:516–526

Besson F (2000) Le paysage pyrénéen dans la littérature de voyage et l'iconographie britannique du dix-neuvième siècle. L'Harmattan, Paris, 463 p

Birot P (1937) Recherches sur la morphologie des Pyrénées orientales franco-espagnoles. Baillière Édit, 318 p

Birot P (1952) Sur quelques contrastes fondamentaux dans la structure et la morphologie des Pyrénées. Actas del Primer Congreso de Estudios Pirenaicos, San Sebastian, 1950, V, Geografia, 17–21

Blanchard R (1914) La morphologie des Pyrénées françaises. Ann Géo 23:303–324

Boissevain H (1934) Étude géologique et géomorphologique d'une partie de la vallée de la haute Sègre (Pyrénées Catalanes). Bull Soc Hist Nat Toulouse 66:33–170

Briffaud S (1994) Naissance d'un paysage. La montagne pyrénéenne à la croisée des regards, XVIe–XIXe siècle. In: Sources et Travaux d'Histoire Haut-Pyrénéenne (Association Guillaume Mauran), vol. 8, 622 p

Calvet M (1996) Morphogenèse d'une montagne méditerranéenne: les Pyrénées orientales. Documents du BRGM, Orléans, n° 255, 3 vols., 1177 p

Calvet M (2004) The Quaternary glaciation of the Pyrenees. In: Ehlers, J., Gibbard, P. (eds), Quaternary Glaciations—Extent and Chronology, part I: Europe. Elsevier, Amsterdam, 119–128

Calvet M, Gunnell Y, Laumonier B (2021) Denudation history and palaeogeography of the Pyrenees and their peripheral basins: an 84-million-year geomorphological perspective. Earth-Sci Rev 215:103436

Canérot J (2008) Les Pyrénées. Tome 1: Histoire géologique, 516 p; Tome 2: Itinéraires de découverte, 127 p. Atlantica, Biarritz & BRGM éditions

Choukroune P (1992) Tectonic evolution of the Pyrenees. Annu Rev Earth Planet Sci 20: 143–158

De Sitter LU (1952) Pliocene uplift of Tertiary mountain chains. Am J Sci 250:297–307

De Sitter LU (1954) Note préliminaire sur la géologie du Val d'Aran. Leidse Geol Med-edelingen 18:272–280

De Sitter LU (1961) La phase tectogénique pyrénéenne dans les Pyrénées méridionales. C R Somm Soc Géol Fr 8:224–225

García Sainz L (1940) Las superficies de erosión que preceden a los glaciares quaternarios del Pirineo central y sus recíprocas influencias. Estudios Geograficos, I:45–73

Goron L (1941a) Les Pré-Pyrénées ariégeoises et garonnaises, essai d'étude morphogénique d'une lisière de montagne. Privat, Toulouse, 886 p

Goron L (1941b) Le rôle des glaciations quaternaires dans le modelé des vallées maîtresses des Pré-Pyrénées ariégeoises et garonnaises et leur avant-pays, Privat, Toulouse, 460 p Abridged in: Rev Géo Pyrén Sud-Ouest, t. XII, pp. 5–61, 147–226, 322–356, 373–430

Margerie E de (1946) Études Pyrénéennes. In: Critique et Géologie, contribution à l'histoire des sciences de la terre (1882–1942), vol. III. A. Colin, Paris, 1157–1714

Martonne E de (1910) Remarque sur la communication de M.L. Carez, «Résumé de la géologie des Pyrénées», Séance du 2 mai 1910, C R Somm Séances Soc Géol Fr, p 425

Mattauer M (1968) Les traits structuraux essentiels de la chaîne pyrénéenne. Rev Géogr Phys Géol Dyn 10:3–12

Nussbaum F (1930) Morphologische studien in den östlichen Pyrenäen. Z Ges Erdkunde, Berlin, 200–210

Nussbaum F (1931) Sur les surfaces d'aplanissement d'âge tertiaire dans les Pyrénées-Orientales et leurs transformations pendant l'époque quaternaire. Comptes Rendus du Congrès International de Géographie, Paris, II, 529–534

Nussbaum F (1943) Orographische und morphologische untersuchungen in den östlichen Pyrenäen. Jahresbericht der Geographischen Gesellschaft von Bern, XXXV, 1942–1943, 1–148 (1945)

Palassou PB (1781) Essai sur la minéralogie des Monts-Pyrénées. Didot-A. Jombert-Esprit éditions, reprint La Découvrance, 2007, 367 p

Palassou PB (1815) Mémoire pour servir à l'histoire naturelle des Pyrénées et des pays adjacents, Imprimerie de Vignancour, Pau, 485 p

Palassou PB (1819) Suite des mémoires pour servir à l'histoire naturelle des Pyrénées et des pays adjacents, Imprimerie de Vignancour, Pau, 428 p

Palassou PB (1823) Nouveau mémoire pour servir à l'histoire naturelle des Pyrénées et des pays adjacents, Imprimerie de Vignancour, Pau, 192 p

Pannekoek AJ (1935) Évolution du bassin de la Têt dans les Pyrénées-Orientales pendant le Néogène. Étude de morphotectonique. Univ. Utrecht, 72 p

Panzer W (1926) Talentwicklung und Eiszeitklima in nord-östlichen Spanien. Abh. der Senckenbergischen Naturforschenden Gesellschaft, Frankfurt 39:141–182

Panzer W (1933) Die Entwicklung der Täler Kataloniens. Géologie des Pays Catalans, III, 21, Barcelona

Pasumot F (1797) Voyages physiques dans les Pyrénées en 1788 et 1789, Histoire naturelle d'une partie de ces montagnes, imp. Le Clère, Paris, 420 p

Penck A (1883) Die Eiszeit in den Pyrenäen. Mitteilungen Ver. für Erdkunde, Leipzig, 163–231. Traduction en Français par L. Braemer (1885), Bull Soc Hist Nat Toulouse, 107–200

Penck A (1894) Studien über das Klima Spaniens während der jungeren Tertiarperiod und Diluvialperiode. Zeitschrift der Geselschaft für Erdkunde zu Berlin 29:109–141

Ramond de Carbonnières LFE (1789) Voyages et observations faites dans les Pyrénées ; pour servir de suite à des observations sur les Alpes, Insérées dans une traduction des Lettres de W. Coxe sur la Suisse. Belin, Paris, 452 p

Roure F, Choukroune P (1998) Contribution of the ECORS seismic data to the Pyrenean geology: crustal architecture and geodynamic evolution of the Pyrenees. In: Damotte B, The ECORS Pyrenean deep seismic surveys, 1985–1994, Mém Soc Géol Fr 173:37–52

Sermet J (1950) Réflexions sur la morphologie de la zone axiale des Pyrénées. Pirineos, VI:323–404

Solé Sabaris L (1951) Los Pirineos, el medio y el hombre. Editorial Martin, Barcelona, 624 p

Sorre M (1913) Les Pyrénées méditerranéennes, étude de géographie biologique. A. Colin, Paris, 508 p

Souquet P, Peybernès B, Bilotte M, Debroas EJ (1977) La chaîne alpine des Pyrénées. Géol Alpine 53: 193–216

Taillefer F (1951) Le piémont des Pyrénées françaises, contribution à l'étude des reliefs de piémont. Privat, Toulouse, 383 p

Taillefer F (1982) Les conditions locales de la glaciation pyrénéenne. Pirineos 116:5–12

Teixell A, Labaume P, Ayarza P, Espurt N, de Saint Blanquat M, Lagabrielle Y (2018) Crustal structure and evolution of the Pyrenean-Cantabrian belt: a review and new interpretations from recent concepts and data. Tectonophysics 724–725:146–170

Vergés J, Fernàndez M, Martínez A (2002) The Pyrenean orogen: pre-, syn-, and post-collisional evolution. In: Rosenbaum, G, Lister, G-S (eds), Reconstruction of the evolution of the Alpine-Himalayan Orogen. J Virtual Explorer, 8:55–74

Viers G (1960) Le relief des Pyrénées occidentales et leur piémont. Pays Basque français et Barétous. Privat, Toulouse, 604 p

Viers G (1962) Les Pyrénées. Presses Univ. de France, collection Que-sais-je n° 995, 128 p

Contents

List of Boxes

Part I
The Eastern Pyrenees in Their Wider Regional Setting: An Overview

1.1 Anatomy of the Crustal Wedge

The Pyrenean orogen is the product of the N–S collision between Eurasia and Iberia and has involved crustal shortening by at least 100 km (Choukroune 1992; Teixell et al. 2018). The mountain belt strikes N110°E, varies in width from 75 to 150 km, and between the Aquitaine Basin in the north and the Ebro Basin in the south is structurally divided into a classic succession of five tectonostratigraphic belts. These are known as the Sub-Pyrenean Zone, North-Pyrenean Zone (NPZ), Axial Zone (AZ), Nogueres Zone, and South-Pyrenean Zone (SPZ). From a more strictly tectonic viewpoint, the orogen presents two major domains (Figs. 1.1, 1.2):

(i) the European domain, where Alpine structures are north-vergent. It consists of most of the NPZ, which overrides the Aquitaine retro-foreland basin along the North-Pyrenean Frontal Thrust (NPFT). The southern edge of the Aquitaine Basin south of the North-Pyrenean Front (Fig. 1.2) is also substantially crumpled over a blind thrust (the so-called Sub-Pyrenean Thrust), which connects at depth with the NPFT and marks the outermost limit of the crustal wedge. This outer tectonic belt is the Sub-Pyrenean Zone and coincides with a fold-and-thrust belt geographically known as the Petites Pyrénées.

(ii) the Iberian domain, where Alpine structures are south-vergent. From north to south, it consists of the southwestern part of the NPZ, the Axial Zone, the Nogueres Zone (a small and controversial unit), and the SPZ; the SPZ overrides the Ebro Basin and terminates at the South-Pyrenean Frontal Thrust (SPFT). The rock sequences of the Ebro Basin itself are variously deformed,

© Springer Nature Switzerland AG 2022
M. Calvet et al., *Geology and Landscapes of the Eastern Pyrenees*, GeoGuide,
https://doi.org/10.1007/978-3-030-84266-6_1

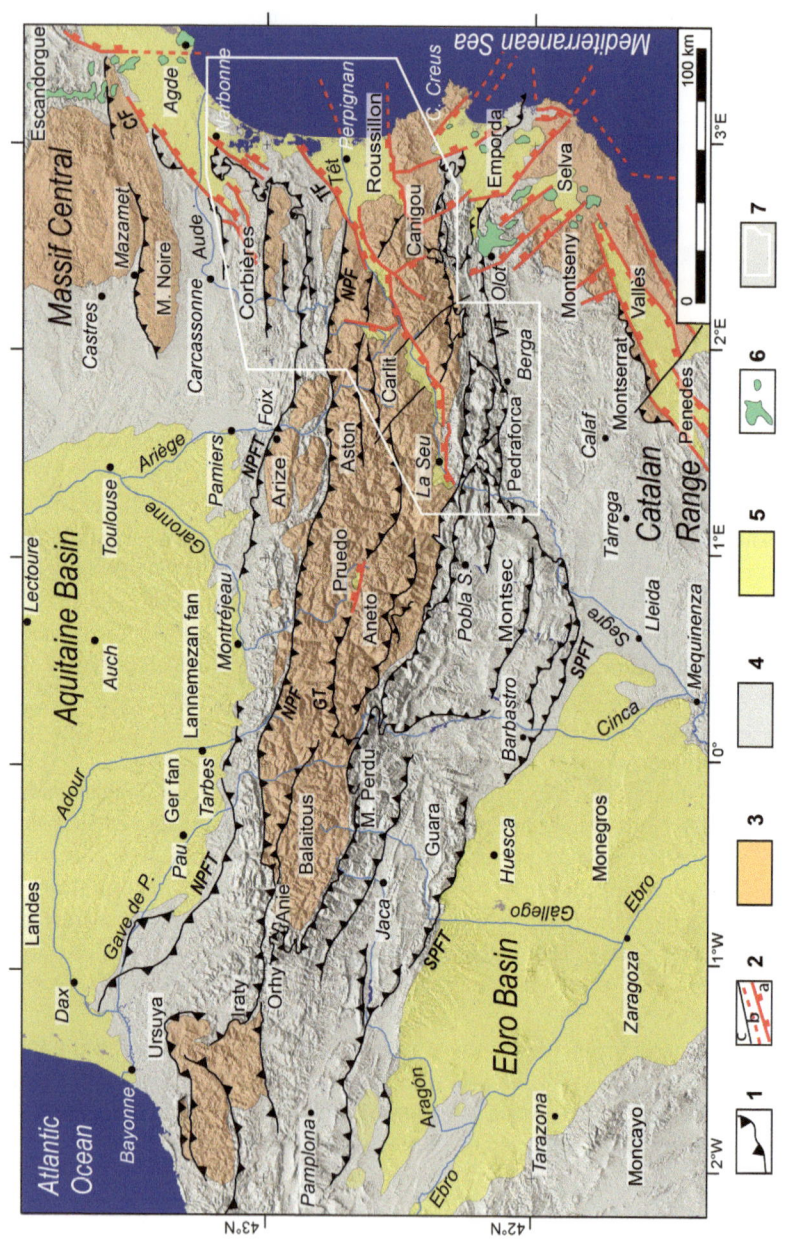

◄**Fig. 1.1** The Pyrenean orogen: topography and first-order geological features. **1**: Paleogene reverse faults and thrusts. **2**: Neogene faults. **2a**: normal faults. **2b**: probable Neogene faults. **2c**: other faults. **3**: Hercynian basement, folded Paleozoic sequences, metasedimentary and granitoid rocks (Axial Zone and related outcrops in the North-Pyrenean and Sub-Pyrenean zones, pro-foreland and retro-foreland basement outcrops, Massif Central and Catalan Range). **4**: folded Mesozoic and Cenozoic cover rocks; Paleogene conglomerate beds. **5**: Neogene sedimentary rocks. **6**: Neogene and Quaternary volcanic rocks. **7**: area documented in greater detail by the nine itineraries. Relief base map: SRTM digital elevation model (3 arcsec)

thus displaying a component of symmetry with the situation in the Sub-Pyrenean Zone of the Aquitaine Basin.

In the central and eastern Pyrenees, cover-rock outcrops in the narrow southern fringe of the NPZ are intensely deformed and underwent low-pressure–high-temperature (LP–HT) metamorphism during the Alpine orogeny. Known as the Internal Metamorphic Zone (IMZ), this belt of metamorphic outcrops can be considered as a sixth structural unit (Fig. 1.2; see Itinerary 5), not recognised as such in earlier literature. It defines the structural centreline of the orogen. It derives directly from a hyperextended mid-Cretaceous rift, on the floor of which the lithospheric mantle (today exposed as small outcrops of lherzolite) became exhumed (Jammes et al. 2009; Lagabrielle et al. 2010). The Axial Zone and IMZ are abruptly separated from one another by the North-Pyrenean Fault (NPF), which has long been considered as the locus of the Europe–Iberia plate boundary (e.g., Le Pichon et al. 1970; Choukroune et al. 1973). The eastern half of the mountain range, examined in greater detail in this book, has undergone extensional tectonics as a consequence of Cenozoic rifting in the Western Mediterranean.

1.2 Pre-Orogenic Evolution of the Pyrenean Domain

Outcrops of the pre-orogenic sedimentary sequences are diversely distributed throughout the orogen because of nondeposition or postdepositional removal by denudation. Some are also buried beneath Cenozoic sequences in the two foreland basins (e.g., Canérot 2008). In the central Pyrenees, the sedimentary sequences of the (future) SPZ tectonic unit (Fig. 1.2) originally rested on the Hercynian basement outcrops of the Axial Zone.

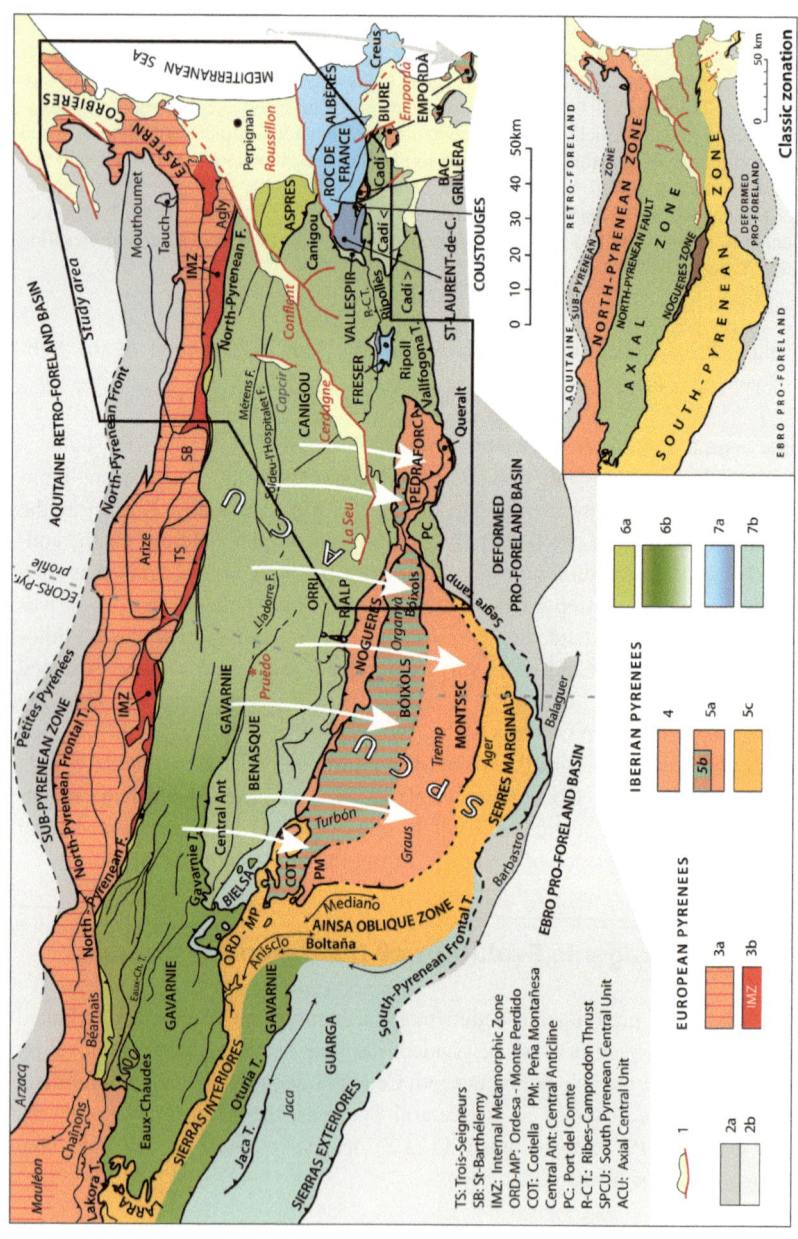

◄**Fig. 1.2**   **Structural map of the Alpine Pyrenees.** Classical structural zones of the Pyrenees are given in the inset, bottom right (North-Pyrenean Zone: NPZ, Axial Zone: AZ, South-Pyrenean Zone: SPZ). The main map emphasises (i) terrane origin—Europe or Iberia; (ii) the root areas of the Iberian nappe units (Upper, Middle and Lower Thrust Sheets, see below), which may involve cover rocks initially resting on the Hercynian basement, basement nappes, or mixed basement and cover-rock nappes. Key to symbols and ornaments—**1**: Neogene normal faults and basins. **2**: pro- and retro-foreland basins; **2a**: tectonically deformed; **2b**: tectonically undeformed. **3**: European Pyrenees (north-vergent structures); **3a**: North-Pyrenean Zone (NPZ, including main Hercynian North-Pyrenean massifs: Arize, etc.). **3b**: Internal Metamorphic Zone (IMZ), limited to the south by the North-Pyrenean Fault. **4 to 7**. **Iberian Pyrenees (south-vergent structures, including the Axial Zone).** **4**: southwest of the NPZ ('Chaînons béarnais'). **5**. *Upper Thrust Sheets*, rooted north of the Axial Zone; **5a**: South-Pyrenean klippen or allochtons (South-Pyrenean Central Unit, or SPCU, and Pedraforca and Empordà nappes). **5b**: Uppermost units with thick Lower Cretaceous (Upper Pedraforca, Organya). White arrows: displacement of the southern upper allochtons from their pre-orogenic locations in the Axial Zone. **5c**: other Upper Thrust Sheets. **6**: *Middle Thrust Sheets*, rooted within the Axial Zone; **6a**: small uppermost Thrust Sheets (Eaux-Chaudes, Aspres). **6b**: main Middle units, especially Gavarnie (Axial Zone and SPZ) and Benasque–Orri (Axial Zone) in the central Pyrenees, and Canigou–Vallespir (Axial Zone)–Cadí (SPZ) in the eastern Pyrenees; the Axial Central Unit (ACU) corresponds to the Canigou–Orri–Gavarnie units. **7**: *Lower Thrust Sheets*, rooted beneath the Axial Zone and forming the base of the southern crustal wedge; **7a**: Albères, Roc de France, Saint-Laurent-de-Cerdans, Freser and Rialp units, exclusively involving Axial Zone rocks; **7b**: Bielsa (Axial Zone), Guarga (SPZ)

The youngest deposits of the Triassic sequence (250–200 Ma) were the Keuper evaporites, which played a major role in the subsequent Pyrenean crustal wedge dynamics as a lubricant on thrust soles and in diapiric structures. The Jurassic and lowermost Cretaceous sedimentation (200–125 Ma) consists of a mudstone and limestone sequence. The lower units were deposited in what was, at the time, the western portion of the young Alpine Tethys; but a rifting episode started during the late Jurassic and early Cretaceous in relation to the opening of the Atlantic and promoted mainly carbonate rocks (e.g., Barremian limestone). Rifting intensified until the Cenomanian (110–95 Ma), i.e., a time when, in the area which was to become the future NPZ, rapidly subsiding grabens were being formed and mantle lherzolites became exposed as a result of extreme crustal thinning. Syn-rift sedimentation, up to 4–6 km thick in some depocentres, ranged from fine-textured clastic deposits (Aptian mudstones and the so-called Black Flysch, i.e., Middle Albian to Lower Cenomanian turbidites) to coarse breccia around the basin margins.

The pre-collisional chronology and geodynamics of the Euro-Iberian plate boundary are still poorly constrained because the strike-slip distance of travel and magnitude of anticlockwise rotation of Iberia are difficult to reconstruct. During early Cretaceous time, the region corresponding to the future Axial Zone was largely a land area, though not a mountain range. From middle Cenomanian time, rifting slackened rapidly, and within a regional context of sea-level rise the North-Pyrenean foredeep widened northward and filled with flysch.

1.3   Plate Convergence and Its Resulting Tectonic Structures in the Eastern Pyrenees

Convergence was relatively rapid during two successive episodes, namely from the Santonian to the Maastrichtian (~84–68 Ma) and subsequently from the early Eocene to the Aquitanian (~55–20 Ma). Convergence rates slackened between ~68 and 55 Ma. The mountain range thus grew in two discrete stages: first during the late Cretaceous, producing the Proto-Pyrenees; then from the Eocene to the Oligocene, producing the Ancestral Pyrenees—which themselves differed from the Modern Pyrenees of today (Calvet et al. 2021). The earlier period of peak crustal convergence caused tectonic inversion of the Cretaceous rift zone, whereas the second period corresponds to the collision itself. This chronology is mostly valid for the central Pyrenees, which have been the focus of much research, and cannot substitute perfectly for other parts of the mountain range. Tectonic collision in the eastern Pyrenees, for example, terminated in early Oligocene time and was thereafter superseded by the new extensional stress field impinging landward from the Western Mediterranean.

1.3.1   A Hercynian Backbone: The Axial Zone

A prominent feature of the Pyrenees is its extensive outcrop of Paleozoic rocks. These basement rocks define the Axial Zone, also at one time named *Haute Chaîne Primaire* ('Paleozoic high range'), but are also exposed in large outcrops forming massifs within the NPZ (so-called North-Pyrenean massifs: Arize, Saint-Barthélemy, Trois-Seigneurs, Agly) and the Mouthoumet massif. The Paleozoic outcrops once belonged to the southern outer zones of the Hercynian orogen (Barnolas and Chiron 1996). They formed during the late Carboniferous (~330–295 Ma) and involved sedimentary and some volcanic rock sequences of Ediacaran to early Carboniferous age (~580–325 Ma), with Ordovician plutonic intrusions (~475–445 Ma). Major crustal deformations (detected through a range of characteristic rock textures, thrusts, large upright and recumbent folds, detachment faults, mylonites, and well mapped mega-antiforms such as the Canigou–Carança dome; see Itinerary 6) also involved LP–HT regional metamorphism, and resulted in the currently observable array of micaschists, marble, paragneiss, orthogneiss, and migmatites. The orogenic cycle ended in Permian time with an abundant production of red continental (volcano-sedimentary) molasse sequences. These are mostly preserved along the southern edge of the Axial Zone. Because depths of crustal denudation were greater in the eastern and central Axial Zone than further west, the eastern half of the Axial Zone displays comparatively deeper facies of upper and middle crustal rocks such as the Ediacaran Canaveilles Group (schists and carbonates), the Cambrian Jujols Group (mainly schists), and the Ordovician Canigou metagranite. Younger Hercynian rocks (Silurian, Devonian, Carboniferous) are rare and confined to narrow synclines (e.g., the Villefranche-de-Conflent syncline, see Itinerary 6). A reverse situation is observed in the western Axial Zone, which is dominated by limestone-rich Devonian and Carboniferous series. Hercynian granites (~310–295 Ma) have intruded all of these series. The North-Pyrenean massifs of the retro-wedge expose complete crustal sections, from migmatites and granulites of the lower crust to Devonian and Carboniferous sedimentary rocks.

1.3.2 The North Pyrenean Fault: Plate Boundary Between Iberia and Europe

The NPF is a major crustal discontinuity separating the Axial Zone from the Internal Metamorphic Zone of the NPZ, and extends laterally for at least 250 km. Whereas, for some authors, the NPF has offset the Moho, the currently prevailing view considers that it is cross-cut by the NPFT, which is a major south-dipping thrust plane over which the NPZ and part of the Axial Zone have been transported northward towards the retro-foreland (Figs. 1.2, 1.3). Reactivation of this fault during the Eocene is held as the main cause for raising the Axial Zone relative to the NPZ and restoring the NPF's prominence in the structure of the crustal wedge. Consistent with this view is that it terminates with the Axial Zone in the west (Fig. 1.2).

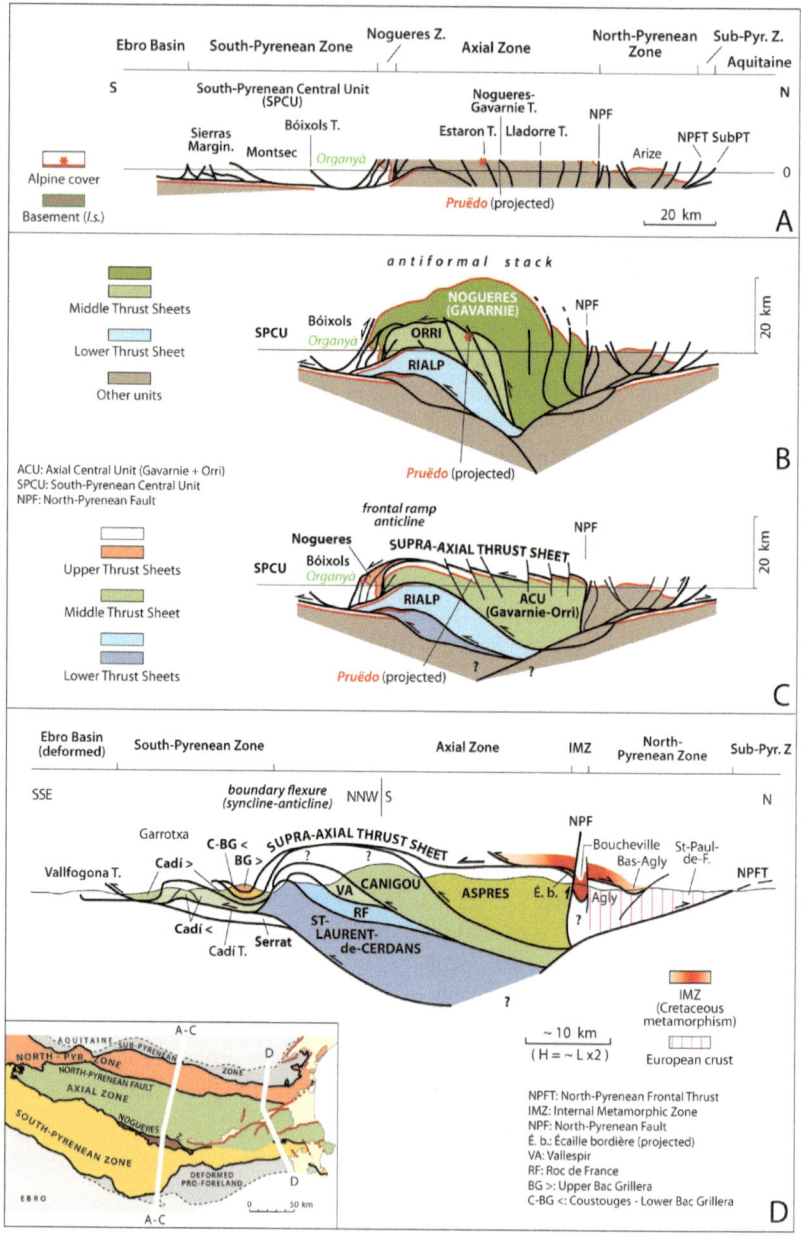

◀**Fig. 1.3** **Tectonostratigraphic cross-sections through the eastern and central Pyrenees. A** Structural units encountered along the ECORS-Pyrenees profile, after Muñoz (1992). Organyà is one of the Lower Cretaceous basins inverted by crustal convergence during the latest Cretaceous, thus forming the Bóixols unit. **B** Cross-section of the Axial Zone along the ECORS-Pyrenees profile according to Muñoz (1992). In this model, the large nappe stack thickens the Axial Zone basement and generates near-vertical dips among the major Alpine thrusts, particularly the Nogueres (Garvarnie) Thrust. **C** ECORS-Pyrenees cross-section in a and b reinterpreted by Laumonier (2015). In this model, a Supra-Axial Thrust connects the uppermost nappes of the Pyrenean orogen (namely the South-Pyrenean Central Unit and the Nogueres units) with the NPZ root area. These overspanning nappes overlie an 'Axial Central Unit', which corresponds to the Gavarnie and Orri basement nappes. Note the contrast in geometry with model B. This alternative model is supported by a number of field constraints, including the preservation of residual and non-metamorphic Triassic deposits at Pruëdo (red star; see location in Fig. 1.3). **D** Schematic cross-section through the eastern Pyrenees. Colour coding identical to Fig. 1.3. Components of the Supra-Axial Thrust Sheet here include the South-Pyrenean klippe of Bac Grillera (at least its upper part) and the North-Pyrenean Internal Metamorphic Zone (southern part); the eroded central part of the overspanning nappe exposes a basement window of Paleozoic rocks commonly defined as the Axial Zone

1.3.3 The Late Cretaceous to Paleocene Mountain Range

The late Cretaceous to Paleocene Proto-Pyrenees rose only in the eastern part of where the Pyrenean mountain range now stands. The south-vergent Bóixols thrust (Figs. 1.2, 1.3) was produced during Campanian time. It results from an inversion of the Cretaceous Organyà graben's southern boundary fault and has generated a fault-bend growth fold (Sant Corneli ramp anticline) (Bond and McClay 1995; Mencos et al. 2015). The Upper Pedraforca thrust sheet (Itinerary 9) is a similar structure. The thrust is covered by unfaulted rocks of the Garumnian series. Other thrusts probably formed during the Campanian in the highly thinned crust of the Cretaceous rift. These thrusts were north-vergent and included the Internal Metamorphic Zone, which overrides the rest of the NPZ. The entire NPZ was extruded partly northward (with a first episode of displacement along the NPFT) and partly southward (e.g., Espurt et al. 2019; Ternois et al. 2019). The NPZ is thus the proposed root area of the SPZ nappes, the latter currently interpreted as the outermost envelope of the central Pyrenean nappe stack (Figs. 1.2, 1.3). By early Paleocene time, the convergent plate motions had resulted in (i) cancelling out the crustal extension generated by Cretaceous rifting, (ii) thickening the crust to near-normal values beneath the NPZ, and (iii) accelerating subsidence in the two range-parallel foredeeps.

After 85 Ma, the foredeep in the north (future North-Pyrenean Zone) functioned as a receptacle for clastic sequences. The marine trough filled sequentially from east to west during Maastrichtian time, in step with the Axial Zone progressively rising above sea level in the east in response to early Pyrenean collision. The foredeep marine sequences and part of the Axial Zone in the eastern Pyrenees were soon covered by continental red beds known regionally as the Paleocene Garumnian series (e.g., Rosell et al. 2001). Meanwhile, the west of the present-day Pyrenees remained below sea level (Calvet et al. 2021).

1.3.4 The Middle Cenozoic Mountain Range

The width of the crustal wedge generated during Paleogene collision ranges between ~ 125 km in the centre and ~ 60 km in the east. Some weakly- to strongly-dipping thrusts and reverse faults from this period carve up the Hercynian basement and cover sequence. Three categories of tectonic unit have resulted from the collision. Some units consist exclusively of pre- to synorogenic Alpine sedimentary sequences. They form cover-rock nappes in which the Keuper evaporites—such as in the central part of the South Pyrenean Zone, known as the South Pyrenean Central Unit, or SPCU—have aided thrust propagation. Other tectonic units, often among the thickest, consist of mixed basement (Axial Zone) and cover-rock packages (pre- to synorogenic strata, now the SPZ): Gavarnie nappe at the centre of the range; Aspres, Canigou, Vallespir and Cadí nappes in the east (see Itinerary 4). Others still are basement nappes situated within the Axial Zone. Those three nappe categories may be termed upper, middle, and lower, respectively (Muñoz et al. 1986; Fontboté et al. 1986; Laumonier 1987, 2015). A major consequence of the late Eocene to early Oligocene tectonic movements (in the east) has been the uplift of the Axial Zone like a large pop-up relative to the NPZ and SPZ on either side of it. As a result, the Axial Zone forms a vast basement window while also hosting the roots of the middle and lower nappe units.

The Iberian pro-wedge is a south-vergent thrust sequence consisting of upper, middle and lower units, respectively. The South Pyrenean Frontal Thrust (SPFT) schematically represents the brow of the middle and/or lower thrusts. This configuration is very clear in the southeastern Iberian domain, where growth of the orogenic pro-wedge ended in early Oligocene time, but less clear in the southwestern Iberian domain, where collisional deformation continued until the early Miocene and where deformed foreland rock sequences became incorporated into the SPZ. The limit between the Axial Zone and the SPZ presents itself as a large range-front flexure—schematically a monocline pertaining to a frontal ramp fold (with an anticline on the southern edge of the Axial Zone and a syncline on the northern edge of the SPZ). Another consequence of the late uplift of the Axial Zone is that the upper units of the SPZ are klippen preserved in the flexural syncline and 'floating' on the middle units of Paleogene age (Fig. 1.3D). This is clearly observable in the case of the Pedraforca nappe resting on the Cadí nappe (see Itinerary 9). The position of their root area, perhaps within but, equally possibly, north of the Axial Zone, is debated (see Laumonier 2015; Calvet et al. 2021).

1.3.4.1 Tectonic Units of the Iberian Domain

In greater detail, the various tectonic units described above display lateral variation along the strike of the orogen (Figs. 1.2, 1.3D). Here we focus on the eastern and central Pyrenees.

Eastern Pyrenees

The nappe pile in the eastern Pyrenees displays from top to bottom (Figs. 1.2, 1.3D): (i) the Aspres upper unit within the Axial Zone; (ii) the SPZ upper units (Pedraforca nappe; Empordá nappe; klippen of Coustouges, Bac Grillera, and Biure), which are preserved in the Ripoll syncline (see Itineraries 4 and 9); (iii) the Canigou nappe (see Itineraries 4 and 6): a thick and extensive middle unit which incorporates a large component of Axial Zone basement rocks (as well as the Amélie-les-Bains Mesozoic sequence, which is the only remaining evidence of pre-orogenic cover rocks in the eastern Axial Zone), and most of the SPZ (Cadí units). The thin and discontinuous Vallespir unit is the lower portion of the Canigou nappe (see Itinerary 4); (iv) some lower units displaying outcrops of their basal thrust planes (Albères and Roc de France units; Freser duplex structure); (v) components of the lowermost unit (mainly the Saint-Laurent-de-Cerdans unit), the floor-thrust of which (no known outcrops of it, however) forms the base of the pro-wedge and functions as the lowermost imbricate for the overlying frontal ramps; and (vi) the entirely concealed South-Pyrenean Serrat unit (Fig. 1.3D).

In the Axial Zone, the lower units are visible beneath the Canigou nappe in the vast Albères and smaller Freser windows. The eastern segment of the SPFT is known as the Vallfogona Thrust, where the lower and middle thrusts converge. The Coustouges–Lower Bac Grillera–Biure klippen have usually been correlated with the Aspres unit of the Axial Zone. The Pedraforca thrust (now essentially a klippe), which appears rootless, probably originated somewhere just north of the Axial Zone. In such a scenario, the Mérens and Hospitalet faults (Fig. 1.2) should be interpreted as Hercynian mylonitic reverse faults (see Itinerary 6), having functioned as normal faults in the context of Cretaceous rifting but subsequently reactivated as low-angle reverse faults in the late Cretaceous and involved in the southward displacement of the upper nappes.

The foregoing structural outline gets locally more complicated in the eastern Pyrenees because of at least two main factors:

(i) the rhomboidal shape of a number of tectonic units. This results from two sets of thrusts striking respectively NW–SE (Aspres) and NE–SW (Vallespir), in each case because they correspond to either left-lateral or right-lateral synorogenic ramps. These ramps were themselves guided by older Hercynian structures such as mylonitic fault zones, late Hercynian (quartz-plugged) normal faults; probably likewise by other pre-orogenic (Cretaceous rift-related) normal faults; and possibly also by synsedimentary normal faults of Ypresian age in the South-Pyrenean Zone;

(ii) the interference with the synorogenic fabric from Neogene normal faults (e.g., Têt Fault, Tech Fault), a number of which appear to be extensionally reactivated components of the NE–SW ramps (see Itineraries 4, 6, 7 and 8).

Central Pyrenees

The central Pyrenees present a classic sequence of thrust units from north to south: (i) within the Axial Zone, the Gavarnie Thrust shears the Paleozoic basement but also overrides the Mesozoic sequences to the south, along the border between the Axial Zone and the SPZ. The thrust plane separates the Gavarnie nappe from the underlying Benasque/Orri nappe, which itself overrides the Bielsa and Rialp nappes (Fig. 1.3); (ii) just south of the Axial Zone, the Nogueres Zone is the leading edge of a nappe consisting of Upper Paleozoic basement units carrying a thin envelope of Permian to Triassic sedimentary rocks; (iii) the South-Pyrenean Central Unit (SPCU)—actually a lateral continuation of the Pedraforca nappe (see above and Itinerary 9)—is a vast thrust-sheet lobe advancing far out over the Ebro Basin (it mainly consists of the Montsec thrust, but also incorporates the late Cretaceous Bóixols unit; Figs. 1.2, 1.3), and effectively also a vast klippe; (iv) lastly, the Serres Marginales unit is the westward continuation into the central Pyrenees of the lowermost sheet of the Pedraforca nappe (Figs. 1.2, 1.3; see also Itinerary 9, Fig. 14.2 therein).

Interpretations of the SPCU have been strongly influenced by results from the ECORS deep seismic (1985–1986), from which emerged a 'standard model' of the central Pyrenean crust (Fig. 1.3B; Muñoz 1992; Beaumont et al. 2000). The standard model advocates the following geometry: (i) the Gavarnie nappe extends eastward via the Nogueres Zone, the Gavarnie Thrust being renamed Nogueres Thrust as a result; (ii) the Axial Zone is interpreted as a vigorously compressed anticlinal nappe stack in which syn-orogenic thrust planes, such as the eastern end of the Gavarnie/Nogueres Thrust, have been tilted almost vertically; (iii) the putative root area of the South-Pyrenean allochtonous units (particularly the Cretaceous Organyà Basin) is inferred to have coincided with the Rialp basement unit, i.e., it was south of the present-day Axial Zone. A partly analogous model has been proposed for the Pedraforca unit further east (Vergès 1999, and others thereafter), examined and discussed in Itinerary 9.

The 'standard model', however, contains a number of structural inconsistencies, shown in Fig. 1.3B. A new model (Fig. 1.3C), based on extensive geological field observations (Soler et al. 1998; Laumonier 2015; Teixell et al. 2018; Angrand 2017; Cochelin et al. 2018; Espurt et al. 2019; Ternois et al. 2019; Calvet et al. 2021) emphasizes the following alternative interpretation:

(i)   an absence in the Axial Zone, between the Gavarnie and Aspres thrusts, of an Alpine thrust, whether upturned or not;

(ii) the need for rooting the SPCU–Pedraforca thrust north of the Axial Zone, implying the existence of a major nappe overspanning the entire Axial Zone and overriding a comparatively more stationary basement unit named Axial Central Unit (ACU, Figs. 1.2, 1.3). The ACU lumps together the eastern parts of the Gavarnie and Orri (Benasque) units and the basement component of the Canigou nappe;

(iii) the need for the Nogueres units to form a duplex structure between the Gavarnie and Benasque–Orri units (Fig. 1.3C).

Crustal shortening within the Axial Zone basement rocks was thus more limited than inferred by the 'standard model', and the pretectonic position of the SPCU–Pedraforca nappe system was not situated south of the current Axial Zone: it was, instead, located above the Axial Zone. Furthermore, unlike the 'standard model', the nappe stack in the revised model does not form a thick anticlinal hump: it is, instead, flatter and thinner (compare cross-sections B and C, Fig. 1.3). Because of its décollement in the Keuper evaporites, the SPCU has propagated much further southward than its eastward fragment (Pedraforca, via the Segre ramp) or its westward analogues (Ordesa, via the Ainsa Oblique Zone) (Fig. 1.2) (Muñoz et al. 2013).

1.3.4.2 Tectonic Units of the European Domain

North of the Axial Zone, the Pyrenean retro-wedge displays outcrops of Hercynian basement and Mesozoic–Cenozoic cover. The NPZ results from the tectonic inversion of older rift sequences during the Cretaceous (~80–66 Ma), and most of all during the Eocene (~56–35 Ma). The NPZ, NPF, and part of the Axial Zone were displaced northward by 25–30 km over the North-Pyrenean Frontal Thrust which, for example, can be inspected in the field at the Pic de Bugarach (Itinerary 5). Note that the IMZ (Internal Metamorphic Zone) is thrust over unmetamorphosed rocks, both northwards over the rest of the NPZ and southward over the Axial Zone (in this case, forming the proximal part of the Supra-Axial nappe, now eroded; Fig. 1.3D). North of the NPZ, the narrow Sub-Pyrenean Zone consists of folded synorogenic late Cretaceous and Eocene sedimentary series (e.g., Mas d'Azil and Plantaurel anticlines). This outermost zone of the retro-wedge corresponds to rock series from the least thinned crustal boundary of the Cretaceous rift zone. North of the Sub-Pyrenean Zone, the synorogenic sequences of the Aquitaine Basin are weakly deformed or undeformed.

The NPZ is narrow (10–15 km) and highly deformed to the east; it widens in the central Pyrenees (15–30 km), locus of the main North-Pyrenean basement massifs of Arize, Saint-Barthélemy, and Trois-Seigneurs. In the west, the NPZ transitions to moderately deformed and inverted versions of the Cretaceous rift basins, which today form the Béarn Ranges and Mauléon Ranges. In the far east, the retro-wedge configuration changes abruptly: (i) the southern part of the NPZ (especially the Internal Metamorphic Zone) disappears as result of the extensional tectonics prevalent since the Oligocene in the Gulf of Lion; (ii) the outer NPZ bends around and forms the NNW-vergent orocline of the eastern Corbières; and (iii) the Sub-Pyrenean Zone widens considerably and includes the basement outcrops of the Mouthoumet massif and Montagne Noire (see Itineraries 2 and 5). In the eastern NPZ and Corbières, salt tectonics have overall played a major role in generating the structural mosaic (Ford and Vergés 2020).

1.4   Lithospheric Processes and First-Order Links with the Modern Topography

Knowledge of the deep structure of the Pyrenees arises from a number of geophysical studies (Souriau and Granet 1995; Pous et al. 1995a, b; Damotte 1998; Vacher and Souriau 2001; Souriau et al. 2008, Chevrot et al. 2014, 2015, 2018; Wehr et al. 2018; Campanyá et al. 2018). Estimated maximum shortening is 165 km in the central Pyrenees (100 km or 147 km, depending on the authors), dropping off to 125 km to the east and~80–90 km to the west (Choukroune 1992; Muñoz 1992; Roure and Choukroune 1998; Vergés et al. 1995, 2002; Sinclair et al. 2005; Beaumont et al. 2000; Vissers and Meijer 2012b; Mouthereau et al. 2014; Teixell et al. 2018). Collision terminated sooner in the eastern portion of the orogen and endured until 20 Ma in the westernmost segment.

A lithospheric root has been detected to depths of >100 km, thus documenting in the central Pyrenees a component of subduction of the Iberian lower crust beneath the European. The crustal root itself exceeds thicknesses of 50 km beneath the central Pyrenees, with a large Moho jump beneath the NPF. These attributes change rapidly east of the Ariège River (ca. 1°30′E), where features of the steeply dipping Iberian subduction slab disappear and the Iberian Moho is almost flat-lying at shallower depths of 40–45 km. West of Andorra, the crustal root is anomalously dense (eclogite) and apparently undergoing gravitational subsidence and/or delamination (Chevrot et al. 2018; Campanyá et al. 2018; Dufréchou et al. 2018).

A major negative Bouguer anomaly detected in the eastern Pyrenees has been attributed to the disappearance of the lower crust or to an anomalously hot lithospheric mantle (Wehr et al. 2018), which in either case would provide dynamic support to the anomalously elevated, isostatically overcompensated topography in this part of the Pyrenees, and would explain the elevated topography of the eastern Pyrenees despite an absence of crustal root (Gunnell et al. 2008, 2009). The hypothesis of an overheated mantle is potentially consistent with a southward flow of asthenosphere from the Massif Central plume (active since the Oligocene) towards the Mediterranean (Barruol and Granet 2002; Barruol et al. 2004). Geophysical evidence also suggests that the lithospheric root of the central Pyrenees is hot and partially melting (Pous et al. 1995b; Campanyá et al. 2018), thus potentially also contributing a component of dynamic uplift (see also Conway-Jones et al. 2019).

The detected contrast in crustal thickness along the strike of the orogen (i) could be the legacy of pre-collisional rift geometry during the Cretaceous, itself controlled by contrasting crustal properties on either side of the Toulouse Fault

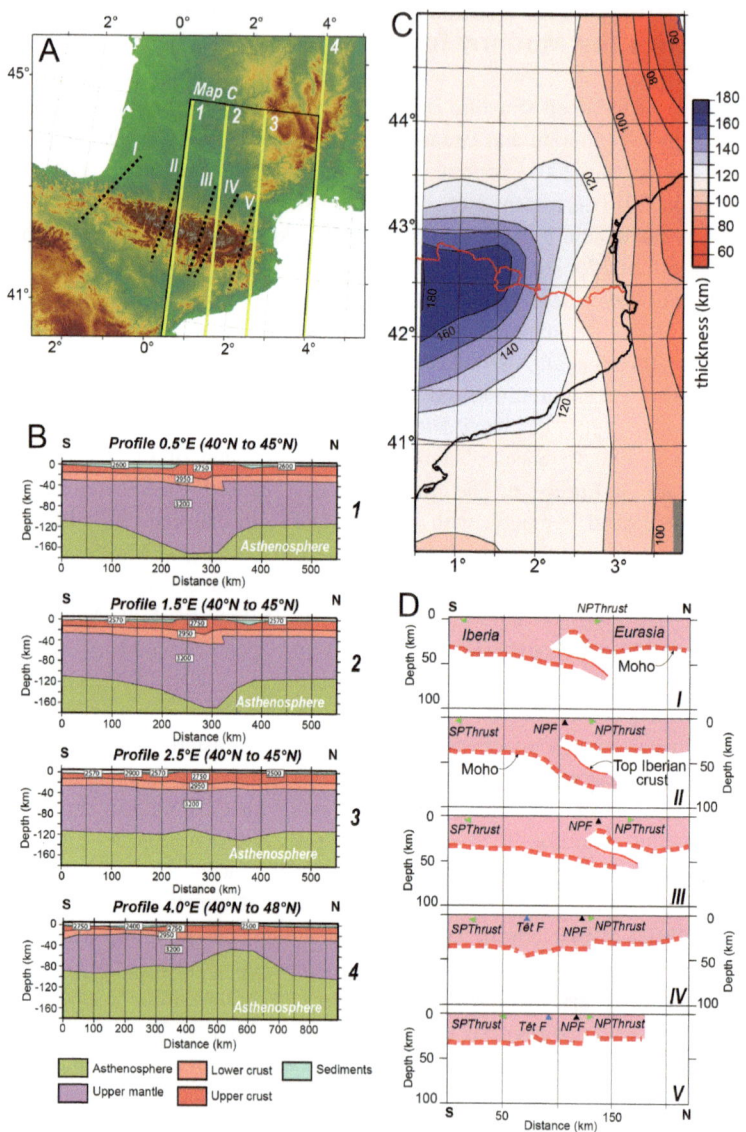

◀**Fig. 1.4 Geophysical imaging of the deep lithospheric structure of the Pyrenees**. **A** Location of geophysical swath profiles (lines 1–4) and seismic profiles (lines I to V). Relief base map: Copernicus EU-DEM, v. 1.1. **B** Lithospheric cross-sections obtained from simultaneous best fits between free-air gravity, geoid, thermal, and topographic data (after Gunnell et al. 2008, modified); note lithospheric taper in the east (shallowing of lithosphere–asthenosphere boundary), along profiles 3 and 4. Profile 4 extends to the Massif Central and reveals the well-documented thinned lithosphere and shallow asthenosphere responsible for late Neogene volcanism dynamic uplift in that region. **C** Map of the lithosphere–asthenosphere boundary interpolated from profiles 1–4 (after Gunnell et al. 2008, modified). **D** Crustal-scale cross-sections through the Pyrenees (data from Chevrot et al. 2018); note the thinner crust along profiles IV and V, immediately east of the ECORS profile (III)

(Chevrot et al. 2018); but it (ii) could also, at least in part, be the result of relief-transforming processes caused by Mediterranean crustal extension impinging on the Pyrenean domain. From the middle Oligocene to the middle Miocene, rifting and southeastward drift of the Corsica–Sardinia block marked the opening of the Western Mediterranean back-arc basin, forming the Gulf of Lion (Séranne 1999; Lacombe and Jolivet 2005). Break-up of the orogen in the Western Mediterranean has been linked to the Catalan Transfer Zone (Rehault et al. 1984; Mauffret et al. 2001), which is a deep crustal fault striking NW–SE along which the Corsica–Sardinia block sheared itself away from the remainder of the Pyrenees during the Neogene as part of the wider process of Mediterranean trench rollback (see Itinerary 1, Fig. 1.3 therein). The Pyrenean crust thins rapidly just 20 km east of the current Roussillon coastline, and the crustal root that would be expected to occur beneath a mountain range as elevated as the eastern Pyrenees is absent (Fig. 1.4; Gunnell et al. 2008; Chevrot et al. 2018; Campanyá et al. 2018; Calvet et al. 2021).

Today, Iberia and Europe are considered to be locked to one another, thus forming a single tectonic plate (Nocquet and Calais 2003, 2004; Fernandes et al. 2007; de Vicente et al. 2008; Asensio et al. 2012). The main focus of continental deformation has transferred to the Betic Cordillera, i.e., to the southern boundary of the Iberian Plate. Seismicity in the Pyrenees appears moderate (Fig. 1.5), with one $M > 5$ event every five years since 1950, and 12 events of a higher intensity (i.e., greater than VIII on the Modified Mercalli damage intensity scale) since the Middle Ages (Souriau and Pauchet 1998). The focal mechanisms are poorly constrained: whereas previous studies supported a compressional stress regime involving strike-slip and reverse faulting (Goula et al. 1999), which is in agreement with geological evidence of Quaternary reverse faulting (Philip et al. 1992; Calvet 1996, 1999; Carbon et al. 1995; Goula et al. 1999; Fleta et al. 2001; Baize

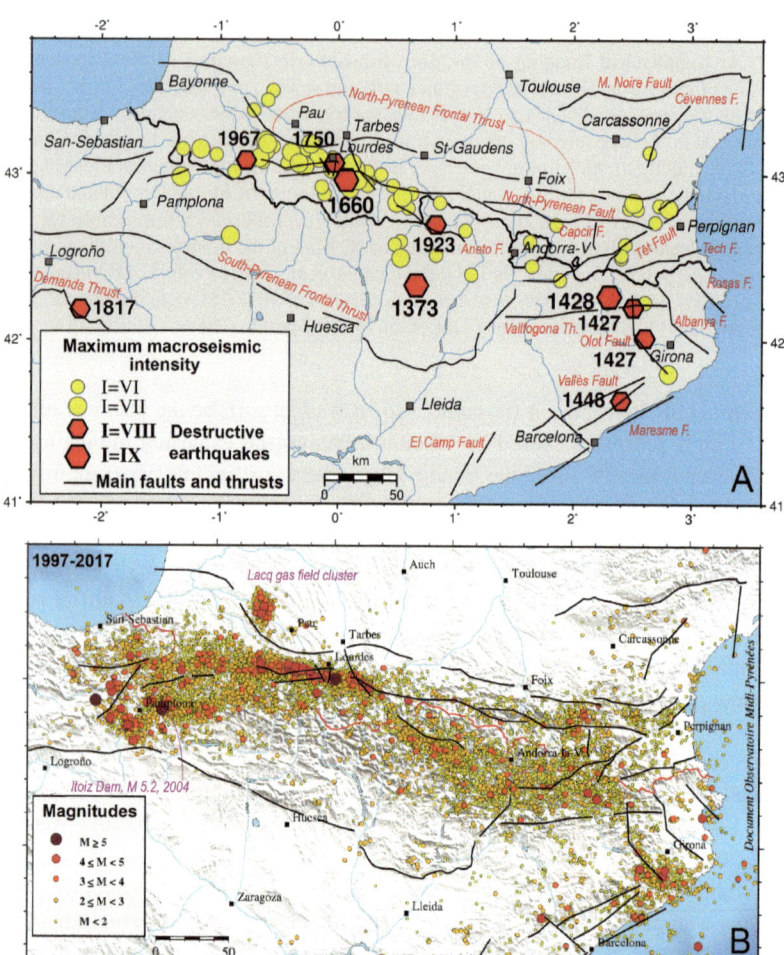

Fig. 1.5 Seismic activity in the Pyrenees. A historic earthquakes since the Middle Ages. **B** Instrumentally measured seismicity from 1997 to 2017. Note similarities in the spatial distribution of historical earthquakes during the last 600 years and instrumental seismicity in the last 20. Earthquake frequencies are highest along the north edge of the Axial Zone in the west of the orogen, then cross the Axial Zone in the vicinity of the Maladeta massif, then rather concentrates along the southern edge of the Axial Zone in the eastern half of the range, where the pattern is nonetheless more diffuse. A cluster of epicentres highlights the westernmost Catalonian extensional basins. The main tectonic lineaments are shown but the exact sources of Pyrenean seismicity are under continuing debate (Source: Réseau de Surveillance Sismique des Pyrénées, Observatoire Midi-Pyrénées, Toulouse)

et al. 2002; Alasset and Meghraoui 2005; Lacan et al. 2012; Lacan and Ortuño 2012), recent revision favours instead an extensional regime for the Pyrenees, i.e., exclusively involving normal faults (Chevrot et al. 2011). The most recent synthesis based on GPS measurements and focal mechanism records (Rigo et al. 2015), however, suggests a more nuanced picture that lays emphasis on regional variation in deformational style and intensity depending on the stress field through time. The weakness of horizontal deformation is nonetheless confirmed, with a gradient from transtension in the west to transpression in the east.

1.5 Synthesis

The eastern Pyrenees, locus of the Proto-Pyrenees and thus region with the longest history as a mountain range, have been affected by multiple orogenic overprints. The long Variscan history of the Pyrenean region ended in the complete destruction of the Hercynian orogen and in the burial of its residual topography beneath a mass of its own debris (initially Permian and Triassic red claystone and sandstone). A new chapter then began with the Alpine orogenic cycle. This started with a long period of marine sedimentation extending from the Triassic (~240 Ma) to the Barremian (~125 Ma). A period of rifting then transformed the European Plate margin, leading to the separation of Iberia from the rest of Eurasia. Convergence, collision and underthrusting of the Iberian Plate beneath the Eurasian Plate ensued in a succession of stages between the late Cretaceous and at least the late Oligocene. A subsequent phase of Neogene crustal extension was recorded, beginning during the Oligocene and subsequently impinging on at least the eastern half of the orogen.

References

Alasset JP, Meghraoui M (2005) Active faulting in the western Pyrénées (France): paleoseismic evidence for late Holocene ruptures. Tectonophysics 409:39–54

Angrand P (2017) Evolution 3D d'un rétro-bassin d'avant-pays: le Bassin d'Aquitaine. Univ, de Lorraine, France, 162 p

Asensio E, Khazaradze G, Echeverria A, King RW, Vilajosana I (2012) GPS studies of active deformation in the Pyrenees. Geophys J Int 190:913–921

Baize S, Cushing M, Lemeille T, Granier B, Grellet B, Carbon D, Combes C, Hibsch C (2002) Inventaire des indices de rupture affectant le Quaternaire, en relation avec les grandes structures connues en France métropolitaine et dans les régions limitrophes. Mém H-S Soc Géol Fr, vol 175, 142 p

Barnolas A, Chiron JC (eds) (1996) Synthèse géologique et géophysique des Pyrénées. Vol. 1: introduction, géophysique, cycle hercynien. Edition BRGM-ITGE, 729 p

Barnolas A, Chiron JC (eds) (2018) Synthèse géologique et géophysique des Pyrénées. Vols 2 & 3: Cycle Alpin. AGSO and BRGM

Barruol G, Granet M (2002) A tertiary asthenospheric flow beneath the southern Massif Central indicated by upper mantle seismic anisotropy and related to the west Mediterranean extension. Earth Planet Sci Lett 202:31–47

Barruol G, Deschamps A, Coutant O (2004) Mapping upper mantle anisotropy beneath SE France by SKS splitting indicates Neogene asthenospheric flow induced by Apenninic slab roll-back and deflected by the deep Alpine roots. Tectonophysics 394:125–138

Beaumont C, Muñoz JA, Hamilton J, Fullsack P (2000) Factors controlling the Alpine evolution of the central Pyrenees inferred from a comparison of observations and geodynamical models. J Geophys Res 105:8121–8145

Bond RMG, McClay KR (1995) Inversion of a lower cretaceous extensional basin, South Central Pyrenees, Spain. In: Buchanan JG, Buchanan PG (eds) Basin inversion. Geol Soc Lond Spec Publ 88:415–431

Calvet M (1996) Morphogenèse d'une montagne méditerranéenne : les Pyrénées orientales. Documents du BRGM, Orléans, n° 255, 3 vols. 1177 p

Calvet M (1999) Régime des contraintes et volumes de relief dans l'Est des Pyrénées. Géomorphol Relief Processus Environ 3:253–278

Calvet M, Gunnell Y, Laumonier B (2021) Denudation history and palaeogeography of the Pyrenees and their peripheral basins: an 84-million-year geomorphological perspective. Earth-Sci Rev 215:103436

Campanyà J, Ledo J, Queralt P, Marcuello A, Muñoz JA, Liesa M, Jones AJ (2018) New geoelectrical characterization of a continental collision zone in the Central–Eastern Pyrenees: constraints from 3-D joint inversion of electromagnetic data. Tectonophysics 742–743:168–179

Canérot J (2008) Les Pyrénées, t 1: Histoire géologique, t2: Itinéraires de découverte. Atlantica and BRGM éds, Biarritz, Orléans, 516 and 127 pp

Carbon D, Combes P, Cushing M, Granier T, Grellet B (1995) Rupture de surface post-pléistocène moyen dans le bassin aquitaine. C R Acad Sci Paris IIa 320:311–317

Chevrot S, Sylvander M, Delouis B (2011) A preliminary catalog of moment tensors for the Pyrenees. Tectonophysics 510:239–251

Chevrot S, Villaseñor A, Sylvander M, Benahmed S, Beucler E, Cougoulat G, Delmas Ph, de Saint Blanquat M, Diaz J, Gallart J, Grimaud F, Lagabrielle Y, Manatschal G, Mocquet A, Pauchet H, Paul A, Péquegnat C, Quillard O, Roussel S, Ruiz M, Wolyniec D (2014) High-resolution imaging of the Pyrenees and Massif Central from the data of the PYROPE and IBERARRAY portable array deployments. J Geophys Res Solid Earth 119:6399–6420

Chevrot S, Sylvander M, Diaz J, Martin R, Mouthereau F, Manatschal G, Masini E, Calassou S, Grimaud F, Pauchet H, Ruiz M (2018) The non-cylindrical crustal architecture of the Pyrenees. Sci Rep 8:9591. https://doi.org/10.1038/s41598-018-27889-x

Chevrot S, Sylvander M, Diaz J, Ruiz M, Paul A and the PYROPE Working Group (2015) The Pyrenean architecture as revealed by teleseismic P-to-S converted waves recorded along two dense transects. Geophys J Int 200:1096–1107

Choukroune P (1992) Tectonic evolution of the Pyrenees. Annu Rev Earth Planet Sci 20:143–158

Choukroune P, Le Pichon X, Séguret M, Sibuet J-C (1973) The Pyrenees: subduction and collision? Earth Planet Sci Lett 18:109–118

Cochelin B, Lemirre B, Denèle Y, de Saint Blanquat M, Lahfid A, Duchêne S (2018) Structural inheritance in the Central Pyrenees: the Variscan to Alpine tectonometamorphic evolution of the Axial Zone. J Geol Soc 175:336–351

Conway-Jones BW, Roberts GG, Fichtner A, Hoggard M (2019) Neogene epeirogeny of Iberia. Geochem Geophys Geosyst 20:1138–1163

Damotte B (Ed.) (1998) The ECORS Pyrenean deep seismic surveys, 1985–1994. Mém Soc Géol Fr 173:108

Dufréchou G, Tiberi C, Martin R, Bonvalot S, Chevrot S, Seoane L (2018) Deep structure of Pyrenees range (SW Europe) imaged by joint inversion of gravity and teleseismic delay time. Geophys J Int 214:282–301

Espurt N, Angrand P, Teixell A, Labaume P, Ford M, de Saint Blanquat M, Chevrot S (2019) Crustal-scale balanced cross-section of the Central Pyrenean belt (Nestes-Cinca transect): highlighting the structural control of Variscan belt and Permian-Mesozoic rift systems on mountain building. Tectonophysics 764:25–45

Fernandes RMS, Miranda JM, Meijninger BML, Bos MS, Noomen R, Bastos L, Ambrosius BAC, Riva REM (2007) Surface velocity field of the Ibero-Maghrebian segment of the Eurasia-Nubia plate boundary. Geophys J Int 169:315–324

Fleta J, Santanach P, Goula X, Martínez P, Grellet B, Masana E (2001) Preliminary geologic, geomorphologic and geophysical studies for the palaeoseismological analysis of the Amer fault, NE Spain. Neth J Geosci 80:243–253

Fontboté JM, Muñoz JA, Santanach P (1986) On the consistency of proposed models for the Pyrenees with the structure of the eastern parts of the belt. Tectonophysics 129:291–301

Ford M, Vergés J (2020) Evolution of a salt-rich transtensional rifted margin, eastern North Pyrenees. France. J Geol Soc Lond. https://doi.org/10.1144/jgs2019-157.

Goula X, Olivera C, Fleta J, Grellet B, Lindo R, Rivera LA, Cisternas A, Carbon D (1999) Present and recent stress regime in the eastern part of the Pyrenees. Tectonophysics 308:487–502

Gunnell Y, Zeyen H, Calvet M (2008) Geophysical evidence of a missing lithospheric root beneath the Eastern Pyrenees: consequences for post-orogenic uplift and associated geomorphic signatures. Earth Planet Sci Lett 276:302–313

Gunnell Y, Calvet M, Brichau S, Carter A, Aguilar JP, Zeyen H (2009) Low long-term erosion rates in high-energy mountain belts: insights from thermo- and biochronology in the Eastern Pyrenees. Earth Planet Sci Lett 278:208–218

Jammes S, Manatschal G, Lavier L, Masini E (2009) Tectonosedimentary evolution related to extreme crustal thinning ahead of a propagating ocean: example of the western Pyrenees. Tectonics 28(TC4012):24

Lacan P, Ortuño M (2012) Active tectonics of the Pyrenees: a review. J Iber Geol 38:9–30

Lacan P, Nivière B, Rousset D, Sénéchal P (2012) Late Pleistocene folding above the Mail Arrouy Thrust, North-Western Pyrenees (France). Tectonophysics 541–543:57–68

Lacombe O, Jolivet L (2005) Structural and kinematic relationships between Corsica and the Pyrenees-Provence domain at the time of the Pyrenean orogeny. Tectonics 24:TC1003. https://doi.org/10.1029/2004TC001673

Lagabrielle Y, Labaume P, de Saint Blanquat M 2010 Mantle exhumation, crustal denudation, and gravity tectonics during Cretaceous rifting in the Pyrenean realm (SW Europe): insights from the geological setting of the lherzolite bodies. Tectonics 29(TC4012). https://doi.org/10.1029/2009TC002588

Laumonier B (1987) Les structures tangentielles alpines de la partie orientale de la chaîne pyrénéenne, en particulier du Vallespir. C R Acad Sci Paris, 304(II):1081–1086

Laumonier B (2015) Les Pyrénées alpines sud-orientales (France, Espagne)—essai de synthèse. Revue Géol Pyr 2:44. http://www.geologie-despyrenees.com/

Le Pichon X, Bonnin J, Sibuet J-C (1970) La faille nord-pyrénéenne: faille transformante liée à l'ouverture du Golfe de Gascogne. C R Acad Sci Paris 271:1941–1944

Mauffret A, Durand de Grossouvre B, Dos Reis AT, Gorini C, Nercessian A (2001) Structural geometry in the eastern Pyrenees and western Gulf of Lion (Western Mediterranean). J Struct Geol 23:1701–1726

Mencos J, Carrera N, Muñoz JA (2015) Influence of rift basin geometry on the subsequent postrift sedimentation and basin inversion: the Organyà Basin and the Bóixols thrust sheet (south central Pyrenees). Tectonics 34:1452–1474

Mouthereau F, Filleaudeau P-Y, Vacherat A, Pik R, Lacombe O, Fellin MG, Castelltort S, Christophoul F, Masini E (2014) Placing limits to shortening evolution in the Pyrenees: role of margin architecture and implications for the Iberia/Europe convergence. Tectonics 33:2283–2314

Muñoz JA, Martínez A, Vergés J (1986) Thrust sequences in the Eastern Spanish Pyrenees. J Struct Geol 8:399–405

Muñoz JA, Beamud E, Fernández O, Arbués P, Dinarès-Turell J, Poblet J (2013) The Ainsa fold and thrust oblique zone of the central Pyrenees: kinematics of a curved contractional system from palaeomagnetic ans structural data. Tectonics 32:1142–1175

Muñoz JA (1992) Evolution of a continental collision belt: ECORS-Pyrenees crustal balanced cross-section. In: McClay KR (ed) Thrust tectonics. Chapman and Hall, New York, pp 235–246

Nocquet JM, Calais E (2003) The crustal velocity field in Western Europe from permanent GPS array solutions, 1996–2001. Geophys J Int 154:72–88

Nocquet JM, Calais E (2004) Geodetic measurements of crustal deformation in the western Mediterranean and Europe. Pure Appl Geophys 161:661–681

Philip H, Bousquet JC, Escuer J, Fleta J, Goula X, Grellet B (1992) Présence de failles inverses d'âge quaternaire dans l'est des Pyrénées: implications sismotectoniques. C R Acad Sci Paris II (314):1239–1245

Pous J, Muñoz JA, Ledo JJ, Liesa M (1995) Partial melting of subducted continental lower crust in the Pyrenees. J Geol Soc Lond 152:217–220

Pous J, Ledo J, Marcuello A, Daignières M (1995) Electrical resistivity model of the crust and upper mantle from a magnetotelluric survey through the central Pyrenees. Geophys J Int 121:750–762

Rehault JP, Boillot G, Mauffret A (1984) The western Mediterranean basin geological evolution. Mar Geol 55:447–477

Rigo A, Vernant P, Feigl KL, Goula X, Khazaradze G, Talaya J, Morel L, Nicolas J, Baize S, Chery J, Sylvander M (2015) Present-day deformation of the Pyrenees revealed by GPS surveying and earthquake focal mechanisms until 2011. Geophys J Int 201:947–964

Rosell J, Linares R, Llompart C (2001) El 'Garumniense' prepirenaico. Rev Soc Geol España 14:47–56

Roure F, Choukroune P (1998) Contribution of the ECORS seismic data to the Pyrenean geology: crustal architecture and geodynamic evolution of the Pyrenees. In: Damotte B (ed) The ECORS Pyrenean deep seismic surveys, 1985–1994. Mém Soc Géol Fr 173:37–52

Séranne M (1999) The Gulf of Lion continental margin (NW Mediterranean) revisited by IBS: an overview. In: Durand B, Jolivet L, Horvath F, Séranne M (eds) The Mediterranean basins: tertiary extension within the Alpine Orogen. Geol Soc Lond Spec Publ 156:15–36

Sinclair HD, Gibson M, Naylor M, Morris RG (2005) Asymmetric growth of the Pyrenees revealed through measurement and modelling of orogenic fluxes. Am J Sci 305:369–406

Soler D, Teixell A, García-Sansegundo J (1998) Amortissement latéral du chevauchement de Gavarnie et sa relation avec les unités sud-pyrénéennes. C R Acad Sci Paris 327:699–704

Souriau A, Granet M (1995) A tomographic study of the lithosphere beneath the Pyrenees from local and teleseismic data. J Geophys Res 100:18117–18134

Souriau A, Pauchet H (1998) A new synthesis of the Pyrenean seismicity and its tectonic implications. Tectonophysics 290:221–224

Souriau A, Chevrot S, Olivera C (2008) A new tomographic image of the Pyrenean lithosphere from teleseismic data. Tectonophysics 460:206–214

Teixell A, Labaume P, Ayarza P, Espurt N, de Saint Blanquat M, Lagabrielle Y (2018) Crustal structure and evolution of the Pyrenean-Cantabrian belt: a review and new interpretations from recent concepts and data. Tectonophysics 724–725:146–170

Ternois S, Odlum M, Ford M, Pik R, Stockli D, Tibari B, Vacherat A, Bernard V (2019) Thermochronological evidence of early orogenesis, eastern Pyrenees, France. Tectonics 38:1308–1336

Vacher P, Souriau A (2001) A 3D-model of the Pyrenean deep structure based on gravity modelling, seismic images and petrological constraints. Geophys J Int 145:460–470

Vergés J, Millán H, Roca E, Muñoz JA, Marzo M, Cirés J, Den Bezemer T, Zoetemeijer R, Cloetingh S (1995) Eastern Pyrenees and related foreland basins: pre-, syn- and post-collisional crustal scale cross-sections. Mar Pet Geol 12:893–915

Vergés J, Fernàndez M, Martínez A (2002) The Pyrenean orogen: pre-, syn-, and post-collisional evolution. In: Rosenbaum G, Lister GS (eds) Reconstruction of the evolution of the Alpine-Himalayan Orogen. J Virtual Expl 8:55–74 de Vicente G, Cloetingh S, Muñoz-Martin A, Olaiz A, Stich D, Vegas R, Galindo-Zaldivar J, Fernandez-Lozano J (2008) Inversion of moment tensor focal mechanisms for active stresses around the microcontinent Iberia: tectonic implications. Tectonics 27:TC1009. https://doi.org/10.1029/2006TC002093

Vissers RLM, Meijer PT (2012) Iberian plate kinematics and Alpine collision in the Pyrenees. Earth-Sci Rev 114:61–83

Wehr H, Chevrot S, Courrioux G, Guillen A (2018) A three-dimensional model of the pyrenees and their foreland basins from geological and gravity data. Tectonophysics 734–735:16–32

The denudation history of mountain belts is recorded in the clastic sedimentary components of their basins, with marine incursions potentially also providing additional clues about palaeogeography and base-level changes. Constraints on the chronostratigraphy of the foreland sequences are usually provided by the fossil record and, when suitable sequences exist, magnetostratigraphy; basin analysis help to reconstruct the tectonic regime and succession of depositional environments through time. These clues yield indirect evidence about magnitudes of energy in the slope system, and thus about the topography and relief of the mountain range. In the case of the Pyrenees, however, the late Cenozoic environmental conditions are paradoxically less well documented than the earlier periods of Mesozoic rifting and Paleogene collision, where marine sedimentary rocks were more ubiquitous and provided more precise chronostratigraphic information. We provide here a synthesis of the evolution of the eastern and central Pyrenees captured through the sedimentary record of erosion in the evolving mountain belt. We emphasise the chronology and spatial distribution of conglomerate sequences, the rock-cooling record of its basement outcrops, and the denudation of the foreland basins themselves in relation to base-level changes. The stratigraphic ages of the continental sedimentary sequences highlight information provided by biochronology (usually fossil mammalian remains), calibrated against the geological time scales of Vandenberghe et al. (2012) for the Paleogene, and of Hilgen et al. (2012) for the Neogene. The scarcity of magnetostratigraphic data on the French side (compared to the Spanish side) does not allow a uniform chronostratigraphic reference frame to be constructed on that basis.

© Springer Nature Switzerland AG 2022

27

M. Calvet et al., *Geology and Landscapes of the Eastern Pyrenees*, GeoGuide,
https://doi.org/10.1007/978-3-030-84266-6_2

2.1   Patterns of Terrigenous Sedimentation in the Ebro Foreland

Because of the semi-arid climate and thin vegetation cover of northern Spain, and because the clastic pro-foreland zone has been more vigorously uplifted and incised by deep valleys that its French retro-foreland counterpart (Fig. 2.1), very good outcrops and exposures have promoted many opportunities for extensive and detailed stratigraphic and sedimentological studies of the Iberian fill sequences. This has also spawned the description and naming of a multitude of lithostratigraphic units of local extent, resulting in a complicated inventory of place-based

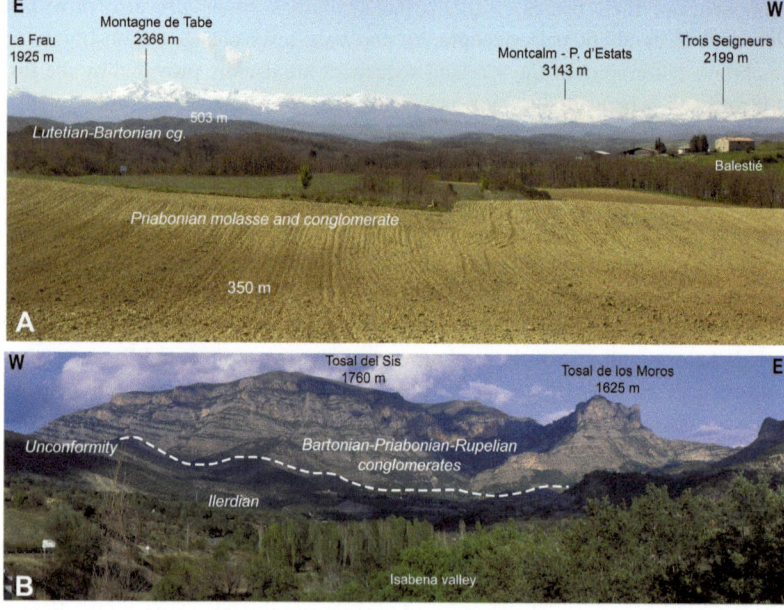

Fig. 2.1 Structural asymmetry of the Pyrenees: evidence in the landscapes. **A** Low hills of the Paleogene retro-foreland piedmont (Priabonian 'Palassou' conglomerate beds, and Bartonian–Lutetian in the background), vicinity of Mirepoix (Ariège); note the elevated, snow-capped massifs of the Axial Zone and their North-Pyrenean Hercynian basement outliers (e.g., Trois Seigneurs). **B** Paleogene pro-foreland piedmont, which has been incorporated into the mountain range by more intense Neogene uplift than in the retro-foreland. Note the thick conglomerate series of Sierra de Sis (Bartonian to Rupelian), resting unconformably on the early Eocene marine outcrops of the Isabena valley

terminologies. Despite appearances of regional uniformity inferred from attempts at correlations based mostly on lithostratigraphic criteria, geochronological constraints acquired progressively over time have revealed that many units displaying similar petrographic facies in different parts of the foreland are in fact substantially diachronous (Calvet et al. 2021). Progress in revising the lithostratigraphic correlations has mainly been gained from sequence stratigraphy, magnetostratigraphy, thermochronological dating of detrital sediments, and from the discovery of various continental fossil assemblages.

Unlike the Aquitaine foreland (from which the sea withdrew rapidly from the Corbières westward to the Béarn in Ypresian time, i.e., before ~48 Ma; see Sect. 2.2), the Iberian foreland remained a marine environment for an additional 10–12 million years, i.e., until the early Priabonian (~36 Ma) (Sanjuan et al. 2012; Costa et al. 2013; Barnolas and Gil-Peña 2001, 2019; Garcés et al. 2020). After ~35 Ma, however, the Ebro Basin became internally drained for the subsequent 20–25 million years—thus ruling out constraints from marine indicator fossils. Unfortunately, mammalian faunal assemblages contained in the post-35 Ma continental deposits have turned out to be scarce, low in species diversity, and very unevenly distributed in the stratigraphy and across the foreland area. Fossil assemblages, for example, are much less abundant in the pro-wedge molasse and conglomerate sequences than in the Ebro Basin itself (Alvarez Sierra et al. 1990; Cuenca et al. 1992; Sudre et al. 1992). The scarcity of biochronological markers is nonetheless compensated by other tools such as magnetostratigraphy (Barberà et al. 2001; Pérez-Rivarés et al. 2018; Garcés et al. 2020).

Despite these constraints and complications, the record of Paleogene clastic sequences in the eastern and central Pyrenees can broadly be divided into two strike-perpendicular segments from east to west (Barnolas et al. 2019): the South Pyrenean Foreland Basin in western Catalonia and onward into Aragón (Fig. 2.2), and the Southeast Pyrenean Foreland Basin, which is confined to eastern and central Catalonia (Fig. 2.3; see Itinerary 9). We emphasise clast provenance and the diachronous patterns of basin fill generated by the rising Pyrenees.

2.1.1 Synorogenic Chronostratigraphy of the Southeast Pyrenean Foreland Basin

The depositional environment of the Proto-Pyrenees, and subsequently of the Ancestral Pyrenees, in the southeastern part of the Iberian foreland is recorded in the Southeast Pyrenean Foreland Basin. It has been well constrained by sequence

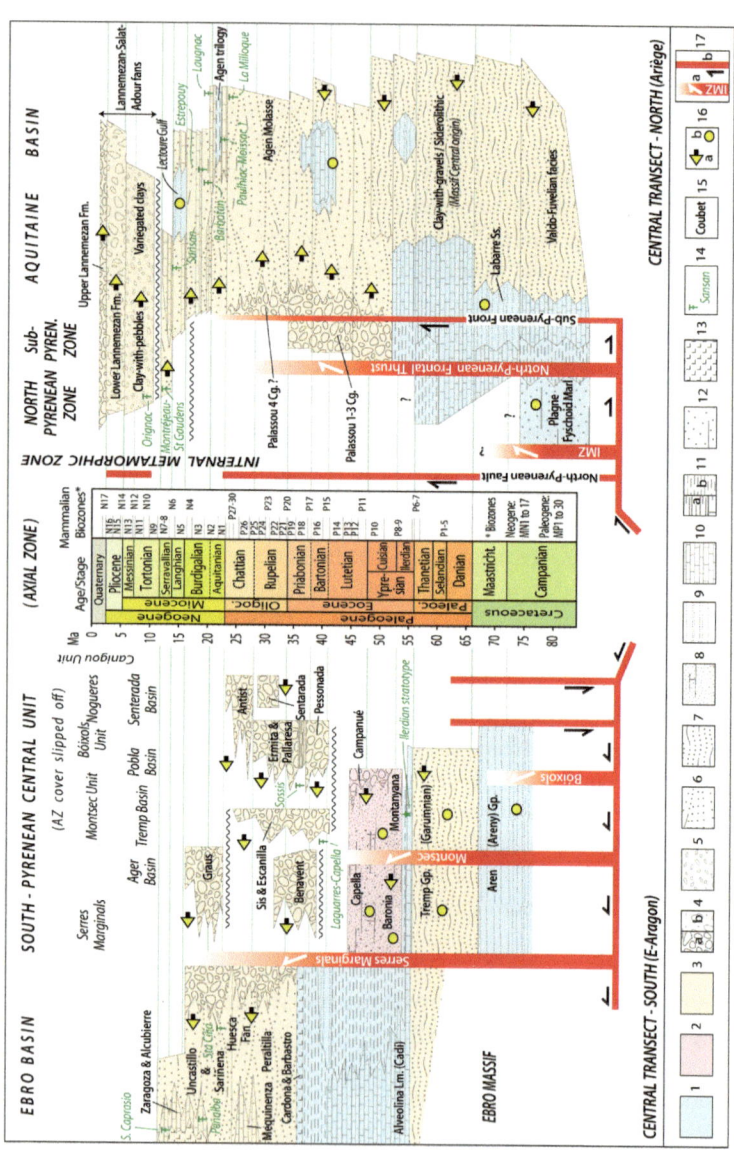

◀**Fig. 2.2** **Lithostratigraphy and tectonic evolution of the central Iberian foreland and European retro-foreland sedimentary basins of the Pyrenees**. These paired diagrams are chronostratigraphic (indicate the age of sedimentary formations, and includes European Land Mammal Age zones) and tectonostratigraphic (show the depositional contents of tectonic units). Horizontal scale is unspecified because the N–S widths of the tectonic domains narrowed progressively over time. Key to symbols, colours and ornaments chronostratigraphy **1**: marine sequences; **2**: deltaic and base-level floodplain sequences; **3**: continental sequences; **4**: coarse clastic formations; **4a**: proximal conglomerates; **4b**: other coarse-textured debris-fan formations; **5**: clay with decalcified pebbles; **6**: sand- and clay-textured, carbonate-rich fluvial molasse; **7**: other sand- and clay-textured molasse occurrences; **8**: sand- and clay-textured deltaic facies; **9**: other marine sandstones; **10**: shelf limestone, and lacustrine or palustrine limestone; **11a**: epipelagic carbonates; **11b**: outer-shelf marls; **12**: slope and abyssal flysch and turbidites; **13**: marine or continental evaporites (gypsum, salt); **14**: ELMA-related data points; **14a**: main mammalian fossil site; **14b**: stratigraphic type section name; **15**: main lithostratigraphic unit names; **16**: Palaeodrainage directions; **16a**: transverse drainage systems; **16b**: strike-parallel drainage systems; **17**: tectonic discontinuities; **17a**: main thrusts; **17b**: main faults

stratigraphy from the Empordà Basin westward to the Ripoll syncline (Puigdefàbregas and Souquet 1986; Burbank et al. 1992; Vergés and Burbank 1996; Costa et al. 1996; Serra-Kiel et al. 2003; Barnolas et al. 2019). Nine depositional units have been identified in this easternmost portion of the foredeep: eight within the Eocene, and one defining the base of the Oligocene. Their names are directly adopted from the regional lithostratigraphic unit names, hence some ambiguity in the literature between the two schemes (i.e., lithostratigraphy vs. sequence stratigraphy)—particularly in the case of Bellmunt (unit VI) and Milany (unit VII) and the respective boundaries of these units in each of the reference frames. Units I to IV (Ypresian–early Lutetian) testify to submarine thrust sheet emplacement, with olistoliths, turbidites, and deltaic facies. Vestigial outcrops of proximal conglomerate beds of Ypresian and Lutetian age occur also at several locations, in Berga–Queralt (Solé Sugrañes and Clavell 1973) and the far east (mapped on the geological sheet of Figueres: Fleta et al. 1994; Pujadas et al. 1989). The younger depositional units (VI, VII, IX) consist of thick conglomerate units, including the fan-deltas of the Bellmunt (48–41 Ma, i.e., Lutetian) and Milany (41–38 Ma, i.e., Bartonian) formations. Units V (Beuda gypsum) and VIII (Cardona salt beds) consist of evaporites and record the growing confinement of the Ebro Basin and the definitive termination of the marine environment, respectively (see Itinerary 9, and Fig. 14.2 therein).

Between the Llobregat and Segre rivers, the Bellmunt and Milany units have been deformed by Pyrenean tectonics and incorporated into the Pedraforca nappe

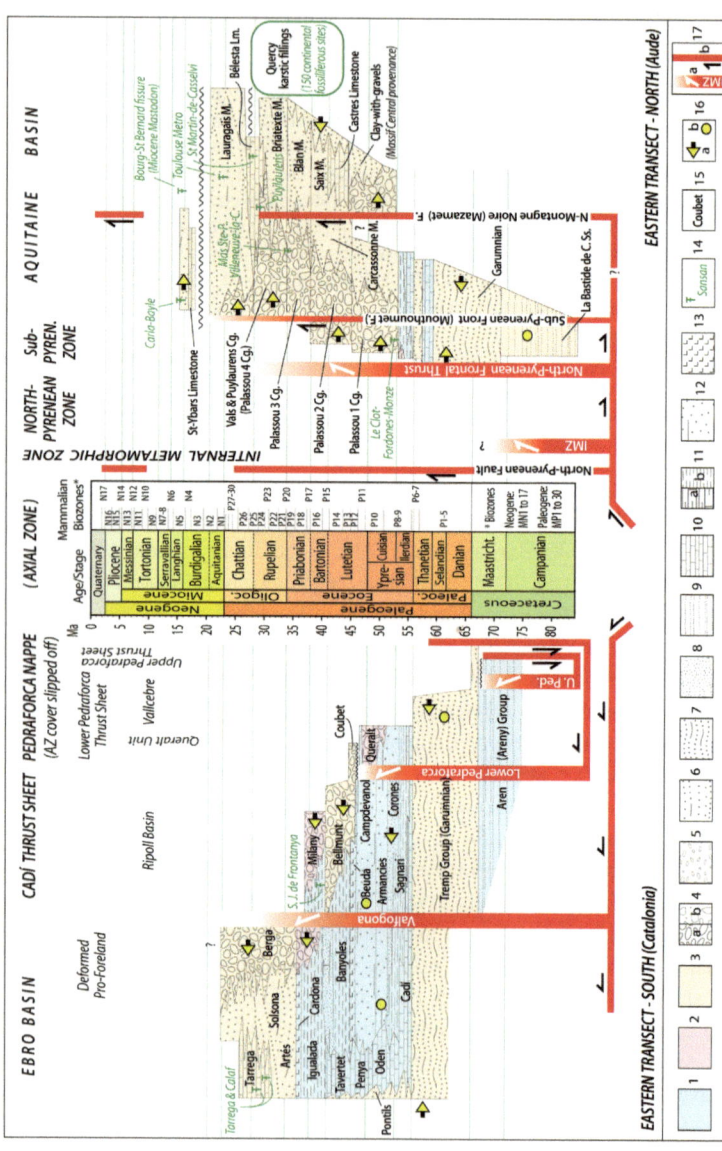

Fig. 2.3 Lithostratigraphy and tectonic evolution of the eastern Iberian foreland and European retro-foreland sedimentary basins of the Pyrenees. Key as in Fig. 2.2

(Fig. 1.2), and thus into the orogenic wedge. South of this tectonostratigraphic unit lie the well-documented Berga and Oliana conglomerate sequences (Priabonian–Rupelian, i.e., ~36–28 Ma), which are 2.5 km thick and have been subdivided into three units based on conspicuous boundaries within the fanning dip structure (Riba 1973, 1976). The clast compositions are lithologically fairly uniform, with debris from the Axial Zone unevenly distributed through the sequence, but nonetheless indicating that rivers had made inroads into the Hercynian core of the mountain range at the time of deposition, i.e., the Priabonian–Oligocene. For example, granite clasts occur almost exclusively in the Priabonian upper unit, which has buried the eastern termination of the Pedraforca nappe at Serrat Negre–Serra de Catllaràs (Itinerary 9).

2.1.2 Synorogenic Chronostratigraphy of the South Pyrenean Foreland Basin

Synorogenic sedimentation in the south-Pyrenean foreland occurred in two contrasting environments separated by the External Sierras: an elongated piggyback basin riding on the back of the SPZ thrust units, and a deformed but undisplaced outer foreland basin south of the External Sierra thrust fronts. Together they form what is sometimes known as the South Pyrenean Foreland Basin (Barnolas et al. 2019), but each setting is described separately below.

2.1.2.1 The Outer Ebro Foreland
In this palaeogeographic domain, the rising sea lapped progressively southward during Bartonian time onto the basement and thin sedimentary cover of the Ebro microcontinent. Shelf limestones, followed by marls, aggraded during the Priabonian as accommodation space increased in response to loading by the advancing Pyrenean nappes. Closure of the marine foredeep in the west, however, occurred eventually as a result of rapid Priabonian mountain building. The transition to continental conditions generated thick evaporite deposits (Balaguer and Barbastro gypsum deposits, Fig. 2.2), homologous to the Cardona evaporites further east.

   The clastic sequences in the South Pyrenean Foreland Basin are much less conspicuous in the landscape than in the Southeast Pyrenean Foreland Basin because of the different tectonic history at these western longitudes. For example, as a consequence of time-transgressive westward growth of the Pyrenees, the Paleogene deposits at the northern margin of the Ebro Basin have been concealed west of the Cinca river by younger Neogene clastic sequences that do not appear to exist farther east in Catalonia.

The history of denudation in the Ancestral Pyrenees is recorded by the continental mud- and sandstone sequences that overlie the Priabonian evaporites and have been deformed into a fairly continuous frontal anticline: the Balaguer–Barbastro anticline. Conglomerate sequences have buried the External Sierras between the Cinca and Segre rivers, generating a clearly identifiable syntectonic unconformity. The Peralta de la Sal Basin (Meigs 1997), SW of the Ager Basin on the outer curve of the SPCU lobe (Fig. 1.2), is a representative example. It displays a very similar chronostratigraphic configuration to that of the Artés area in the Southeast Pyrenean Foreland Basin, where the 1800-m-thick molasse sequence can be subdivided into three Priabonian to Chattian (35 and 25 Ma) syntectonic units (magnetostatigraphic ages after Meigs et al. 1996). As in the Artés analogue, extensive outcrops of the 1000-m-thick Peraltilla Formation, which consists of red mudstone deposits and sandstone-filled palaeochannels, occur west of the Cinca River along the Barbastro anticline. Based on a micromammalian fossil assemblage labelled as MP 23, the lower units (marl and lacustrine limestones) of the Peraltilla Formation are early Oligocene (Reille 1967, 1971; Alvarez-Sierra et al. 1990; Cuenca et al. 1992). The Peraltilla sequence has been folded and is unconformably covered by an early Neogene terrigenous sequence (Uncastillo and Sariñena formations; Fig. 2.2). The fine-textured sedimentological features of the Peraltilla sequence would suggest a relatively low-energy environment at the front of the outer sierras at the time. Mineralogical provenance analysis nonetheless documents some sediment transfer from basement outcrops in the rising Axial Zone (Yuste et al. 2004).

2.1.2.2 The Tremp–Jaca–Pamplona Piggyback Basin

From the Segre River to Jaca and as far as Pamplona, the Paleogene sequences are continuous and contained within a strike-parallel furrow sometimes called South Pyrenean Foreland Basin (Barnolas et al. 2019), or Tremp–Jaca Basin, or Aragón syncline. Its connection and continuity with the Southeast Pyrenean Foreland Basin remains unclear (Nijman 1989), which explains why they are invariably treated as separate domains.

In this segment of the foreland zone, the Paleogene sequences themselves cover older clastic sequences from the earlier period of denudation in the Proto-Pyrenees. The Jaca–Tremp furrow had thus previously received a 1-km-thick discharge of Garumnian continental molasse (Rosell et al. 2001). These deposits lie conformably over the Maastrichtian Aren Formation (marine and littoral sandstone; Figs. 2.2, 2.3), which represent somewhat distal submarine facies of depositional systems generated by the eroding Cretaceous Proto-Pyrenees. For a detailed presentation, see Calvet et al. (2021); here we focus on the area situated east of the Cinca River, which is important for understanding the geodynamic evolution of the eastern Pyrenees.

The presence of an extensive Lower Ypresian shelf-carbonate platform documents a distinctive but brief respite in the chronology of clastic output from the rising orogen, suggesting that the foredeep was still underfilled in earliest Eocene time. These pelagic episodes (which in the Tremp Basin define the Ilerdian Stage of the Lower Ypresian as an international stratotype: see Pujalte et al. 2009) subsequently yielded to a system of piedmont, alluvial plain, deltaic, and marine shelf facies. The Upper Ypresian to Lutetian succession of thick sandstone and conglomerate beds, all prograding from east to west along the strike of the orogen's axis (Nijman 1998), marks the transition from an underfilled marine foredeep to the overfilled foreland environment of a lengthening and widening successor orogen to the Proto-Pyrenees: the Ancestral Pyrenees (see Itinerary 9).

The conglomerate and sandstone units of the South Pyrenean Foreland Basin were supplied by geological structures situated to the northeast of the depositional area, i.e., where the Proto-Pyrenees had formed in the early Paleocene and were presently expanding westward and producing the Ancestral Pyrenees. A secondary contribution to clastic output was nonetheless also made by the South Pyrenean Central Unit fold systems and by the Ebro microcontinent (which at the time stood above sea level: Puigdefábregas 1975). The source rocks in the SPCU were mainly sedimentary cover sequences, with nonetheless quartz- and feldspar-rich (arkose) deposits and a minority presence of basement clasts from the Axial Zone. The pebbles are well rounded, with median sizes ranging from 10–20 cm to 40–50 cm in the Campanué Formation (Isábena–Ésera Basin; thickness: 500–1000 m), the top of which could be of early Lutetian age (~45 Ma; Fig. 2.2).

The fluvial Capella Formation (Tremp–Ager Basin: thickness: 600 m) records the tipping moment when the ancestral Isábena River catchment became lastingly cut off from marine influence. The Capella Formation is reported to connect laterally to the upper section of the Campanué Formation. The continental character of the foreland environment was accentuated at the time of the Escanilla Formation (Fig. 2.2), which rests unconformably over the Capella–Campanué sequences in the south and southwest and displays a much more accentuated angular unconformity over the Triassic and Cretaceous outcrops of the Mediano anticline (western part of the Sierras Exteriores; see Fig. 1.2). The Escanilla Formation consists of 500–1000 m of red- and ochre-coloured mudstone, and of grey conglomerate containing some pebbles from the Axial Zone. This formation also displays two levels of limestone and lignite containing two taxonomically diverse fossil assemblages at Capella (MP 14, ~42 Ma, not to be confused with the name of the underlying formation) and Laguarres (200 m further up-sequence: MP 15, i.e., ~40.5 Ma according to Sudre et al. 1992; but MP 16, i.e., ~39 Ma according to Antunes et al. 1997 and Badiola et al. 2009). The top of the sequence, which has not been preserved, could thus potentially be extrapolated into the Priabonian, and perhaps even into the Oligocene. This biochronology is corroborated by magnetostratigraphic data from that area (Bentham and Burbank 1996).

2.1.2.3 A Note on the Paleogene Nogueres Conglomerates

The Nogueres conglomerates have long been singled out as unique within the Pyrenean orogenic environment, and therefore enigmatic (Ashauer 1934; Birot 1937; Crusafont et al. 1956; De Sitter 1961; Rosell and Riba 1966, for a literature review; Rosell 1967; Reille 1971; Mellere and Marzo 1992; Coney et al. 1996; Vincent and Elliot 1996; Vincent 2001; Beamud et al. 2003, 2011). They occur exclusively on the SPCU and are recessed in a comparatively more northerly position of the pro-wedge between the modern Isábena and Segre rivers, i.e., much closer to the Axial Zone than any other conglomerate deposit of the Iberian foreland (a few small outliers have been mapped on the Pedraforca tectonic unit east of the Segre; see Itinerary 9). The Nogueres sequence occurs as four discrete outcrops, from west to east: Sierra de Sis, Sierra de Gurp, La Pobla de Segur; and, offset to the north: the Senterada outcrop. Cumulative thicknesses attain 3500 m, subdivided into 5 allostratigraphic units altogether forming a stack of 20 alluvial fans as well as a backfill sequence lapping far into the Axial Zone (Melleres and Marzo 1992; Beamud et al. 2003, 2011). Unit 2 and the base of Unit 3 contain the fossil mammalian remains of Sossis, which belong to biozone MP 17a (Crusafont et al. 1963; Sudre et al. 1992; Antunes et al. 1997; Sigé 1997; López-Martínez et al. 1998; Cuesta et al. 2006), i.e., early Priabonian (37.5 Ma: Escarguel et al. 1997; Legendre et al. 1997; Aguilar et al. 1997; Vandenberghe et al. 2012), and constitute the richest assemblage of the South Pyrenean Foreland. The slope of the basal angular unconformity decreases from north to southwest, with the conglomerate beds resting successively on the Hercynian basement, on the folded Cretaceous cover sequence of the SPCU (Fig. 1.2), and on the Paleocene and Lower Eocene deposits of the Aragón syncline (e.g., the upper Ypresian, i.e., MP 10 at Cajigar, in the Sierra de Sis). Then, just a few kilometres to the SW, they connect to the Escanilla Formation, which itself rests disconformably on the underlying Lutetian (Beamud et al. 2003). The Nogueres conglomerate sequence is considered to represent the proximal eastern tract of the fluvial systems that supplied the Escanilla Formation and, farther to the west, the Campodarbe Formation in the Jaca Basin (Michael et al. 2014; Fig. 2.2). These conglomerate sequences and their significance for understanding the palaeogeography and relief of the Ancestral Pyrenees are discussed in greater detail in Itinerary 9, and in Calvet et al. (2021).

2.1.3 The Neogene Record

2.1.3.1 Early to Middle Miocene: The Last Synorogenic Conglomerate Beds

The Neogene sediments of the Ebro Basin were deposited within an enclosed, internally drained basin, and probably as a result are much thicker than in the Aquitaine Basin. The five Neogene tectono-sedimentary units (TSUs 4 to 8) reach a total thickness of ~1000 m, and TSU 4 straddles the Oligocene and the Miocene (Villena et al. 1992; Arenas and Pardo 1999; Arenas et al. 2001; Pérez Rivarés et al. 2018). The top of the Miocene sequence at the centre of the basin corresponds to biozone MN 6 (San Caprasio deposit, i.e., ~13.5 Ma; Agustí et al. 2011) and to biozones MN 7–MN 8 to the south of the Ebro River near Tarazona (Murelaga et al. 2008; Pérez-Rivarés et al. 2018), i.e., ~12.5–12 Ma. The carbonates of TSU 8 are devoid of fossils but could be early Vallesian (Vázquez-Urbez et al. 2013). The eastern edge of the Miocene outcrops does not extend beyond the Cinca–Segre drainage divide. This interruption is most likely an erosional boundary, but whether a megafan the size of the Huesca fan ever existed east of the Segre River remains currently a matter of speculation.

Late Neogene denudation of the Ebro Basin, after its reconnection to the Mediterranean, has resulted in preservation of the harder lithologies such as the central carbonate and peripheral conglomerate masses (Huesca-Mallos de Riglos conglomerates), but has stripped a large proportion of the softer marl and sandstone outcrops typical of mid-fan to distal alluvial depositional systems. Added to the previously mentioned relative scarcity of fossil remains among the Cenozoic rocks of the Iberian foreland, stratigraphic correlations between distal and proximal alluvial deposits, as well as between proximal deposits along the strike of the orogen, have remained partly speculative. Age constraints are precise at some localities within the mountain-front conglomerates, owing for example to the mammalian fossil deposit at Santa Cilia, near Huesca, of Aquitanian to early Burdigalian age (Alvarez Sierra et al. 1990; Cuenca et al. 1992), but overall variations through time in the intensity of post-orogenic sediment delivery to the basins by the Pyrenean orogen remain difficult to reconstruct regionally. Many alluvial fans (Nichols 2005, 2018) were fed with gravel and boulders from local outcrops in the External Sierras, which often became locally buried as a result of piedmont backfilling; but far-field source areas, with abundant quartzite, schist and granite pebbles from the Axial Zone, are definite features of the Huesca Fan, which extends from Graus to west of Barbastro. In the far west, the Luna Fan appears in contrast to mostly contain rock facies fed and redistributed from the piggyback Jaca Basin (i.e., sandstone from the flysch or the Campodarbe outcrops, and reworked Paleogene conglomerates), which was undergoing folding and erosion at the time. Few inputs from the Axial Zone, which still today is buried beneath Mesozoic cover sequences west of the Pic d'Anie, are reported from the Luna deposits (Arenas 1993; Arenas et al. 2001).

2.1.3.2 Late Miocene: Post-Orogenic Return to External Drainage

When exactly the Ebro Basin ceased to be internally drained, particularly in relation to the possible impact of the Messinian salinity crisis in reconnecting the Ebro River with the Mediterranean Sea, has been a source of intense speculation (Coney et al. 1996). A late Cenozoic transition from internal to external drainage was a general feature among the major rivers of the Iberian Meseta, whether or not these join up with the Mediterranean (the Tagus and Duero, for example, connect to the Atlantic). There should thus be no reason for the abrupt and short-lived Messinian base-level fall (~5.9–5.3 Ma) to have been a key cause. Accordingly, some authors have ruled it out and advocate instead a progressive reconnection of the Ebro Basin beginning in the Pliocene, i.e., somewhat later than the Messinian (Babault et al. 2006).

A consensus has nonetheless progressively emerged in favour of an early reconnection of the Ebro to the Mediterranean, i.e., at some time before rather than after the Messinian. The inferences are model-based and consistently generate a best estimate falling between 13 and 8.5 Ma (García-Castellanos et al. 2003) or between 12 and 7.5 Ma (García-Castellanos et al. 2015). This time window (late Serravalian–Tortonian) coincides with a strong influx of sediment into the Valencia Basin (Castellón Group) (Cameselle et al. 2014), and seems compatible with thermochronological data obtained for the Nogueres conglomerate beds, into which fluvial incision began ca. 9.5 Ma (Fillon and Van der Beek 2012, Fillon et al. 2013). Note that these studies postulate palaeoelevations of either ~1000 m (García-Castellanos et al. 2003; Babault et al. 2006) or 530–750 m (García-Castellanos and Larrasoaña 2015) for the top of the Ebro Basin fill prior to the switch to external drainage. These are substantial elevations above sea level, but as boundary conditions for the numerical models they do not allow for evidence that the Ebro Basin has itself undergone regional surface uplift in the last 12 Ma (Soto et al. 2016; Pérez-Rivarés et al. 2018; Conway-Jones et al. 2019).

2.2   Patterns of Terrigenous Sedimentation in the Aquitaine Foreland

A comparative advantage of the Aquitaine foreland over the Ebro is that its chronostratigraphy of continental deposits is constrained at some localities by the occurrence of marine interlayers, which recur between ~56 (Ypresian) and ~14 Ma (middle Miocene). Existing syntheses have tended to focus on the marine deposits or the sub-Cenozoic stratigraphy (e.g., Kieken 1973; Biteau et al. 2006; Ortiz et al. 2020). The body of work devoted to the Eocene marine record is the largest (e.g., Plaziat 1981, 1984; Sztrakos et al. 1998), but a few overviews also exist for the Oligocene and Miocene (Cahuzac et al. 1995, 1996, Cahuzac and Turpin 1999, Cahuzac and Janssen 2010; Sztrakos and Steurbaut 2017). Note that the Aquitaine Basin is the locus of two type sections of the International Stratigraphic Scale: the Burdigalian (Latin name for the city of Bordeaux; Depéret 1892) and the Aquitanian (Mayer Eymar 1858) (see also Parize et al. 2008; Londeix 2014). The Aquitaine Basin is also a rich repository of continental fossils—whether in alluvial and lacustrine deposits or in a karst setting. Among its several hundred sites, some are type localities used in Europe's Paleogene (MP) and Neogene (MN) Land Mammal Age scale, or ELMA (see Richard 1948; Crouzel 1957; Bergounioux and Crouzel 1960; Sudre et al. 1992; Duranthon 1991, 1993; Muratet et al. 1992; Muratet and Cavelier 1992; Antoine et al. 1997, 2006, 2011; Legendre et al 1997; Duranthon and Cahuzac 1997; Astruc et al. 2003, etc.). The key sites for these biozones, however, are often situated in the more distal, usually lacustrine molasse deposits of the basin. As a result, the more proximal conglomerate sequences (i.e., closer to the mountain front) are often dated by stratigraphic correlation rather than by direct dating.

2.2.1   Paleogene Conglomerates: The Palassou and Jurançon Series

The Palassou Series has been well studied in the eastern segment of the retro-foreland (Crochet 1991), where a band of almost vertically upturned conglomerate beds extends between the Arize and Ariège rivers. The outcrop broadens substantially to the east of the Ariège and into the Aude catchment, where it was folded during the Pyrenean orogeny (see Itinerary 5). The Palassou alluvial sequence is >2 km thick and commonly subdivided into four units interpreted as four successive intervals of tectonic activity, each separated by synsedimentary unconformities that display fanning dip angles and grade into sharper angular unconformities nearer the mountain front. The petrographic composition indicates that the river catchments sampled most of the Ancestral Pyrenees in varying proportions. Inputs from the Axial Zone are in a minority (<10%) in the older Palassou Unit 1 and at the base of middle Unit 2, but become dominant or exclusive in the remainder of the Unit 2 (90%, including granite and metamorphic rocks). Uppermost Unit 3 almost entirely lacks debris from the Axial Zone.

The age of Palassou units 1 to 3 has been established on the basis of (i) their stratigraphic continuity with the marine and coastal Lower Ypresian (i.e., Ilerdian) beds, (ii) a series of mammalian fossil deposits (particularly in the Aude syncline and in the distal molasse deposits between Villefranche-de-Lauragais and Castres), and (iii) a selection of freshwater and terrestrial mollusc assemblages. The lower Palassou (Unit 1) lies conformably over the Lower Ypresian marine units and contains in its basal beds some mammalian assemblages ascribable to MP 7 and MP 8–MP 9 (Marandat 1991; Marandat et al. 2012), i.e., 54–52 Ma (according to Legendre and Lévêque 1997; Escarguel et al. 1997; Vandenberghe et al. 2012). This lower unit continues up to at least the base of the Lower Lutetian. The middle Palassou (unit 2) spans the Upper Lutetian to Bartonian interval. The upper Palassou (unit 3) contains at its top the mammalian fossils of Mas-Saintes-Puelles (MP 19, ~35 Ma) and Villeneuve-la-Comtal (MP 20, ~34 Ma), and is thus compatible with a Priabonian age.

The Palassou Series continues upward into the Rupelian, but this Unit 4 was not described by Crochet (1991). Unit 4 has been mapped between the Hers valley and Lauragais ('Poudingues de Vals'), and continues northward towards Castres ('Poudingues de Puylaurens' and 'Molasses de Briatexte': Mouline 1967, 1978), where biozones MP 24–25 at the fossil site of Puylaurens provide an age of 30–29 Ma for the top of the sequence. This early Oligocene component of the clastic wedge, which contains a minority of pebbles from the Axial Zone, may have reached these quite distal areas because of a decline in subsidence rate and accommodation space in the retro-wedge foredeep, forcing the rivers to prograde northwards across the overfilled foreland towards an increasingly lacustrine environment (Astruc et al. 2003). The lacustrine environment itself indicates conditions of diminished clastic output from the eastern Pyrenees in the middle Oligocene ('Calcaires de Bélesta'), at a time when mountain topography may have attained a state of subdued relief and moderate energy. The Chattian age of the Lauragais and Toulousain molasse deposits, which grade conformably to the Lower Miocene, have been well constrained by several mammalian fossil deposits. The molasse is fine-textured and consists of limestone-rich silty clay with sand interlayers (Astre 1959, 1964; Antoine et al. 2006).

Outcrops of proximal Paleogene conglomerate sequences, which are otherwise mostly covered by Neogene deposits west of the Ariège, occur around the city of Pau, where they are known as 'Poudingues de Jurançon' after their type locality. The 1200 m-thick molasse sequences that fill the Arzacq Basin between the Gave de Pau and the Adour River have been interpreted as distal, and thus coeval, continuations of the Jurançon conglomerates. Belief since the early twentieth century was that the 'Poudingues de Jurançon' were of Neogene age (Douvillé 1924; Crouzel 1957), but more recent investigations have established that both they and the 'Molasses d'Arzacq' are westward continuations of the synorogenic Paleogene units of the Palassou Series (Hourdebaigt et al. 1986; Hourdebaigt 1988). Two mammalian fossil sites in the Arzacq molasses have also provided an Eocene–Oligocene boundary age (Stehlin 1910, in Capdeville 1997) and a Rupelian age (MP 23) (Sudre et al. 1992; Viret 1938; Glangeaud 1938; Schoeffler 1969).

2.2.2 Early to Middle Neogene Molasse

By comparison with the Paleogene foreland deposits, their Neogene successors are thin and exhibit shallow dips. On seismic profiles, the Neogene cover sequence never exceeds thicknesses of a few hundred metres (Schoeffler 1971,1973)—on average 200–300 m, with local maxima attaining 600 m at the Landes coastline or below the middle Adour valley, 500 m north of Tarbes, 540 m north of Auch, and 560 m at Lézat (i.e., 15 to 30 km north of the NPF). As on the Spanish side, however, early to middle Neogene continental deposits thin out substantially and disappear in the eastern Pyrenees, i.e., east of the Ariège River.

During early and middle Miocene time, marine sedimentation prevailed in the west and produced interlayers with the ongoing continental deposit accumulations farther east. Three marine transgressions from the west have been recorded, each successively forming gulfs extending as far as 150 km into the interior of the Aquitaine Basin, and with depocentres situated between 100 and 75 km north of the mountain front. The most extensive transgression occurred in Aquitanian time (Londeix 2014), reaching the vicinity of Agen. Such a regional pattern suggests that Pyrenean tectonics were not directly driving the chronology of sea-level changes. The easternmost paralic marls contain fossil oyster beds and are sandwiched between two lacustrine limestone beds: 'Calcaires blancs' below, and 'Calcaires gris' above. The limestones contain two mammalian sites (Paulhiac and Laugnac) emblematic of biozones MN 1 and MN 2, respectively (i.e., Aquitanian and early Burdigalian). The sandwich is known as the 'Trilogie agenaise'. The existence of biochemical deposits of this nature capping the Chattian molasse at the centre of the Aquitaine Basin suggest a sharp decline in detrital input from the Pyrenees, and thus probably a decline in relief, mean slope gradients, and stream energy. The Bélesta limestones (mentioned in Sect. 2.2.1), situated in a proximal position closer to the mountain front, were already foreshadowing this decline in catchment steepness and river loads during the early Chattian. The geometry of the Burdigalian marine incursion was almost similar to its Aquitanian precursor, whereas the Langhian to Serravallian transgression ('Sables fauves') lapped southward a little closer to the Pyrenees, generating the Gulf of Lectoure (Cahuzac et al. 1995; Cahuzac and Poignant 1996, 2004; Cahuzac and Turpin 1999; Cahuzac and Janssen 2010; Gardère 2002, 2005; Gardère et al. 2002).

Continental deposits throughout the entire Neogene sequence are all fine-textured (molasse and lacustrine limestone), indicative of subdued relief in the Pyrenees in Neogene time. Deposit lithology ranges from marl to fine sand and greywacke, cross-cut by mica-rich sand-filled palaeochannels (Crouzel 1957). Deposit ages range from biozones MN 1 to MN 8 (i.e., 23–11 Ma) (Antoine et al. 1997), and have been confirmed in the west by the palaeontological ages of marine interlayers (Magné et al. 1985; Rey et al. 1997). Unlike earlier periods of the Paleogene, when fine-textured debris accumulated mostly in the more distal, northern parts of the Aquitaine Basin, lacustrine marl and limestone deposits were now forming in a proximal position, in direct, nonconformable contact with the Sub-Pyrenean fold structures buried beneath them (Petites Pyrénées, etc.). This applies for example to the Lower Burdigalian (?) cover rocks at La Lèze (Saint-Ybars limestone), or to the Serravallian–Tortonian basal beds at Montréjeau and Saint-Gaudens.

The Miocene depositional area extends eastward slightly beyond the Ariège valley. Outcrops just above the town of Pamiers show that these Miocene deposits comprise of an Aquitanian conglomerate sequence at their base, which fines rapidly through the middle and upper beds and was almost exclusively fed by basement rock formations from the Axial Zone. The stratigraphy of the cover sequence exhibits a number of lacunae. For example, all of biozone MN 3, and perhaps the base of MN 4, are missing, i.e., 3 Ma of Burdigalian biostratigraphy (Duranthon 1991). The geodynamic cause of this stratigraphic lacuna in the eastern part of the Aquitaine retro-foreland is uncertain, but a eustatic fall is ruled out because a Burdigalian transgression was occurring at the same time farther west in the Aquitaine Basin. Erosion caused by regional uplift close to the Mediterranean rift zone is a more likely explanation for the observed gaps in the stratigraphic record (Calvet et al. 2021).

2.2.3 Abrupt Changes During the Late Neogene

2.2.3.1 Stratigraphic Aspects

The Aquitaine Basin did not undergo any major drainage reorganisation on a scale or magnitude comparable with events recorded in the Ebro Basin ca. 10 Ma, and yet the late Neogene stratigraphic sequence in the Aquitaine foreland is radically different from anything that preceded it in at least two respects: (i) its sedimentology, which consists of almost pure siliciclastic debris always devoid of any postdepositional carbonate matrix; and (ii) its stratigraphic identity, which consistently displays a sharp basal ravinement surface involving networks of fluvial channels vertically incised into the underlying substrate (Crouzel 1957). This highly irregular erosional boundary is well documented throughout the Aquitaine foreland where its decalcified, pebble-bearing clay deposits form a disconformable marker horizon across every existing geological structure and stratigraphic formation from the Neogene molasse (whether marine or continental) to the Paleogene Palassou conglomerates and the Cretaceous bedrock of the Sub- and North-Pyrenean zones. The basal unit of the late Neogene cover sequence is varyingly named Clay-with-pebbles ('argile à galets') or Variegated clays ('argiles bariolées') depending on the map sheet concerned and on its proximal or distal position. Its reported thickness ranges from 20 to 60 m.

2.2.3.2 Growth of Piedmont Megafans

The late Neogene cover sequences extend as far out from the Pyrenees as the Landes plateau and the Entre-deux-Mers (between Dordogne and Garonne) near Bordeaux. Closer to the Pyrenean mountain front, the late Neogene clastic output formed large alluvial fans at the mouths of the main Pyrenean valleys. Among these: the Lannemezan megafan (coaxial with the Neste River) (Boule 1894; Patin 1967; Icole 1968, 1969), the Ger megafan (Gave de Pau), and lesser range-front fans such as the Adour, Gave d'Oloron, Lasserre–Lahitère (Salat River), and residual strips of formerly much larger fans coaxial with the Arbas, Arize and Ariège rivers. In the Basque country, the extensive limestone catchment lithology was a limiting factor on the potential for clastic supply. The Ariège and Garonne rivers likewise failed to produce a fan of Lannemezan dimensions—probably because, owing to their larger watersheds and higher discharges, they were comparatively more efficient at exporting most of the bedload beyond the piedmont zone. Fan thickness attains 60–80 m in the Ger and Adour (up to 120 m for the Ger), 90–110 m in the central and eastern Lannemezan, and 170 m in the western Lannemezan.

Four sedimentary units have been identified in some of those clastic sequences, with evidence of internal disconformities. The three uppermost units are clearly defined within the Ger megafan (Karnay and Dubreuilh 1998; Capdeville and Darboux 1998) and broadly of Pliocene age. They each contain 2 to 4 upward-fining depositional sequences, but mean pebble size increases from Unit p1 (2–5 cm) to Unit p3 (15 cm; maximum: 25–30 cm). The base of the basal unit ('Clay-with-pebbles') is well dated in a proximal position at Orignac, on the Adour Fan, where a lignite bed has yielded a rich assemblage of mammalian fossils (known since 1865) containing *Hipparion* (Astre 1932; Richard 1948; Crouzel 1957). It can thus be assigned a Vallesian age at Orignac (MN 10; Antoine et al. 1997), i.e., no older than 9.9 Ma. This constraint fixes the development of the ravinement surface between the times of biozone MN 8 (Montréjeau–Saint-Gaudens sites) and biozone MN 10 (Orignac). Biozone MN 9, which is characterised by the unprecedented presence of Hipparion in the Vallès Basin of Catalonia ca. 11.1 Ma (Garcés et al. 1997; Agustí et al. 2001), has never been reported from the Aquitaine Basin (Antoine et al. 1997). Nonetheless, it can be inferred that the MN 9 hiatus and the period of fluvial incision lasted ~1 Myr (11.1 to 9.9 Ma). The age of the terminal units of the retro-foreland megafans is currently unknown, but their surfaces are cut and filled by shallow alluvial channel deposits containing even coarser debris than in underlying Unit p3, and traditionally attributed to the 'Donau' (Alimen 1964; Icole 1974) and thus roughly the Calabrian.

The size of the late Neogene megafans and large calibre of their constituent clasts have been correlated by most authors with a tectonic pulse in the Pyrenees. The distal tracts of these megafans correlate stratigraphically with the sedimentary sequence of the Landes, which consists of a stack of five disconformable units. The 'Variegated clays' and lignite beds of the basal unit contain at Arjuzanx a late Miocene mammalian fossil assemblage and warmth-loving plant remains (Huard and Lavocat 1963) similar to those encountered closer to the mountain front at Orignac and Capvern (Sauvage 1969; Bugnicourt et al. 1988). The four fluvial to deltaic units above the 'Variegated clays' contain lignite deposits and document a progressive decline in warmth-loving biota. The two lower units (Arengosse and Onesse–Beliet formations) are Pliocene to Gelasian, the third (Belin Fm.) is early Pleistocene, and the fourth (Castet Fm.) middle Pleistocene and mainly of aeolian origin (Dubreuilh et al. 1995).

At the same time as the piedmont megafans were aggrading on land, some very thick clastic wedges were likewise aggrading offshore in the Bay of Biscay, with accelerating sedimentation rates after the middle Miocene (Kieken 1973; Bellec 2003; Bellec et al. 2009; Ortiz et al. 2020).

2.3   Extensional Intermontane Basins of the Eastern Pyrenees

From late Oligocene time, the Mediterranean became the eastern boundary of the Pyrenean mountain range (Bache et al. 2010; Calvet et al. 2021). In addition to an increasing intensity of tectonic fragmentation of the Paleogene mountain range by Neogene extensional faulting, the new Gulf of Lion marine base level thus controlled the entire geomorphological evolution of the eastern segment of the Pyrenees from Aquitanian time to the present. These basins are presented in detail in Itineraries 1, 2, 6, 7, and 8.

The Aquitanian sea-level rise, already documented in the Aquitaine Basin as entering eastward from the Atlantic (see Sect. 2.2.2) but in the present case advancing from the east (Calvet et al. 2021), did not, however, extend westward past the longitude of Montpellier. The back-arc related extensional faulting produced a population of half-grabens striking NE–SW and NNE–SSW, but initially these evolved in a continental environment. A dozen extensional basins were generated on the Mediterranean seaboard and in the eastern Pyrenees (Fig. 1.1), fragmenting the orogenic wedge, its topography, and thus progressively zipping the orogen open from east to west along the Axial Zone.

The first extensional episode affecting the Mediterranean back-arc basin occurred during the late Oligocene–early Miocene; the resulting onshore basins were the Narbonne–Sigean, Roussillon, Conflent, Vallès-Penedès, and several other smaller basins in the Corbières (Tuchan–Paziols, Lapalme, Fabrezan–Tournissan). The clastic fill of these basins documents the geomorphological evolution of their immediate surroundings, often diachronous and with complex histories involving phases of rift-flank uplift that typically produced coarse deposits ranging from slope breccia to boulder fans or fluvial conglomerate. These tectonically active periods alternated with more quiescent intervals reflected in finer-textured clastic deposits and even lacustrine limestone (see Itinerary 1). The eustatic high was reached during the Langhian and Serravallian intervals in all the coastal basins from Languedoc (including the Roussillon) to Catalonia (Vallès-Penedès).

The second extensional episode made inroads into the core of the Axial Zone during the late Miocene, producing intermontane basins such as the Cerdagne and Seu d'Urgell, but also the Val d'Aran (Pruëdo basin: Ortuño et al. 2013). It also impacted the Mediterranean onshore area of Catalonia, where the older basins closest to the coast (Vallès, Roussillon) were reactivated during the late Miocene and Pliocene and two new basins—Selva and Empordà—were generated. The Catalan passive margin and Valencia Trough share a tectonic and sedimentary history similar to that of the Gulf of Lion. Extension in the southern Catalan margin has even inverted to grabens some Oligocene piggyback basins that were previously produced by convergent tectonics in the Catalan Coastal Range (Roca 1996a,b; Roca et al. 1999). After the middle Miocene eustatic maximum, shorelines during Tortonian time receded seaward of the current coastline in Languedoc and Roussillon, but in Catalonia the sea advanced into the Empordà area and deposited coastal sediments around Barcelona. The dramatic late Messinian sea-level fall caused deep canyons to cut into the continental margin and onshore, but such canyons have not been observed or detected at all the region's river mouths, and do not appear to concern either the Aude or the Ter. Rapid drowning of these deep fluvial valleys by the subsequent sea-level rise ca. 5.4 Ma generated a landscape of rias, which soon became filled with Pliocene sediment (Clauzon 1990; Clauzon et al. 1987, 1990).

Another unprecedented feature of the late Miocene was the onset of protracted episodes of intraplate volcanism. Although eruptions did not occur in the mountain range itself, a trail of volcanic outcrops occurs from the Massif Central to the Languedoc coastline (Gillot 1974), with a possible continuation on the continental shelf. The youngest alignment of volcanoes, from the Escandorgue hills to the coastal town of Agde, was emplaced between 2 and 0.5 Ma (Dautria et al. 2010). Traces of volcanic activity reappear on the south side of the Axial Zone in the Selva and Empordà basins, in Catalonia after ~10 Ma (Donville 1973a, b, c, 1976; Marti 2004; Marti et al. 1992).

The facies of the late Neogene basin sedimentary fills indicates growing relief in the surrounding footwall uplands, particularly in Turolian time when thick alluvial-fan deposits containing huge boulders suggest tectonic activity along the southern master fault of the Cerdagne Basin (see Itineraries 7 and 8), as likewise in Empordà. Unlike the Cerdagne and Seu d'Urgell basins, the Capcir Basin strikes N–S and is unique in that respect (see Itinerary 7). Perhaps even younger than the other two, it may have formed in response to a new regional stress regime but its clastic fill has until now never been dated.

Similar to processes documented in the shelf zone of the Aquitaine Basin (Sect. 2.2.3.2), depositional siliciclastic wedges began to thicken substantially along the Western Mediterranean seaboard after the middle Miocene, particularly during the Pliocene and Quaternary when enhanced clastic influx buried the Messinian ravinement surface from the Gulf of Lion (Roussillon, Empordà) to the Ebro delta (Medialdea et al. 1994; Tassone et al. 1994; Martínez del Olmo 1996; Guennoc et al. 2000; Mauffret et al. 2001; Duvail et al. 2005; Bache et al. 2010; Cameselle et al. 2014; Granado et al. 2016).

References

Aguilar JP, Legendre S, Michaux J (eds) (1997) Synthèses et tableaux de corrélations. Actes du Congrès BiochroM'97, Biochronologie mammalienne du Cénozoïque en Europe et domaines reliés, Mém Trav EPHE Inst Montpellier 21:769–805

Agustí J, Cabrera L, Garcés M, Krijgsman W, Oms O, Parés JM (2001) A calibrated mammal scale for the Neogene of Western Europe. State of the Art. Earth-Sci Rev 52:247–260

Agustí J, Pérez-Rivarés FJ, Cabrera L, Garcés M, Pardo G, Arenas C (2011) The Ramblian-Aragonian boundary and its significance for the European Neogene continental chronology. Contributions from the Ebro Basin record (NE Spain). Geobios 44:121–134

Alimen H (1964) Le Quaternaire des Pyrénées de Bigorre. Mém Serv Carte Géol, Paris, 394 p

Alvarez-Sierra M, Daams R, Lacomba JI, López-Martínez N, Meulen A van der, Sesé C, de Visser J (1990) Paleontology and biostratigraphy (micromammals) of the continental Oligocene-Miocene deposits of the North-Central Ebro Basin (Huesca, Spain). Scr Geol 94:1–77

Antoine PO, Duranthon F, Hervet S, Fleury G (2006) Vertébrés de l'Oligocène terminal (MP30) et du Miocène basal (MN1) du métro de Toulouse (Sud-Ouest de la France). C R Palevol 5:874–884

Antoine PO, Métais G, Orliac MJ, Peigné S, Rafaÿ S, Solé F, Vianey-Liaud M (2011) A new late early Oligocene vertebrate fauna from Moissac, South-West France. C R Palevol 10:239–250

Antoine PO, Duranthon F, Tassy P (1997) L'apport des grands mammifères (Rhinocerotidés, Suoïdés, Proboscidiens) à la connaissance des gisements du Miocène d'Aquitaine (France). Actes du Congrès BiochronM'97, Aguilar JP, Legendre S, Michaux J (eds) Mém Trav EPHE Inst Montpellier 21:581–590

Antunes MT, Casanovas ML, Cuesta MA, Checa Ll, Santafé JV, Agustí J (1997) Eocene mammals from Iberian Peninsula. In: Actes du Congrès BiochroM'97, Aguilar JP, Legendre S, Michaux J (eds) Mém Trav EPHE Inst Montpellier 21:337–352

Arenas C, Pardo G (1999) Latest Oligocene-late Miocene lacustrine systems of the north-central part of the Ebro Basin (Spain): sedimentary facies models and palaeogeographic synthesis. Palaeogeogr Palaeoclim Palaeoecol 151:127–148

Arenas C, Millán H, Pardo G, Pocoví A (2001) Ebro Basin continental sedimentation associated with late compressional Pyrenean tectonics (north–eastern Iberia): controls on basin margin fans and fluvial systems. Basin Res 13:65–89

Arenas C (1993) Sedimentología y paleogeografía del Terciario del margen pirenaico y sector central de la cuenca del Ebro (zona aragonesa occidental). Tesis, Universidad de Zaragoza 858 p

Ashauer H (1934) Die östliche Endigung der Pyrenäen. Abh Ges Wiss Göttingen Math Phys Kl 10:2–115

Astre G (1932) Mammifères des lignites pontiens d'Orignac. Bull Soc Hist Nat Toulouse 64:581–584

Astre G (1959) Terrains stampiens du Lauragais et du Tolosan. Bull Soc Hist Nat Toulouse 94:8–168

Astre G (1964) Le problème des aires d'affleurement du Stampien terminal au sommet des marno-molasses Tolosannes. Bull Soc Hist Nat Toulouse 99:229–234

Astruc JG, Hugueney M, Escarguel G, Legendre S, Rage JC, Simon-Coiçon R, Sudre J, Sige B (2003) Puycelci, nouveau site à vertébrés de la série molassique d'Aquitaine. Densité et continuité biochronologique dans la zone Quercy et bassins périphériques au Paléogène. Geobios 36:629–648

Babault J, Loget N, Van den Driessche J, Castelltort S, Bonnet S, Davy P (2006) Did the Ebro basin connect to the Mediterranean before the Messinian salinity crisis. Geomorphology 81:155–165

Bache F, Olivet JL, Gorini C, Aslanian D, Labails C, Rabineau M (2010) Evolution of rifted continental margins: the case of the Gulf of Lions (Western Mediterranean Basin). Earth Planet Sci Lett 292:345–356

Badiola A, Checa L, Cuesta MA, Quer R, Hooker JJ, Astibia H (2009) The role of new Iberian finds in understanding European Eocene mammalian paleobiogeography. Geol Acta 7:243–258

Barbera X, Cabrera L, Marzo M, Parés JM, Agustí J (2001) A complete terrestrial Oligocene magnetobiostratigraphy from the Ebro Basin, Spain. Earth Planet Sci Lett 187:1–16

Barnolas A, Gil-Peña I (2001) Ejemplos de relleno sedimentario multiepisódico en una cuenca de antepaís fragmentada: la Cuenca Surpirenaica. Bol Geol Min Esp 112:17–38

Barnolas A, Pujalte V, Schmitz B (2019) South Pyrenean Foreland and Basque–Cantabrian Paleogene Basins. In: Quesada C, Oliveira JT (eds) The Geology of Iberia: a Geodynamic Approach, Springer, Berlin, pp 7–40

Beamud E, Garcés M, Checa LI, Muñoz JA, Almar Y (2003) A new middle to late Eocene continental chronostratigraphy from NE Spain. Earth Planet Sci Lett 216:501–514

Beamud E, Muñoz JA, Fitzgerald PJ, Baldwin SL, Garcés M, Cabrera L, Metcalf JR (2011) Magnetostratigraphy and detrital apatite fission track thermochronology in syntectonic conglomerates: constraints on the exhumation of the South-Central Pyrenees. Basin Res 23:309–331

Bellec V, Cirac P, Faugères JC (2009) Formation and evolution of paleo-valleys linked to a subsiding canyon, North Aquitaine shelf (France). C R Géosci 341:36–48

Bellec V (2003) Évolution morphostructurale et morphosédimentaire de la plate-forme aquitaine depuis le Néogène. Thèse Univ. Bordeaux I, 268 p

Bentham P, Burbank DW (1996) Chronology of Eocene foreland basin evolution along the western oblique margin of the South-Central Pyrenees. In: Friend PF, Dabrio CJ (eds) Tertiary Basins of Spain. The stratigraphic record of crustal kinematics. Cambridge University Press, Cambridge, pp 144–152

Bergounioux FM, Crouzel F (1960) Mastodontes du Miocène du Bassin d'Aquitaine. Bull Soc Hist Nat Toulouse 48:232–286

Birot P (1937) Recherches sur la morphologie des Pyrénées orientales franco-espagnoles. Baillière Édit., 318 p

Biteau JJ, Le Marrec A, Le Vot M, Masset JM (2006) The Aquitaine Basin. Pet Geosci 12:247–273

Boule M (1894) Le plateau de Lannemezan et les alluvions anciennes des hautes vallées de la Garonne et de la Neste. Bull Serv Carte Géol Fr VI(43):447–469

Bugnicourt D, Claracq P, Dupéron J, Privé-Gil C, Sauvage J (1988) Sédimentologie, bois fossiles et palynologie d'une couche à lignites de Capvern (plateau de Lannemezan, Hautes-Pyrénées). Bull Centre Rech Explor Prod Elf Aquitaine, Pau 12:739–757

Burbank DW, Puigdefabregas C, Muñoz JA (1992) The chronology of the Eocene tectonic and stratigraphic development of the eastern Pyrenean foreland basin, northeast Spain. Geol Soc Am Bull 104:1101–1120

Cahuzac B, Janssen AW (2010) Eocene to Miocene holoplanktonic Mollusca (Gastropoda) of the Aquitaine Basin, southwest France. Scripta Geol 141:1–193

Cahuzac B, Poignant A (1996) Foraminifères benthiques et Microproblematica du Serravallien d'Aquitaine. Géol Fr 3:35–55

Cahuzac B, Poignant A (2004) Les foraminifères du Burdigalien moyen à supérieur de la région sud-aquitaine (golfe de Saubrigues, SW France). Rev Micropaléontol 47:153–192

Cahuzac B, Turpin L (1999) Stratigraphie isotopique du Strontium dans le Miocène marin du Bassin d'Aquitaine (SW France). Rev Soc Geológica España Madrid 12:37–56

Cahuzac B, Janin MC, Steurbaut E (1995) Biostratigraphie de l'Oligo-Miocène du bassin d'Aquitaine fondée sur les nannofossiles calcaires. Implications paléogéographiques. Géol Fr 2:57–82

Calvet M, Gunnell Y, Laumonier B (2021) Denudation history and palaeogeography of the Pyrenees and their peripheral basins: an 84-million-year geomorphological perspective. Earth-Sci Rev 215:103436

Cameselle AL, Urgeles R, De Mol B, Camerlenghi A, Canning JC (2014) Late Miocene sedimentary architecture of the Ebro Continental Margin (Western Mediterranean): implications to the Messinian Salinity Crisis. Int J Earth Sci (Geol Rundsch) 103:423–440

Capdeville JP, Gineste MC, Turq A, Vergain P (1997) Handbook, Carte géol. France (1:50,000 scale), sheet Hagetmau (978). Orléans : BRGM, 70 p. Geological map by J.P. Capdeville (1997)

Capdeville JP, Darboux F (1998) Carte géol. France (1/50 000), sheet Aire-sur-l'Adour (979). Orléans: BRGM. Handbook by J.P. Capdeville, D. Millet, F. Millet (1998), 51 p

Clauzon G, Aguilar JP, Michaux J (1987) Le bassin pliocène du Roussillon (Pyrénées-Orientales, France): exemple d'évolution géodynamique d'une ria méditerranéenne consécutive à la crise de salinité messinienne. C R Acad Sci Paris II 304:585–590

Clauzon G, Suc JP, Aguilar JP, Ambert P, Capetta H, Cravatte J, Drivaliari A, Domenech R, Dubar M, Leroy S, Martinell J, Michaux J, Roiron P, Rubino JL, Savoye B, Vernet JL (1990) Pliocene geodynamic and climatic evolutions in the french mediterranean region. In: Iberian Neogene Basins, Paleontologia i Evolució, Mem. Especial n° 2, Sabadell, pp 131–186

Clauzon G (1990) Restitution de l'évolution géodynamique néogène du bassin du Roussillon et de l'unité adjacente des Corbières d'après les données écostratigraphiques et paléogéographiques. Paléobiol contin (Montpellier) XVII:125–155

Concy PJ, Muñoz AJ, McClay KR, Evenchik CA (1996) Syntectonic burial and post-tectonic exhumation of the southern Pyrenees foreland fold-thrust belt. J Geol Soc Lond 153:9–16

Conway-Jones BW, Roberts GG, Fichtner A, Hoggard M (2019) Neogene epeirogeny of Iberia. Geochem Geophys Geosyst 20:1138–1163

Costa JM, Maestro-Maideu E, Betzler C (1996) The Paleogene basin of the Eastern Pyrenees. In: Friend PF, Dabrio CJ (eds) Tertiary Basins of Spain. The stratigraphic record of crustal kinematics. Cambridge University Press, Cambridge, pp 106–113

Costa E, Garcés M, López-Blanco M, Serra-Kiel J, Bernaola G, Cabrera L, Beamud E (2013) The Bartonian-Priabonian marine record of the eastern South Pyrenean foreland basin (NE Spain): a new calibration of the larger foraminifers and calcareous nannofossil biozonation. Geol Acta 11:177–193

Crochet B (1991) Molasses syntectoniques du versant nord des Pyrénées: la série de Palassou. Documents du BRGM 199, 387 p

Crouzel F (1957) Le Miocène continental du bassin d'Aquitaine, Bull Serv Carte Géol Fr. 248, LIV: 264 p

Crusafont M, Hartenberger JL, Thaler L (1963) Sur des nouveaux restes de mammifères du gisement Eocène supérieur de Sossis, au nord de Tremp (Lérida, Espagne). C R Acad Sci Paris 257:3014–3017

Crusafont M, Villalta JF, de Truyols J (1956) Caracterización del Eoceno continental en la cuenca de Tremp y edad de la orogenesis pirenaica. Actes du deuxième Congrès international d'études pyrénéennes, Pau 1954, 2(I):39–53

Cuenca G, Canudo JI, Laplana C, Andres JA (1992) Bio y cronoestratigrafia con mamíferos en la Cuenca Terciaria del Ebro: ensayo de síntesis. Acta Geol Hisp 27:127–143

Cuesta MA, Checa LI, Casanovas ML (2006) Los artiodáctilos del yacimiento ludiense de Sossís (Cuenca Prepirenaica, Lleida, España). Rev Española Paleontología 21:123–144

Dautria JM, Liotard JM, Bosch D, Alard O (2010) 160 Ma of sporadic basaltic activity on the Languedoc volcanic line (Southern France): A peculiar case of lithosphere–asthenosphere interplay. Lithos 120:202–222

De Sitter LU (1961) La phase tectogénique pyrénéenne dans les Pyrénées méridionales. C R Somm Soc Géol Fr 8:224–225

Depéret C (1892) Sur la classification et le parallélisme du système miocène. C R Somm Séances Soc Géol Fr 3:170–266

Donville B (1973a) Ages K-Ar des vulcanites du Bas Ampurdan. C R Acad Sci Paris 276:3253–3256

Donville B (1973b) Ages K-Ar des vulcanites du Haut Ampurdan (NE de l'Espagne). Implications stratigraphiques. C R Acad Sci Paris 276:2497–2500

Donville B (1973c) Ages K-Ar des roches volcaniques de la depression de La Selva (NE de l'Espagne). C R Acad Sci Paris 277:1–4

Donville B (1976) Géologie néogène de la Catalogne orientale. Bull Bur Rech Geol Min III 2:177–210

Douvillé H (1924) A propos du poudingue de Palassou. C R Soc Géol Fr 15:160–162

Dubreuilh J, Capdeville JP, Farjanel G, Karnay G, Platel JP, Simon-Coinçon R (1995) Dynamique d'un comblement continental néogène et quaternaire: l'exemple du bassin d'Aquitaine. Géol Fr 4:3–26

Duranthon F (1993) Nouveaux gisements à rongeurs dans les molasses oligo-miocenes de la région toulousaine. Paleovertebrata 22:113–136

Duranthon F, Cahuzac B (1997) Eléments de corrélation entre échelles marines et continentales: les données du bassin d'Aquitaine au Miocène. Actes du Congrès BiochronM'97. In: Aguilar JP, Legendre S, Michaux J (eds) Mémoire and Travaux de l'Ecole Pratique des Hautes Etudes, Montpellier, vol 21, pp 591–608

Duranthon F (1991) Biozonation des molasses continentales oligo-miocènes de la région toulousaine par l'étude des mammifères. Apports à la connaissance du Bassin d'Aquitaine. C R Acad Sci Paris 313:965–970

Duvail C, Gorini C, Lofi J, Le Strat P, Clauzon G, Dos Reis T (2005) Correlation between onshore and offshore Pliocene-Quaternary systems tracks below the Roussillon basin (Eastern Pyrenees, France). Mar Pet Geol 22:747–756

Escarguel G, Marandat B, Legendre S (1997) Sur l'âge numérique des faunes de mammifères du Paléogène d'Europe occidentale, en particulier celles de l'Eocène inférieur et moyen. Actes du Congrès BiochronM'97, Aguilar JP, Legendre S, Michaux J (eds) Mémoire and Travaux de l'Ecole Pratique des Hautes Etudes, Montpellier, 21, 441–460

Fillon C, Van der Beek P (2012) Post-orogenic evolution of the southern Pyrenees: constraints from inverse thermo-kinematic modelling of low-temperature thermochronology data. Basin Res 23:1–19

Fillon C, Gautheron C, Van der Beek P (2013) Oligocene-Miocene burial and exhumation of the Southern Pyrenean foreland quantified by low-temperature thermochronology. J Geol Soc Lond 170:67–77

Fleta J, Vergés J, Escuer J, Pujadas J (1994) Mapa geologico de España, scale 1:50,000, Figueras (258). Memoria, 83 p, IGME, Madrid.

Garcés M, Cabrera L, Agustí J, Parés JM (1997) Old World first appearance datum of "Hipparion" horses: late Miocene large mammal dispersal and global events. Geology 25:19–22

Garcés M, López-Blanco M, Valero L, Beamud E, Muñoz JA, Oliva-Urcia B, Vinyoles A, Arbués P, Cabello P, Cabrera L, 2020. Paleogeographic and sedimentary evolution of the south-Pyrenean foreland basin. Mar Pet Geol 113, 104105

Garcia-Castellanos D, Vergés J, Gaspar-Escribano J, Cloetingh S (2003) Interplay between tectonics, climate, and fluvial transport during the Cenozoic evolution of the Ebro Basin (NE Iberia). J Geophys Res 108:2347. https://doi.org/10.1029/2002jb002073

García-Castellanos D, Larrasoaña JC (2015) Quantifying the post-tectonic topographic evolution of closed basins: The Ebro basin (Northeast Iberia). Geology 43:663–666

Gardère P (2005) La Formation des Sables Fauves: dynamique sédimentaire au Miocène moyen et évolution morpho-structurale de l'Aquitaine (SW France) durant le Néogène. Eclog Geol Helv 97:201–217

Gardère P, Rey J, Duranthon F (2002) Les «Sables Fauves», témoins de mouvements tectoniques dans le basin d'Aquitaine au Miocène moyen. C R Geosci 334:987–994

Gardère P (2002) Les Sables Fauves: dynamique sédimentaire et évolution morpho-structurale du Bassin d'Aquitaine au Miocène moyen. Strata, Toulouse, 40, 264 p

Gillot PY (1974) Chronométrie par la méthode K/Ar des laves des Causses et du Bas-Languedoc. Interprétation. PhD thesis, université d'Orsay, 61 p

Glangeaud P (1938) Sur la découverte d'un gisement stampien à Anthracothérium dans les argiles à lignites de Nassiet (Landes). P.V. Soc. Linn. Bordeaux, XC, 16–22

Granado P, Urgeles R, Sàbat F, Albert-Villanueva E, Roca E, Muñoz JA, Mazzuca N, Gambini R (2016) Geodynamical framework and hydrocarbon plays of a salt giant: the NW Mediterranean Basin. Pet Geosci 22:309–321

Guennoc P, Gorini C, Mauffret A (2000) Histoire géologique du golfe du Lion et cartographie du rift oligo-aquitanien et de la surface messinienne. Géol Fr 3:67–97

Hilgen FJ, Lourens LJ, Van Dam JA, with contributions by Beu AG, Boyes AE, Cooper RA, Krijgsman W, Ogg JG, Piller WE, Wilson DS (2012) The Neogene period. In: Gradstein et al. (eds), The Geologic Time Scale, Elsevier, chap. 29, 923–978

Hourdebaigt ML, Villatte J, Crochet B (1986) Le poudingue de Jurançon du Sud de Pau appartient à la série syntectonique de Palassou : preuve par la découverte d'une malacofaune éocène. C R Acad Sci Paris II 303:10, 951–955

Hourdebaigt ML (1988) Stratigraphie et sédimentologie des molasses synorogéniques en Béarn et en Bigorre. PhD thesis, Univ. Paul Sabatier, Toulouse, 240 p

Huard J, Lavocat R (1963) Sur la découverte de fossiles dans les formations à lignite d'Arjuzanx (Landes) et leur signification stratigraphique. C R Acad Sci Paris 257:3979–3980

Icole M (1968) Données nouvelles sur la formation de Lannemezan. C R Acad Sci Paris 267:687–689

Icole M (1969) Age et nature de la formation dite «de Lannemezan». Revue Géogr Pyrénées Sud-Ouest 40:157–170

Icole M (1974) Géochimie des altérations dans les nappes d'alluvions du piémont occidental nord-pyrénéen. Essai de paléopédologie quaternaire. Sciences Géologiques, mémoire n°40, Université Louis Pasteur, Strasbourg, 201 p

Karnay G, Dubreuilh J (1998) Carte géol. France (1:50,000), sheet Lembeye (1005). Orléans: BRGM. Handbook by G. Kamay, B. Mauroux, J.J. Châteauneuf (1998), 50 p

Kieken M (1973) Évolution de l'Aquitaine au cours du Tertiaire. Bull Soc Géol Fr XV: 51–60

Legendre S, Lévêque F (1997) Étalonnage de l'échelle biochronologique mammalienne du Paléogène d'Europe occidentale : vers une intégration à l'échelle globale. In: Aguilar et al (eds) Actes du Congrès BiochroM'97, Mém Trav EPHE Inst Montpellier 21:461–473

Legendre S, Sigé B, Astruc JG, De Bonis L, Crochet J-Y, Denys C, Godinot M, Hartenberger J-L, Leveque F, Marandat B, Mourer-Chauvire C, Rage J-C, Remy JA, Sudre J, Vianey Liaud M (1997) Les phosphorites du Quercy: 30 ans de recherche. Bilan et perspectives. [The phosphorites of Quercy: 30 years of investigations. Results and prospects]. Geobios M.S. 20:331–345

Londeix L (Ed) (2014) Stratotype Aquitanien. Muséum national d'Histoire naturelle, Paris; Biotope, Mèze, 416 p

López-Martínez N, Civis J, Casanovas L, Daams R (eds) (1998) Geología y Paleontología del Eoceno de la Pobla de Segur (Lleida). Edicions Universitat de Lleida, Lleida, 267 p

Magné J, Baudelot S, Crouzel F, Gourinard Y, Wallez MJ (1985) La mer du Langhien inférieur a envahi le centre du bassin d'Aquitaine : arguments biostratigraphiques et géochronologiques. C R Acad Sci Paris 300(II):19, 961–964

Marandat B (1991) Mammifères de l'Ilerdien moyen (Eocène inférieur) des Corbières et du Minervois, systématique, biostratigraphie, corrélations. Palaeovertebrata 20(2–3):55–144

Marandat B, Adnet S, Marivaux L, Martinez A, Vianey-Liaud M, Tabuce R (2012) A new mammalian fauna from the earliest Eocene (Ilerdian) of the Corbières (Southern France): palaeobiogeographical implications. Swiss J Geosci https://doi.org/10.1007/s00015-012-0113-5

Martí J (2004) La región volcánica de Gerona. In: Vera JA (ed) Geología de España. SGE-IGME, Madrid, pp 672–675

Martí J, Mitjavila J, Rocá E, Aparicio A (1992) Cenozoic magmatism of the Valencia trough (western Mediterranean): relationship between structural evolution and volcanism. Tectonophysics, 203:145–165

Martínez del Olmo W (1996) Depositional sequences in the Gulf of Valencia Tertiary basin. In: Friend PF, Dabrio CJ (eds) Tertiary basins of Spain. The Stratigraphic Record of Crustal Kinematics. Cambridge University Press, Cambridge, pp 55–67

Mauffret A, Durand de Grossouvre B, Dos Reis AT, Gorini C, Nercessian A (2001) Structural geometry in the eastern Pyrenees and western Gulf of Lion (Western Mediterranean). J Struct Geol 23:1701–1726

Mayer-Eymar K (1858) Versuch einer neuen klassifikation der Tertiär-Gebilde Europa's. Réunion de Trogen (16–19 août 1857). Verhandlungen der allgemeinen Schweizerischen Gesellschaft für die gesammten Naturwissenschaften 70–71, 165–199

Medialdea T, Vasquez JT, Vegas R (1994) Estructura y evolución geodinámica del extremo noreste del margen continental catalán durante el Neógeno. Acta Geol Hisp, 29, 4 (Pub. 1996):9–53

Meigs AJ, Vergés J, Burbank DW (1996) Ten-million-year history of a thrust sheet. Geol Soc Am Bull 108:1608–1625

Meigs A (1997) Sequential development of selected Pyrenean thrust faults. J Struct Geol 19:481–502

Mellere D, Marzo M (1992) Los depósitos aluviales sintectónicos de la Pobla de Segur: alogrupos y su significado tectonoestratigráfico. Acta Geol Hisp 27:145–159

Michael NA, Carter A, Whittaker AC, Allen PA (2014) Erosion rates in the source region of an ancient sediment routing system: comparison of depositional volumes with thermochronometric estimates. J Geol Soc Lond 171:401–412

Mouline MP (1967) Etude des poudingues dits de Puylaurens, leurs conditions de mise en place, les conséquences paléoclimatiques de ces phénomènes. Actes Soc Linn Bordeaux, 104B:3–15

Mouline M-P (1978) Les épandages conglomératiques de l'Eocène inférieur à l'Oligocène dans le Castrais et l'Albigeois: l'importance de l'orographie et ses conséquences climatiques dans une des principales manifestations de la rhexistasie paléopyrénenne d'origine tectonique. Bull Soc Géol Fr XX:215–219

Muratet B, Cavelier C (1992) Caractère séquentiel discontinu des molasses oligocènes à la bordure orientale du Bassin aquitain; signification des conglomérats bordiers (Tarn, Tarn et Garonne, Sud-Ouest de la France). Géol Fr 1:3–14

Muratet B, Duranthon F, Lange-Badré B, Riveline J (1992) Discontinuité d'origine eustatique dans les molasses oligocènes de l'Est du Bassin Aquitain (SW France). Apport de la biochronologie. C R Acad Sci Paris II 315:1113–1118

Murelaga X, Pérez-Rivarés FJ, Vázquez-Urbez M, Zuluaga MC (2008) Nuevos datos bioestratigráficos y paleoecológicos del Mioceno medio-superior (Aragoniense) del área de Tarazona de Aragón (Cuenca del Ebro, provincia de Zaragoza, España). Ameghiniana, 45:393–406

Nichols GJ (2005) Tertiary alluvial fans at the northern margin of the Ebro Basin. In: Harvey AM, Mather AE, Stokes M (eds) Alluvial fans: geomorphology, sedimentology, dynamics. Geol Soc Lond Spec Publ 251:187–206

Nichols GJ (2018) High-resolution estimates of rates of depositional processes from an alluvial fan succession in the Miocene of the Ebro Basin, northern Spain. In: Ventra D, Clarke LE (eds) Geology and geomorphology of alluvial and fluvial fans: terrestrial and planetary perspectives. Geol Soc Lond Spec Publ 440:159–173

Nijman W (1989) Thrust sheet rotation? The South Pyrenean Tertiary basin configuration reconsidered. Geodin Acta 3:17–42

Nijman W (1998) Cyclicity and basin axis shift in a piggyback basin: towards modelling of the Eocene Tremp–Ager Basin, South Pyrenees, Spain. In: Mascle A, Puigdefàbregas C, Luterbacher HP, Fernàndez M (eds), Cenozoic foreland basins of Western Europe. Geol Soc Lond Spec Publ 134, pp. 135–162

Ortiz A, Guillocheau F, Lasseur E, Briais J, Robin C, Serrano O, Fillon C, (2020) Sediment routing system and sink preservation during the post-orogenic evolution of a retro-foreland basin: the case example of the North Pyrenean (Aquitaine, Bay of Biscay) basins. Mar Pet Geol 112:104085

Ortuño M, Marti A, Martin-Closas C, Jimenéz-Moreno G, Martinetto E, Santanach P (2013) Palaeoenvironments of the Late Miocene Pruëdo Basin: implications for the uplift of the Central Pyrenees. J Geol Soc Lond 170:79–92

Parize O, Mulder T, Cahuzac B, Fiet N, Londeix L, Rubino JL (2008) Sedimentology and sequence stratigraphy of Aquitanian and Burdigalian stratotypes in the Bordeaux area (southwestern France). C R Géosci 340:390–399

Patin J (1967) L'évolution morphologique du plateau de Lannemezan. Rev Géogr Pyrén Sud-Ouest 38:325–337

Pérez-Rivaré FJ, Arenas C, Pardo G, Garcé M (2018) –Temporal aspects of genetic stratigraphic units in continental sedimentary basins: examples from the Ebro basin, Spain. Earth-Sci Rev 178:136–153

Plaziat JC (1981) Late Cretaceous to late Eocene paleogeographic evolution of southwest Europe. Palaeogeogr Palaeoclimatol Palaeoecol 36:263–320

Plaziat JC (1984) Stratigraphie et évolution paléogéographique du domaine pyrénéen de la fin du Crétacé (phase Maastrichtienne) à la fin de l'Éocène (phase Pyrénéenne). Université Paris Sud, 1362 p

Puigdefábregas C (1975) La sedimentación molásica en la cuenca de Jaca. Monografías del Instituto de Estudios Pirenaicos, 104, Pirineos, Jaca, 188 p

Puigdefàbregas C, Souquet P (1986) Tecto-sedimentary cycles and depositional sequences of the Mesozoic and Tertiary from the Pyrenees. Tectonophysics 129:173–203

Pujadas C, Casas JM, Muñoz JA, Sabat F (1989) Thrust tectonics and Paleogene syntectonic sedimentation in the Empordà area, Southeastern Pyrenees. Geodin Acta 3:195–206

Pujalte V, Baceta JI, Schmitz B, Orue-Etxebarria X, Payros A, Bernaola G, Apellaniz E, Caballero F, Robador A, Serra-Kiel J, Tosquella J (2009) Redefinition of the Ilerdian Stage (early Eocene). Geol Acta 7:177–194

Reille JL (1967) Subdivisions stratigraphiques et phases de plissement dans le Paléogène continental sud-pyrénéen (region de Barbastro, province de Huesca). C R Acad Sci Paris D 265:852–854

Reille JL (1971) Les relations entre tectogénese et sédimentation sur le versant sud des Pyrénées centrales d'après l'étude des formations tertiaires essentiellement continentales. Unpubl. PhD thesis, Université de Montpellier, 330 p

Rey J, Duranthon F, Gardère P, Gourinard Y, Magné J, Feinberg H, Muratet B (1997) – Découverte d'un encaissement entre dépôts de sables fauves dans la région de Sos (Miocène centre-aquitain). Géol Fr 2:23–29

Riba O (1973) Las discordancias syntectónicas del Alto Cardener (Prepirineo catalán), ensayo de interpretación. Acta Geol Hisp:1–6

Riba O (1976) Syntectonic unconformities of the Alto Cardener, Spanish Pyrenees: a genetic interpretation. Sed Geol 15:213–233

Richard M (1948) Contribution à l'étude du bassin d'Aquitaine. Les gisements de mammifères tertiaires. Mém Soc Géol Fr 380 p

Roca E (1996a) The Neogene Cerdanya and Seu d´Urgell intramontane basins (eastern Pyrenees). In: Friend PF, Dabrio CJ (eds) Tertiary Basins of Spain. The Stratigraphic Record of Crustal Kinematics. Cambridge University Press, Cambridge, pp 114–118

Roca E (1996b) La evolución geodinámica de la Cuenca Catalano-Balear y áreas adyacentes desde el Mesozoico hasta la actualidad. Acta Geol Hisp 29:3–25

Roca E, Sans M, Cabrera L, Marzo M (1999) Oligocene to Middle Miocene evolution of the central Catalan margin (northwestern Mediterranean). Tectonophysics 315:209–233

Rosell J, Riba O (1966) Nota sobre la disposicion sedimentaria de los conglomerados de La Poble de Segur. Actas V Cong. Int. Est. Pyr., Jaca, Pamplona. Pirineos 81–82:61–74

Rosell J, Linares R, Llompart C (2001) El 'Garumniense' prepirenaico. Rev Soc Geol Esp 14:47–56

Rosell J (1967) Estudio geológico del sector del Prepirineo comprendido entre los ríos Segre y Noguera Ribagorzana. Pirineos XXI:9–214

Sanjuan J, Martín-Closas C, Serra-Kiel J, Gallardo H (2012) Stratigraphy and biostratigraphy (charophytes) of the marine-terrestrial transition in the Upper Eocene of the NE Ebro Basin (Catalonia, Spain). Geol Acta 10:19–31

Sauvage J (1969) Etude sporo-pollinique des formations miocènes d'Orignac (Pyrénées centrales françaises). Doc Labo Géol Fac Sci Lyon 31:1–19

Schoeffler J (1969) L'âge des molasses de Pau à Peyrehorade. Bull Soc Géol Fr 7:28–34

Schoeffler J (1971) Étude structurale des terrains molassiques du piémont nord des Pyrénées de Peyrehorade à Carcassonne. PhD thesis (unpubl.), Université de Bordeaux 1

Schoeffler J (1973) Étude structurale des terrains molassiques du piémont nord des Pyrénées de Peyrehorade à Carcassonne. Rev Inst Fr Pétrole, 28:515–549 and 639–665

Serra-Kiel J, Mato E, Saula E, Travé A, Ferrandez-Cañadell C, Busquets P, Samso JM, Tosquella J, Barnolas A, Alvarez-Perez G, Franquès J, Romero J (2003) An inventory of the marine and transitional Middle/Upper Eocene deposits of the Southeastern Pyrenean Foreland Basin (NE Spain). Geol Acta 1:201–229

Sigé B (1997) Les mammifères insectivores des nouvelles collections de Sossis et sites associés (Éocène supérieur, Espagne). Geobios 30:91–113

Solé Sugrañes L, Clavell E (1973) Nota sobre la edad y posición tectónica de los conglomerados eocenos de Queralt (Prepirineo oriental, Prov. de Barcelona). Acta Geol Hisp VIII:1–6

Soto R, Larrasoaña JC, Beamud E, Garcés M (2016) Early–Middle Miocene subtle compressional deformation in the Ebro foreland basin (northern Spain); insights from magnetic fabrics. C R Géosci 348:213–223

Stehlin HG (1910) Die Säugetiere des schweizerischen Eocaens. Mém Soc paléont Suisse XXXVI:838–1164

Sudre J, de Bonis L, Brunet M, Crochet JY, Duranthon F, Godinot M, Hartenberger JL, Jehenne Y, Legendre S, Marandat B, Remy JA, Ringeade M, Sigé B, Vianey-Liaud M (1992) La biochronologie mammalienne du Paléogène au Nord et au Sud des Pyrénées: état de la question. C R Acad Sci Paris (II) 314:631–636

Sztrakos K, Steurbaut E (2017) Révision lithostratigraphique et biostratigraphique de l'Oligocène d'Aquitaine occidentale (France). Geodiversitas 39:741–781

Sztrakos K, Gély JP, Blondeau A, Müller C (1998) L'Éocène du bassin sud-aquitain: lithostratigraphie, biostratigraphie et analyse séquentielle. Géol Fr 4:57–105

Tassone A, Roca E, Muñoz JA, Cabrera L, Canals M (1994) Evolución del sector septentrional del margen continental catalán durante el Cenozoico. Acta Geol Hisp 29:3–37

Vázquez-Urbez M, Arenas C, Pardo G, Pérez-Rivarés J (2013) The effect of drainage reorganization and climate on the sedimentologic evolution of intermontane lake systems: the final fill stage of the Tertiary Ebro Basin (Spain). J Sed Res 83:562–590

Vandenberghe N, Hilgen FJ, Speijer RP, with contributions by Ogg JG, Gradstein FM, Hammer O (2012) The Paleogene period. In: Gradstein et al (eds) The Geologic Time Scale, Elsevier, chap. 28, 855–921

Vergés J, Burbank DW (1996) Eocene–Oligocene thrusting and basin configuration in the eastern and central Pyrenees (Spain). In: Friend PF, Dabrio CJ (eds), Tertiary basins of Spain, the stratigraphic record of crustal kinematics. Cambridge University Press, Cambridge, pp. 120–133

Villena J, Gonzalez A, Muñoz A, Pardo G, Perez A (1992) Síntesis estratigráfica del Terciario del borde Sur de la Cuenca del Ebro: unidades genéticas. Acta Geol Hisp 27:225–245

Vincent SJ (2001) The Sis paleovalley a record of proximal fluvial sedimentation and drainage basin development in response to Pyrenean mountain building. Sedimentology 48:1235–1276

Vincent SJ, Elliott T (1996) Long-lived fluvial palaeovalleys sited on structural lineaments in the Tertiary of the Spanish Pyrenees. In: Friend PF, Dabrio CJ (eds) Tertiary Basins of Spain. The Stratigraphic Record of Crustal Kinematics. Cambridge University Press, Cambridge, pp 161–165

Viret J (1938) Sur l'âge des argiles ligniteuses de Nassiet près Amou (Landes). C R Acad Sci Paris 207:500–501

Yuste A, Luzón A, Bauluz B (2004) Provenance of Oligocene-Miocene alluvial and fluvial fans of northern Ebro Basin (NE Spain): an XRD, petrographic ans SEM study. Sed Geol 172:251–268

Cenozoic Transformations of the Mountain Range: Evidence from Denudation Chronology and Landforms

In the case of the Ancestral Pyrenees the record reveals a state of transience in which rates of crustal deformation and crustal denudation during the Paleogene were neither steady through time at any given location, nor spatially uniform (see also Rahl et al. 2011; Whitchurch et al. 2011; Parsons et al. 2012; Ford et al. 2016). Given these persistent signatures of transience in the sediment record, it is reasonable to expect that a record of unsteady or transient landscape evolution should also be encoded in landform assemblages within the mountain range, whether in the Axial Zone or the outer fold belts. Slope systems, interfluve summits, and perhaps valleys at times responded in characteristic ways to relative lulls in the history of crustal deformation, and/or were at certain places buffered from base-level changes. A rapid overview of the mountain landscape tends to confirm this hypothesis, with its (i) mosaic of range-top, low-gradient erosion surfaces of varying sizes and states of conservation, often in abrupt juxtaposition with (ii) a population of younger incisional landforms, themselves displaying evidence of successive stages of downcutting and drainage reorganization such as dry valleys, wind gaps, mountain-flank pediments and rock benches, all situated above (iii) staircases of Quaternary alluvial terraces, fans and debris cones. We focus first on the population of older, pre-Quaternary erosional landforms, and address the Quaternary landforms subsequently.

3.1 A Distinctive Population of Vestigial Erosion Surfaces

Low-gradient mountain tops occur in the glaciated as well as non-glaciated parts of the orogen (Calvet and Gunnell 2008; Calvet et al. 2015a) and are among the oldest surviving landforms of the mountain range (Fig. 3.1). Many of these elevated, low-gradient landscapes are invisible to roadside geologists, but early

© Springer Nature Switzerland AG 2022
M. Calvet et al., *Geology and Landscapes of the Eastern Pyrenees*, GeoGuide,
https://doi.org/10.1007/978-3-030-84266-6_3

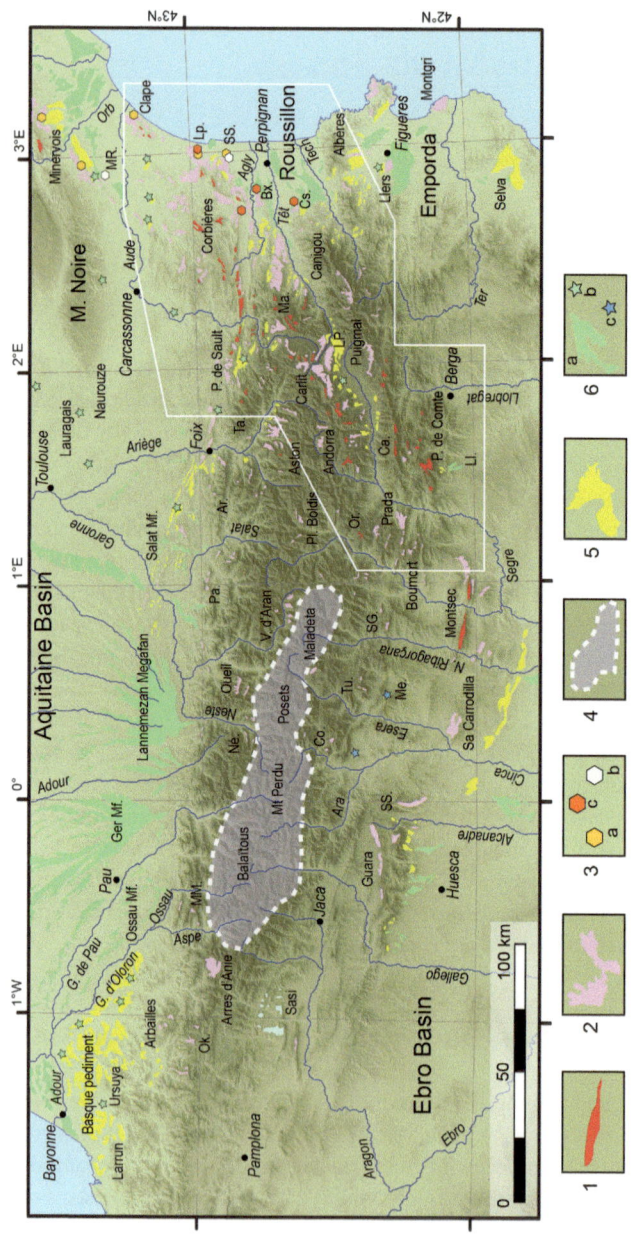

◄**Fig. 3.1** **Elevated erosion surfaces and other diagnostic landforms in the Pyrenees, with corresponding sedimentary deposits**. The map is based on a systematic map and field survey of each individual topographic unit. Key to symbols and ornaments—**1**: range-top surface S (late Oligocene to Aquitanian). **2**: range-flank pediment P1 (mid-Miocene); in the western Pyrenees S and P1 are indistinguishable. **3**: base-level connections for P1; **3a**: Langhian–Serravallian coastal deposits; **3b**: Serravallian and Tortonian lacustrine limestones; **3c**: micromammalian fossil deposits of Neogene age preserved in the limestone karst. **4**: elevated massifs of the western Axial Zone apparently never bevelled by S or P1. **5**: late Neogene local pediments P2 and P3, also represented as karstic poljes when in limestone outcrops. **6**: deposits relating to the generation of P2 or P3; **6a**: Mediterranean Pliocene sediments, Lannemezan Formation (ancient range-front fluvial megafans), calcrete-capped upper wash pediments of the Ebro Basin; **6b**: isolated vestiges of siliciclastic lag gravels capping P2; **6c**: upper slope breccia deposits of the Cinca watershed (Merli Fm.). Key to place name abbreviations—MR: Montredon, LP: Lapalme, SS: Serre du Scorpion, Bx: Baixas, Cs: Castelnou, Ma: Madrès, Ta: Tabe, Ar: Arize, Pa: Paloumère, Né: Néouvielle, MM: Mailh Massibé, Ca: Cadi, Or: Orri, Ll: Lladurs, Ok: Okabé, SG: Sant Gervas, Tu: Turbon, Me: Merli, Co: Cotiella, SS: Sierra de Sevil

hillwalking geomorphologists have been reporting them from various parts of the mountain range for over 100 years (de Martonne 1910; Blanchard 1914; Sorre 1913; Nussbaum 1930, 1931; Boissevain 1934; Pannekoek 1935; Birot 1937; Goron 1941a; Sermet 1950; De Sitter 1952; Zandvliet 1960). The presence of low-energy landforms in a high-energy mountain landscape has never been fully explained, but generates five threads of inquiry, all narrowly intertwined:

(i) the spatial distribution of these low-gradient surfaces, e.g., whether they are isolated or occur in clusters, whether they form accordant levels or topographic staircase patterns, and whether they can be mapped along the entire strike of the mountain range or are instead confined to its outer fold belts or to the eastern segment of the orogen;

(ii) the age of the erosion surfaces with respect to the Proto-, Ancestral and Modern Pyrenees, respectively, i.e., whether they are pre-, syn-, or post-tectonic;

(iii) the relative ages of the different surfaces within any one of the given broad time windows (e.g., corresponding to the Ancestral or the Modern orogenic states)—i.e., whether there is just one generation that perhaps underwent subsequent tectonic offsets, of whether several generations can be distinguished and mapped based on a variety of criteria;

(iv) the geodynamic conditions under which the low-energy topographic end-state that they document could be attained;

(v) the elevation at which the erosion surfaces were formed, i.e., whether the mountain tops of the ancestral Pyrenees underwent some form of levelling in situ (perhaps by altiplanation), implying a subsequent static evolution of the mountain range; or whether the transition from the synorogenic Ancestral to the post-orogenic Modern Pyrenees involved some form of transient peneplanation prior to regional uplift and deep re-incision by rivers and glaciers.

3.1.1 Erosion Surfaces in the Axial Zone

The spatial pattern and morphology of erosion surfaces is well established in the east of the Modern Pyrenees, where they occupy ~15% of the surface area (Calvet 1996; Calvet and Gunnell 2008; Calvet et al. 2021). Two distinct generations can be mapped in some massifs (e.g., eastern Corbières, Carlit, see Itineraries 1, 6, 7, 8), and this bimodal feature is extendable to southern Andorra, the Aston and the Madrès. The upper, mountain-top surface (noted S_0 in Calvet 1996, subsequently S in Calvet and Gunnell 2008) is either a peneplain or a pediplain: there are no discriminating criteria other than the occasional presence of monadnocks such as, for example, Pic Carlit (2921 m; see Itinerary 7). These residual summit surfaces are flanked at lower elevations by a population of range-flank pediments, some of which are quite extensive (noted S_1 in Calvet 1996; P1 in Calvet and Gunnell 2008). The elevation difference between S and P1 is typically 300–500 m at the most, regardless of whether they occur in the limestone thrust-and-fold belt of the eastern Corbières (Itinerary 1), or the crystalline Carlit or Aston massifs of the Axial Zone (Itineraries 7, 8). In any given massif, S is always in a range-top position, often forming a drainage divide between the Atlantic and Mediterranean base levels—whether at 600–900 m in the Corbières or at 2400 m in the Madrès, 2800–2900 m in the Carlit massif, and likewise along the borders with Andorra (Hartevelt 1970). These widespread relics of low-gradient, low-energy topography are legacies of some advanced state of orogenic downbearing of the ancestral Pyrenees in at least the eastern third of the modern mountain range. In the central Pyrenees, the Axial Zone only displays vestiges of P1, e.g., in the upper Noguera Pallaresa watershed (Zandvliet 1960), in Val d'Aran (Kleinsmiede 1960; Ortuño et al. 2008, 2013, 2018), the Neste watershed (Monod et al. 2016), the Montagne d'Oueil; and arguably around the edges of the Néouvielle massif, where disproportionately wide and flat-floored glacial cirques are possibly seated in the upper reaches of pre-Quaternary pediments (Barrère 1952a). In the western

Pyrenees erosion surfaces reappear (Uzel et al. 2020) among the higher massifs around the base of Pic d'Anie, and bevel the Chaînons Basques at decreasing altitudes towards the west (Viers 1960).

3.1.2 Erosion Surfaces in the Pro- and Retro-Wedges

Along the retro-wedge, the S/P1 pair is mappable as far west as the Aude valley and Pays de Sault (see Itinerary 5). When it occurs on massive Jurassic and Lower Cretaceous limestone, surface S forms elevated plateaus that typically rise towards the SW and additionally exhibit a diverse array of mesoscale exo- and endokarstic features: Roc Paradet (900 m), Fanges (1000 m), Bac d'Estable (1450 m), Forêt de la Serre (1330 m), etc. Wide erosional corridors ascribable to P1 in this context present themselves as wind gaps and dry valleys between these massifs, i.e., intermediate landforms between bedrock straths and bedrock pediments (see Itinerary 8, Box 8.1). These ancient valleys usually contain quartz gravel trains. Also at similar elevations to the pediments and dry valleys, and thus similarly ascribable to P1, are a large number of subhorizontal endokarstic cave systems, often filled with allochtonous gravel deposits (now deeply weathered) supplied by the Axial Zone (e.g., Clat cave in Clat palaeovalley 1130 m; Paradet-Malabrac caves; and on the plateau de Sault; see Itinerary 5).

From Pays de Sault westwards, however, any clear distinction between S and P1 becomes impossible. The only evidence is a generally bevelled appearance of massif summits, but at an apparently random range of mean elevations suggesting either the existence of several generations of surfaces, or differentiated magnitudes of tectonic movements responsible for offsetting a single original surface (see Goron 1931, 1937, 1941a for a descriptive inventory; cyclic interpretations therein, however, are obsolete). Range-flank erosional benches are also noted in some basement massifs such as Tabe, Arize (Le Planel, 1067 m; Plaine d'Uscla, 1274), and Bouirex. They are nonetheless best preserved on massive limestone such as, from east to west: La Frau (1650 m, which contains a very ancient network of horizontal endokarstic galleries at la Caunha de Montségur); Pech de Foix (700–1000 m); the Sourroque (1200 m) and Paloumère massifs (1200–1600 m) in the Salat catchment; a sequence of very smooth ridgetop flats around 1200 m between the Garonne, Neste and Adour rivers; ridgetops of the NPZ between the Aspe and Ossau valleys (Uzel et al. 2020) around 1600 and 1900 m (Ourdinse plateau, Montagnon); and onward to the Arbailles massif, where upturned and folded geological structures are truncated at an elevation of 1000–1200 m (Vanara 2000).

On the pro-wedge side of the orogen, Birot (1937) had ruled out the presence of any erosion surfaces mostly by default, i.e., because of an absence of adequate topographic and geological maps at the time. The accepted default view was that any low-gradient land surfaces were structural. Since then, however, it has been possible to document the existence of erosional bevels on the Pedraforca nappes (Calvet 1996; see Itinerary 9). Likewise, east of the Segre valley two generations of erosion surfaces have been identified on that basis (Itinerary 9). Erosion surfaces also occur on the massive limestone outcrops of the Internal Sierras (Boumort, Turbon) and External Sierras (Montsec and other 'Sierras Marginales' of Catalonia, and Sierra de Guara still farther west; Barrère 1952b; Peña 1983; Rodriguez Vidal 1984; Calvet et al. 2021).

3.1.3 Synthesis

The major items of evidence arising from the inventory of erosion surfaces in the Pyrenean orogenic wedge are (i) the widespread occurrence of a pair of surfaces—a range-top and a range-flank surface—throughout the eastern third of the Pyrenees; (ii) the regional extent of the range-top surface within this eastern segment of the mountain range; (iii) the absence of a well-defined summit paleoplain in the elevated Axial Zone massifs of the central Pyrenees, i.e., over a distance of 160 km from Encantats westward to Pic d'Anie; and (iv) the occurrence, often very clear but sometimes more hypothetical, of erosion surfaces (sometimes several generations) throughout the fold-and-thrust belts of the pro- and retro-wedges, as well as in the higher ranges of the western Pyrenees (Basque country: Arres d'Anie, Occabé, and Nives region; Viers 1960). All of these features require interpretation.

3.2   Age and Origin of the Pyrenean Erosion Surfaces

3.2.1   Products of Late- to Post-Orogenic Landscape Evolution

It seems justified to rule out that the surfaces on basement rocks in the Axial Zone are pre-orogenic. Such a view was once held on the basis of a small vestige of marine Cretaceous caprock at the top of Pic de Balaïtous (3144 m; García Sainz 1940). It is now clear, however, that the isolated character of this outlier confirms on the contrary that all other potential vestiges of this sub-Cretaceous unconformity were soon destroyed by deep and vigorous syntectonic denudation. Everywhere else, and particularly along the boundaries of the basement-cored Axial Zone, the surfaces of sub-Triassic or sub-Cretaceous unconformities have been so steeply up- and overturned by plate collision that the random exposure of these exhumed surfaces has produced occurrences at best a few tens to a few hundreds of metres in extent.

From rock-cooling studies based on thermochronology, we know now that Paleogene denudation depths were much too large for any pre-tectonic erosion surfaces to have survived the orogeny of the Ancestral Pyrenees, Axial Zone included. The geomorphological history of the eastern Pyrenees thus really begins with the peak of Paleogene continental collision, which involved the denudation (driven by subaerial erosion but also by nappe displacement) of 6–10 km of crust. These values are documented by, among others, zircon FT data in the Axial and North-Pyrenean zones, which often have not been entirely reset by synorogenic burial and have instead preserved signatures of the pre-collisional Mesozoic rifting period (Garwin 1985; Yelland 1990, 1991; Sère 1993; Morris et al. 1998; Fitzgerald et al. 1999; Maurel 2003; Maurel et al. 2002, 2008; Sinclair et al. 2005; Gibson et al. 2007; Metcalf et al. 2009; Jolivet et al. 2007; Meresse 2010; Beamud et al. 2011; Filleaudeau et al. 2011; Rahl et al. 2011; Whitchurch et al. 2011; Fillon et al. 2013; Rushlow et al. 2013; Michael et al. 2014; Vacherat et al. 2014, 2016, 2017; Bosch et al. 2016a, b). The thermochronological data mostly document the pre- and synorogenic histories, particularly the Paleogene denudation maximum (accelerated denudation began ca. 50 Ma and peaked around 35–30 Ma), and have revealed a north-to-south time lag in the timing of peak denudation. They also detect some effects of Oligocene to early Miocene crustal extension in the east, and locally pick up recent valley incision in the Central Pyrenees (AFT and (U–Th)/He cooling between 10 and 20 Ma).

Gunnell et al. (2009) produced low-temperature thermochronological data directly documenting the denudation histories of erosion surfaces S and P1 (eastern Pyrenees, including Andorra and Ariège). Results showed that denudation rates declined after the Paleogene maximum (Fig. 3.3). By the time of the Oligocene–Miocene transition, a regional land surface of low relief prevailed (Fig. 3.2). This is evidenced by AFT and (U–Th)/He rock-cooling histories, where each sampled range-top or range-flank site, after an initial period of very rapid cooling, reached a steady and definitive low-temperature plateau ca. ~25 Ma. This low-temperature state has endured to the present day (Fig. 3.3). Such a plateau in the cooling curve of a low-relief landform in a high-energy mountain range is a clear signal of post-orogenic topography having reached a state of low energy and, by reasonable inference, of low relief. This pattern has been observed in

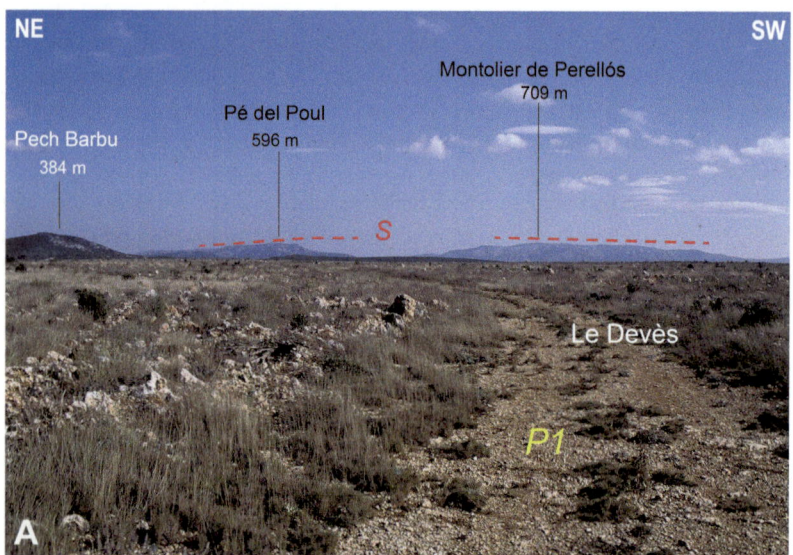

◄**Fig. 3.2**   **Range-top and range-flank erosion surfaces in the east-Pyrenean landscape**. A Range-flank pediment P1 (mid-Miocene) in the eastern Corbières, here seen looking south from the Devès plateau. The tread of this low-gradient erosional land surface cross-cuts the dips of Mesozoic limestone beds of the Corbières nappe (North-Pyrenean Zone) as well as folded Eocene beds of the Sub-Pyrenean Zone. Note residual topography on the horizon, bearing vestiges of the range-top surface, S. **B** Range-top surface, S, here seen from the top of the Campcardós massif looking east. Younger range-flank pediments P1 and P2 are highlighted in the background, around the edge of the late Neogene Cerdagne Basin

many mountain ranges around the world (for a review: Calvet et al. 2015), and in the Pyrenees is fully consistent with the fact that the erosion surfaces cross-cut the many syntectonic structures of the orogenic wedge such as thrust sheets, nappes and folds—whether in the Axial Zone or among the crumple belts of the pro- and retro-wedge.

3.2.2   Evidence of Episodic Surface Uplift

In the eastern Pyrenees, rifting during the early Miocene initiated a cycle of topographic rejuvenation in response to the newly forming Mediterranean base level, and was largely responsible for the tectonic regime that generated the pairs of S–P1 surfaces. Rifting produced substantial local relief around certain footwall uplifts such as Mt. Canigou, where surface S has been entirely destroyed by denudation, but this was an exception. Other massifs (Carlit, Campcardós, Aston, Madrès, Corbières…) were not endowed with comparable magnitudes of relative relief. As a result, the S–P1 pairs of land surfaces in these massifs are preserved and vertically separated by 250–500 m at most (Itineraries 1, 6–8).

3.2.2.1 The Summit Surface: Generation S

The age of the summit surface, S, is imprecise, and probably diachronous according to the AFT and AHe results reported above and in Fig. 3.3. The sedimentological inflexion observed during Chattian time in the eastern Aquitaine Basin (Figs. 2.2, 2.3), when conglomerate was replaced in the stratigraphy by lacustrine limestone, provides a relatively robust palaeoenvironmental indication of declining topographic energy in the Ancestral Pyrenees at the time, and thus of relatively gentle gradients compatible with the widespread existence of a Pyrenean plateau or paleoplain. Thick lacustrine limestone also occurs in the (initially Oligocene) Narbonne Basin, and likewise (with interlayers

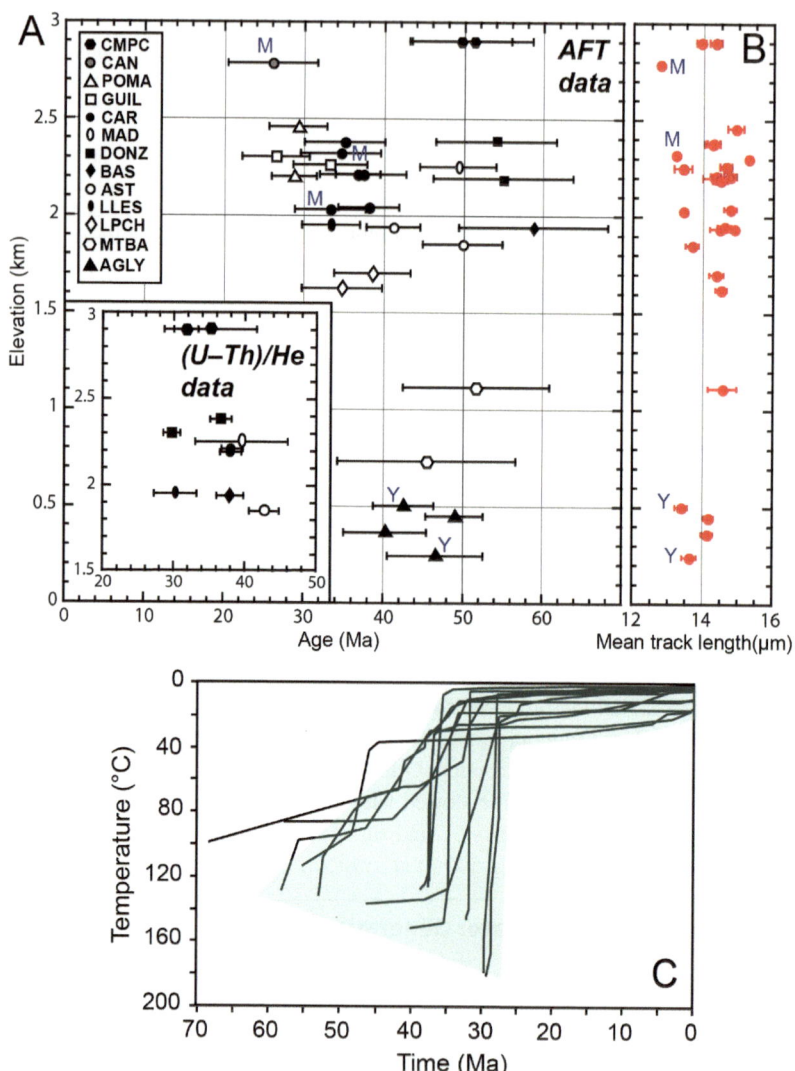

◄**Fig. 3.3** **Rock-cooling histories of range-top and range-flank erosion surfaces of the eastern Pyrenees**. **A** AFT and AHe age–elevation plot (data from Gunnell et al. 2009). Massif names: CMPC–Campcardós; CAN–Canigou; POMA–Pla de Pomarola; GUIL–Pla Guillem; CAR–Carlit; MAD–Madres; DONZ–Donnezan; BAS–Bassiès; AS–Aston; LLES–Tossa Plana de Llès; LPCH–Coll de la Perxa; MTBA–Plateau de Montalba; AGLY–Agly massif. AFT ages produced by Yelland (1991) and Maurel et al. (2008) (marked in blue: Y and M, respectively) show good agreement between independent workers. **B** Mean fission-track length–topographic elevation plot. **C** Best-fitting cooling histories obtained by forward thermal modelling of the AFT and AHe (or U–Th/He) data presented in A and B. The curves highlight the residence times of individual samples at the surface or at least in the shallow crust since 35–25 Ma irrespective of their current elevations (after Gunnell et al. 2009, modified)

of lignite) in the eastern Ebro Basin at Calaf, Tarrega, and Mequinenza (Cabrera and Saez 1987; Gomis et al. 1997). Aquitanian marl is often sandwiched between beds of lacustrine limestone and is a deposit also diagnostic of low-energy environments; it likewise overlies the molasse in the centre of the Aquitaine Basin (the 'Trilogie agenaise', see Sect. 2.2.2) and in the eastern extensional basins (Narbonne Basin, and base of the Roussillon Basin fill sequence at depths of −1200 to −1500 m in Canet borehole; Berger et al. 1988). Consistent with the thermochronological data of Gunnell et al. (2009), the most likely age of range-top palaeosurface, S, is thus late Oligocene to early Miocene. It records the transitional state between a deeply eroded, low-relief Ancestral Pyrenees and the soon-to-rise-again Modern Pyrenees.

3.2.2.2 The Range-Flank Pediments: Generation P1

Age constraints on generation P1 are good in the eastern Pyrenees (Calvet 1992, 1996; Calvet and Gunnell 2008; Gunnell et al. 2009; Calvet et al. 2021). Depending on location, different occurrences of pediment P1 (i) cross-cut uptilted Upper Oligocene and Aquitanian beds within the Narbonne and Tuchan–Paziols basins (Itinerary 1); (ii) grade to the top of the middle Miocene shoreline outcrops of the Mediterranean seaboard (Itinerary 1); and (iii) display across their treads a very large number of fossil micromammalian assemblages preserved in regolith-filled cracks of the limestone pavements, with ages ranging between 20 and 10 Ma (Itineraries 1, 2). The preservation of these fossil assemblages suggests a low-energy surface environment. In the Canigou massif, P1 formed after the intense period of rifting and deep unroofing of the Paleozoic gneiss dome, which (based on the inverse stratigraphy and age of correlated deposits in the adjacent intermontane Conflent Basin, Itinerary 6) occurred during early Aquitanian to

Burdigalian time. In the Corbières and Minervois retro-wedge nappes, P1 continued to evolve into the late Miocene, coevally with the lacustrine limestone deposits of Montredon (biozone MN 10, 10–9 Ma). In the Ariège, the upward-fining molasse sequence from the Aquitanian through to the late Burdigalian, as likewise the onlap of lacustrine limestone beds southward up to the outermost folds of the Pyrenean retro-wedge, suggest a similar chronology, thus also valid for the planar range-flank surfaces of the Aston massif.

Along the Iberian pro-wedge, the Sierras Interiores and their overlying (unconformable and often north-dipping) conglomerate sequences between the Segre and Ésera valleys have been truncated by a single erosion surface. This population of occurrences is at least post-Chattian, but perhaps belongs to generation S rather than to P1. In the External Sierras, the main surface on the Guara massif (Barrère 1952b), which slopes from 1700 to 1200 m, belongs to generation P1: large masses of residual topography have been preserved on the Guara limestones, but P1 cuts across the entire Campodarbe sequence, and grades to the top of the Miocene conglomerate beds. In the Montsec thrust belt, vestiges of summit surface S have undergone warping and below a pediment ascribable to P1 can be mapped truncating Paleogene conglomerates in the east while grading to the top of the Sariñena Formation in the west.

3.2.3 Influence of Multiple Base Levels

The short geological time to completion of surfaces belonging to generations S and P1, respectively, can be explained by three complementary causes, all of which are particularly relevant to the eastern Pyrenees.

(i) The narrowness of the Ancestral Pyrenees at the end of the Paleogene is a first factor. Whether in the Aquitaine or the Ebro foreland, the Pyrenean fold-and-thrust belts were in large part buried beneath the molasse and conglomerate sequences described in Sects. 2.1 and 2.2, so that the Paleogene mountain range was effectively a backbone consisting mainly of the Axial Zone. Paleogene mountain width did not exceed ~80 km (Calvet et al. 2021).

(ii) The interruption of tectonic convergence during the Oligocene in the eastern Pyrenees, and its fragmentation into discrete tectonic blocks generated by extensional tectonics, is a second factor. Horst-and-graben relief multiplied the total length of tectonic mountain fronts vulnerable to erosion by steep, high-energy catchments. Tectonic fragmentation limited the total volume and height of positive relief exposed to denudation, and thus (compared to the central Pyrenees) conspired to promote conditions favourable to relief decline and to the development of a population of erosion surfaces during relative lulls in base-level change.

(iii) A third factor is the possibility of a dense lithospheric root pulling downward and counteracting or delaying the full potential for isostatic rebound. Rebound is a function of the elasticity of the underthrust Iberian Plate and of orogen width (the orogen being assumed here to behave as a line load). In the case of the Ancestral Pyrenees, i.e., an orogen narrower and shorter than the Modern Pyrenees, isostatic rebound in the Pyrenees was thus perhaps 15% of what Airy isostasy would otherwise predict (Montgomery 1994).

The planar land surfaces were thus formed by denudation among competing catchments grading to the Atlantic and Mediterranean base levels of the time, making unnecessary any speculation that the summit surface in the Axial Zone was formed by altiplanation processes at palaeoelevations of 2000 m or more, i.e., at their currently observed altitudes of occurrence—for example as a consequence of Ebro Basin overfilling (Babault et al. 2005; Gunnell and Calvet 2006; Bosch et al. 2016). Such a theory would imply that the narrow, Ancestral Pyrenees somehow remained vertically static during the last ~25 million years and that the externally drained Aquitaine Basin somehow was also an overfilled basin.

No evidence, however, exists in favour of such north–south symmetry (see also arguments given by Sinclair et al. 2009). The scenario of a raised base level caused by an Ebro Basin backfilled to the Axial Zone would imply that the summit erosion surface graded, for example, to the top of the Nogueres conglomerate sequence. There is no direct evidence, however, in support of this scenario on a basis comparable with, for example, the clastic 'gangplanks', or ramps, that today still connect the Colorado Front Range to the High Plains of Texas, Colorado or Wyoming (see Calvet et al. 2015, for an overview). The Nogueres conglomerate beds are steeply backtilted (up to 30° in the Sierra de Sis) towards the Axial Zone, including the uppermost formation known as the Antist Formation (Fig. 2.2), as can be clearly observed north of the town of La Pobla de Segur. Moreover, erosion surfaces along this piedmont zone cross-cut the tilted clastic beds themselves, suggesting that the erosion surfaces are actually younger than the clastic sequences, and not coeval with their formation (Calvet et al. 2021). In the Aquitaine foreland, the Palassou conglomerates, which are homologous to their Nogueres counterparts, are likewise folded and, at many places, upturned vertically. In the Catalonian foreland, the gently north-dipping Bartonian to Lower Priabonian marine beds have been uplifted to altitudes of ~1300 m east of the

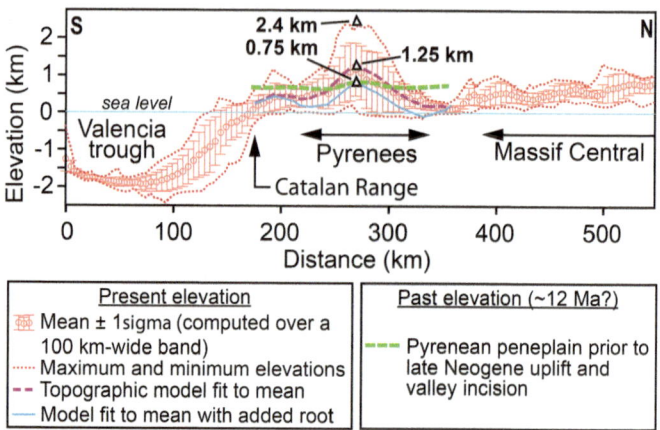

Fig. 3.4 Initial altitude of Pyrenean peneplain formation inferred from multi-proxy geophysical modelling (see Fig. 1.4). Current mean elevation (1.25 km in the Axial Zone, purple line) is shown here with dispersion envelopes along the 100 km-wide swath profile (centred on 2.5°E). Mean elevation with an added root (blue line) mimics the mean topography that would be obtained if a lithospheric load identical to that still present beneath the central Pyrenees (ca. 0.5°E) were instantaneously added beneath the present-day eastern Pyrenees (corresponding topographic valleys and peaks are maintained). The peak altitude on the blue line in the crest zone (0.75 km) defines the initial elevation of the paleoplain (range-top surface S) at a time when the lithospheric root in the eastern Pyrenees was still present. The paleoplain is now uplifted to ~2.4 km. Note that because of uplift and incision, dispersion (dotted lines) around the present-day mean elevation curve (purple dashes) is much higher than at the time of peneplain completion when values of mean, maximum and minimum elevation were almost identical (i.e., low topographic relief). Topography displayed as thick green dashes is meant to represent the paleoplain prior to the onset of thermally-driven lithospheric thinning and associated topographic uplift (after Gunnell et al. 2008, modified)

Ter valley, emphasising the magnitude of post-depositional uplift of these Eocene marine beds. Such geometric relations do not conform with the simple model of a mountain range getting buried beneath a backfill of its own debris (Coney et al. 1996). Connecting (by imaginary lines) vestiges of generation S or P1 at their current altitudes in the Axial Zone to clastic wedge-tops in the Aquitaine or Roussillon basins would require 'gangplank' gradients of 4%. Such a high value is totally unrealistic for putative alluvial megafans of the sizes observed or reconstructed along the Pyrenean range fronts (Calvet and Gunnell 2008; Miall 2016; Ventra and Clarke 2018). By logical elimination, it thus seems far more likely to consider that the summit surfaces are vestiges of a regional paleoplain that formed across part of the Ancestral Pyrenees at an elevation lower than its current elevations would suggest. A variety of geophysics- and geomorphology-based inferences (based on alluvial terrace tilts and more realistic 'gangplank' gradients) have suggested that the initial altitude of this paleoplain in the eastern Pyrenees would have not exceeded 800 m, even in the most internal parts of the orogen's catchments (Fig. 3.4) (Gunnell et al. 2008, 2009; Delmas et al. 2018).

3.3 Intermittent Surface Uplift in Late Neogene Time: Evidence from Landforms

The late Neogene (last ~10 million years) is the history of a second rising, or resurrection, of the Pyrenees. The evidence in support of this phenomenon derives primarily from geomorphological clues, further corroborated by geophysical, tectonic and stratigraphic observations. The alternative hypothesis to this view is that the modern mountain mass is an unchanged legacy of the Ancestral Pyrenees, simply bevelled off at the top by altiplanation (see discussion above) and recently dissected by rivers and glaciers as a result of late Neogene climatic change (Babault et al. 2005a; b; Bosch et al. 2016). Prior to the advent of plate tectonics, some mountain ranges used to be interpreted as 'raised peneplains' and thereby implied recent uplift (and concomitant valley re-incision) of an older orogenic structure (e.g., Davis 1911; de Sitter 1952). This model appears to be gaining new relevance globally within the theoretical framework of plate tectonics (Calvet et al. 2015), and is enhanced by the growing recognition that a clutch of relief-enhancing lithospheric and sublithospheric processes has been occurring since the late Neogene beneath a number of plateau regions and mountain ranges (Potter and Szatmari 2009; Baran et al. 2014; Leroux et al. 2018), including the Pyrenees, their foreland basins, and the Iberian peninsula more generally (Conway-Jones et al. 2019).

3.3.1 Proxies of Topographic Uplift Throughout the Mountain Belt

3.3.1.1 Thermal Relaxation and Lithospheric Thinning Since 10 Ma

From the summit surface having formed at a likely mean palaeoelevation of 0.8 km it follows that ~60% of the topographic relief we see today was produced during the late Neogene, i.e., after 12–10 Ma and in several stages involving regional surface uplift. The inferred mean topographic uplift rate is 0.2 mm/yr, i.e., sufficiently slow to lie within the uncertainty envelope of present-day detection of vertical crustal motion by GPS (0.1 ± 0.2 mm/year: Nguyen et al. 2016). Assemblages of late Miocene warmth-loving plant remains in the sedimentary deposits of the Cerdagne Basin (currently at 1200 m a.s.l., see Itinerary 8) independently confirm the notion of recent regional uplift (Suc and Fauquette 2012), and even more so in the case of the warmth-loving plant assemblages reported from the Val d'Aran, which currently occur at ~2000 m (Ortuño et al. 2013). Oxygen isotope ratios extracted from late Miocene fossil rodent teeth and used as gauges of palaeoaltimetric change corroborate at least 500 m of uplift in the Cerdagne Basin (Huyghe et al. 2020).

The driving mechanism of crustal uplift in the eastern Pyrenees could be ascribed to partial melting or partial detachment of a dense lithospheric root beneath the orogen (Fig. 1.4), accompanied by sustained crustal extension along the Mediterranean back-arc margin (Gunnell et al. 2008, 2009; Calvet et al. 2021). Maintaining the buoyancy of the crust to the east of Andorra almost certainly implies a thermal contribution from the asthenospheric mantle, which in turn causes thermal or mechanical thinning of the mantle lithosphere (discussion in Jolivet et al. 2020). The resulting dynamic uplift supports the regional topography of the modern Pyrenees (Chevrot et al. 2018)—a process also consistent with the situation inferred for the neighbouring Catalan Coastal Range (Lewis et al. 2000) and the Massif Central (Malcles et al. 2020), with and incidence of Late Neogene volcanism in both those areas.

The most recent deep tomography of the mantle based on teleseismic and gravity data has imaged a low-velocity anomaly located between the surface and a depth of 100 km in the central Pyrenees. It coincides with a strong density anomaly interpreted as a mass of eclogite in the deep crust, also consistent with the depth-distribution of seismic activity in the crust. The conjunction of these independent components of evidence support the notion of a crustal drip or delamination process occurring beneath the Pyrenees (Dufréchou et al. 2018). Magnetotelluric tomography along the strike of the ECORS PYRENEES profile had previously inferred partial melting of the Iberian lithospheric mantle beneath the Pyrenees, with a thermal relaxation time compatible with the cessation of crustal convergence ca. 25 Ma. The reason why magmatism in relation to this process has not yet produced volcanic eruptions farther west than eastern Catalonia in the last 10 Ma has been attributed to delays in an overall extensional collapse of the orogenic edifice (Pous et al. 1995a, b). Campanya et al. (2018) recently detected a low-resistivity zone between 20 and 70 km beneath the central Pyrenees, interpreted as a region of partial melting, and in agreement with

Gunnell et al. (2008) and Chevrot et al. (2018) concluded on the current absence of a lithospheric root beneath the mountain range anywhere east of Andorra.

3.3.1.2 Clastic Output: A Response to Relief Growth

The increase in clastic output since the late Miocene, both in terms of volume of debris and average clast size, is recorded in piedmont deposits such as the Lannemezan megafan and its other coeval retro-foreland formations, among the eastern extensional basins such as the Roussillon, Empordà, Cerdagne, and in Mediterranean offshore clastic wedges. A climatic cause for accelerated denudation is theoretically possible, but the time period of regional uplift spans palaeoclimatic regimes as diverse as the warm subtropical conditions alternating between super-humid and semi-arid of the Tortonian, Messinian and Pliocene; and the cold, glaciated and temperate oscillations of the Quaternary. Tectonically-driven relief growth and steepening of hillslopes seems a more plausible explanation for such widespread delivery of terrigenous sediment by the mountain range, and is consistent with the lithospheric and mantle processes mentioned above. Another strong indication of neotectonic activity is the occurrence of well-preserved triangular-faceted spurs, for example along the southern boundary-fault scarp faces of the Roussillon and Cerdagne basins (Albères massif, Itinerary 2; north face of Mt. Canigou, Itinerary 6; Cerdagne Basin, Itineraries 7 and 8). These are in each case associated with debris-cone sequences several hundred metres thick, of Pliocene and Turolian age, respectively, containing giant boulders supplied by very steep mountain-front catchments (Itineraries 2 and 8).

3.3.1.3 Evidence of Fault Activity

Many Pyrenean deposits of know Miocene to Quaternary age exhibit tectonic strain indicators, from tilting to fault-controlled offsets. The known spatial distribution of these tectonic indicators, however, is uneven: in the Roussillon Basin, for example, faults in the Miocene fill sequences are widespread, but faults in younger Pliocene beds are more uncommon and restricted to a few documented sites. At the scale of the entire Pyrenees, the inventory of tectonically offset Quaternary deposits does not exceed ~20 sites (Fig. 3.5). The intensity of deformation is variable depending on the sequence: Upper Miocene beds are steeply upturned in the Cerdagne, La Seu, and Empordà basins, and syn- and post-sedimentary fault throws through the graben fills attain ~1 km (see Itinerary 8). Strata in Pliocene sequences exhibit shallower dips, and fault throws do not exceed a few hundred metres. Tectonic offsets through Quaternary deposits are an order of magnitude less. Deformational style is also variable, and includes extensional, strike-slip, and thrust faults (Viers 1960, 1961a; Ellenberger and Gottis 1967;

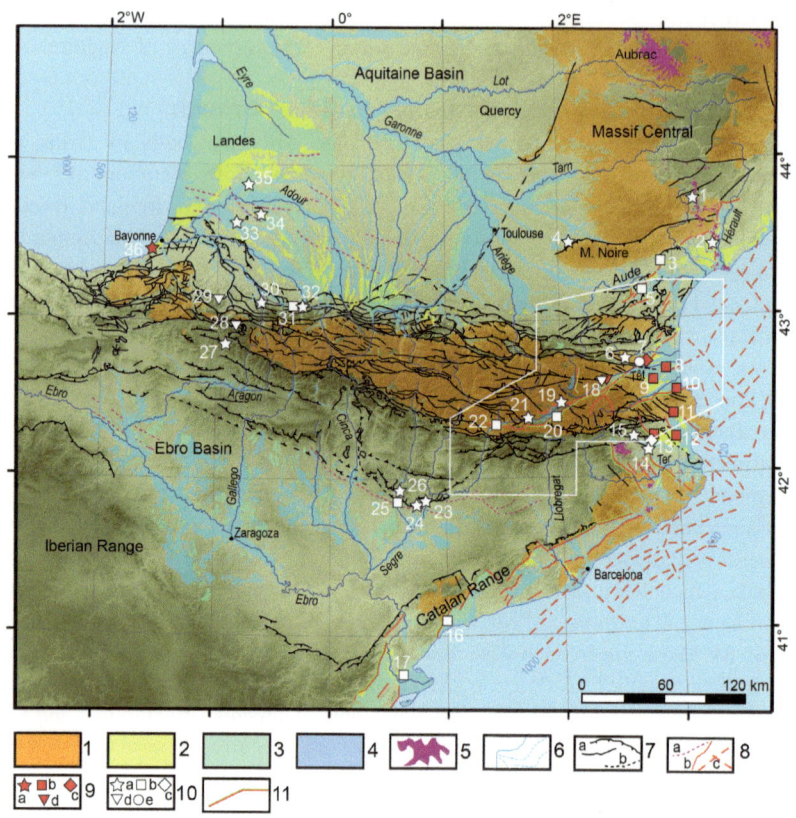

Birot 1969; Mouline et al. 1969; Ambert 1977; Pous et al. 1986; Cabrera et al. 1988; Briais et al. 1990; Philip et al. 1992; Masana 1994; Saula et al. 1994; Carbon et al. 1995; Roca 1996a; Genna et al. 1997; Goula et al. 1999; Calvet 1996, 1999; Fleta et al. 2001; Baize et al. 2002; Alasset and Meghraoui 2005; Dubos-Sallée et al. 2007; Ortuño et al. 2008, 2018; Lacan et al. 2012; Lacan and Ortuño 2012; Philip 2018). Inside the limestone cave systems, endokarstic galleries offset by tectonic throws of up to 10 m have been reported in the Villefranche syncline (Coronat massif, Calvet et al. 2015; Hez et al. 2015), but the ages and fault-plane solutions of these tectonic fractures are difficult to establish (see Itinerary 6). As a result, the chronology of stress regimes during the late Neogene is difficult to

◄**Fig. 3.5** **Neotectonic record in the Pyrenees and foreland areas.** Key to symbols and ornaments—**1**: outcrops of Hercynian basement; **2**: Pliocene continental and marine deposits; **3**: early Pleistocene alluvial deposits, uppermost terraces; **4**: middle and late Pleistocene and Holocene alluvial terraces; **5**: late Miocene to late Pleistocene volcanic rocks; **6**: continental shelf isobaths; **7**: Faults—**7a**: main faults and thrusts of Hercynian and Pyrenean age; **7b**: blind or probable faults and thrusts of Hercynian and Pyrenean age; **8**: other tectonic features—**8a**: main fold axis in the outer fold belts; **8b**: Neogene normal fault. **9**: neotectonic deformations affecting Pliocene deposits—**9a**: reverse fault, **9b**: normal fault; **9c**: strike-slip fault; **9d**: faulted karstic landform of deposit (unspecified). **10**: neotectonic deformations affecting Pleistocene deposits—**10a**: reverse fault, fold; **10b**: normal fault; **10c**: strike-slip fault; **10d**: faulted karstic landform of deposit (unspecified); **10e**: fault (unspecified); **11**: triangular faceted spurs associated with Neogene normal faults. Place names where deformation indices have been recorded (some are logged under a single number given their close proximity); 1: Escandorgue, 2: Le Rièges; 3: Bize-Minervois; 4: Saint Chipoli–Dourgne; 5: Fabrezan; 6: Caramany–Le Mas; 7: Millas, Néfiach, Ille-sur-Têt; 8: Perpignan–Serrat d'en Vaquer; 9: Trouillas, Fourques; 10: Sorède, Laroque des Albères, Villelongue; 11: St Climent Sescebes; 12: Pontos, Fluvia; 13: Pedrinya, Incarcal; 14: Serinya; 15: Tortella, Rajolins, Burro; 16: Camp de Tarragona; 17: Baix Ebre; 18: Villefranche karst (Notre-Dame de Vie Cave, Faubourg Cave); 19: Estavar; 20: Osséja; 21: Martinet; 22: La Seu d'Urgell–Montferrer; 23: Balaguer NE–Rio Sio; 24: Balaguer NW; 25: Alfarras; 26: Canelles; 27: Isaba; 28: Pierre-St-Martin karst; 29: Arbailles karst; 30: Lurbe–Asasp; 31: Capbis; 32: Arcizac; 33: Bastenne-Gaujac, Heugas; 34: Horsarieux; 35: Meilhan; 36: Côte Basque, Arcangue-Castagnet

reconstruct. It is likely that it was neither uniform across the orogen nor constant through time at any given site.

Present-day vertical motions have been documented only by levelling methods, with debatable values of up to 1 mm/year at Mt. Canigou and in the Têt valley (Lenôtre and Fourniguet, unpublished BRGM report, 1987, cited in Calvet 1996; see also Philip 2018) and 1–4 mm/year in NE Catalonia (Gimenez et al. 1996). The limitations associated with such methods have been revealed in the context of events such as the St-Paul-de-Fenouillet (1996, M_L 5.2, see Itinerary 5) and Arudy (1980, M_L 5.1) earthquakes (Rigo and Cushing 1999). With very low measured values of 0.1 ± 0.2 mm/year, GPS measurements of vertical displacement currently remain within the background noise of analytical error (Nguyen et al. 2016).

3.3.2 Geomorphological Evidence of Recent Uplift

3.3.2.1 Late Neogene Rock Pediments (Generation P2)

Late Neogene uplift was not steady. Pauses or relative lulls allowed pediments (i.e., partial planation surfaces, see Itinerary 8, Box 8.1) to expand across the edges of the Neogene extensional basins in the east, but also into more interior areas of the Axial Zone and all the way around the Pyrenees generally, in the form of flat-floored erosional corridors, bedrock straths or erosional topographic basins (usually in soft or weatherable rock outcrops such as marl, flysch, deeply weathered granite), but also limestone basins. As a consequence of uplift, these poljes today are no longer within reach of seasonal water table fluctuations and have thus become fossil landforms. The age of these landforms is often difficult to establish, and sometimes they display successions of multiple benches. The oldest occurrences grade to (and thus seem coeval with) the Vallesian deposits of the Cerdagne (Itineraries 7 and 8) and La Selva basins.

The pediment generation most indicative of unsteady post-orogenic uplift regime is P2. Its residual occurrences grade to the depositional tops of retro-wedge megafans such as the Lannemezan, and are thus estimated to be of late Pliocene or earliest Pleistocene age (Fig. 3.6; Delmas et al. 2018). These partial pediments are several kilometres across and may form deep, flat-floored embayments and make inroads into the mountain topography, typically hanging today 200–400 m above the modern valley floors (see Itineraries 4, 5, 6, 7, 8, 9). Smaller erosional benches of a younger generation (labelled here P3) grade laterally to the oldest Pleistocene alluvial terrace deposits, and may thus qualify as bedrock straths. These successive cohorts of planar landforms are relatively ubiquitous and formed prior to the deep vertical valley incision and widespread slope steepening events of the Pleistocene. Quaternary glaciation has often failed to erase these late Neogene features from the landscape (see Itineraries 7 and 8).

Occurrences of these characteristically low-gradient landforms have been widely mapped in the eastern Pyrenees—particularly among granitic outcrops but also in schist (Goron 1931, 1937, 1941a; Zandvliet 1960; Hartevelt 1970; Lagasquie 1969, 1984a, 1984b, 1987; Calvet 1996). Around the edges of the Cerdagne Basin, a good example of P2 is the Plateau de la Perche (Calvet 1996; Delmas et al. 2018), where the low-gadient topography bevels the basement as well as a steeply dipping sequence of Vallesian and Turolian Miocene fill deposits (see Itineraries 7 and 8). Other landforms belonging to this generation include the Plateau de Sault (Itinerary 5), the surface of which is covered by a veneer of siliciclastic lag deposits displaying facies identical to the constituent depositional units of the Lannemezan megafan (Fig. 3.6). Occurrences of pediment P2 are

Fig. 3.6 Quaternary geology of the piedmont environments: some landscape features.
A Lannemezan megafanhead (Neste River), consisting of the oldest and most elevated early
Pleistocene alluvial deposit in the Pyrenean Quaternary record. **B** General profile view of
the Lannemezan megafan. **C** Patinated quartz ventifacts on the oldest alluvial terrace of the
Roussillon Basin (probably coeval with the Lannemezan Fm.)

also observed in the Neste watershed, in the weathered granite outcrops around the town of Bordères (Monod et al. 2016). Around the Empordà Basin (Catalonia), the pediment surfaces grade to the top of Pliocene gravel beds forming the Llers–Figueres plateau. Similar configurations are reported on either side of the lower Aude valley, e.g., in the Corbières (at Les Vals) and in Minervois (Montouliers–Montplo) (Birot 1969; Ambert 1994; Larue 2008). Farther west, from the Ariège River to the Gave rivers and the Basque country, the P2 piedmont ramp cuts across almost all of the Sub- and North-Pyrenean fold sequences—particularly folds in the softer Cretaceous and Eocene flysch—and is even locally embayed into its regional P1 predecessor (Calvet et al. 2021).

On the pro-wedge side of the eastern Pyrenees, similar suites of planar landforms extend across outcrops of Eocene, Oligocene and Cretaceous flysch and shale, such as at Pla de Lladurs (~400 m above the Cardener River, Itinerary 9). Often sealed by cemented limestone breccia, they also occur further west in the Cinca watershed (Stange et al. 2018), such as at the south-facing Guara thrust front, and on the flysch outcrops of Navarre (Barrère 1952b, 1962, 1975). All of these soft-rock pediments belong to the same generation P2.

3.3.2.2 Dry Valleys and Drainage Capture

The population of abandoned valleys in the Pyrenean landscape consists mostly of broad, shallow furrows which document a time interval during the earlier stages of uplift of the Modern Pyrenees, i.e., when fluvial incision into the stumps of the Ancestral Pyrenees was still limited. These ancient valleys sometimes display well preserved entrenched meander belts (e.g., the dry valley at Col de Saint-Louis, in the Fenouillèdes massif; see Itinerary 5).

Different generations of ancient valleys have been identified. The most elevated are rare and correlate with pediment surface P1. They often connect to the upper portion of these pediments. The largest population of dry valleys is coeval with generation P2 and occurs systematically at elevations lower than P1. In the piedmont zones, some of the valleys are of Quaternary age. Whatever their exact age, they all document a diachronous process of river piracy and progressive drainage reorganisation during the late Neogene regrowth of the Pyrenees. The drainage captures have mostly benefited the northern watersheds of the orogen. In the eastern Pyrenees, the Têt River has also lost part of its left bank watershed to the Aude as a consequence of tectonic downthrow in the Capcir graben; the shallow cradle of abandoned hanging valleys can still be observed along the skyline of the N–S boundary fault scarp to the east of the Capcir graben (Itinerary 7). In the central part of the Axial Zone, Pla de Beret is the abandoned upper valley of the south-flowing Noguera Pallaresa, its upper catchment having been conquered by the north-flowing Garonne River.

Most of the other dry valleys have been preserved by dint of the drainage becoming subterranean and flowing through extensive belts of massive lime-stone. A series of dry valleys, for example, has been preserved in the Aptian and Barremian outcrops that extend from the Fenouillèdes to the Pays de Sault (Itinerary 5). All of these ancient valleys currently hang 300–400 m above the modern active valley floors. In the Sub-Pyrenean massifs of the pro-wedge, the Llinars–Pla de la Llacuna dry valley, which must be very ancient given its relative elevation of 875 m above the Segre River, is a landmark of similar importance (Itinerary 9), as likewise the dry valleys incising the eastern termination of the Montsec thrust-front scarp in Catalonia (Calvet et al. 2021).

3.3.2.3 Quaternary Alluvial Terraces

Vertical successions of alluvial terraces are commonly used as records of crustal uplift (Bridgland 2000; Kiden and Törnqvist 1998; Bridgland and Westaway 2014; Demoulin et al. 2017), particularly when the elevation offset between two given generations of alluvial deposit increases upstream but tends to level off in the vicinity of the coastline, and eventually become a stack of conformable strati-graphic units on the continental shelf. This configuration is valid in the case of the Roussillon Basin, where five generations of Quaternary alluvial deposits fit this scenario (Delmas et al. 2018), and where the steep seaward tilt of the stratigraphic boundary between Pliocene marine and Pliocene continental deposits also con-firms a rise of the east-Pyrenean Axial Zone during the last 4–5 Ma (Itineraries 2 and 6). The hinge zone at the coast remained relatively stationary over that time period, which explains why the Corbières and Albères coastlines fail to display the abundance of uplifted Pleistocene marine terraces that might be expected in other active tectonic settings. Alluvial terrace sequences similar to those of the Têt are also recognised along the Orb (Larue 2008) and Aude rivers in Languedoc, the Fluvia and the Ter rivers in Catalonia, and equally the Garonne, Gave rivers, and lower Adour. Alluvial terraces have similarly been used in the Ebro Basin to document regional uplift, with patterns likewise suggesting increasing magni-tudes of uplift towards the mountain belt and an acceleration of uplift during the late Neogene (Stange et al. 2013a, 2016; Lewis et al. 2017).

The vertical successions of alluvial terraces also record a number of strike-perpendicular tectonic upwarps that have cause the rivers to migrate laterally across their foreland topography. In the Aquitaine Basin, the middle segment of the Garonne has steadily drifted eastward by at least 25 km, forming the extensive left-bank staircase of terraces in the Toulouse area and undercutting the upper terrace system of the Tarn River in the process (Enjalbert 1960; Barrère et al. 2009). The lower Ariège River has undergone a similar evolution. The magnitude of tectonic deformation is appreciated by comparing the maximum altitude of the highest terrace of the Garonne in the Forêt de Bouconne (330 m), west of Toulouse, with the heights of interfluve summits in the Lauragais area, which are capped by coeval alluvial gravels from the Montagne Noire and never exceed 280 m between the Ariège and Hers Mort valleys. Likewise, the Comminges area, situated between Saint-Gaudens and Tarbes, was in early Tortonian time the last surviving fluvio-lacustrine depocentre of the Aquitaine Basin. As a result of subsequent uplift it now holds a commanding position over the piedmont at an altitude of ~600 m.

3.3.2.4 Groundwater Karst and Cave Systems

During periods of crustal stability, the altitude of major drainage conduits within the karst is adjusted to the upper surface of water tables, the position of which is itself dictated by local topographic base levels. Vadose entrenchment forms (which typically consist of vertical shafts and entrenched meandering canyons) develop at times of base-level fall, i.e., in response to crustal uplift and valley incision (Audra and Palmer 2011, 2013). Discontinuous uplift has accordingly been recorded by the groundwater karst system of some Pyrenean limestone massifs, where several levels of horizontal endokarstic cave systems have been mapped over vertical heights of 1 km (Maire and Vanara 2008). Occurrences in the Axial Zone and the retro-wedge tectonic belts include the Arbailles massif (5 major levels over a vertical distance of 800 m; Vanara et al. 1997; Vanara 2000), the Pierre-Saint-Martin–Arres d'Anie (8 levels between 1950 and 450 m; Maire 1990), and the Arbas massif (6 major levels over a vertical distance of 900 m; Bakalowicz 1988) in the west. Further east, occurrences also include the Tarascon syncline (middle Ariège valley, with 10 levels over a vertical distance of 600 m; Sorriaux et al. 2016), the upper Aude valley (9 levels over a vertical distance of 600 m), and the Villefranche-de-Conflent syncline (10–12 levels over a vertical distance of 1100 m; Hez et al. 2015; Calvet et al. 2015b; Calvet et al. 2019) (see Itineraries 5 and 6). Among the pro-wedge tectonic units, the Cotiella massif records at least 3 main levels (Belmonte 2014), and probably 5–6 levels occurring at elevations between 2300 and 800 m.

When such karstic cavities are entered by allocthonous rivers transporting quartz-rich debris supplied by Hercynian basement outcrops located upstream, the alluvium deposited in the caves can be subjected to cosmogenic-nuclide burial dating in order to obtain its residence time. A burial age calculation relies on the radioactive decay of the nuclide because, once confined to the cave environment, the debris are shielded from exposure to cosmic rays and thus from further nuclide accumulation. This method usually relies on measuring the concentrations of two nuclides with different half-lives, most commonly ^{26}Al and ^{10}Be, and allows residence times typically between 0.2 and 5.5 Ma to be detected (Granger et al. 1997; Granger and Muzikar 2001). This kind of work in the Pyrenees is still in its infancy, but should provide clues to the chronology of Pyrenean uplift. Data from the Villefranche karst have yielded a mean incision rate of 0.06 mm/year since the beginning of the Pliocene and 0.11 m/year since 1 Ma (Calvet et al. 2015; Itinerary 6).

3.3.3 Synthesis

The extensional and pull-apart tectonics prevalent during the Oligocene and Miocene in the eastern Pyrenees progressively gave way to regional uplift of the Pyrenees, elevating the gently undulating surfaces S and P1 to maxima of 2900 m (e.g., Campcardós massif); and likewise raising their younger successor population P2, such as the Plateau de la Perche pediment (eastern Pyrenees) or the Pla de Beret palaeovalley (central Pyrenees) as well as the intermontane basins themselves (Pruëdo, La Seu, Cerdagne), to altitudes exceeding 1500 m. This late regional uplift caused extremely rapid incision of V-shaped valleys during the latest Pliocene and Quaternary. At locations where glaciation has failed to alter the longitudinal profiles of the valleys, some large fluvial knickpoints have been preserved in the watershed headwaters. Major knickzones thus occur on the Têt River near Mont-Louis (see Itinerary 6), on the Segre River at Martinet (Itinerary 8), and on the Aude at Escouloubre (Itinerary 7), thereby preserving above them some substantial expanses of relatively unrejuvenated, pre-Quaternary landscape.

3.4 Quaternary Landscapes

The landscapes of the Modern Pyrenees are dominated by fluvial and glacial landforms. Aeolian, periglacial, and karstic environments have left a lesser or more localised imprint (Fig. 3.7).

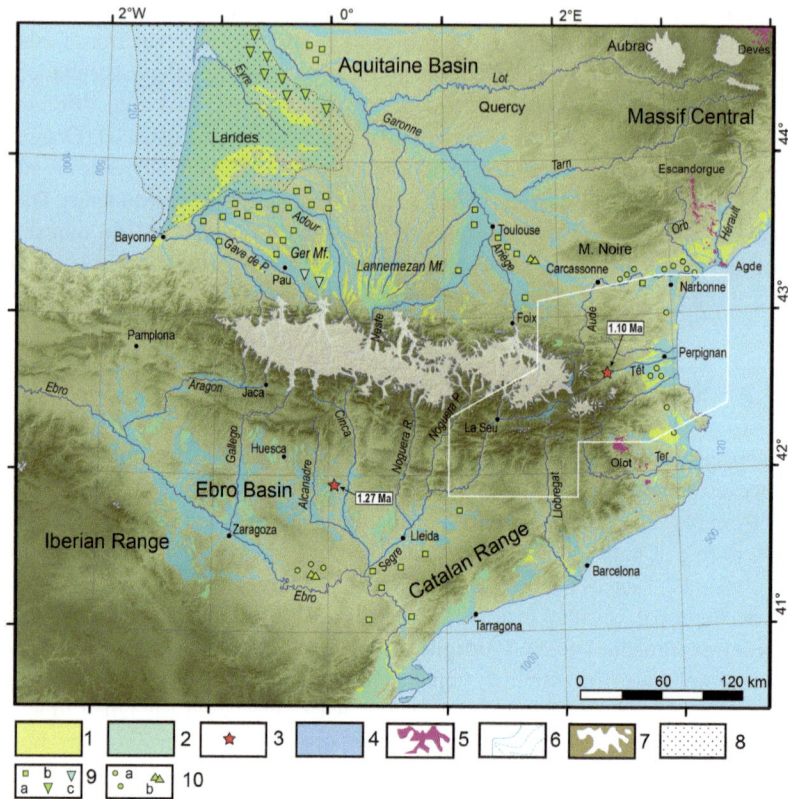

Fig. 3.7 Quaternary geology of the Pyrenees and their piedmont zones: a simplified map. Key to symbols and ornaments—**1**: Pliocene deposits (marine and continental); **2**: early Pleistocene (upper alluvial and glacifluvial terraces); **3**: dated deposits of the uppermost terrace (ESR age); **4**: middle and late Pleistocene alluvial deposits, Holocene alluvium; **5**: Quaternary volcanism (late Pliocene to early Pleistocene); **6**: offshore isobaths, including LGM isobath (–120 m); **7**: outline of most Late Pleistocene extensive glaciation (differences with most extensive Quaternary glaciation are small); **8**: middle and late Pleistocene coversands (erg of the Landes); **9**: other periglacial features—**9a**: loess, **9b**: sand wedges, **9c**: cryoturbation features; **10**: aeolian landforms—**10a**: deflation hollows, **10b**: yardangs. Alluvial deposits after 1:1,000,000 geological map of France, Barrère et al. (2009), and 1:50,000 scale IGME sheets, with additional information from Mensua et al. (1977)

3.4.1 Alluvial Landforms

Stairways of fluvial and glaciofluvial terraces, which are ubiquitous in the Ebro and Aquitaine piedmont zones as well as along the Mediterranean seaboard, have been the most widely used tool for understanding late Quaternary landscape evolution (Fig. 3.7). Alluvial terrace treads have been mapped in the pro-wedge valleys outward of the Internal Sierras, and in some areas as far back as the southern edge of the Axial Zone. Occurrences are also widespread throughout the eastern Pyrenees, where glaciers were mostly confined to the more elevated portions of the Axial Zone and where, additionally, the alluvial terrace systems of rivers such as the Têt, Aude and Segre have met with opportunities for wide floodplain development in the intermontane grabens of Cerdagne, Conflent and Capcir (Delmas et al. 2018).

3.4.1.1 Alluvial Stratigraphy
Stages of fluvial incision can be reconstructed from five main generations of alluvial terraces in the eastern basins (Cerdagne, Conflent, Roussillon, Aude catchment) and the Aquitaine foreland (Ariège, Garonne, Adour, various Gave rivers). This alluvial chronostratigraphy has benefited from several generations of geological studies underpinned by relative dating criteria such as terrace-tread altitude, palaeontological content, archaeological artefacts (Boule 1894; Depéret 1923; Chaput 1927; Denizot 1928), weathering indices, and topsoil characteristics (Alimen 1964; Tricart et al. 1966; Icole 1968, 1969, 1974; Hubschman 1973, 1975a, 1975b; Calvet 1981, 1996; Debals 2000). Relatively robust correlations between watersheds have been established on that basis for most of the retro-foreland and east-Pyrenean systems (Calvet et al. 2008; Barrère et al. 2009)—at least in the case of alluvial deposits that contain suitable proportions of granite or gneiss debris, as these lend themselves fairly readily to the elaboration of a relative weathering intensity scale.

Generations of alluvial deposits in the French geomorphological mapping system are named T0–T5, upward from the modern floodplain to the oldest vestige. On 1:50,000 geological sheets, the ranking is similar but with an alphabetical scheme, i.e., Fz (youngest) to Fu (most ancient). In Spain, the earlier systematic inventories tended to follow the French approach of numbering generations of alluvial deposits from bottom to top (Mensua et al. 1977; Bomer 1979; Peña 1983; Peña and Sancho 1988), but Iberian studies have operated on strictly watershed-based inventories of terrace treads, with distinctions between terrace generations usually restricted to altitudinal criteria and topographic continuity.

The synthesis for the central part of the Ebro Basin, for example, records 6 terrace levels (from TI, i.e., the modern floodplain, to TVI) spread over an elevation range of 250 m (Mensua et al. 1977). The uppermost level was given a 'Pliocene to Quaternary' age. The overview produced for the Segre River basin by Peña and Sancho (1988) identified up to 11 terrace levels within a vertical elevation bracket of 200 m in the Cinca watershed, and along similar lines: 8 in the Noguera Ribagorçana and 6 along the Segre itself (modern floodplain included). For the Aragón River, Bomer (1979) reported a minimum of 6 levels. Subsequent investigations maintained the tradition of catchment-scale inventories but elected to invert the labelling scheme, thereafter numbering alluvial vestiges from the most elevated / oldest vestige down to the floodplain. This is a source of confusion for correlating terrace systems in different valleys of the Iberian foreland, and even more so for correlations between different watersheds on the French and Spanish sides of the mountain range. Appreciation also varies between analysts. In the lower Gàllego catchment, 12 levels were identified above the active floodplain by Benito et al. (1998, 2010), but only 9 by Lewis et al. (2009, 2017)—who also reported 10 in the Cinca watershed. A study by Stange et al. (2013a) reports 8 levels for the Segre (noted TQ0 down to TQ7), 8 for the Cinca (noted in that study Qt2–Qt10), and 7 for the Noguera Ribagorçana (noted T8 to T2—but numbering for this particular watershed, unlike the others, starts from the bottom). A correlation of the full sequence of terraces was attempted for the Segre catchment by Peña et al. (2011).

Soil and alluvium weathering criteria in Iberian research output have not been emphasised, except by Lewis et al. (2009). Basing correlations between the pro- and retro-foreland river systems on such criteria is nonetheless of limited value (i) given the overwhelming prevalence of limestone debris in the Quaternary alluvium of the Iberian foreland; and (ii) given further the aridity of the Ebro Basin, which has promoted the development of calcrete on all the terrace levels, thus making distinctions tenuous (caprock indurations nonetheless tend to be substantially thicker and harder among the older generations of alluvial deposits). Carbonate duricrusts are entirely absent from the more humid Aquitaine piedmont, and even from the Mediterranean environment of the Roussillon Basin, where soil eluviation and carbonate dissolution are the rule except locally along limestone scarp-foot settings where hydrological conditions can favour $CaCO_3$ precipitation.

The coarse, often clast-supported gravel texture of all the Quaternary terrace levels indicates a prevalence of braided channel belts forming floodplains sometimes up to 10 km wide. The depositional sequences are relatively thin (mean thicknesses between 5 and 10 m), and tend to thicken upstream towards the mountain front or near the terminal moraines of outlet glaciers. In the Iberian foreland and around the edges of the extensional basins in the eastern Pyrenees, the alluvial deposits almost always grade laterally to wash pediments covered by a thin mantle of colluvium. These low-gradient and typically concave-up slope systems become laterally quite extensive in soft rock outcrops such as marl and molasse, where different generations of these pediments form staircase topography and the valleys become correspondingly very wide (Barrère 1975, 1981; Bomer 1979; Peña 1983; Stange et al. 2018). Such landscapes are typical of the Iberian drylands, thus displaying strong affinities with landform systems of the Maghreb. Apart from the Aude valley, no such landscapes exist in the Aquitaine Basin, where hillslope profiles are more typically convexo-concave. Colluvial deposits are also thicker and were emplaced by solifluction and cryogenic transfer rather than by hillwash processes. These sharp contrasts in slope morphology and regolith characteristics emphasise the profound climatic differences between the pro- and retro-foreland environments throughout the Quaternary and to the present day.

3.4.1.2 Fluvial Terraces: A Chronology

Several of the terraces—mainly the lower levels—connect directly with terminal moraines. Terrace T1, whether along the Garonne or the Ariège, contains numerous Pleistocene faunal remains—typically *Mammuthus primigenius* (Pouech 1873; Harlé 1893; Astre 1926; Clot and Duranthon 1990), a species that became extinct after the Magdalenian period (i.e., after ~14 ka), but also *Mammuthus trogontherii*, which is understood to have become extinct during the early Late Pleistocene (Astre 1967). The first radiometric ages produced in the Pyrenees with the aim of obtaining indirect constraints on the glaciation chronology were ^{14}C ages from peat and lake levels adjacent to some terminal moraines. Results delivered an early Late Pleistocene age, whether for these landforms or for alluvial-terrace generation T1 (Andrieu et al. 1988; Jalut et al. 1992). The continuation of these alluvial units offshore in the Mediterranean also yielded Late Pleistocene (MIS 3, MIS 2) radiocarbon ages obtained from intraformational shell specimens (Monaco et al. 1972).

A Pyrenean-scale synthesis of alluvial chronostratigraphy was produced by Delmas et al. (2018). In the Iberian foreland, only the two uppermost dated alluvial levels are older than the Bruhnes–Matuyama limit, i.e., >780 ka. The topmost alluvial deposits in the Cinca and Têt watersheds have yielded ESR ages of 1.27 Ma and 1.1 Ma, respectively (Fig. 3.6). These depositional ages are probably coeval with the upper units of the Lannemezan Formation (noted Fu on French geological maps), which are distinguished by the exceptionally large calibre of individual boulders in their boulder beds (Fig. 3.6; Icole 1968, 1969, 1974). In the extensional basins of the Mediterranean seaboard, Fu (i.e., T5) is inset in the top of the Pliocene depositional sequence, which is ~2 Ma old according to micromammalian assemblages contained in coeval fluvial deposits trapped in karstic cavities among the limestone plateaus situated along the edge of the Roussillon Basin (Delmas et al. 2018; see Itinerary 1). For the Middle Pleistocene, existing [10]Be profiles have mostly provided minimum ages for the alluvial terraces, with some well dated by ESR (Delmas et al. 2018), discussed in Itineraries 1 and 2.

In Aquitaine, existing [10]Be profiles indicate that glacifluvial outwash terrace T1 aggraded continuously until the Last Glacial Maximum (LGM, ca. 19–21 ka), perhaps extending into the Last Glacial–Interglacial Transition (i.e., post-LGM period before the Holocene) (Delmas et al. 2015; Stange et al. 2014). On the Iberian side, the last glacial cycle generated in contrast a sequence of 4 terraces along the left-bank tributaries of the Ebro River, with just the lowermost tier correlating with the LGM. Fluvial incision rates thus appear to have accelerated everywhere during the Middle Pleistocene, but more so in the Iberian foreland probably because of larger magnitudes of regional uplift across the Iberian Plate.

3.4.2 Glacial Landforms

3.4.2.1 Patterns of Erosion and Deposition

Despite their southerly latitude, the Pyrenees were glaciated at every stage of the Pleistocene. The Pyrenean icefield typically extended uninterrupted for 250 km from the Capcir Basin in the east to Pic d'Orhy in the west (Fig. 3.8). The spatial distribution of glaciers is now well established and has been synthesised and updated repeatedly over time (Penck 1883, 1894; Taillefer 1957, 1967, 1969; Hérail et al. 1987; Martí Bono and García Ruiz 1994; Calvet 2004, with detailed references therein; Barrère et al. 2009; Calvet et al. 2011; Delmas et al. 2021a, b, c).

The spatial distribution of Quaternary glaciers was dictated by dual E–W and N–S asymmetries in the climatic and topographic configuration of the orogen. The N–S asymmetry is the sharpest, with the northern mountain front open to Atlantic influence and concentrating 75% of the glaciated surface area (Figs. 3.7 and 3.8). The mean Equilibrium Line Altitude (ELA) for the Quaternary, when reconstructed from isolated cirque-floor altitudes, lay between 1200 and 1600 m among the massifs of the North-Pyrenean Zone, rising a little into the core of the Axial Zone. The largest outlet glaciers reached lowland altitudes of 350 m

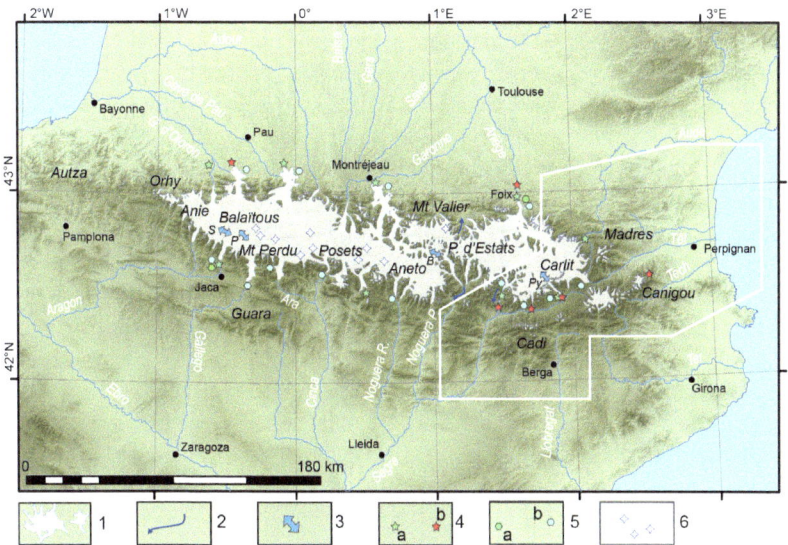

Fig. 3.8 **Glaciation and glacial deposits of the Pyrenees**. Key to symbols and ornaments—**1**: most extensive glaciation during the Late Pleistocene (numerous nunataks not shown); **2**: possible Late Pleistocene maximum advance of some valley glaciers; **3**: main transfluence cols between the north and south sides of the icefield divide (Py: Puymorens, B: Bonaigue, P: Pourtalet, S: Somport); **4**: pre-Late Pleistocene moraines and till occurrences—**4a**: MIS 6 moraines (locally based on exposure ages); **4b**: middle to early Pleistocene glacial till; **5**: exposure-dated frontal moraines and ice-marginal deposits—**5a**: MIS 6 frontal moraines (OSL- or [10]Be-dated); **5b**: MIS 4 to LGM frontal moraines ([14]C–, OSL–, or [10]Be-dated); **6**: massifs currently hosting residual glaciers (the population of massifs with a record of Little Ice Age glaciers is larger). After Calvet (2004), Calvet et al. (2011), and Delmas et al. (2021a; b, c)

(Ariège) and 450 m (Garonne), with glacier lengths attaining 37 km along the Gave d'Ossau, 50 km along the Gave de Pau, 70 km for the Garonne, 65 km for the Ariège, with ice thicknesses in each case 0.8–1 km. Limiting factors of ice extent have been the narrowness of the Pyrenees (and therefore of the accumulation zone), the predominance of transverse drainage, and the limited opportunities for valley confluence that such a parallel drainage network imposes on ice flow patterns. The most elevated ridges of the mountain range always stood above the top of the icefield, and transfluence cols between parallel valleys are uncommon (Col de Lhers, Col du Portillon). Despite some fairly thin, localised plateau icefield occurrences on summit surfaces such as at the Arres d'Anie and in the Aston and Carlit massifs, the Pleistocene icefield was thus by no means an icecap. The larger ice accumulation on the north side of the range nonetheless contributed to spill over to the southern side via a number of transfluence cols across the main divide, each situated at increasingly lower elevations from east to west (Col de Puymorens: 1917 m; Port de Bonaigua: 2072 m; Pla de Beret: 1870 m; Col du Pourtalet: 1795 m; Col du Somport: 1631 m), with an additional number of more minor divide breaches at higher altitudes.

The Iberian domain contained comparatively shorter and thinner (400–600 m) valley glaciers, with outlet glaciers terminating at elevations between 750 and 940 m. Glacier lengths rarely exceeded 30 km. The Pallaresa and Valira trunk glaciers may have attained maximum lengths of 60 and 42 km, respectively, but mainly by virtue of the inputs from tributary valley glaciers that could feed into the main stem at points quite far down the trunk valley (Serrat et al. 1994; Turu et al. 2007; Turu et al. 2011a, b; Turu 2011). The Late Pleistocene ELA rose rapidly southward to elevations above 2100–2200 m in the outermost massifs of the Axial Zone and Sierras Interiores, and to even higher altitudes in the case of south-facing slopes.

The glaciation gradient along the strike of the range was more gradual than across it. The icefield thinned eastward as a combined result of diminishing Atlantic moisture advection from the west and of the increase in aggregate sunshine hours under Mediterranean influence. Among the outlet glaciers along the northern mountain front, only the Gave de Pau at Lourdes and Gave d'Ossau at Arudy formed piedmont glacier lobes. The ELA was particularly low in the Basque Country (1100–1200 m; Viers 1960). From there, it rose progressively along strike, reaching 1300–1400 m among the outermost massifs in the Ariège. The ELA attained 1600 m in the upper catchments of the Hers, Aude and Boulzane (Dourmidou massif), i.e., ~60 km from the Mediterranean coast (Itinerary 6). In the eastern Pyrenees, the greater fragmentation of relief resulting from Neogene extensional tectonics and relative aridity of the sheltered intermontane basins conspired to a confinement of glaciation to the most elevated massifs of the Axial Zone. Here, the ELA extended between 2000 and 2300 m, the valley glaciers were short (Têt: 18 km, Querol: 25 km) and never extended below the 1000–1500 m contour band. At these easterly longitudes, the icefield was often little more than a population of cirque glaciers (Itineraries 4, 6, 7, 8).

These Quaternary glacial gradients were essentially an exaggerated version of present-day climatic contrasts, also reflected in the pattern of the modern winter snowline. It can thus be reasonably inferred that average climatic conditions and average atmospheric circulation patterns in the region remained similar throughout the Quaternary (Barrère 1954; Taillefer 1982; Calvet 1996). These conditions include: (i) permanent air flow from the W to NW, bringing snow but also favouring its local redistribution over ridgetops, thereby supplying east- and southeast-facing cirques; (ii) the interference of Mediterranean air flow from the southeast, which is also a source of abundant snowfall in present-day conditions in the eastern part of the range; and (iii) the considerably greater dryness and warmth of the southern and eastern Pyrenees compared to anywhere else—with negative consequences on the thermal budget of glaciers in those areas.

The geomorphological legacy of Quaternary glaciation on Pyrenean landscapes and slope systems is widespread but not often intense. The glacial imprint is strongest in the cirque belt (Crest et al. 2017), which in some massifs displays characteristic arêtes and a few pyramidal peaks. In the eastern Pyrenees, however, the limited erosive power of the Pleistocene glaciers has, for example, failed to eradicate the erosion surfaces. Very deep mantles of saprolite, which at many places cover these elevated residuals of Neogene topography, have been preserved (Delmas et al. 2009; Itineraries 7 and 8). Most of the larger valleys nonetheless exhibit large bedrock steps, e.g., along the Ariège at Tarascon or at Les Cabannes. None of the wider glacial troughs are calibrated to a characteristic U shape; V-shaped gorge sections are frequent and even include entrenched fluvial meanders (such as between Ax-les-Thermes and Mérens in the Ariège). This relatively light erosional imprint of warm-based glaciers also explains the indecision among scholars as to the true terminal positions of valley glaciers in some V-shaped valleys such as the Noguera Pallaresa, Cinca, Valira, and Salat.

3.4.2.2 Age of Glacial Deposits

Middle Pleistocene till exists in most outlet valleys of the Aquitaine foreland, whether stratigraphically beneath the Late Pleistocene till or slightly forward of Late Pleistocene terminal moraines. The earliest evidence of this was provided by the mapping and analysis of alluvial fill in limestone caves adjacent to ice-filled valleys. Depending on the ice thickness at each successive glaciation, certain cave levels would be supplied by glacifluvial deposits. When sandwiched between different generations of speleothems, the age of the gravel or varved silt units in the caves could be determined by U–Th dating the encasing calcite concretions. The Niaux–Lombrives cave system, which is situated at the junction between the Vicdessos and Ariège valleys, has in this way produced evidence of 4 glacial periods in the last 400 ka (Sorriaux 1981, 1982; Bakalowicz et al. 1984; Sorriaux et al. 2016).

Producing a more detailed chronology of the last glacial cycle is still work in progress. The first evidence was obtained by ^{14}C dating of ice-marginal lake sediments for the Late Pleistocene (Würmian Stage; Andrieu et al. 1988); and from caves, where U–Th dating of speleothems has placed the last glaciation between 90 and 20 ka (Bakalowicz et al. 1984). The most extensive glaciation was inferred from this evidence to have occurred during the first half of the Late Pleistocene, with deglaciation beginning at some time during MIS 3 (57–29 ka). Systematic OSL dating of deposits in the pro-wedge valleys (e.g., Lewis et al. 2009; Garcia Ruiz et al. 2013; Guerrero et al. 2018; Sancho et al. 2018); [10]Be, [36]Cl or [21]Ne exposure dating at many sites across the range (Pallàs et al. 2006, 2010; Delmas et al. 2008, 2011, 2012; Turu et al. 2011, 2017; Palacios et al. 2015a, 2015b, 2017; Crest et al. 2017); systematic ^{14}C dating (Turu et al. 2017); and Schmidt hammer proxy dating based on [10]Be calibrations (Tomkins et al. 2018) have refined the chronology substantially and bring out some important components of regional variation (see Itineraries 7 and 8). Note that, in the case of cosmogenic exposure ages, some allowance concerning apparent age discrepancies must be made for uncertainties relating to rock shielding by snow cover, and also for the older ages published before 2015 because the physical calibration of nuclide production rates has since been officially revised to lower values.

3.4.2.3 LGM and Post-LGM Deglacial Chronology

At the time of the LGM, a paradoxically reverse climatic contrast between the eastern and western Pyrenees appears to have prevailed: in the eastern Pyrenees, glaciers advanced at some places as far as their earlier (MIS 4) maximum positions, with the two moraine generations often bunching into a tight single mass in the case of the Têt, Querol, Malniu, Duran, Llosa, Aranser, and Noguera Ribagorçana glaciers. The LGM ice front stood 7 km upstream of the MEG position in the Ariège valley, and at most 9 km in the Valira. In the west (Gàllego, Aragon, Ossau), the positions of LGM frontal moraines are unknown but LGM glaciers were probably also several kilometres shorter than their most extensive MIS 3/4 predecessors because of colder and drier conditions in the nearby Bay of Biscay than in the western Mediterranean (Delmas et al. 2011, 2021a; b, c). Deglaciation nonetheless appears to have been almost universally rapid at the end of the LGM, particularly in the eastern Axial Zone where the valley glaciers had receded all the way to the cirques by 20 ka (Delmas 2005; Delmas et al. 2008, Delmas 2015) (see Itinerary 7). A major glacial readvance during the Oldest Dryas, ca. 18 ka and persisting until 16–15 ka, has been documented. It produced glaciers ~20 km long in the upper Ariège, 7 km long in the Bassiès valley, 5 km in the Têt, 6 km in the Noguera Ribagorçana, 4 km in the Ésera, and 9 and 15 km, respectively, in two tributary valleys of the upper Gàllego catchment. By the end of the Allerød Interstadial (~13 ka), ice had disappeared even from the highest ranges of the eastern Pyrenees, with evidence of the tree line having risen to above 1800 m in the NE ranges of the orogen by that time (Reille and Andrieu 1993). The last documented readvance during the Younger Dryas did not descend below the 2000 m contour in the eastern Pyrenees, remaining confined to the cirques and uppermost valley areas. Even at that time, the tree line stood around 1300 m. Many cirques of the eastern ranges became populated at this time with rock glaciers.

3.4.3 Other Landforms

3.4.3.1 Limestone Karst

Surface karst landforms are good indicators of climate, and in the present case confirm the relatively long-term stability of present-day climatic patterns in the Quaternary past. The plateau limestone surfaces of the Corbières, for example, display very few dolines and other solutional features, suggesting they have long resided in a dry environment with a relatively benign potential for rock solution. This also explains the excellent state of preservation of the micromammalian fossil assemblages in this area, some of which have remained trapped in surface fissures of limestone pavements for up to 20 million years (Aguilar et al. 1986b; Faillat et al. 1990; Calvet 1992, 1996) (see Itineraries 1 and 2). The pattern of doline distribution along the belt of retro-wedge limestone outcrops displays a trend from east to west where rare, isolated dolines in the east (e.g., around Roc Paradet, in the Fenouillèdes massif) give way to increasingly higher doline densities (e.g., in the Pays de Sault) (Itinerary 5), and eventually to a chaotic topography produced by high densities of intersecting dolines closer to the Atlantic seaboard (e.g., in the Arbailles massif).

3.4.3.2 Periglacial Features

Colluvium, talus deposits, and blockstreams are widespread in the unglaciated highland areas of the Pyrenees, but have also populated most deglaciated valley sides given the context of ice retreat during various stadial episodes during and soon after MIS 3–MIS 2, and on each occasion under palaeoclimatic conditions that remained suitably cold and dry. The oldest generation of periglacial stratified talus deposits is roughly contemporaneous with terrace T5, which clearly operated as a local base level for talus cones at several type localities (see Itinerary 6). Such slope deposits in limestone are widespread in the unglaciated eastern part of the orogen, but induration as a result of debris cementation has also allowed them at other locations, such as the Roc de Sédour in the middle Ariège valley, to survive the repeated passage of valley glaciers. Features indicative of discontinuous permafrost, e.g., frost-wedge structures and other active-zone features, have been reported from the Aquitaine piedmont (Fig. 3.5; Gangloff et al. 1994; Sitzia et al. 2015). A climatic interpretation that still endures has been given to the asymmetric valley sides of the various parallel or fanning rivers of the Armagnac region: their low-angle western slopes would be the result of gelifluction exacerbated by snow accumulation downwind of the prevailing Westerlies (Taillefer 1951), and as a result of being pushed eastward by the advancing active layer, the rivers have undercut and steepened their eastern valley slopes.

3.4.3.3 Aeolian Features

Evidence of periglacial environments is also supported by a number of aeolian deposits. By far the most regionally prominent is the sand sea, or erg, on the Landes plateau—a vast, triangular-shaped outcrop of sand between the Garonne and Adour rivers. This major dunefield has been active since at least the Middle Pleistocene (Sitzia 2014; Sitzia et al. 2015). Further inland, the sand deposits give way to occurrences of loess, e.g., as far east as the Lauragais (transition area into Languedoc; Hubschman 1975c; Revel et al. 1978, 1979; Revel and Bourgeat 1981; Bertran et al. 2011), locally in the Aude River catchment (Birot 1969), and in the SE Ebro Basin at the foot of the Catalan Coastal Range (Boixadera et al. 2015).

The eastern termination of the Pyrenees lacks coversand and aeolian silt deposits because prevailing winds blow from the NW, exporting the products of deflation out to the Mediterranean. However, the high frequency and intensity of these steady NW-wind regimes, which result from a Venturi effect through the Aude valley and air-flow deflection by the Pyrenean mountain mass (today one of the highest wind potentials in France: >500 W/m^2), explain the presence in Languedoc, the Corbières and Roussillon of hydro-aeolian deflation pans. These bowl-shaped topographic depressions are up to 1 km wide and several tens of metres deep, and result from a conjunction of phreatic corrosion during seasonal water-table highstands and aeolian deflation of fine particles produced in the process (Calvet 1981; Ambert and Clauzon 1992; David and Carozza 2013; Carozza et al. 2016, 2017). The deflation pans occur on interfluves in softer rocks such as molasse, and at a range of lowland altitudes down to below the tread of alluvial terrace T2, which implies continuing activity during the Late Pleistocene (Itineraries 1 and 2). A trail of pans can be mapped in the Aude valley downstream of Carcassonne, in Languedoc, across the Roussillon plains, in Empordà, and in the SE Ebro Basin (Fig. 3.7), where yardangs have also been described (Gutiérrez-Elorza et al. 2002). These landforms also occur on lowland granitic outcrops where deep, in situ saprolite is vulnerable to piping, scouring and win-nowing processes, such as on the northern edge of the Roussillon (Montalba plateau, in the Millas granite, see Itinerary 5) and in Empordà on the south-ern backslope of the Albères footwall uplift (Calvet 1996). Such landscape-scale aeolian blasting is consistent with the occurrence in the region of ventifacts such as dreikanter, i.e., alluvial quartz pebbles transformed by the abrasive action of blown sand (Fig. 3.6). These occur exclusively on the treads of alluvial terraces T3, T4 and T5, and suggest extremely intense wind regimes during the Middle Pleistocene (Calvet 1981, 1996; see Itineraries 2 and 8).

References

Aguilar JP, Calvet M, Michaux J (1986b) Découvertes de faunes de micromammifères dans les Pyrénées-Orientales (France) de l'Oligocène supérieur au Miocène supérieur ; espèces nouvelles et réflexions sur l'étalonnage des échelles continentales et marines. C R Acad Sci Paris II 303:755–760

Alasset JP, Meghraoui M (2005) Active faulting in the western Pyrénées (France): paleoseis-mic evidence for late Holocene ruptures. Tectonophysics 409:39–54

Alimen H (1964) Le Quaternaire des Pyrénées de Bigorre. Mém Serv Carte Géol, Paris, 394 p

Ambert P (1977) Déformation tectonique d'une terrasse quaternaire de la Cesse à Bize (Aude). Bull Soc Hist Nat Toulouse 113:147–151

Ambert P, Clauzon G (1992) Morphogénèse éolienne en ambiance périglaciaire: les dépres-sions fermées du pourtour du Golfe du Lion (France méditerranéenne). Z Geomorph Suppl Bd 84:55–71

Ambert P (1994) L'évolution géomorphologique du Languedoc central depuis le Néogène (Grands Causses méridionaux-Piémont Languedocien). Document du BRGM 231, Ed. BRGM, Orléans, 210 p

Andrieu V, Hubschman J, Jalut G, Hérail G (1988) Chronologie de la déglaciation des Pyrénées françaises. Bull Assoc Fr Étude Quat 34–35, 55–67

Astre G (1926) répartition stratigraphique des deux types de Mammouths. Bull Soc Hist Nat Toulouse 183–188

Astre G (1967) Elephas trogontherii dans des graviers de Palaminy. Bull Soc Hist Nat Toulouse 19–29

Audra P, Palmer AN (2013) The vertical dimension of karst: controls of vertical cave pattern. In: Shroder JF, Frumkin A (eds) Treatise on Geomorphology, vol 6, Karst Geomorphol-ogy. Academic Press, San Diego, pp 186–206

Audra P, Palmer AN (2011) The pattern of caves: controls of epigenic speleogenesis. Géo-morph Rel Proc Environ 4:359–378

Babault J, Bonnet S, Crave A, Van den Driessche J (2005) Influence of piedmont sedimenta-tion on erosion dynamics of an uplifting landscape: an experimental approach. Geology 33:301–304

Babault J, Van den Driessche J, Bonnet S, Castelltort S, Crave A (2005b) Origin of the highly elevated Pyrenean peneplain. Tectonics, 24, TC2010. 10.1029 /2004TC001697

Baize S, Cushing M, Lemeille T, Granier B, Grellet B, Carbon D, Combes C, Hibsch C (2002) Inventaire des indices de rupture affectant le Quaternaire, en relation avec les grandes structures connues en France métropolitaine et dans les régions limitrophes. Mém H S Soc Géol Fr 175:142 p

Bakalowicz M, Sorriaux P, Ford DC (1984) Quaternary glacial events in the Pyrenees from U. series dating of speleothems in the Niaux-Lombrives-Sabart Caves, Ariège France. Norsk Geogr Tidsskr 38:193–197

Bakalowicz M (1988) L'évolution paléohydrologique et morphologique des Pyrénées centrales: l'exemple du massif karstique d'Arbas (Pyrénées garonnaises). Actes des Journées F. Trombe, 8–10 mai 1987, Moulis, CNRS, pp 43–57

Baran R, Friedrich AM, Schlunegger F (2014) The late Miocene to Holocene erosion pattern of the Alpine foreland basin reflects Eurasian slab unloading beneath the western Alps rather than global climate change. Lithosphere 6:124–131

Barrère P (1952a) Le relief des massifs granitiques de Néouvielle, de Cauterets et de Panticosa. Rev Géogr Pyrén Sud-Ouest 23:69–98

Barrère P (1952b) La morphologie des Sierras Oscences. Actas del Primer Congreso de Estudios Pirenaicos, San Sebastian, 1950, V, iv Geografia, pp 51–79

Barrère P (1954) Equilibre glaciaire actuel et quaternaire dans l'Ouest des Pyrénées centrales. Rev Géogr Pyrén Sud-Ouest 24:116–134

Barrère P (1962) Reliefs murs perchés de la Navarre orientale. Rev Géogr Pyrén Sud-Ouest 33:309–323

Barrère P (1975) Terrasses et glacis d'érosion en roches tendres dans les montagnes du Haut-Aragon. In: Etudes géographiques, Mélanges offerts à G. Viers, Université de Toulouse-le-Mirail, vol 2, 29–43

Barrère P (1981) Le bassin de Sangüesa, articulation majeure du versant sud des Pyrénées. In: Estudios de Geografía, Homenaje a Alfredo Floristan, Instituto Principe de Viana, pp 1–39

Barrère P, Calvet M, Courbouleix S, Gil Peña I, Martin Alfageme S (2009) In: Courbouleix S, Barnolas A (eds) Carte géologique du Quaternaire des Pyrénées, 1:400,000 scale. BRGM and ITGM

Beamud E, Muñoz JA, Fitzgerald PJ, Baldwin SL, Garcés M, Cabrera L, Metcalf JR (2011) Magnetostratigraphy and detrital apatite fission track thermochronology in syntectonic conglomerates: constraints on the exhumation of the South-Central Pyrenees. Basin Res 23:309–331

Belmonte A (2014) Geomorfología del macizo de Cotiella (Pirineo oscense): cartografía, evolución paleoambiental y dinámica actual. PhD thesis, Universidad de Zaragoza, 581 p

Benito G, Pérez-González A, Gutiérrez F, Machado MJ (1998) River response to Quaternary large-scale subsidence due to evaporite solution (Gállego River, Ebro Basin, Spain). Geomorphology 22:243–263

Benito G, Sancho C, Peña JL, Machado MJ, Rhodes EJ (2010) Large-scale karst subsidence and accelerated fluvial aggradation during MIS6 in NE Spain: climatic and paleohydrological implications. Quat Sci Rev 29:2694–2704

Berger G, Clauzon G, Michaux J, Suc JP, Aloïsi JC, Monaco A, Got H, Augris C, Gadel F, Buscail R (1988) Carte géologique de la France, 1:50,000 scale, sheet Perpignan (1091), Orléans, BRGM. Handbook by Clauzon G, Berger G, Aloïsi JC, Got H, Monaco A, Buscail R, Gadel F, Augris C, Marchal JP, Michaux J, Suc JP (1989), 40 p

Bertran P, Bateman MD, Hernandez M, Mercier N, Millet D, Sitzia L, Tastet JP (2011) Inland aeolian deposits of south-west France: facies, stratigraphy and chronology. J Quat Sci 26:374–388

Birot P (1937) Recherches sur la morphologie des Pyrénées orientales franco-espagnoles. Baillière Édit., 318 p

Birot P (1969) Le Quaternaire de la basse vallée de l'Orbieu. Livret guide excursion A6, Pyrénées orientales et centrales, Roussillon, Languedoc occidental, VIIIe Congrès INQUA, Paris, pp 101–105

Bosch GV, Van den Driessche J, Babault J, Robert A, Carballo A, Le Carlier C, Loget N, Prognon C, Wyns R, Baudin T (2016a) Peneplanation and lithosphere dynamics in the Pyrenees. C R Geosci 348:194–202

Bosch GV, Teixell A, Jolivet M, Labaume P, Stockli D, Doménech M, Monié P (2016b) Timing of Eocene-Miocene thrust activity in the western Axial Zone and Chaînons Béarnais (west-central Pyrenees) revealed by multi-method thermochronology. C R Geosci 348:246–256

Blanchard R (1914) La morphologie des Pyrénées françaises. Ann Géogr 23:303–324

Boissevain H (1934) Étude géologique et géomorphologique d'une partie de la vallée de la haute Sègre (Pyrénées Catalanes). Bull Soc Hist Nat Toulouse 66:33–170

Boixadera J, Poch RM, Lowick SE, Balasch JC (2015) Loess and soils in the eastern Ebro Basin. Quat Int 376:114–133

Bomer B (1979) Les Piedmonts du Bassin de l'Ebre (Espagne). Méditerranée 36:19–25

Boule M (1894) Le plateau de Lannemezan et les alluvions anciennes des hautes vallées de la Garonne et de la Neste. Bull Serv Carte Géol Fr 43(VI):447–469

Briais A, Armijo R, Winter T, Tapponnier P, Herbecq A (1990) Morphological evidence for Quaternary normal faulting and seismic hazard in the Eastern Pyrenees. Ann Tecton IV:19–42

Bridgland DR (2000) River terrace systems in north-west Europe: an archive of environmental change, uplift and early human occupation. Quat Sci Rev 19:1293–1303

Bridgland DR, Westaway R (2014) Quaternary fluvial archives and landscape evolution: a global synthesis. Proc Geol Assoc 125:600–629

Cabrera L, Saez A (1987) Coal deposition in carbonate-rich shallow lacustrine systems: the Calaf and Mequinenza sequences (Oligocene, eastern Ebro Basin, NE Spain). J Geol Soc Lond 144:451–461

Cabrera L, Roca E, Santanach P (1988) Basin formation at the end of a strike-slip fault: the Cerdanya basin (Eastern Pyrenees). J Geol Soc Lond 145:261–268

Calvet M (1981) Nappes alluviales et niveaux quaternaires du bas-Vallespir. Implications néotectoniques et paléoclimatiques. Rev Géogr Pyr Sud-Ouest 52:125–159

Calvet M (1992) Aplanissements sur calcaire et gîtes fossilifères karstiques. L'exemple des Corbières orientales. Tübinger Geogr Studien H 109:37–43

Calvet M (1996) Morphogenèse d'une montagne méditerranéenne: les Pyrénées orientales. Documents du BRGM, Orléans, n° 255, 3 vols., 1177 p

Calvet M (1999) Régime des contraintes et volumes de relief dans l'Est des Pyrénées. Géomorph Rel Proc Environ 3:253–278

Calvet M (2004) The Quaternary glaciation of the Pyrenees. In: Ehlers J, Gibbard P (eds) Quaternary glaciations—extent and chronology, part I: Europe. Elsevier, Amsterdam, pp 119–128

Calvet M, Delmas M, Gunnell Y, Braucher R, Bourlès D (2011) Recent advances in research on Quaternary glaciations in the Pyrenees. In: Ehlers J, Gibbard PL (eds) Quaternary glaciations, extent and chronology, a closer look, Part IV, Elsevier édit, Developments in Quaternary Science, vol 15, pp 127–139

Calvet M, Gunnell Y (2008) Planar landforms as markers of denudation chronology: an inversion of East Pyrenean tectonics based on landscape and sedimentary basin analysis. In: Gallagher K, Jones SJ, Wainwright J (eds) Landscape evolution: denudation, climate and tectonics over different time and space scales, Geolo Soc Lond Spec Publ 296: 147–166

Calvet M, Gunnell Y, Delmas M (2008) Géomorphogenèse des Pyrénées. In: Canérot J, Colin J-P, Platel J-P, Bilotte M (eds) Pyrénées d'Hier et d'Aujourd'hui. TOTAL, BRGM, AGSO, AIPT, CNRS, Atlantica, Biarritz, pp 129–143

Calvet M, Gunnell Y, Farines B (2015a) Flat-topped mountain ranges: their global distribution and value for understanding the evolution of mountain topography. Geomorphology 241:255–291

Calvet M, Gunnell Y, Braucher R, Hez G, Bourles D, Guillou V, Delmas M, Aster Team (2015b) Cave levels as proxies for measuring post-orogenic uplift: evidence from cosmogenic dating of alluvium-filled-cave in the French Pyrenees. Geomorphology, 246:617–633

Calvet M, Gunnell Y, Laumonier B (2021) Denudation history and palaeogeography of the Pyrenees and their peripheral basins: an 84-million-year geomorphological perspective. Earth-Sci Rev 215:103436

Calvet M, Hez G, Gunnell Y, Jaillet S (2019) Le karst du synclinal de Villefranche, enregistreur de l'incision de la vallée de la Têt. Bol Soc Esp Speleol Sciencias del Karst 14:15–32

Campanyà J, Ledo J, Queralt P, Marcuello A, Muñoz JA, Liesa M, Jones AJ (2018) New geoelectrical characterization of a continental collision zone in the Central–Eastern Pyrenees: constraints from 3-D joint inversion of electromagnetic data. Tectonophysics 742–743:168–179

Carbon D, Combes P, Cushing M, Granier T, Grellet B (1995) Rupture de surface postpléistocène moyen dans le bassin aquitain. C R Acad Sci Paris, IIa 320:311–317

Carozza JM, Llubes M, Danu M, Faure E, Carozza L, David M, Manen C (2016) Geomorphological evolution of Mediterranean enclosed depressions in the Late glacial and Holocene: the example of Canohès (Roussillon, SE France). Geomorphology 273:78–92

Carozza JM, Llubes M, Carozza L, Danu M, David M (2017) Les processus de formation et évolution des dépressions fermées du golfe du lion au cours du pléistocène supérieur et du tardiglaciaire. Quaternaire 28:323–336

Chaput E (1927) Recherches sur l'évolution des terrasses de l'Aquitaine. Bull Soc Hist Nat Toulouse **LVI**:17–84

Chevrot S, Sylvander M, Diaz J, Martin R, Mouthereau F, Manatschal G, Masini E, Calassou S, Grimaud F, Pauchet H, Ruiz M (2018) The non-cylindrical crustal architecture of the Pyrenees. Sci Rep 8:9591. https://doi.org/10.1038/s41598-018-27889-x

Clot A, Duranthon F (1990) Les mammifères fossiles du Quaternaire dans les Pyrénées. Muséum d'Histoire Naturelle de Toulouse and Accord Ed., 80 p

Coney PJ, Muñoz AJ, McClay KR, Evenchik CA (1996) Syntectonic burial and post-tectonic exhumation of the southern Pyrenees foreland fold-thrust belt. J Geol Society Lond 153:9–16

Conway-Jones BW, Roberts GG, Fichtner A, Hoggard M (2019) Neogene epeirogeny of Iberia. Geochem Geophys Geosyst 20:1138–1163

Crest Y, Delmas M, Braucher R, Gunnell Y, Calvet M, ASTER Team (2017) Cirques have growth spurts during deglacial and interglacial periods: evidence from [10]Be and [26]Al nuclide inventories in the central and eastern Pyrenees. Geomorphology 278:60–77

David M, Carozza JM (2013) Les dépressions fermées du Languedoc central et du Roussillon: inventaire, caractérisation géomorphométrique et essai de typologie. Géomorph Rel Proc Environ 19:407–424

Davis WM (1911) The Colorado Front Range, a study on physiographic presentation. Ann Assoc Am Geogr 1:21–83

De Sitter LU (1952) Pliocene uplift of Tertiary mountain chains. Am J Sci 250:297–307

Debals B (2000) Mise au point sur la chronostratigraphie des dépôts alluviaux quaternaires de la Plaine du Roussillon: exemple de la vallée de la Têt (France). Quaternaire 11:31–39

Delmas M (2005) La déglaciation dans le massif du Carlit (Pyrénées orientales): approches géomorphologique et géochronologique nouvelles. Quaternaire 16:45–55

Delmas M, Gunnell Y, Braucher R, Calvet M, Bourles D (2008) Exposure age chronology of the last glaciation in the eastern Pyrenees. Quat Res 69:231–241

Delmas M, Calvet M, Gunnell Y (2009) Variability of Quaternary glacial erosion rates—a global perspective with special reference to the Eastern Pyrenees. Quat Sci Rev 28:484–498

Delmas M, Calvet M, Gunnell Y, Braucher R, Bourlès D (2011) Palaeogeography and 10Be exposure-age chronology of Middle and Late Pleistocene glacier systems in the northern Pyrenees: implications for reconstructing regional palaeoclimates. Palaeogeogr Palaeoclimatol Palaeoecol 305:109–122

Delmas M, Calvet M, Gunnell Y, Braucher R, Bourlès D (2012) Les glaciations quaternaires des Pyrénées ariégeoises. Approche historiographique, données paléogéographiques et chronologiques nouvelles. Quaternaire 23:61–85

Delmas M, Braucher R, Gunnell Y, Guillou V, Calvet M, Bourlès D (2015) Constraints on Pleistocene glaciofluvial terrace age and related soil chronosequence features from vertical [10]Be profiles in the Ariège River catchment (Pyrenees, France). Global Planet Change 132:39–53

Delmas M, Gunnell Y, Calvet M, Reixach T, Oliva M (2021a) Glacial landscape of the Pyrenees (chapter 16). In: Palacios D, Hughes P, García-Ruiz JM, Andrés A (eds) European glacial landscapes: maximum extent of glaciations. Elsevier, pp 123–128

Delmas M, Gunnell Y, Calvet M, Reixach T, Oliva M (2021b) The Pyrenees: glacial landforms prior to the last glacial maximum (chapter 40). In: Palacios D, Hughes P, García-Ruiz JM, Andrés A (eds) European glacial landscapes: maximum extent of glaciations. Elsevier, pp 295–307

Delmas M, Gunnell Y, Calvet M, Reixach T, Oliva M (2021c) The Pyrenees: glacial land-forms from the last glacial maximum (chapter 59). In: Palacios D, Hughes P, García-Ruiz JM, Andrés A (eds) European Glacial Landscapes: Maximum Extent of Glaciations. Elsevier, 461–472

Delmas M (2015) The last maximum ice extent and subsequent deglaciation of the Pyrenees: an overview of recent research. Cuad Inv Geográfica 41:109–137

Delmas M, Calvet M, Gunnell Y, Voinchet P, Manel C, Braucher R, Tissoux H, Bahain JJ, Perrenoud C, Saos T, Aster Team (2018) Terrestrial ^{10}Be and Electron Spin Resonance dating of fluvial terraces quantifies Quaternary tectonic uplift gradients in the eastern Pyrenees. Quat Sci Rev 193:188–211

Demoulin A, Mather A, Whittaker A (2017) Fluvial archives, a valuable record of vertical crustal deformation. Quat Sci Rev 166:10–37

Denizot G (1928) Note sur la morphologie, sur l'évolution et sur l'âge des terrasses toulou-saines. Bull Soc Hist Nat Toulouse 346–356

Depéret C (1923) Les glaciations des vallées pyrénéennes françaises et leurs relations avec les terrasses fluviales. C R Hebdo Séances Acad Sci 176:1519–1524

Dubos-Sallée N, Nivière B, Lacan P, Hervouët Y (2007) A structural model for the seismicity of the Arudy (1980) epicentral area (Western Pyrenees, France). Geophys J Int 171:259–270

Dufréchou G, Tiberi C, Martin R, Bonvalot S, Chevrot S, Seoane L (2018) Deep structure of Pyrenees range (SW Europe) imaged by joint inversion of gravity and teleseismic delay time. Geophys J Int 214:282–301

Ellenberger F, Gottis M (1967) Sur les jeux de failles pliocènes et quaternaires dans l'arrière-pays narbonnais. Rev Géogr Phys Géol Dyn IX:153–159

Enjalbert H (1960) Les Pays aquitains, le modelé et les sols. Imp. Bière, Bordeaux, 618 p

Faillat JP, Aguilar JP, Calvet M, Michaux J (1990) Les fissures à remplissages fossilifères néogènes du plateau de Baixas (Pyrénées orientales, France), témoins de la distension oligo-miocène. C R Acad Sci Paris II(311):205–212

Filleaudeau P-Y, Mouthereau F, Pik R (2011) Thermo-tectonic evolution of the south-central Pyrenees from rifting to orogeny: insights from detrital zircon U/Pb and (U–Th)/He thermochronometry. Basin Res 23:1–17

Fillon C, Gautheron C, van der Beek P (2013) Oligocene-Miocene burial and exhumation of the Southern Pyrenean foreland quantified by low-temperature thermochronology. J Geol Soc Lond 170:67–77

Fitzgerald PG, Muñoz JA, Coney PJ, Baldwin SL (1999) Asymmetric exhumation across the Pyrenean orogen: implications for the tectonic evolution of a collisional orogen. Earth Planet Sci Lett 173:157–170

Fleta J, Santanach P, Goula X, Martínez P, Grellet B, Masana E (2001) Preliminary geo-logic, geomorphologic and geophysical studies for the palaeoseismological analysis of the Amer fault, NE Spain. Neth J Geosci 80:243–253

Ford M, Hemmer L, Vacherat A, Gallagher K, Christophoul F (2016) Retro-wedge foreland basin evolution along the ECORS line, eastern Pyrenees, France. J Geol Soc 173:419–437

Gangloff P, Hétu B, Courchesne F, Richard PJH (1994) Présence d'un pergélisol würmien sur le piémont des Pyrénées Atlantiques. Géog Phys Quatern 48:169–178

García Sainz L (1940) Las superficies de erosión que preceden a los glaciares quaternarios del Pirineo central y sus recíprocas influencias. Estudios Geograficos I:45–73

García-Ruiz JM, Martí-Bono C, Peña-Monné JL, Sancho C, Rhodes E, Valero B, Gonzalez Samperiz P, Moreno A (2013) Glacial and fluvial deposits in the Aragón Valley, central western Pyrenees: chronology of the Pyrenean late Pleistocene glaciers. Geogr Ann Ser A Phys Geogr 95:15–32

Garwin LG (1985) Fission-track dating and tectonics in the Eastern Pyrenees. PhD thesis (unpubl.), University of Cambridge

Genna A, Lenotre N, Capdeville JP (1997) Proposition d'un modele d'inversion tectonique au Plio-Quaternaire dans les Corbieres et le Minervois (France). Consequences morphologiques et hydrologiques. C R Acad Sci Paris II 325:807–813

Gibson M, Sinclair HD, Lynn GJ, Stuart FM (2007) Late- to post-orogenic exhumation of the Central Pyrenees revealed through combined thermochronological data and modelling. Basin Res 19:323–334

Giménez J, Surinach E, Fleta J, Goula X (1996) Recent vertical movements from high-precision leveling data in northeast Spain. Tectonophysics 263:149–161

Gomis Coll E, Parés JM, Cabrera L (1997) Nuevos datos magnetoestratigráficos del tránsito Oligoceno-Mioceno en el sector SE de la Cuenca del Ebro (provincias de Lleida, Zaragoza y Huesca, NE de España). Acta Geol Hisp 185–199 (Pub. 1999)

Goron L (1931) Un type de vallée pyrénéenne: La Barguillère (Pyrénées Ariègeoises). Rev Géogr Pyrén Sud-Ouest 2:59–94

Goron L (1937) Les unités topographiques du Pays ariégeois: Le rôle des cycles d'érosion tertiaires et des glaciations quaternaires dans leur morphologie. Rev Géogr Pyrén Sud-Ouest 8:300–334

Goron L (1941a) Les Pré-Pyrénées ariégeoises et garonnaises, essai d'étude morphogénique d'une lisière de montagne. Privat, Toulouse, 886 p

Goula X, Olivera C, Fleta J, Grellet B, Lindo R, Rivera LA, Cisternas A, Carbon D (1999) Present and recent stress regime in the eastern part of the Pyrenees. Tectonophysics 308:487–502

Granger DE, Muzikar P (2001) Dating sediment burial with cosmogenic nuclides: theory, techniques, and limitations. Earth Planet Sci Lett 188:269–281

Granger DE, Kirchner JW, Finkel RC (1997) Quaternary downcutting rate of the New River, Virginia, measured from differential decay of cosmogenic ^{26}Al and ^{10}Be in cave-deposited alluvium. Geology 25:107–110

Guerrero J, Gutiérrez F, García-Ruiz JM, Carbonel D, Lucha P, Arnold LJ (2018) Landslide-dam paleolakes in the Central Pyrenees, Upper Gállego River Valley, NE Spain: timing and relationship with deglaciation. Landslides 15:1975–1989

Gunnell Y, Zeyen H, Calvet M (2008) Geophysical evidence of a missing lithospheric root beneath the Eastern Pyrenees: consequences for post-orogenic uplift and associated geomorphic signatures. Earth Planet Sci Lett 276:302–313

Gunnell Y, Calvet M, Brichau S, Carter A, Aguilar JP, Zeyen H (2009) Low long-term erosion rates in high-energy mountain belts: insights from thermo- and biochronology in the Eastern Pyrenees. Earth Planet Sci Lett 278:208–218

Gunnell Y, Calvet M (2006) Comment on "Origin of the highly elevated Pyrenean peneplain", by J Babault et al. Tectonics, 25, TC3003. https://doi.org/10.1029/2005TC001849

Gutiérrez-Elorza M, Desir G, Gutiérrez-Santolalla F (2002) Yardangs in the semiarid central sector of the Ebro Depression (NE Spain). Geomorphology 44:155–170

Harlé E (1893) Restes d'éléphants du sud-ouest de la France. Bull Soc Hist Nat Toulouse 28–34.

Hartevelt JJA (1970) Geology of the Upper Segre and Valira valleys, central Pyrenees, Andorra-Spain. Leidse Geol Meded 45:167–236

Hérail G, Hubschman J, Jalut G (1987) Quaternary glaciation in the French Pyrenees. Quat Sci Rev 5:397–402

Hez G, Jaillet S, Calvet M, Delannoy JJ (2015) Un enregistreur exceptionnel de l'incision de la vallée de la Têt: le karst de Villefranche, Pyrénées-orientales–France. Karstologia 65:9–32

Hubschman J (1973) Etablissement par l'étude des faciès d'altération, d'un schéma stratigraphique du Quaternaire garonnais et ariégeois. C R Acad Sci D 277:753–755

Hubschman J (1975a) Morphogenèse et pédogenèse dans le piémont des Pyrénées garonnaises et ariégeoises. H Champion, Paris, 745 p

Hubschman J (1975b) Conclusion, évolution pédo-géochimique et interprétation paléobioclimatique du piémont quaternaire garonnais. Bull Assoc Fr Étude Quat 3–4:211–216

Hubschman J (1975c) Modelés et formations quaternaires du terrefort molassique au sud de Toulouse. Bull Assoc Fr Étude Quat 12:125–136

Huyghe D, Mouthereau F, Ségalen L, Furió M (2020) Long-term dynamic topographic support during post-orogenic crustal thinning revealed by stable isotope (δ^{18}O) paleoaltimetry in eastern Pyrenees. Sci Rep. https://doi.org/10.1038/s41598-020-58903-w

Icole M (1968) Données nouvelles sur la formation de Lannemezan. C R Acad Sci Paris 267:687–689

Icole M (1969) Age et nature de la formation dite «de Lannemezan». Rev Géogr Pyrén Sud-Ouest 40:157–170

Icole M (1974) Géochimie des altérations dans les nappes d'alluvions du piémont occidental nord-pyrénéen. Essai de paléopédologie quaternaire. Sci Géol, Mém 40, 201 p

Jalut G, Montserrat J, Fontugne M, Delibrias G, Vilaplana JM, Julia R (1992) Glacial to interglacial vegetation changes in the northern and southern Pyrenees: deglaciation, vegetation cover and chronology. Quat Sci Rev 11:449–480

Jolivet L, Romagny A, Gorini C, Maillard A, Thinon I, Couëffé R, Ducoux M, Séranne M (2020) Fast dismantling of a mountain belt by mantle flow: late-orogenic evolution of Pyrenees and Liguro-Provençal rifting. Tectonophysics 776(228312):15 p

Jolivet M, Labaume P, Monié P, Brunel M, Arnaud N, Campani M (2007) Thermochronology constraints for the propagation sequence of the south Pyrenean basement thrust system (France–Spain). Tectonics 26:TC5007.10.1029 /2006 TC002080

Kiden P, Törnqvist TE (1998) Can river terrace flights be used to quantify Quaternary tectonic uplift rates? J Quat Sci 13:573–574

Kleinsmiede WFJ (1960) Geology of the Valle de Aran (Central Pyrenees). Leidse Geol Meded 25:129–245

Lacan P, Ortuño M (2012) Active Tectonics of the Pyrenees: A review. J Iber Geol 38:9–30

Lacan P, Nivière B, Rousset D, Sénéchal P (2012) Late Pleistocene folding above the Mail Arrouy Thrust, North-Western Pyrenees (France). Tectonophysics 541–543:57–68

Lagasquie JJ (1969) Le Bassin de Saint-Girons et la vallée du Baup (Pyrénées du Couserans). Etude morphologique. Rev Géogr Pyrén Sud-Ouest 40:267–286

Lagasquie JJ (1984a) Géomorphologie des granites, les massifs granitiques de la moitié orientale des Pyrénées françaises. CNRS, Toulouse, 374 p

Lagasquie JJ (1984b) Les relations piémont-montagne en Couserans et Bas-Salat. Actes du colloque «Montagnes et piémonts» (12-15 mai 1982) Hommage à F. Taillefer. Rev Géogr Pyr Travaux I:239–245

Lagasquie JJ (1987) Signification géodynamique des formes et dépôts de piémont dans la moitié orientale des Pyrénées. Rev Géomorph Dyn XXXVI:85–86

Larue JP (2008) Effects of tectonics and lithology on long profiles of 16 rivers of the southern Central Massif border between the Aude and the Orb (France). Geomorphology 93:343–367

Leroux E, Aslanian D, Rabineau M, Pellen R, Moulin M (2018) The late Messinian event: A worldwide tectonic revolution. Terra Nova 30:207–214

Lenôtre N, Fourniguet J (1987) Mouvements verticaux actuels: comparaison des nivellements. Synthèse géologique des Pyrénées, Rapport dugroupe de travail Tectonique récente et actuelle, 7 p. and 2 maps, Doc. BRGM-IGME (unpubl).

Lewis CJ, Vergès J, Marzo M (2000) High mountains in a zone of extended crust: Insights into the Neogene-Quaternary topographic development of northeastern Iberia. Tectonics 19:86–102

Lewis CJ, McDonald EV, Sancho C, Peña JL, Rhodes EJ (2009) Climatic implications of correlated Upper Pleistocene and fluvial deposits on the Cinca and Gállego Rivers (NE Spain) based on OSL dating a soil stratigraphy. Glob Planet Change 67:141–152

Lewis CJ, Sancho C, McDonald EV, Peña-Monné JL, Pueyo EL, Rhodes E, Calle M, Soto R (2017) Post-tectonic landscape evolution in NE Iberia using staircase terraces: combined effects of uplift and climate. Geomorphology 292:85–103

Maire R, Vanara N (2008) Les karsts du domaine pyrénéen, témoins des paléoenvironnements depuis le Paléozoïque (transect Zone axiale-Zone nord pyrénéenne dans les Pyrénées atlantiques). In: Canérot J, Colin J-P, Platel J-P, Bilotte M (eds) Pyrénées d'Hier et d'Aujourd'hui, International Year of Planet Earth, TOTAL, BRGM, AGSO, AIPT, CNRS, ed. Atlantica, Biarritz, pp 109–127

Maire R (1990) La haute montagne calcaire. Karsts, cavités, remplissages, Quaternaire, paléoclimats. Karstologia Mém 3, 731 p

Malcles O, Vernant P, Chéry J, Camps P, Cazes G, Ritz JF, Fink D (2020) Determining the Plio-Quaternary uplift of the southern French Massif Central; a new insight for intraplate orogen dynamics. Solid Earth 11:241–258

Martí Bono CE, García Ruiz JM (eds) (1994) El glaciarismo surpirenaico: nuevas aportaciones, Logroño, Geoforma édit., 142 p

de Martonne E (1910) Remarque sur la communication de ML Carez, «Résumé de la géologie des Pyrénées», Séance du 2 mai 1910, C R Séances Soc Géol Fr, p 425

Masana E (1994) Neotectonic features of the Catalan Coastal Ranges, Northeastern Spain. Acta Geol Hisp 29:107–121 (Publ. in 1996)

Maurel O, Brunel M, Monie P (2002) Exhumation cénozoïque des massifs du Canigou et de Mont-Louis (Pyrénées orientales, France). C R Geosci 334:941–948

Maurel O, Monié P, Pik R, Arnaud N, Brunel M, Jolivet M (2008) The Meso-Cenozoic thermo-tectonic evolution of the Eastern Pyrenees: an $^{40}Ar/^{39}Ar$ fission track and (U–Th)/He thermochronological study of the Canigou and Mont-Louis massifs. Int J Earth Sci 97:565–584

Maurel O (2003) L'exhumation de la Zone Axiale des Pyrénées orientales: une approche thermo-chronologique multi-méthodes du rôle des failles. PhD thesis, Université de Montpellier 2, 218 p

Mensua S, Ibañez J, Yetano M (1977) Sector central de la depresion del Ebro, mapa de terrazas fluviales y glacis. Departamento de Geografia, Universidad de Zaragoza, 5 sheets 1:100,000 scale, Handbook, 18 p

Meresse F (2010) Dynamique d'un prisme orogénique intracontinental: évolution thermochronologique (traces de fission sur apatite) et tectonique de la Zone Axiale et des piémonts des Pyrénées centro occidentales. PhD thesis, Université de Montpellier 2, 277 p

Metcalf JR, Fitzgerald PJ, Baldwin SL, Muñoz JA (2009) Thermochronology of a convergent orogen: constraints on the timing of thrust faulting and subsequent exhumation of the Maladeta Pluton in the Central Pyrenean Axial Zone. Earth Planet Sci Lett 287:488–503

Miall A (2016) Fluvial depositional systems. Springer, 316 p

Michael NA, Carter A, Whittaker AC, Allen PA (2014) Erosion rates in the source region of an ancient sediment routing system: comparison of depositional volumes with thermochronometric estimates. J Geol Soc Lond 171:401–412

Monaco A, Thommeret JY (1972) L'âge des dépôts quaternaires sur le plateau continental du Roussillon (Golfe du Lion). C R Acad Sci Paris D 274:2280–2283

Monod B, Regard V, Carcone J, Wyns R, Christophoul F (2016) Postorogenic planar palaeosurfaces of the central Pyrenees: Weathering and neotectonic records. C R Geosci 348:184–193

Montgomery DR (1994) Valley incision and the uplift of mountain peaks. J Geophys Res 99:13913–13921

Morris RG, Sinclair HD, Yelland AJ (1998) Exhumation of the Pyrenean orogen: implications for sediment discharge. Basin Res 10:69–85

Mouline M, Birot P, Paquereau M (1969) Le rebord NE de la Montagne Noire dans la région de Revel. Livret guide excursion A6, Pyrénées orientales et centrales, Roussillon, Languedoc occidental, VIIIe Congrès INQUA, Paris, pp 106–109

Nguyen HN, Vernant P, Mazzotti S, Khazaradze G, Asensio E (2016) 3D GPS velocity field and its implications on the present-day postorogenic deformation of the Western Alps and Pyrenees. Solid Earth 7:1349–1363

Nussbaum F (1930) Morphologische studien in den östlichen Pyrenäen. Z Gesellschaft Erdkunde, Berlin, pp 200–210

Nussbaum F (1931) Sur les surfaces d'aplanissement d'âge tertiaire dans les Pyrénées-Orientales et leurs transformations pendant l'époque quaternaire. C R Congrès Int Géogr Paris II:529–534

Ortuño M, Viaplana-Muzas M (2018) Active fault control in the distribution of elevated low relief topography in the Central-Western Pyrenees. Geol Acta 16:499–518

Ortuño M, Queralt P, Martí A, Ledo J, Masana E, Perea H, Santanach P (2008) The North Maladeta Fault (Spanish Central Pyrenees) as the Vielha 1923 earthquake seismic source: recent activity revealed by geomorphological and geophysical research. Tectonophysics 453:246–262

Ortuño M, Marti A, Martin-Closas C, Jimenéz-Moreno G, Martinetto E, Santanach P (2013) Palaeoenvironments of the Late Miocene Pruëdo Basin: implications for the uplift of the Central Pyrenees. J Geol Soc Lond 170:79–92

Palacios D, Andrés N, López-Moreno JI, García-Ruiz JM (2015) Late Pleistocene deglaciation in the upper Gállego Valley, central Pyrenees. Quat Res 83:397–414

Palacios D, Gómez-Ortiz A, Andrés N, Vázquez-Selem L, Salvador-Franch F, Oliva M (2015) Maximum extent of Late Pleistocene glaciers and last deglaciation of La Cerdanya mountains, Southeastern Pyrenees. Geomorphology 231:116–129

Palacios D, García-Ruiz JM, Andrés N, Schimmelpfennig I, Campos N, Léanni L, ASTER Team (2017) deglaciation in the central Pyrenees during the PleistoceneeHolocene transition: timing and geomorphological significance. Quat Sci Rev 162:111–127

Pallàs R, Rodés A, Braucher R, Carcaillet J, Ortuño M, Bordonau J, Bourlès D, Vilaplana JM, Masana E, Santanach P (2006) Late Pleistocene and Holocene glaciation in the Pyrenees: a critical review and new evidence from ^{10}Be exposure ages, south-central Pyrenees. Quat Sci Rev 25:2937–2963

Pallàs R, Rodès A, Braucher R, Bourlès D, Delmas M, Calvet M, Gunnell Y (2010) Small, isolated glacial catchment as priority targets for cosmogenic surface exposure dating of Pleistocene climate fluctuations, southeastern Pyrenees. Geology 38:891–894

Pannekoek AJ (1935) Évolution du bassin de la Têt dans les Pyrénées-Orientales pendant le Néogène. Étude de morphotectonique. University of Utrecht, 72 p

Parsons AJ, Michael NA, Whittaker AC, Duller RA, Allen PA (2012) Grain-size trends reveal the late orogenic tectonic and erosional history of the south–central Pyrenees, Spain. J Geol Soc Lond 169:111–114

Peña Monné JL, Turu V, Calvet M (2011) Les terrasses fluvials del Segre i afluents principals: descripció d'afloraments i assaig de correlació. In: Turu V, Constante A (eds) El Cuaternario en España y areas afines, avances en 2011, XIII Reunión Nacional de Cuaternario, Andorra, 4–7 juillet, Asociación Española para el Estudio del Cuaternario (AEQUA), pp 51–55

Peña Monné JL (1983) La conca de Tremp y sierras prepirenaicas comprendidas entre los ríos Segre y Noguera Ribagorzana, estudio geomorfológico. Instituto de Estudios Ilerdenses, SCIC, vol 2, 373 p

Peña JL, Sancho C (1988) Correlación y evolución cuaternaria del sistema fluvial Segre-Cinca en su curso bajo (Provincias de Lerida y Huesca). Cuatern Geomorf 2:77–83

Penck A (1894) Studien uber das klima spaniens während der jungeren Tertiärperiod und Diluvialperiode. Z Gesellschaft Erdkunde Berlin 29:109–141

Penck A (1883) Die Eiszeit in den Pyrenäen. Mitteil Ver Erdkunde, Leipzig 163–231. Traduction en Français par L Braemer (1885), Bull Soc Hist Nat Toulouse 107–200

Philip H, Bousquet JC, Escuer J, Fleta J, Goula X, Grellet B (1992) Présence de failles inverses d'âge quaternaire dans l'est des Pyrénées: implications sismotectoniques. C R Acad Sci, Paris II 314:1239–1245

Philip H (ed) (2018) Tectonique récente et actuelle, Synthèse géophysique et géologique des Pyrénées-volume 3: Cycle alpin: Phénomènes alpins, chap. 23, 361–404, AGSO and BRGM, 483 p

Potter PE, Szatmari P (2009) Global Miocene tectonics and the modern world. Earth-Sci Rev 96:279–295

Pouech (1873) Note au sujet des restes d'un éléphant fossile découvert à Pamiers (Ariège). Bull Soc Géol Fr 2:8–14

Pous J, Julià R, Sole SL (1986) Cerdanya basin geometry and its implication on the Neogene evolution of the eastern Pyrenees. Tectonophysics 129:355–365

Pous J, Muñoz JA, Ledo JJ, Liesa M (1995a) Partial melting of subducted continental lower crust in the Pyrenees. J Geol Soc Lond 152:217–220

Pous J, Ledo J, Marcuello A, Daignières M (1995b) Electrical resistivity model of the crust and upper mantle from a magnetotelluric survey through the central Pyrenees. Geophys J Int 121:750–762

Rahl JM, Haines SH, van der Pluijm BA (2011) Links between orogenic wedge deformation and erosional exhumation: evidence from illite age analysis of fault rock and detrital thermochronology of syn-tectonic conglomerates in the Spanish Pyrenees. Earth Planet Sci Lett 307:180–190

Reille M, Andrieu V (1993) Variations de la limite supérieure des forêts dans les Pyrénées (France) pendant le Tardiglaciaire. C R Acad Sci Paris II 316:547–551

Revel JC, Bourgeat F, Crouzel F, Puissegur JJ (1979) Pédogenèse et morphogenèse sur les loess wurmiens de la région toulousaine. Bull Soc Hist Nat Toulouse 3–4:293–315

Revel JC, Bourgeat F (1981) Sols fossiles du Terrefort toulousain. Leur signification paléoclimatique. Bull Assoc Fr Étude Quat 18:149–158

Revel JC, Bourgeat F, Paquet H (1978) Pédogenèses quaternaires dans la région toulousaine. Les loess et leurs colluvions comme marqueur chronologique. Bull Assoc Fr Étude Quat 15:179–185

Rigo A, Cushing M (1999) Effets topographiques sur les comparaisons de profils de nivellement: cas français de Saint-Paul-de-Fenouillet (Pyrénées-Orientales) et d'Arudy (Pyrénées-Atlantiques). C R Acad Sci Paris 329:697–704

Roca E (1996) The Neogene Cerdanya and Seu d´Urgell intramontane basins (eastern Pyrenees). In: Friend PF, Dabrio CJ (eds) Tertiary Basins of Spain. The Stratigraphic Record of Crustal Kinematics. Cambridge University Press, Cambridge, pp 114–118

Rodriguez-Vidal J (1984) Geomorfología de las sierras exteriores oscenses y su piedemonte. Tesis Universidad de Zaragoza (1983), Instituto de Estudios Alto Aragoneses (SCIC), 172 p

Rushlow CR, Barnes JB, Ehlers TA, Vergés J (2013) Exhumation of the southern Pyrenean fold-thrust belt (Spain) from orogenic growth to decay. Tectonics 32:843–860

Sancho C, Arenas C, Pardo G, Peña-Monné JL, Rhodes EJ, Bartolomé M, García-Ruiz JM, Martí-Bono C (2018) Glaciolacustrine deposits formed in an ice-dammed tributary valley in the south-central Pyrenees: New evidence for late Pleistocene climate. Sed Geol 366:47–66

Saula E, Picart J, Mató E, Llenas M, Losantos M, Berástegui X, Agustí J (1994) Evolución geodinámica de la fosa del Emporda y las Sierras Transversales. Acta Geol Hisp 29:55–75

Sère V (1993) Analyse cinématique et évolution thermotectonique des mylonites de la faille de la Têt (versant nord du Canigou, P-O). MPhil Dissertation (unpubl.), Université de Montpellier II

Sermet J (1950) Réflexions sur la morphologie de la zone axiale des Pyrénées. Pirineos VI:323–404

Serrat D, Bordonau J, Bru J, Furdada G, Gomez A, Marti J, Marti M, Salvador F, Ventura J, Vilaplana JM (1994) Síntesis cartográfica del glaciarismo surpirenaico oriental. In: Martí Bono CE, García Ruiz JM (eds) El glaciarismo surpirenaico: nuevas aportaciones, Logroño, Geoforma Édit., pp 9–15

Sinclair HD, Gibson M, Naylor M, Morris RG (2005) Asymmetric growth of the Pyrenees revealed through measurement and modelling of orogenic fluxes. Am J Sci 305:369–406

Sinclair HD, Gibson M, Lynn G, Stuart F (2009) The evidence for a Pyrenean resurrection: a response to Babault et al. (2008). Basin Res 21:143–145

Sitzia L, Bertran P, Bahain JJ, Bateman MD, Hernandez M, Garon H, de Lafontaine G, Mercier N, Leroyer C, Queffelec A, Voinchet P (2015) The quaternary coversands of southwest France. Quat Sci Rev 124:84–105

Sitzia L (2014) Chronostratigraphie et distribution spatiale des dépôts éoliens quaternaires du bassin Aquitain. PhD thesis, Univ. Bordeaux 1, 341 p. https://tel.archives-ouvertes.fr/tel-01009617

Sorre M (1913) Les Pyrénées méditerranéennes, étude de géographie biologique. A Colin, Paris, 508 p

Sorriaux P, Delmas M, Calvet M, Gunnell Y, Durand N, Pons-Branchu E (2016) Relations entre karst et glaciers depuis 450 ka dans les grottes de Niaux-Lombrives-Sabart (Pyrénées ariégeoises). Nouvelles datations U/Th dans la grotte de Niaux. Karstologia 67:3–16 (Publ. 2018)

Sorriaux P (1981) Étude et datation de remplissages karstiques: nouvelles données sur la paléogéographie quaternaire de la région de Tarascon (Pyrénées ariégeoises). C R Acad Sci Paris II(293):703–706

Sorriaux P (1982) Contribution à l'étude de la sédimentation en milieu karstique. Le système de Niaux-Lombrives-Sabart (Pyrénées Ariégeoises). PhD thesis, Université Paul Sabatier, Toulouse, 255 p.

Stange KM, Van Balen RT, Vandenberghe J, Peña JL, Sancho C (2013) External controls on Quaternary fluvial incision and terrace formation at the Segre River, southern Pyrenees. Tectonophysics 602:316–331

Stange KM, Van Balen RT, Kasse C, Vandenberghe J, Carcaillet J (2014) Linking morphology across the glaciofluvial interface: a [10]Be supported chronology of glacier advances and terrace formation in the Garonne River, northern Pyrenees, France. Geomorphology 207:71–95

Stange KM, Van Balen T, Garcia-Castellanos D, Cloetingh S (2016) Numerical modelling of Quaternary terrace staircase formation in the Ebro foreland basin, southern Pyrenees, NE Iberia. Basin Res 28:124–146

Stange KM, Midtkandal I, Nystuen JP, Kuss HJ, Spiegel C (2018) Direct response of small non-glaciated headwater catchments to Late Quaternary climate change: The Valle de la Fueva, southern Pyrenees. Geomorphology 318:187–202

Suc JP, Fauquette S (2012) The use of pollen floras as a tool to estimate palaeoaltitude of mountains: The eastern Pyrenees in the Late Neogene, a case study. Palaeogeogr Palaeoclim Palaeoecol 321–322:41–54

Taillefer F (1951) Le piémont des Pyrénées françaises, contribution à l'étude des reliefs de piémont. Privat, Toulouse, 383 p

Taillefer F (1957) Le glaciaire pyrénéen: versant nord et versant sud. Rev Géogr Pyrén Sud-Ouest 28:221–244

Taillefer F (1967) Extent of Pleistocene glaciation in the Pyrenees. In: Osborne W (eds) Arctic and Alpine Environments, Indiana University Press, pp 255–266

Taillefer F (1969) Les glaciations des Pyrénées. Actes du VIII° Congrès international de l'INQUA, supplément au Bulletin de l'Association Française pour l'Étude du Quaternaire, Paris, 19–32

Taillefer F (1982) Les conditions locales de la glaciation pyrénéenne. Pirineos 116:5–12

Tomkins MD, Dortch JM, Hughes PD, Huck JJ, Stimson A, Delmas M, Calvet M, Pallas R (2018) Rapid age assessment of glacial landforms in the Pyrenees using Schmidt hammer exposure dating (SHED). Quat Res 1–12. https://doi.org/10.1017/qua.2018.12

Tricart J, Hirsch AR, Griesbach JC (1966) La géomorphologie du bassin du Touch (Haute-Garonne), ses implications pédologiques et hydrologiques. Rev Géogr Pyrén Sud-Ouest 37:5–46

Turu V (2011) Los complejos morrenicos terminales del Valira (Andorra-Alt Urgell). Resúmenes XIII Reunion Nacional de Cuaternario, Andorra 2011, Simposio de glaciarismo, guía de campo, pp 1–8

Turu V, Calvet M, Bordonau J, Gunnell Y, Delmas M, Vilaplana JM, Jalut G (2017) Did Pyrenean glaciers dance to the beat of global climatic events? Evidence from the Würmian sequence stratigraphy of an ice-dammed palaeolake depocentre in Andorra. In: Hughes PD, Woodward JC (eds) Quaternary Glaciation in the Mediterranean Mountains, Geol Soc Lond, Spec Publ 433: 111–136

Turu V, Ventura J, Ros X, Pèlachs A, Vizcaino A, Soriano JM (2011) Geomorfologia glacial del tram final de la Noguera Pallaresa i Riu Flamicell (Els Pallars). Resúmenes XIII Reunion Nacional de Cuaternario, Andorra 2011, Simposio de glaciarismo, pp 37–43

Turu V, Vidal Romani JR, Fernández Mosquera D (2011) Dataciones con isótopos cosmogénicos (^{10}Be): el "LGM" (Last Glacial Maximum) y el "Last Termination" en los valles del Gran Valira y la Valira del Nord (Principado de Andorra, Pirineos orientales). Resúmenes XIII Reunion Nacional de Cuaternario, Andorra 2011, Simposio de glaciarismo, 19–23

Turu V, Boulton GS, Ros X, Peña Monné JL, Martí C, Bordonau J, Serrano-Cañadas E, Sancho-Marcén C, Constante-Orrios A, Pous J, Gonzalez-Trueba JJ, Palomar J, Herrero R, Garcia-Ruiz JM (2007) Structure des grands bassins glaciaires dans le nord de la péninsule ibérique: comparaison entre les vallées d'Andorre (Pyrénées orientales), du Gallego (Pyrénées centrales) et du Trueba (Chaîne cantabrique). Quaternaire 18:309–325

Uzel J, Nivière B, Lagabrielle Y (2020) Fluvial incisions in the North-Western Pyrenees (Aspe Valley): dissection of a former planation surface and some tectonic implications. Terra Nova 32:11–22

Vacherat A, Mouthereau F, Pik R, Bernet M, Gautheron C, Masini E, Le Pourhiet L, Tibari B, Lahfid A (2014) Thermal imprint of rift-related processes in orogens as recorded in the Pyrenees, Earth Planet Sci Lett 408:296–306

Vacherat A, Mouthereau F, Pik R, Bellahsen N, Gautheron C, Bernet M, Daudet M, Balansa J, Tibari B, Pinna JR, Rada J (2016) Rift-to-collision transition recorded by tectonothermal evolution of the northern Pyrenees. Tectonics 35:907–933

Vacherat A, Mouthereau F, Pik R, Huyghe D, Paquette J-L, Christophoul F, Loget N, Tibari B (2017) Rift-to-collision sediment routing in the Pyrenees: a synthesis from sedimentological, geochronological and kinematic constraints. Earth-Sci Rev 172:43–74

Vanara N, Maire R, Lacroix J (1997) La surface carbonatée du massif des Arbailles (Pyrénées Atlantiques): un example de paléoréseau hydrographique néogéne déconnecté par la surrection. Bull Soc Géol Fr 168:255–265

Vanara N (2000) Le Karst des Arbailles. Karstologia Mémoire n° 3, 320 p

Ventra D, Clarke LE (2018) Geology and geomorphology of alluvial and fluvial fans: current progress and research perspectives. In: Ventra D, Clarke LE (eds) Geology and geomorphology of alluvial and fluvial fans: terrestrial and planetary perspectives. Geol Soc Lond Spec Publ 440:1–21

Viers G (1961) La tectonique post-pliocène sur le littoral atlantique entre Biarritz et Hendaye. Actes VI° congrès INQUA. Warsaw 1:539–544

Viers G (1960) Le relief des Pyrénées occidentales et leur piémont. Pays Basque français et Barétous. Privat, Toulouse, 604 p

Whitchurch AL, Carter A, Sinclair HD, Duller RA, Whittaker AC, Allen PA (2011) Sediment routing system evolution within a diachronously uplifting orogen: insights from detrital zircon thermochronological analyses from the South-Central Pyrenees. Am J Sci 311:442–482

Yelland A (1990) Fission track thermotectonics in the Pyrenean orogen. Nucl Tracks Radiat Meas 17:293–299

Yelland A (1991) Thermo-tectonics of the Pyrenees and Provence from fission-track studies, PhD thesis (unpubl.), University of London

Zandvliet J (1960) The geology of the upper Salat and Pallaresa valleys, Central Pyrenees, France/Spain. Leidse Geol Meded 25:1–127

Present-Day Environments

4

4.1 Climate and Vegetation

4.1.1 A Bridge Between Atlantic and Mediterranean Climates

The modern Pyrenees form a major climatic barrier between western Europe and Iberia (Fig. 4.1). This climatic asymmetry has prevailed for at least the entire duration of the Quaternary, and probably much longer. The contrast appears clearly on climate diagrams across the study area (Figs. 4.2, 4.3). In the Aquitaine Basin, the climate is temperate oceanic, with mild winters and evenly distributed precipitation throughout the year. Precipitation nonetheless declines eastward with distance from the Atlantic: thus, for the period 1981–2010, mean annual precipitation values were: Hendaye, 1483 mm; Tarbes: 1047 mm; Toulouse, 638 mm; Carcassonne, 648 mm; Perpignan, 557 mm; with 140 days of precipitation at Biarritz, 120 at Tarbes, 96 at Toulouse, 87 at Carcassonne, 54 at Perpignan, and 175 at the Pic du Midi de Bigorre. Restricting the rainy days criterion to precipitation >5 m/24 h, the number of rainy days declines from 70 in the west to <30 in the east. To the south and east, the climatic regime is Mediterranean, with at least 2 months of summer dry season at the coast and up to 5 in the Ebro Basin interior, the latter being also colder in winter. Annual precipitation rarely exceeds 600 mm in the Roussillon and 400 mm in the Ebro Basin (Zaragoza: 322 mm/yr, data for 1981–2000), concentrated in a short run of 50 to 80 days with large year-to-year variability. These climatic features are altered by altitude, but the mountain climate retains a sharp contrast between the humid and snow-prone north-facing range front (1.5 m to >2 m of precipitation in the outer ranges) and the much drier and sunnier east- and south-facing sides (Fig. 4.1). For example, the elevated

© Springer Nature Switzerland AG 2022
M. Calvet et al., *Geology and Landscapes of the Eastern Pyrenees*, GeoGuide,
https://doi.org/10.1007/978-3-030-84266-6_4

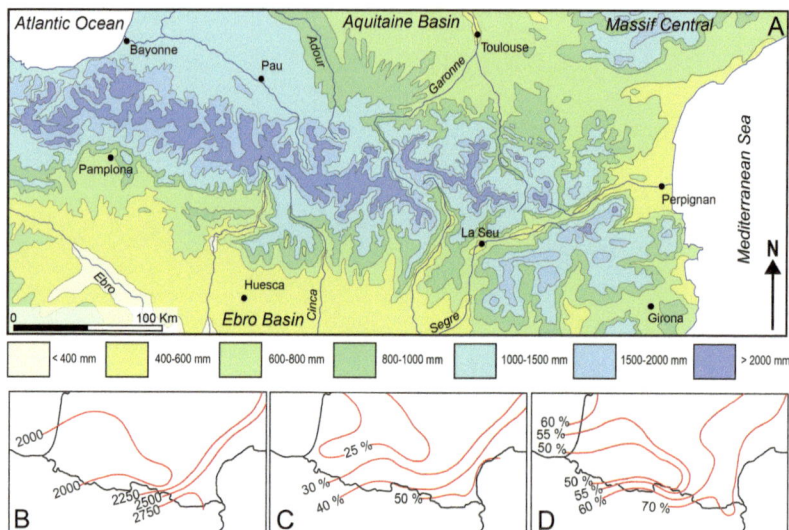

Fig. 4.1 Rainfall and sunshine hours in the Pyrenees. **A** Simplified rainfall distribution map. Isohyet contours across the southern half of the range are uncertain, as likewise rainfall contours and values in excess of >2 m in the more elevated massifs. After Gaussen (1934), in Solé Sabaris (1951), redrawn and modified. **B** Annual number of sunshine hours. **C** Sunshine hours in December (%). **D** Sunshine hours in July (a value of 100% would correspond to continuous sunshine from sunrise to sunset every day of the month). After maps by Kessler and Chambraud (1986), redrawn and modified

tectonic basins of Cerdagne and La Seu (elevations: 1000–1400 m) only record 600–700 mm of annual precipitation (Fig. 4.3); winter snow cover on south-facing slopes is often patchy, and its lower boundary lies generally between 1500 and 2000 m. The prevailing wind in the eastern Pyrenees is the tramontane (meaning 'from over the mountain' in Catalan). At Perpignan, this strong, dry NW wind blows for a yearly average of 127 days at velocities >57 km/h (data sources: http://www.met eofrance.com, https://www.infoclimat.fr; see also Kessler and Chambraud 1986).

4.1.2 Distribution of Montane Vegetation Belts

The vegetation cover (Fig. 4.4) reflects these regional climatic contrasts (Gaussen 1926; Dupias 1985). By virtue of their location where the Atlantic and Mediterranean biogeographical regions meet, with important alpine flora and even

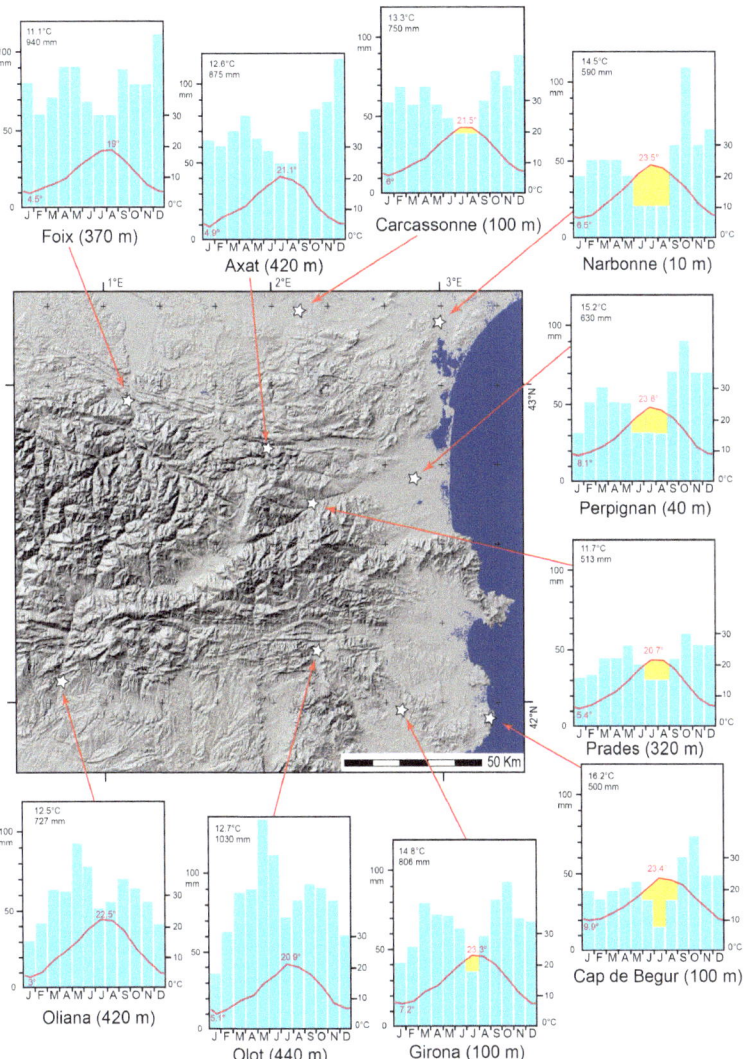

Fig. 4.2 Climatic characteristics of the eastern Pyrenees: piedmont zones and Mediterranean seaboard. All rainfall diagrams scaled to P (mm) = 2T (°C). Mediterranean conditions record at least 2 dry summer months in between two rainy seasons. The northern piedmont features an oceanic climate (e.g., Foix). Note the unique profile of northeastern Catalonia (e.g., Olot), in the lee of the high range but with high spring and summer precipitation despite its proximity to the Mediterranean. Data sources: Météo-France, and Servicio Meteorologico Nacional de España

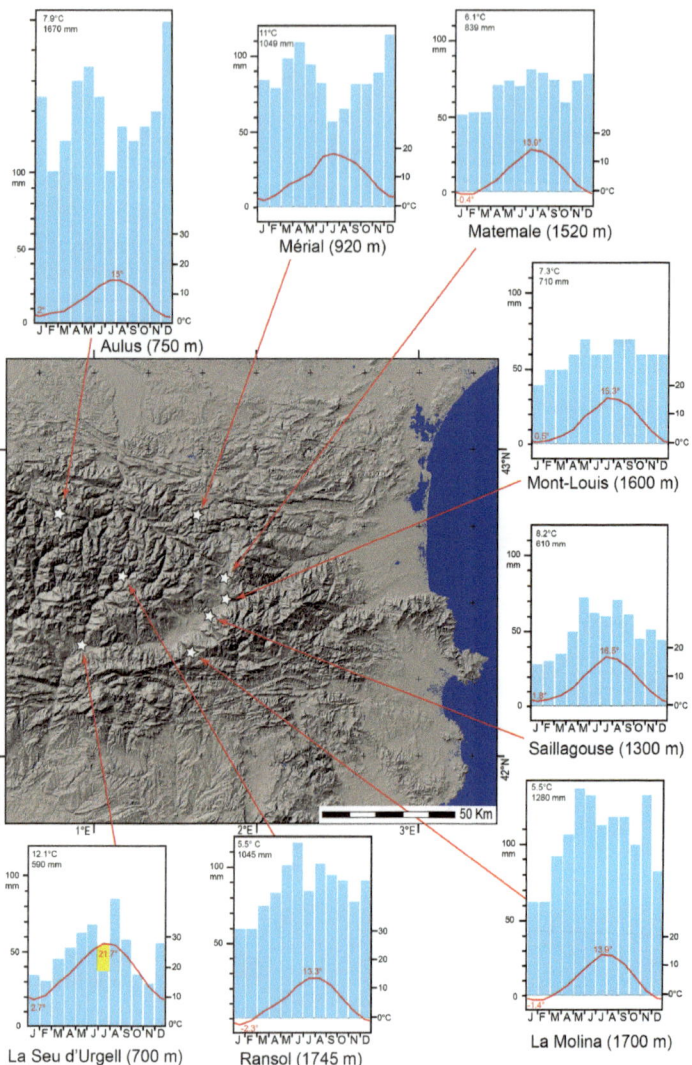

Fig. 4.3   Climatic characteristics of the eastern Pyrenees: elevated massifs and interior montane basins. Note high precipitation in the northern massifs under Atlantic influence (e.g., Mérial, Aulus), and dry environment of the intermontane basins, e.g., (Mont-Louis, Matemale, Saillagouse). The Seu d'Urgell Basin is unique among the population of interior basins in presenting a hint of Mediterranean climate, with one summer dry month

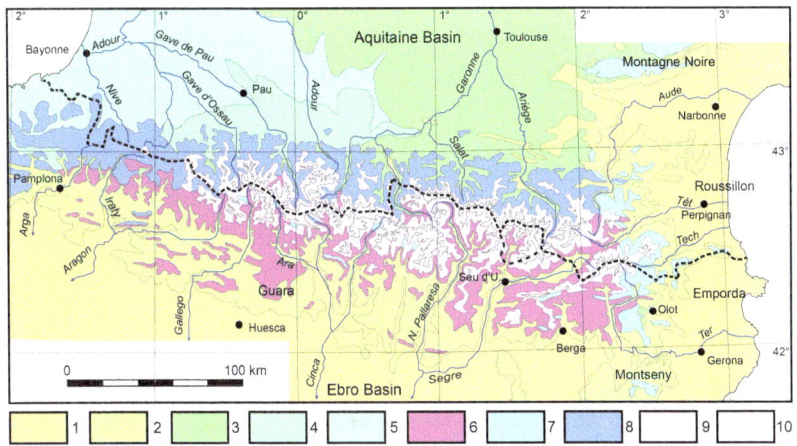

Fig. 4.4   Vegetation belts of the Pyrenees based on geobotanical subdivisions (phytosociological nomenclature after Dupias, 1985; redrawn and augmented). Key to vegetation belts and their plant communities (defined here as vegetation series, i.e., mosaics of successional units) based on elevation boundaries (altitudinal zones): *Foothill woodland communities*— **1**: *Meso-Mediterranean zone*, a vegetation series with mature successional stage comprising green oak, cork oak, stone pine; Aleppo pine and downy oak optional. Degraded states range from kermès oak or strawberry tree and tree heath shrubland, juniper and herb-rich scrub (rosemary, thyme, etc.), to open grassland (*Brachypodium ramosum*, a bunch grass). **2**: *Supra-Mediterranean foothills zone*, a series comprising downy oak (green oak on south-facing rocky slopes), locally with Salzmann pine. **3**: *Mid-European foothills zone*, where sessile oak is the indicator tree species (hornbeam absent), transitioning eastward (east of Toulouse longitude) to French, sessile, and downy oak. This series is also present in some upper valleys of the south (e.g., in Navarra) and in the humid valleys and basins of NE Catalonia. **4**: *Subatlantic foothills zone*, a vegetation series dominated by French oak (Pyrenean oak absent). **5**: *Atlantic zone*, where indicator tree species are French and Pyrenean oak (*Quercus pyrenaica*); degrades to Atlantic heathland. *Montane woodland communities*—**6**: *Montane zone (Continental–Mediterranean and Continental–Circum-Mediterranean variants)*, a series comprising Scots pine with enclaves of fir at more humid sites (southern flank of the Pyrenees and interior basins); this series is also hosted by the more sheltered valleys of the northern massifs (Val d'Aran, Neste, Cauterets) and by the eastern intermontane basins (Capcir, Cerdagne, Conflent). **7**: *Montane zone (Mid-European variant)*, with beech and fir forests of the eastern Pyrenees (Aude to Ter valleys), also forming enclaves in the inner south-Pyrenean valleys (Llobregat, Noguera Ribagorçana, Cinca, Gàllego). **8**: *Montane zone (humid Atlantic variant)*, consisting of beech and fir–beech associations with montane Atlantic undergrowth species. *High-altitude Subalpine woodland and alpine grassland communities*—**9**: *Subalpine zone*, where the dominant tree is mountain pine (*Pinus uncinata*), associated with birch, rowan, and fir at lower elevations. **10**: *Alpine zone*, where *Festuca supina* dominates in the eastern ranges, and *Carex curvula* in the central and western Pyrenees. Numerous endemic grass and flowering species in the higher massifs, particularly in the central and more locally the eastern Pyrenees (Tejero et al. 2017)

Pleistocene relics at higher altitudes, the eastern Pyrenees are home to great species richness and to a large proportion of endemic plant species. These natural landscapes have nonetheless undergone a long history of resource management and land-cover manipulation by human communities. The earliest record of this goes back to 7500–7800 year BP and is related to the settlement along the Mediterranean seaboard and its close hinterland by a Neolithic culture associated with Cardium pottery (Guilaine 2003). This human presence soon involved cattle stocking in high-altitude alpine meadows of the Axial Zone (Jalut 1977; Galop 1998; Rendu 2003; Galop et al. 2008). From Antiquity, the landscape on hillslopes and valley floors underwent lasting changes caused by forest clearance, wood harvesting, and charcoal production to supply the local needs of mining and metallurgy. As early as the tenth century CE, most valleys of the eastern Pyrenees were home to high population densities, already at that time comparable to those later recorded during the nineteenth century demographic maximum (Bonnassie 1990). As a result, Mediterranean forest originally dominated by green oak (*Quercus ilex*), cork oak (*Q. suber*, restricted to silica-rich rocks), Aleppo pine (*Pinus halepensis*) and strawberry tree (*Arbutus unedo*) progressively gave way to varyingly dense shrubland on crystalline soils ('maquis'), sparse Mediterranean scrub on limestone ('garrigue'), and even steppe in the drier areas of the Ebro Basin. This Mediterranean belt does not extend westward beyond the Aude valley in the NW, but spreads across the Iberian foreland as far west as Navarre. Above 500–800 m, the transitional vegetation belt gives way to downy oak (*Quercus pubescens*) and, more locally, to Pyrenean pine (*Pinus nigra subsp. salzmannii var. salzmannii*). Among the massifs of the orogen's retro-wedge, French (*Quercus pedunculata*) and sessile oak (*Qercus petraea*) mix at somewhat higher altitudes in the outer ranges with beech (*Fagus sylvatica*), and with downy oak in the eastern tracts. In the montane belt, and up to 1600–1800 m, beech and silver fir (*Abies alba*) prevail when under oceanic influence but also occur on the more humid slope aspects of the Mediterranean domain. Here, however, in drier locations fir and beech are usually replaced by Scots pine (*Pinus silvestris*). The most elevated forest belt favours mountain pine (*Pinus uncinata*). This vegetation belt is relatively uniform on both sides of the Pyrenean orogen, with occurrences up to 2400 m. Above it, alpine *Festuca* and *Carex* meadows tend to be dense and lush under oceanic climate, but degraded and stonier south of the drainage divide.

4.2   The Legacy of Holocene and Recent Erosional Events

4.2.1   Geomorphological Processes in the Frost Belt

Active glaciation in the Pyrenees is confined to the west-central portion of the range, where only 9 massifs still host a population of small cirque glaciers. The only surviving glacier in the eastern half of the range is at Mont Valier, in the shadow of its unusually high NE rock face (René 2008, 2011, 2013, 2014). Two to three glacial advances have been documented for the Little Ice Age (LIA) during the 17th and early nineteenth centuries. At that time, more that 100 small cirque glaciers, including among massifs as low as 2800–2900 m, existed on both sides of the central and western Pyrenees. The ELA among pro-wedge massifs descended to 2500 m in Aragón, 2700 m in the upper Gàllego watershed (Tendeñera and Panticosa), but was confined to 2900 m in the Posets and Maladeta massifs (Copons and Bordonau 1994; Lampre 1994; Julian and Chueca 1998; Serrano and Agudo 2004; for a synoptic location map of LIA glaciers, see Serrano et al. 2018). East of the Noguera Ribagorçana (south side) and Garonne (north side) rivers, legacies of LIA glaciation are uncertain, with attested easternmost vestiges in the upper Vicdessos valley at Pic Montcalm and Pic de Médecourbe. In 1870, i.e., just after the LIA, small cirque glaciers covered a total surface area of 23 km^2. These glaciers had shrunk to 5 km^2 in 2000, 3.5 km^2 in 2007, and 2.42 km^2 in 2016 (Rico et al. 2017).

Evidence of earlier Holocene glaciation has been until now very scarce. The Maladeta massif, for example, lacks evidence of glacial landforms located between the voluminous LIA frontal moraines and the Younger Dryas terminal moraine situated ~ 500 m further downslope. By default, it is thus assumed that all putative Holocene ice advances reached the same position as during the LIA, the latter thus being the latest episode of a recurring pattern (Crest et al. 2017), as if climatic oscillations had never exceeded a certain upper limit. Farther to the west, however, radiometric ages show that the Holocene advances overreached the subsequent LIA ice fronts: in the Munia–Troumouse massif, for example, a first cirque-glacier advance occurred before 5200 BP and indeed perhaps during the early Holocene, whereas the second occurred between 4900 and 4600 ^{14}C BP (uncalibrated ages; Gellatly et al. 1992). Likewise, moraines at the north base of Mt. Perdu document a glacial readvance ca. 5.1 ka, followed by another between 1.4 and 1.2 ka ([36]Cl exposure-dating results; García Ruiz et al. 2014).

Today, periglacial scree deposits and rock glaciers are widespread in the alpine landscape, but most are also legacies of the Last Glacial–Interglacial Transition rather than of Holocene age (Fernandes et al. 2018). In the Central Pyrenees of Aragón, only 7 rock glaciers among an inventory of 87 are active, and discontinuous permafrost occurs above 2700 m (Serrano et al. 2001, 2006; Serrano and Agudo 2004). Patterned ground has formed locally on LIA glacial deposits (Feuillet 2011; Feuillet and Mercier 2012). No evidence exists of permafrost in the eastern Pyrenees (Huc 2010). All the rock glaciers are fossil occurrences, but some may have been reactivated during the LIA (see Itineraries 4, 6 and 7). A range of minor processes and landforms are controlled by seasonal frost action (Soutadé 1980).

4.2.2 Geomorphological Processes Below the Tree Line

The landscape in the thickly vegetated north-Pyrenean massifs has long been known for the scarcity of modern erosional scars, quite unlike similar environments in the Alps (Blanchard 1914). Torrential activity, which has been partly curbed by major reforestation works carried out by the Service de Restauration des Terrains en Montagne (RTM) since the end of the nineteenth century, is a feature mostly confined to the eastern and southern parts of the mountain belt. In the eastern Pyrenees, debris torrents are triggered by exceptionally intense Mediterranean rainstorms, typically events with a 200–300-year return period such as the 'aiguat' (Catalan word for extreme rainstorm) that occurred on October 17–20, 1940, when 1000 mm of rain fell in 24 h (see Itineraries 4 and 6). During this meteorological event of historical importance, peak discharges on the Têt and Tech rivers exceeded 3000 m^3/s, and specific discharges in small catchments of the upper Tech valley attained 47 m^3/s/km^2 (Pardé 1941; Soutadé 1993). Similar events have been reported from historical records for the years 1763, 1632, 1553, 1421, 1264, and 878 (Bénech 1993). Lesser events, nonetheless highly erosive, also occurred in 1848, in 1907 in Roussillon, in 1963 in the southern Pyrenees, in 1982 across the Axial Zone from Mt. Perdu to Mt. Canigou, in 1999 (Calvet 2001; Calvet and Lemartinel 2002) and 2005 from Roussillon to the Montagne Noire across the Corbières and Minervois (see Itinerary 1), and in 2013 throughout the central Pyrenees.

Existing data on recent erosion rates have been compiled by Calvet (1996) for the eastern Pyrenees. Limestone solution rates reach 81–110 mm/ka in the Arbailles massif and Ariège River catchment, only 39 mm/ka in pro-wedge sierras such as Port de Comte, and 20–37 mm/ka in the western Corbières. Rates fall to ~10 mm/ka along the drier Mediterranean coastline (Maire 1990; Vanara et al. 1997; Vanara 2000; Fabre 1981). Catchment-wide denudation based on river sediment loads is ~200 mm/ka in the Têt, 300–400 mm/ka in the Tech, and 74 mm/ka in the Agly catchments (Calvet 1996; Serrat 1999), falling to ~56 mm/ka in the upper Garonne (Probst and Bazerbachi 1986). Estimates for the Têt (0.1 mm/yr) are corroborated by inferences made from the volume of sediment accumulated over 4 years of below-average discharge in Vinça Dam reservoir (Mussot et al. 1990), although such measurements are skewed by anthropogenic erosion. Other results based only on suspended loads provide much lower values: 14 mm/ka in the Adour and various Gave rivers; 17 mm/ka in the Aude and other Roussillon rivers (Delmas et al. 2012); 15 mm/ka in the Têt (Serrat et al. 2001). An estimate of Holocene sediment volumes stored in the Roussillon Basin and offshore has provided a basis for inferring an average denudation rate of 161–201 mm/ka for the tributary catchments. Catchment-wide denudation rates inferred from ^{10}Be concentrations in river sediment and supposed to filter out anthropogenic erosion have provided values of 60.7 ± 21.7 mm/ka for the Agly River, 132.1 ± 47.4 mm/ka for the Têt, and 104 ± 24.3 mm/ka for the Tech (Molliex et al. 2016). Preliminary results for 8 small catchments situated in the montane and alpine vegetation belts of the central Pyrenees have yielded a denudation range from 42 to 484 mm/ka (mean: 195 mm/ka; Genti 2015). Likewise, 21 other small- to medium-sized montane catchments in the central and eastern Pyrenees document rates of 22 to 404 mm/ka (mean: 170 mm/ka) (Crest 2018).

References

Bénech C (1993) Estimation des périodes de retour de 'l'aiguat' d'octobre 1940 dans quelques vallées des Pyrénées–Orientales. In: l'Aiguat del 40, actes du congrès de Vernet-les-Bains, Generalitat de Catalunya Édit 297–313

Blanchard R (1914) La morphologie des Pyrénées françaises. Ann Géogr 23:303–324

Bonnassie P (1990) La Catalogne au tournant de l'an mil, croissance et mutation d'une société. Albin Michel, Paris, 498 p

Calvet M (1996) Morphogenèse d'une montagne méditerranéenne: les Pyrénées orientales. Documents du BRGM, Orléans, 255:1177 p

Calvet M (2001) La catastrophe exemplaire: premiers enseignements géomorphologiques de la crue de novembre 1999 dans les Corbières. In: "Au chevet d'une catastrophe", actes du colloque de Perpignan, 26–28 juin 2000, 63–86, Presses Universitaires de Perpignan, coll. Etudes

Calvet M, Lemartinel B (2002) Précipitations exceptionnelles et crues éclair dans l'aire pyrénéo-méditerranéenne. Géomorph Rel Proc Environ 1:35–50

Copons R, Bordonau J (1994) La Pequeña Edad del Hielo en el macizo de La Maladeta (Alta cuenca del Ésera, Pirineos centrales). In: Martí Bono CE, García Ruiz JM (eds) El glaciarismo surpirenaico: Nuevas aportaciones. Geoforma Ediciones, Logroño, 111–124

Crest Y (2018) Quantification de la dénudation glaciaire et postglaciaire dans l'orogène pyrénéen. Bilans comparés parmi des cirques et des petits bassins versants en contexte climatique océanique et méditerranéen à l'aide des nucléides cosmogéniques produits in-situ et de mesures topométriques sous SIG. PhD thesis (unpubl.), Université de Perpignan Via Domitia, 426 p

Crest Y, Delmas M, Braucher R, Gunnell Y, Calvet M, ASTER Team (2017) Cirques have growth spurts during deglacial and interglacial periods: evidence from ^{10}Be and ^{26}Al nuclide inventories in the central and eastern Pyrenees. Geomorphology 278:60–77

Delmas M, Cerdan O, Cheviron B, Mouchel JM, Eyrolle F (2012) Sediment export from French rivers to the sea. Earth Surf Proc Land 37:754–762

Dupias G (1985) Végétation des Pyrénées, Notice détaillée de la partie pyrénéenne des feuilles 69 Bayonne-70 Tarbes-71 Toulouse-72 Carcassonne-76 Luz-77 Foix-78 Perpignan, carte de la végétation de la France au 1:200,000. CNRS, Paris, 209 p

Fabre G (1981) Dissolution spécifique actuelle dans les karsts du sud méditerranéen de la France. Bull Assoc Géogr Fr 482:3–7

Fernandes M, Palma P, Lopes L, Ruiz-Fernández J, Pereira P, Oliva M (2018) Spatial distribution and morphometry of permafrost-related landforms in the Central Pyrenees and associated paleoclimatic implications. Quatern Int 470:96–108

Feuillet T (2011) Statistical analyses of active patterned ground occurrence in the Taillon Massif (Pyrénées, France/Spain). Permafrost Periglac Process 22:228–238

Feuillet T, Mercier D (2012) Post-Little Ice Age patterned ground development on two Pyrenean proglacial areas: from deglaciation to periglaciation. Geogr Ann Ser B 94:363–376

Galop D, Cugny C, Rius D (2008) Rythmes et ruptures dans l'histoire de l'anthropisation du massif pyrénéen à partir des données polliniques. In: Canérot J, Colin J-P, Platel J-P, Bilotte M (eds) Pyrénées d'Hier et d'Aujourd'hui. TOTAL, BRGM, AGSO, AIPT, CNRS, ed. Atlantica, Biarritz, pp 181–190

Galop D (1998) La forêt l'homme et le troupeau dans les Pyrénées. 6000 ans d'histoire de l'environnement entre Garonne et Méditerranée. GEODE, Laboratoire d'Ecologie Terrestre et FRAMESPA, Toulouse, 303 p

García-Ruiz JM, Palacios D, de Andrés N, Valero-Garcés BL, López-Moreno J, Sanjuán Y (2014) Holocene and 'Little Ice Age' glacial activity in the Marboré Cirque, Monte Perdido Massif, Central Spanish Pyrenees. The Holocene 24:1439–1452

Gaussen H (1926) Végétation de la moitié orientale des Pyrénées, sol, climat, végétation. Carte prod. végét., série Pyrénées, vol I, 564 p

Gaussen H (1934) Carte de la pluviosité annuelle du Sud-Ouest de la France et des Pyrénées. 4 sheets, Paris

Gellatly AF, Grove JM, Switsur VR (1992) Mid-Holocene glacial activity in the Pyrenees. The Holocene 2–3:266–270

Genti M (2015) Impact des processus de surface sur la déformation actuelle des Pyrénées et des Alpes. PhD thesis, Université de Montpellier, 247 p

Guilaine J (2003) De la vague à la tombe. La conquête néolithique de la Méditerranée. Le Seuil, Paris, 375 p

Huc S (2010) Éboulis mobiles et marqueurs biogéographiques : le cas de la haute montagne des Pyrénées orientales, PhD thesis (unpubl.), Université de Perpignan, 2 vols., 481 and 155 p

Jalut G (1977) Végétation et climat des Pyrénées Méditerranéennes depuis quinze mille ans. Archives d'Ecologie Préhistorique, 2 vols., 141 p

Julián A, Chueca J (1998) Le Petit Âge Glaciaire dans les Pyrénées Centrales méridionales: estimation des paléotempératures à partir d'inférences géomorphologiques. Sud-Ouest Eur 3:79–88

Kessler J, Chambraud A (1986) La météo de la France. Tous les climats localité par localité. J C Lattès, Paris, 312 p

Lampre Vitaller F (1994) La línea de equilibrio glacial y los suelos helados en el macizo de la Maladeta (Pirineo Aragonés): evolución desde la Pequeña edad de hielo y situación actual. In: Martí Bono CE, García Ruiz JM (eds) El glaciarismo surpirenaico: Nuevas. Geoforma Ediciones, Logroño, 125–142

Maire R (1990) La haute montagne calcaire. Karsts, cavités, remplissages, Quaternaire, paléoclimats. Karstologia Mém 3, 731 p

Molliex S, Rabineau M, Leroux E, Bourlès DL, Authemayou C, Aslanian D, Chauvet F, Civeta F, Jouët G (2016) Multi-approach quantification of denudation rates in the Gulf of Lion source-to-sink system (SE France). Earth Planet Sci Lett 444:101–115

Mussot R, Kuzucuoglu C, Grisoni J-M (1990) L'impact d'un barrage et de la construction de seuils d'arrêt sur la sédimentation et la morphologie au fond de la retenue. Rev Géogr 30:149–167

Pardé M (1941) La formidable crue d'octobre 1940 dans les Pyrénées-Orientales. Rev Géogr Pyrén Sud-Ouest, XII:237–279

Probst J-L, Bazerbachi A (1986) Transports en solution et en suspension par la Garonne supérieure. Bull Sci Géol 39:79–98

Rendu C (2003) La montagne d'Enveitg, une estive pyrénéenne dans la longue durée. Trabucaire, Canet-en-Roussillon, 606 p

René P (2011) Régression des glaciers pyrénéens et transformation du paysage depuis le Petit Âge Glaciaire. Sud-Ouest Eur 32:5–19

René P (2008) Les glaciers actuels des Pyrénées. In: Canérot J, Colin J-P, Platel J-P, Bilotte M (eds) Pyrénées d'Hier et d'Aujourd'hui. TOTAL, BRGM, AGSO, AIPT, CNRS, ed. Atlantica, Biarritz, pp 163–176

René P (2013) Glacier des Pyrénées–Le réchauffement climatique en images. Cairn, Pau and Parc national des Pyrénées, 168 p

René P (2014) Le suivi des glaciers dans les Pyrénées françaises. La Météorologie 85:27–34

Rico I, Izagirre E, Serrano E, López-Moreno JI (2017) Current glacier area in the Pyrenees: an updated assessment 2016. Pirineos 172:e029. https://doi.org/10.3989/Pirineos.2017.172004

Serrano E, Agudo C (2004) Glaciares rocosos y deglaciación en la alta montaña de los Pirineos aragoneses (España). Bol Real Soc Esp Historia Nat (sección Geología) 99:159–172

Serrano E, Agudo C, Delaloye R, Gonzalez-Trueba JJ (2001) Permafrost distribution in the Posets massif, central Pyrenees. Norsk Geogr Tidsskr 55:245–252

Serrano E, San José JJ, Agudo C (2006) Rock glacier dynamics in a marginal periglacial high mountain environment: Flow, movement (1991–2000) and structure of the Argualas rock glacier, the Pyrenees. Geomorphology 74:285–296

Serrano E, Oliva M, González-García M, López-Moreno JI, González-Trueba J, Martín-Moreno R, Gómez-Lende M, Martín-Díaz J, Nofre J, Palma P (2018) Post-little ice age paraglacial processes and landforms in the high Iberian mountains: a review. Land Degrad Dev 29:4186–4208

Serrat P (1999) Dynamique sédimentaire actuelle d'un système fluvial méditerranéen : l'Agly (France). C R Acad Sci Paris IIa 329:189–196

Serrat P, Wolfgang L, Navarro B, Blazi JL (2001) Variabilité spatio-temporelle des flux de matières en suspension d'un fleuve côtier méditerranéen: la Têt (France). C R Acad Sci 333:389–397

Solé Sabaris L (1951) Los Pirineos, el medio y el hombre. Editorial Martin, Barcelona, 624 p

Soutadé G (1980) Modelés et dynamiques actuelles des versants supraforestiers des Pyrénées orientales. Imprimerie Coopérative du Sud-Ouest, Albi, 452 p

Soutadé G (1993) Les inondations d'octobre 1940 dans les Pyrénées-Orientales. Conseil Général, Direction des Archives départementales, Perpignan, 351 p

Tejero P, García MB, Gómez D (2017) Spatial distribution and environmental description of the endemic flora of the Pyrenees. Pirineos 172. https://doi.org/10.3989/pirineos.2017.172006

Vanara N (2000) Le karst des Arbailles. Karstologia Mém 8, 320 p

Vanara N, Maire R, Lacroix J (1997) La surface carbonatée du massif des Arbailles (Pyrénées Atlantiques): un example de paléoréseau hydrographique néogéne déconnecté par la surrection. Bull Soc Géol Fr 168:255–265

Synthesis and Overview

5

5.1   History of the Mountain Range

From the K–T boundary to the present, the Pyrenees as a mountain range rising between Europe and Iberia went through three successive states: the Proto-Pyrenees, the Ancestral Pyrenees, and the Modern Pyrenees. Throughout this period, the mountain range changed in length, width, relief and elevation.

At the time of the K–T boundary, a mountain range defined here as Proto-Pyrenees began to emerge, but it coincided with the eastern third of the modern Pyrenees and extended at the time farther east into the area now occupied by the Gulf of Lion. The west-central and western Pyrenes did not exist as a mountain range at the time. The Garumnian sequences constitute the clastic fore-land deposits of these eroding Proto-Pyrenees, and they have been preserved in the north in the Corbières (Itinerary 1) as well as in the south between Tremp and Pedraforca (Plaziat 1981, 1984; Rosell et al. 2001; Itinerary 9). High-energy denudation ca. 78 Ma, probably driven by high relief, has been inferred from fission-track evidence in U–Pb-dated detrital zircons in the Iberian fore-land sequences (Whitchurch et al. 2011), and likewise between 80 and 68 Ma from helium ages on detrital zircon crystals (Filleaudeau et al. 2011). This record of rapid denudation has been confirmed by late-Cretaceous to Paleocene zircon cooling ages in bedrock in the NE Axial Zone and North Pyrenean Zone. Denuda-tion of the Proto-Pyrenees was sufficiently advanced by Paleogene time that part of the orogen became drowned by sea level rise and covered by the Ilerdian (~56–52 Ma) carbonate platform.

These Proto-Pyrenees extended westward during the Eocene as a consequence of continuing tectonic inversion from east to west of the Cretaceous flysch-filled

© Springer Nature Switzerland AG 2022
M. Calvet et al., *Geology and Landscapes of the Eastern Pyrenees*, GeoGuide,
https://doi.org/10.1007/978-3-030-84266-6_5

pro- and retro-foredeeps. Still by the end of Lutetian time, the emergent successor mountain range, i.e., the Ancestral Pyrenees, nonetheless does not appear to
have extended westward beyond the longitude of Pic d'Anie (0°43' W). Denudation of the thickening crust supplied thick clastic mountain-front megafans. The
debris were delivered in a terrestrial environment to the retro-foreland (Palassou
sequences) and in a deltaic environment to the pro-foreland. Crustal denudation
of the Axial Zone resulted in part from the north- and southward transport of the
cover nappes, but also as a consequence of subaerial denudation. The Hercynian
basement core of the orogen first gained exposure during the Bartonian, after
which time the delivery of granite and gneiss pebbles to the piedmonts occurred
regularly, often in periodic pulses recorded in the Paleogene conglomerate stratigraphies of both forelands. The Ancestral Pyrenees appear to have attained their
peak energy and peak topography between the Priabonian and the Oligocene, at
a time when the Ebro Basin also was entering its ~20–25-million-year period of
internally drained confinement and when the mountain-front fan systems on both
sides of the orogen became ubiquitous.

5.2   Similarities Between the Pro- and Retro-Foreland Basins

The Ebro and Aquitaine basins have nearly always been dealt with separately
in the literature, and yet a number of rarely noted correlations are interesting to
emphasise. For example, it is intriguing to note that the eastern limit of Neogene
outcrops in the Ebro Basin occurs almost exactly at the same distance from the
Mediterranean shoreline as in the Aquitaine Basin (Fig. 1.1). The consistency of
this region-wide feature is perhaps a far-field response to Mediterranean rifting
farther east and to the ensuing uplift, which most likely resulted in displacing both
the foreland and retro-foreland depocentres towards the west (Gaspar-Escribano
et al. 2001).

Likewise it is striking to highlight the synchronised onset of fluvial incision
in the Ebro after 10 Ma with similar processes in the Aquitaine basins. The
moment when the Ebro returned to external drainage after 25 Ma of hydrological confinement coincided with (i) the major episode of fluvial incision recorded
in the Aquitaine Basin by the ravinement surface beneath the widespread Vallesian 'Clay-with-pebbles' formation; and with (ii) the resumption of coarse clastic
delivery to the piedmont, of which the Lannemezan and Ger megafans are the
best-preserved legacies (Figs. 2.2 and 3.6). Much rather than a chance or contingent event such as drainage piracy restricted to the Ebro River, such symmetry

in the response of the two foreland basins to fluvial incision strongly suggests a regional episode of crustal uplift occurring not just in the Pyrenees but also more widely in Iberia and across to the Massif Central (Boschi et al. 2010; Malcles et al. 2020). The hypothesis of mantle-supported dynamic topography (Casas-Sainz and de Vicente 2009; Gunnell et al. 2008, 2009; Boschi et al. 2010; Calvet et al. 2015; Conway-Jones et al. 2019; Calvet et al. 2021) contributing to re-energise not just the Pyrenean mountain belt during the Neogene and Quaternary period, but also the pro- and retro-foreland basins themselves, showcases the Pyrenees as a transient orogen that has undergone a complex history of topographic uplift, decay, and resurgence, and fits the scenario of the Modern Pyrenees as a youthful Neogene mountain range successor to an older, Paleogene ancestor (the Ancestral Pyrenees) (Calvet et al. 2021).

Some uncertainty remains over other features of the geological record of the Ancestral Pyrenees. For example, no chronostratigraphic correspondence has ever been established between the four generations of Palassou (and Jurançon) clastic pulses in the retro-wedge and the conglomeratic sequences of the Iberian pro-wedge, which have instead received a collection of local names in the literature. This, however, does not mean that the two forelands did not respond synchronously to common drivers. Another apparent asymmetry between the pro- and retro-foreland basins lies in the sharp difference in cumulative thickness of the proximal conglomeratic (and more widely: detrital) sequences, which turn out to be considerably greater in the Ebro foreland. This contrast, however, may just be a spurious paradox. Four explanations would support this caveat: (i) the pro-wedge watersheds have been considerably larger that their retro-wedge counterparts, thus with more potential for sediment delivery; (ii) in the Aquitaine Basin, the rivers were afforded uninterrupted opportunities to export at least their finer clastic loads to the Atlantic, whereas the Ebro Basin was denied all opportunities of sediment bypass for 25 million years after the Priabonian; (iii) the outward growth of the orogenic wedge during the Paleogene was much more extensive on the south side of the orogen (e.g., producing the South-Pyrenean Central Unit), with associated crustal thickening thus promoting greater potential energy and denudation; (iv) the Aquitaine foreland has undergone limited uplift, and its Paleogene conglomerate sequences accordingly feature few good outcrops, whereas the Ebro foreland has risen to elevations of up to 2000 m and been incised by valleys 1 km deep, thereby affording an exceptionally rich display of stratigraphic exposures.

The multiple and abrupt breaks in the geological record ca. 10 Ma in all three piedmont environments of the orogenic wedge (Aquitaine, Mediterranean,

Ebro) turned a definitive page in the history of the Ancestral Pyrenees. The characteristics of the successor mountain range were heralded by major changes in morphotectonic regime, involving: (i) a relative relaxation of Cenozoic crustal convergence intensity; (ii) a return to external drainage in the Ebro Basin; (iii) the opening of new extensional basins far into the Axial Zone, i.e., in locations progressively more remote from the Mediterranean back-arc basin; (iv) the impact of continuous volcanic activity in the eastern part of the mountain belt; (v) a symmetric interruption in the stratigraphy of the Ebro Basin (which became exorheic after ~10 Ma) and in the Aquitaine Basin (major unconformity beneath the Lannemezan megafan and its coeval analogues along the strike of the mountain range); and (vi) the construction of increasingly voluminous silicilastic wedges on the continental shelves of the Aquitaine, Valencia, and Gulf of Lion marine domains.

Contrary to the Aquitaine Basin, fluvial incision of the Ebro Basin after ~10 Ma was interrupted by only very few more quiescent moments of alluvial aggradation and terrace formation, which explains why the Ebro Basin displays few late Neogene and Quaternary aggradational terraces compared to the Aquitaine Basin (Fig. 3). The Ebro Basin likewise displays no analogues of the Lannemezan Formation (Fig. 2) or of the multiple other late-Neogene megafans that at one time buried considerable portions of the retro-wedge. This contrast is likely explained by regional uplift to >800 m a.s.l. of the Ebro Basin itself, following a different style and greater magnitude than the more simple basinward tilt motion recorded in Aquitaine, where the piedmont grades more progressively to its marine base level.

This broad sketch of Pyrenean palaeogeography gains further detail and finer resolution from (i) a documentation by thermochronological methods of rock-cooling patterns within various massifs and valleys of the modern mountain range; and from (ii) the geomorphological information encoded in various suites of mountain and piedmont landforms, which constitute critical building blocks for producing plausible scenarios of long-term landscape evolution. These scenarios are detailed in the following nine itineraries.

References

Boschi L, Faccena C, Becker TW (2010) Mantle struture and dynamic topography in the Mediterranean. Geophys Res Lett 37:L20303. https://doi.org/10.1029/2010GL045001
Calvet M, Gunnell Y, Farines B (2015) Flat-topped mountain ranges: their global distribution and value for understanding the evolution of mountain topography. Geomorphology 241:255–291

Calvet M, Gunnell Y, Laumonier B (2021) Denudation history and palaeogeography of the pyrenees and their peripheral basins: an 84-million-year geomorphological perspective. Earth-Sci Rev 215:103436

Casas-Sainz AM, de Vicente G (2009) On the tectonic origin of Iberian topography. Tectonophysics 474:214–235

Conway-Jones BW, Roberts GG, Fichtner A, Hoggard M (2019) Neogene epeirogeny of Iberia. Geochem Geophys Geosyst 20:1138–1163

Filleaudeau P-Y, Mouthereau F, Pik R (2011) Thermo-tectonic evolution of the south-central pyrenees from rifting to orogeny: insights from detrital zircon U/Pb and (U-Th)/He thermochronometry. Basin Res 23:1–17

Gaspar-Escribano J, van Wees JD, Ter Voord M, Cloetingh S, Roca E, Cabrera L, Muñoz JA, Ziegler PA (2001) Lithospheric structure of the Ebro Basin (NE Spain): constraints from 3D flexural modeling. Geophys J Int 145:349–367

Gunnell Y, Calvet M, Brichau S, Carter A, Aguilar JP, Zeyen H (2009) Low long-term erosion rates in high-energy mountain belts: insights from thermo- and biochronology in the eastern pyrenees. Earth Planetary Sci Lett 278:208–218

Gunnell Y, Zeyen H, Calvet M (2008) Geophysical evidence of a missing lithospheric root beneath the eastern pyrenees: consequences for post-orogenic uplift and associated geomorphic signatures. Earth Planetary Sci Lett 276:302–313

Malcles O, Vernant P, Chéry J, Camps P, Cazes G, Ritz JF, Fink D (2020) Determining the plio-quaternary uplift of the southern French massif central; a new insight for intraplate orogen dynamics. Solid Earth 11:241–258

Plaziat JC (1981) Late Cretaceous to late Eocene paleogeographic evolution of southwest Europe. Palaeogeogr Palaeoclimatol Palaeoecol 36:263–320

Plaziat JC (1984) Stratigraphie et évolution paléogéographique du domaine pyrénéen de la fin du Crétacé (phase Maastrichtienne) à la fin de l'Éocène (phase Pyrénéenne). Univ, Paris Sud, 1362 p

Rosell J, Linares R, Llompart C (2001) El 'Garumniense' prepirenaico. Rev Soc Geol España 14:47–56

Whitchurch AL, Carter A, Sinclair HD, Duller RA, Whittaker AC, Allen PA (2011) Sediment routing system evolution within a diachronously uplifting orogen: insights from detrital zircon thermochronological analyses from the South-Central Pyrenees. Am J Sci 311:442–482

Part II
Field Itineraries

From Perpignan to Narbonne by the coast, return through the Corbières (170 km, excluding optional loops; allow 1–3 days depending on the number of stops and optional excursions). See Fig. 6.1.

Itinerary 1 showcases regional tectonic and sedimentary evolution during the Oligocene and Miocene (see Box 6.1). The successive stops track landscape evolution and base-level changes during the Neogene by examining fault systems, sediment lithostratigraphy and provenance, and erosion surface vestiges cross-cutting folds and structures of the North-Pyrenean and Sub-Pyrenean zones. Stops include sites where the planar land surfaces have been dated by various methods, and where their geodynamic and palaeoclimatic context can be reconstructed from a variety of environmental archives. Insights are also gained into the Pleistocene and Holocene evolution of the coastal plains, highlighting the recurrence of catastrophic flooding in the valleys.

Box 6.1 The Western Mediterranean rift system
The tectonic basins of Roussillon, Lapalme, and Sigean–Narbonne (see also Box 6.2) are onshore examples of the Cenozoic Western Mediterranean rift system, a southern extension of the European Cenozoic Rift System (Sissingh 2006) which underwent rapid crustal thinning and extension during the Oligocene and lower Miocene when Sardinia and Corsica rifted away from this coastline (Bache et al. 2010; Calvet et al. 2021). As the lithospheric slab currently beneath the Italian peninsula rolled back from the Iberian continent (Fig. 6.2), the Gulf of Lion developed as a back-arc basin in which a NW to SE gradient of intensified crustal extension

© Springer Nature Switzerland AG 2022   129
M. Calvet et al., *Geology and Landscapes of the Eastern Pyrenees*, GeoGuide,
https://doi.org/10.1007/978-3-030-84266-6_6

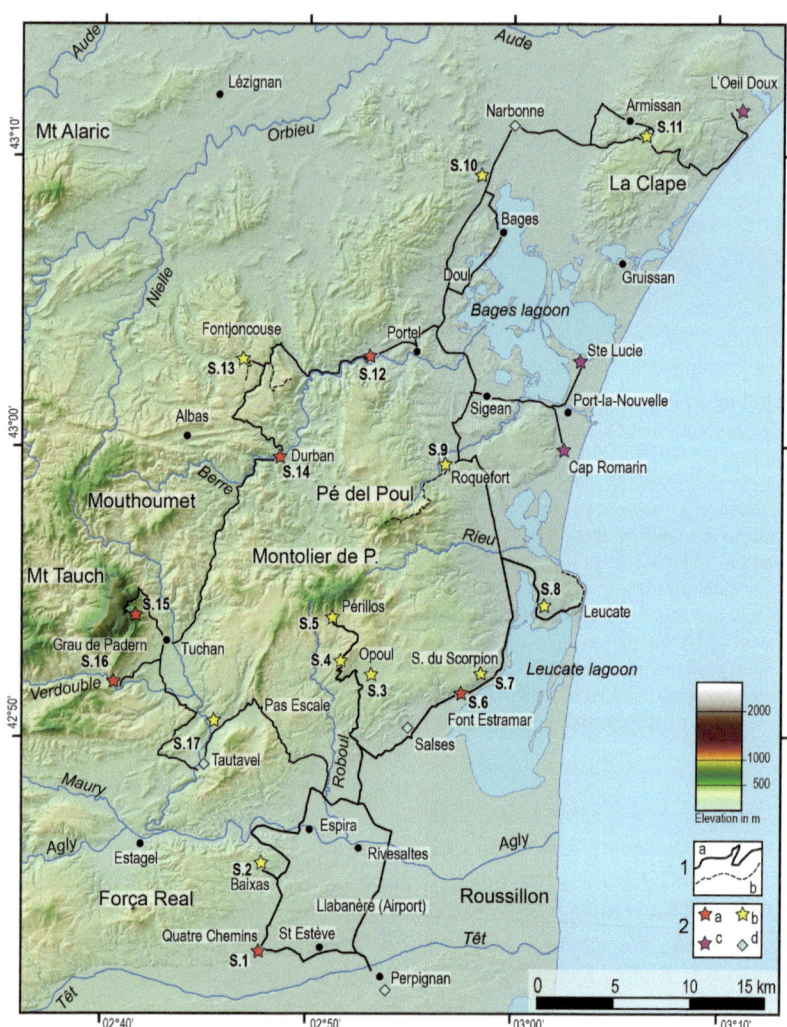

Fig. 6.1 Itinerary 1: route map and overview. Key to symbols—**1**: roads and trails. **1a**: roadways, irrespective of road rank, size, or quality. **1b**: trailwalks. **2**: Stopping points (outcrops, exposures, landforms, or landscapes). **2a**: roadside stops. **2b**: roadside stops involving a short walk, or marking the start of a more major trailwalk. **2c**: subsidiary stops. **2d**: cultural stops. Labels S.1 to S.17 locate the sequence of roadside stops

has been inferred; i.e., thinned upper crust in the eastern Pyrenees yields progressively offshore to lower crustal necking, then to exhumed lower crust and serpentinised upper mantle in the Provençal Basin (the presence of an aborted oceanic ridge between Provence and Corsica is currently speculative; Granado et al. 2016; Canva et al. 2020; Jolivet et al. 2020). Mediterranean extension began with intracontinental rifting and continued with marine flooding. Onlap of the Miocene transgression eventually sealed the basin boundary fault lines and established base-level controls on denudation in the hinterland. The sedimentary fill of the Corbières grabens consists of continental and paralic deposits (lacustrine limestone, shale, evaporites, with conglomerate and breccia along the faulted boundaries). These sequences have been dated on the basis of late Oligocene and early Miocene mammalian faunas remains contained in the sediments (Aguilar 1977) and are covered by Burdigalian to Langhian marine, sandstone-textured molasse.

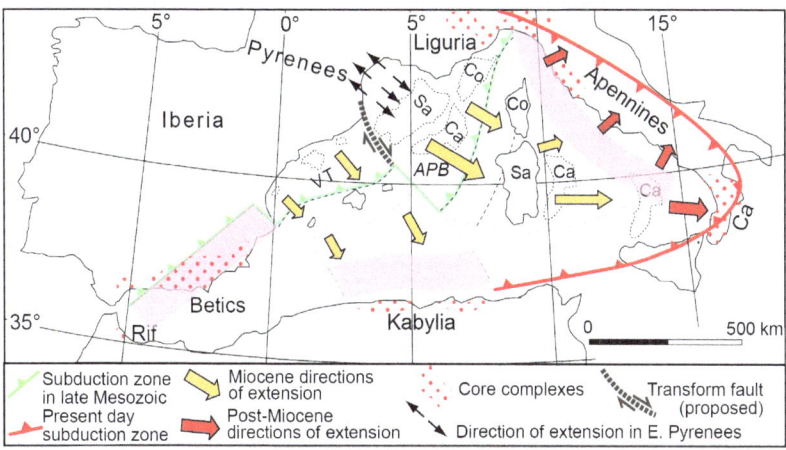

◄**Fig. 6.2** **Schematic interpretation of the Western Mediterranean subduction system and some kinematic consequences of slab rollback: rifting and formation of a back-arc basin** (after Gunnell et al. 2008, modified). APB: Algero-Provençal Basin; VT: Valencia Trough; TYR: Tyrrhenian basin. In Middle Cenozoic time, the subduction zone of Mesozoic oceanic Tethyan lithosphere beneath the eastward-migrating Iberian microplate was segmented by two transform fault zones. The leading edge of Mesozoic lithosphere has since migrated eastward to the Italian peninsula, with slab rollback causing back-arc basins to open in its wake (APB during the Miocene, and TYR after Miocene time) and the Corsica–Sardinia–Calabria blocks (Co, Sa, Ca) to rotate counter-clockwise. Areas interpreted to have been generated by slab breakoff during rollback (pink shading) exhibit rapidly formed metamorphic core-complexes (after Zeck 1999; Séranne 1999). Note the logic requiring the existence of a transform fault separating the Tethyan subduction domain from the eastern edge of the Pyrenean continental subduction zone (see Part I)

The itinerary proceeds initially through the floodplain gravel deposits of the Têt River (Fig. 6.3), which overlie the Pliocene continental sequences of the Roussillon Basin (see Itinerary 2). From Perpignan to Saint-Estève, road D616 follows the lowermost alluvial terrace tread of the Têt, with land use dominated here by irrigated market gardens. This Holocene and modern alluvial plain, labelled T0, consists of pebble-filled channels and a thick layer of sandy silt (Fig. 6.4B). Several generations of inset alluvial terraces of early Holocene to historical age occur here, but agricultural land use and urbanisation make these difficult to discern. Still today, a proportion of these deposits gets displaced and redistributed by the largest floods (e.g., in October 1940; see also Itineraries 3 and 4). The lower Late Pleistocene terrace, T1, upstream of Baho (Fig. 6.4A), is barely distinguishable in the landscape because of its small vertical offset, of 5 m at most, above the younger levels.

Between Saint-Estève and Baho, the road cuts into the edge of T3 (Fig. 6.3) and exposes its rubified, deeply weathered materials among which red or ochre patinated quartz pebbles form the only distinctively alluvial signature (Fig. 6.5A, B). These materials occur as channel fill in the silt- and sand-rich continental Pliocene deposits, effectively defining a ravinement surface. Road D616, which subsequently continues as the D614, stays on the middle terraces T3c and T3d (Middle Pleistocene); the tread of T3 actually consists of 4 inset gravel beds (from top to bottom: T3d, c, b, a), one of which (T3b) has yielded an Electron Spin Resonance age of 374 ± 47 ka (Delmas et al. 2018). This alluvial landscape is entirely covered in vineyards.

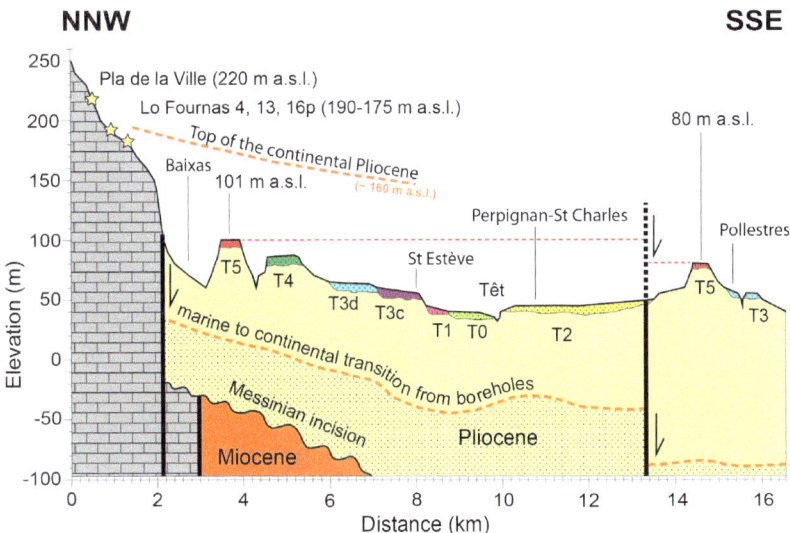

Fig. 6.3 Cross-profile of the Têt terrace sequence in the Roussillon Basin. Yellow stars locate late Pliocene karst fissure fillings. Alluvial sequence thicknesses and dips of the marine-to-continental Pliocene stratigraphic boundary are inferred from the French national subsoil data bank (http://infoterre.brgm.fr/) (after Delmas et al. 2018)

6.1 Baixas Plateau

Stop 1. Quatre-Chemins crossroads (Quatre Camins, roads D616 and D614)
Ascend the terrace situated to the south of the road cut. From there you gain a panoramic view over the modern alluvial plain as well as of terrace T3d. Sections into T3d are visible in both road embankments. Note the relative enrichment in quartz pebbles near the surface.

The D614 towards Baixas initially crosses the tread of T3d, then climbs up a low riser to T4—a well preserved, flat, but currently undated terrace (estimated age: Gelasian to early Middle Pleistocene). Baixas lies at the boundary between the Pliocene deposits of the Roussillon Basin and the Jurassic limestone of the Cor-bières. From the village, follow the Chemin de Sainte-Catherine, then access the plateau (Pla de la Vila) on foot (30 min round trip) along a trail behind the chapel, then up a road track which passes a small, disused quarry. This quarry (Fig. 6.6A; 42°45′27.9″N, 02°47′41.3″E) mines the Paleogene limestone breccia (noted 'eBr'

Fig. 6.4 **Lower alluvial deposits near Pézilla-la-Rivière.** **A** Section through T1 north of the D614; this exposure shows braided channel deposits and evidence of northward channel migration. **B** Section through the modern floodplain sequence (T0) south of the D614; gravel-filled channels at the base and a thick sand and silt deposit at the top; this upper level has yielded charcoal debris with calibrated radiocarbon ages of 414–544 CE (Lyon-8924(OxA), see Delmas et al. 2018)

Fig. 6.5 Terrace T3. A Panoramic view of terrace T3c at La Llabanère (Perpignan airport); T4 terrace riser and tread beneath the pines in the background. **B** Section through terrace T3c near La Llabanère; note strongly eluviated chromic Luvisol and dominance of quartz pebbles in the A horizon, to the exclusion of any other surviving lithology

*on the geological map), and its western face displays the unconformable (and probably karst-enhanced) contact with the Jurassic dolomite. The clean-cut wall in the breccia reveals a number of vertical fissures filled with micromammalian fossils of Miocene age (*Fig. 6.6B, C).

Stop 2. Baixas Plateau

Baixas Plateau is a typical vestige of Miocene erosion surface in the eastern Corbières. It formed and evolved under the same conditions as the other vestiges encountered later along this itinerary (see Figs. 6.12, 6.13 and 6.14) and in Itinerary 2. As in the case of Pied du Poul and Périllos, Baixas Plateau is overlooked in the north by a razorback ridge in limestone, and further to the west by Força Réal—an inselberg consisting of Lower Paleozoic schist (metagreywacke) and quartzite (see itineraries 2 and 5).

Beyond these landscape-scale similarities, the site here at Baixas (Figs. 6.6 and 6.7) is nonetheless exceptional from a palaeontological as well as a geomorphological perspective. Down to maximum depths of ~50 m (Fig. 6.8C), fissures in the limestone and karren at the surface have yielded more than sixty Neogene sedimentary fill deposits containing rich assemblages of micromammalian remains (Figs. 6.6, 6.7, 6.8 and 6.9). At this site of international importance, these fossil-filled fissures, or FFFs, have contributed substantially to improving the methods of biochronology in general, and to documenting the Neogene continental deposits of southern Europe in particular (Fig. 6.10; Aguilar et al. 1986a,

Fig. 6.6   Fossil-bearing Miocene deposits at Baixas. **A** Disused quarry of Sainte-Catherine. The quarry face displays Paleocene breccia ('Brèche de Baixas') drowning an irregular fossil topography in Lower Jurassic dark limestone; the vertical fissures contain Miocene micro-mammalian fossils. **B** Example of an open fissure formed under the Neogene extensional regime (no. 202) and filled here by three generations of rodent fossil assemblages attributed to MN 3. **C** Fragment of deposit (clay, sand) from the karren surface containing an extremely rich assemblage of micromammalian bone debris (Lo Fournas 3)

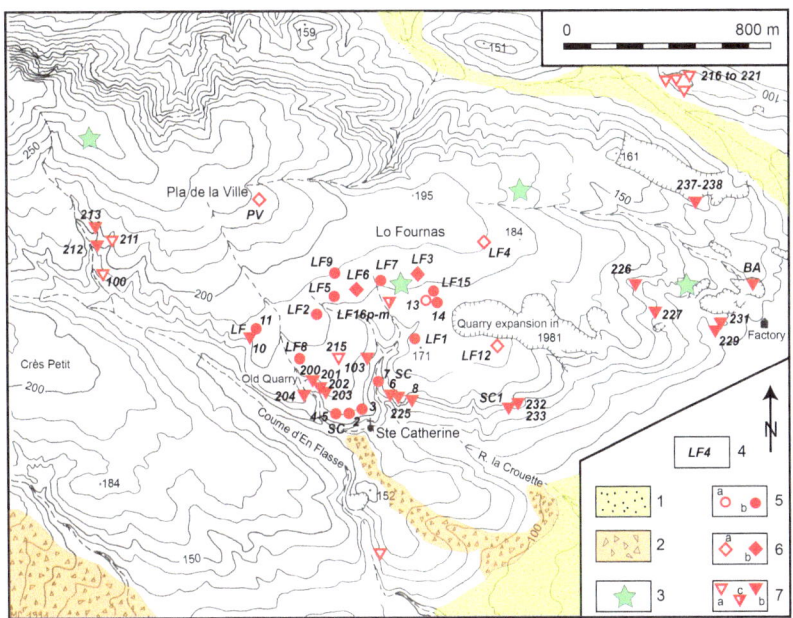

Fig. 6.7 Map of Baixas Plateau and its Neogene fossil-filled fissures, or FFFs (after Calvet 1996, modified and updated). Key to symbols and ornaments—**1**: Pliocene alluvial deposits composed of allochtonous gravels (Agly River palaeovalley on north side of the plateau), sand, and silt (in the south). **2**: locally sourced limestone breccia, rubified and sealed by calcrete; pink-coloured calcrete facies at the entrance to the Sainte-Catherine gully formed during the Messinian Salinity Crisis; the breccia is pre-Pliocene but some of it is Pliocene as it grades laterally to the silt- and sand-textured Pliocene beds that prevail in the Roussillon Basin. **3**: residual alluvium consisting of quartz sand and sparse exogenous pebbles in a red, cemented matrix. **4**: Code name of fossil-bearing sites. **5**: scattered clumps of fossil-bearing fragments derived from the shallow dismantling of surficial accumulations; **5a**, Pliocene; **5b**, Miocene. **6**: horizontal rafts of fossiliferous sediment, sometimes covering the clints of the limestone pavement (e.g., LF3); **6a**, Pliocene, **6b**, Miocene. **7**: Subvertical karstic fissure; **7a**, Pliocene to Quaternary; **7b**, Miocene; **7c**, Pliocene and Miocene. This map, originally produced in 1991, has been updated with newly discovered sites, particularly a 15 m-deep fissure containing two distinct generations of sediment and fossil assemblage—one of Tortonian and the other of mid-Pliocene age (Aguilar et al. 2007)—named LF16a and LF16b, respectively (noted LF16p-m on the geological map). Growth of the quarry has destroyed most of the sites previously named 'Lo Fournas' in the literature (disused quarry in Fig. 6.6A)

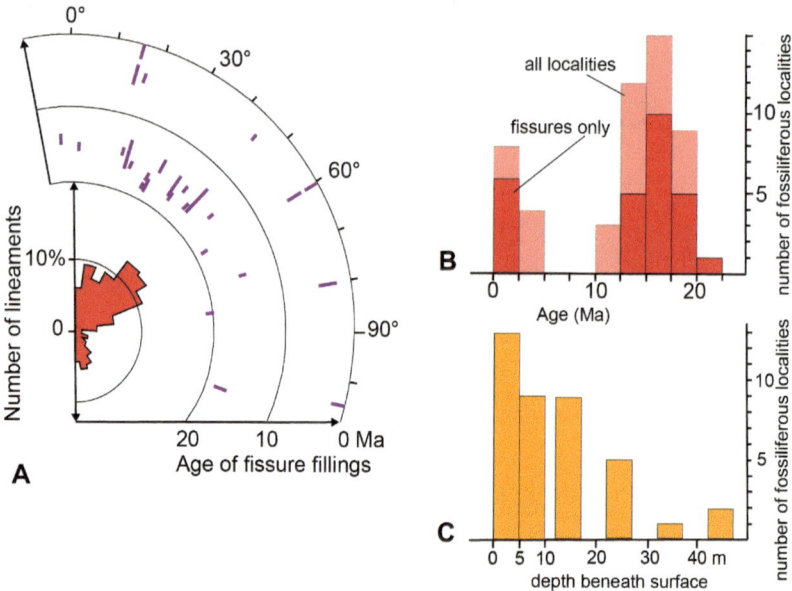

Fig. 6.8 Quantitative attributes of the FFFs (after Calvet 1996 and Faillat et al. 1990, modified). **A** Comparison between lineament directions obtained from photogeological analysis and the age and direction of fossil-filled fissures. The Miocene fissures cluster between N20°E and N40°E, i.e., they show the same direction as the lineaments. These are directly linked to the extensional stress regime prevalent at the time in the Western Mediterranean. Younger fissures exhibit a more dispersed array of directions. **B** FFF age distribution. Note here the lacuna between 5 and 10 Ma (this period is documented at Castelnou in a similar setting; see Itinerary 2). **C** FFF depths beneath the tread of the Miocene pediment; note that the vast majority of sites occur near the topographic surface. Assemblages nearer the surface contain large numbers of bone fragments and well-preserved teeth, forming bone breccia interpreted as barely disintegrated raptor rejection pellets. In the deeper fissures, the fossil debris is more scattered

1986b, 1991, 2007, 2010; Aguilar and Michaux 1987, 1990, 1997; Faillat et al. 1990; Aguilar 2002; etc.). At some FFF sites, the sediment plugs tectonic fissures that did not undergo subsequent widening by solutional processes (Fig. 6.8B); such features can be observed, for example, in the disused quarry of Sainte-Catherine (sites no. 200–203 in the nomenclature of Faillat et al. 1990; Fig. 6.6). Other FFF sites consist of sediment smears and coatings on exhumed karren walls. Many occurrences of this kind have, however, been destroyed by the quarrying; some are nonetheless still visible at the surface, either in situ (LF8, LF10) or as red, fossil-rich blocs in dry stone walls (LF9). The Miocene biochronology obtained from the FFFs records the NW–SE extensional tectonics of the Western Mediterranean (Faillat et al. 1990), whereas the younger fissures exhibit much less uniform directions (Fig. 6.8A).

Despite some uncertainty regarding the reconstruction of Cenozoic palaeoclimates (Aguilar and Michaux 1990; Aguilar et al. 1999), the palaeoecology of the fossil taxa indicates warm subtropical conditions. The evolutionary sequences that form the basis of the regional biochronology suggest alternating phases of drier steppe environments and more humid woodland habitats (Fig. 6.11). From a geomorphological perspective, these sites date back to the shallow denudation that produced the Corbières erosion surfaces (Fig. 6.9). The large proportion of quartz-rich clastic sand and fine gravel admixture to the *terra rossa* matrix that contains the fossil remains supports the idea of limestone pediments displaying shallow surface fissures and forming a sediment transfer zone between the eroding Axial Zone and the aggrading coastal area (Figs. 6.12, 6.13 and 6.14).

The juxtaposition of contrasting FFF biozones across the Baixas Plateau surface suggests extremely shallow denudation, and thus an extraordinarily stable land surface since the beginning of the Miocene. As a result, the ancestral surface, S (which is preserved, e.g., at Pied du Poul and Périllos), the Miocene erosion surface (in this region a collection of limestone pediments labelled P1), and even late Neogene pediment P2, remain largely indistinguishable at Baixas (Calvet 1992, 1996). The impression that the area has operated as a relatively passive sediment transfer zone, perhaps a 'pavement barren' ecosystem, is confirmed by the existence not just of FFF surface sites of widely contrasting ages merely a few metres apart (e.g., Lo Fournas 13a, 13b, 14), but also the occurrence of polygenetic outcrops or fissures—i.e., single cavities presenting their own internal stratigraphy of faunal remains of Tortonian (9.5 Ma) to Pliocene age (3 to 2 Ma) (e.g., Lo Fournas 16: Aguilar et al. 2007). Deeper denudation, whether at or since that time, would have destroyed these palaeontological remains. Pliocene FFFs typically contain quartz-rich alluvial sand, sometimes abundant, and may containing fish teeth (LF 16). These Pliocene occurrences coincide chronostratigraphically and altitudinally with the top of the Pliocene fill in the adjacent Roussillon Basin (elevation: 200–230 m; age: 2–3 Ma; Delmas et al. 2018).

Other fossiliferous deposits of this kind occur on the fringes of the Roussillon Basin (Figs. 6.12 and 6.13). Isolated Miocene sites have been studied on its northern side: Tautavel–Blanquatère, Estagel, Serre de Verges (see Itinerary 5),

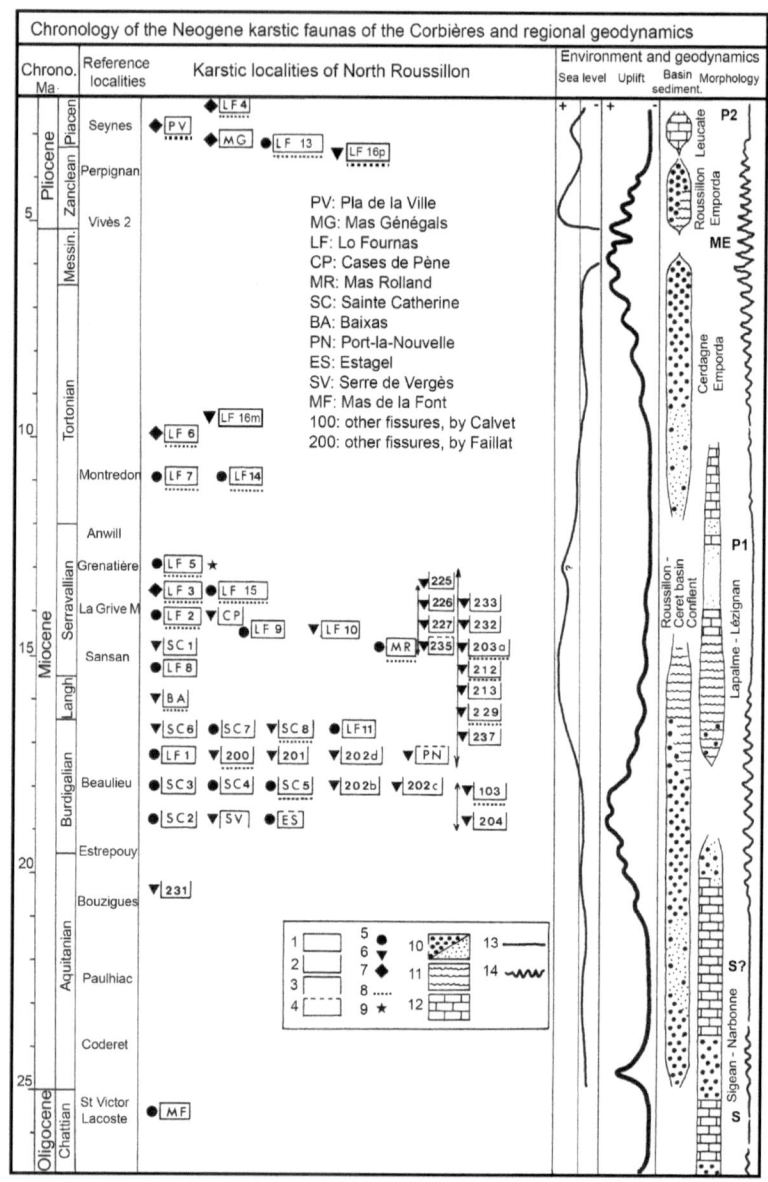

◀**Fig. 6.9 Biochronology of FFFs in the Corbières in relation to regional geodynamics** (after Calvet 1996, modified). Key to symbols and ornaments: Topographic position of the FFFs: **1**: on the surface of pediment P1. **2**: on a shallow valleyside incising P1. **3**: on the slopes of a residual hill rising above P1 (e.g., at Serre de Vergès). **4**: on the boundary fault scarp buried by marine Miocene sediments (e.g., at Port-La-Nouvelle—PN; Font Estramar—no. 235), and at the foot of a hillslope buried by continental Miocene beds (at Estagel—ES). Field setting of fossil remains: **5**: scattered clods of soil lying on the pediment surface. **6**: soil-filled fissures, solution pits, pans, or grikes. **7**: soil coatings on clints in the limestone pavement. **8**: exogenous quartz-rich sand or gravel deposits forming the matrix of the fossil assemblages. **9**: beachrock (marine sand) associated with LF5. Neogene basin lithology: **10**: coarse clastics prevail (large dots); sandy silt prevails (fine stipple). **11**: marine molasse. **12**: continental limestone and marl. Landscape evolution: **13**: planation phase (diffuse hillslope processes dominant). **14**: fluvial incision phase. P1, P2: pediments; S: summit surface; ME: Messinian sequence boundary. Chronology after Aguilar et al. (1986b), Aguilar and Michaux (1987, 1990), and Faillat et al. (1990). This chart does not fully conform to the most recently updated IUGS stratigraphic subdivisions, particularly in the case of the Oligocene and Aquitanian. The position of the FFFs and key sites may also have evolved (see Fig. 6.10)

Cases de Pène, Mas Rolland, Font Estramar, La Nouvelle. To the west, the Thuir–Castelnou plateau surface, which is situated at the eastern termination of the Axial Zone and cross-cuts the dips of Devonian limestone strata, also exhibits a rich cluster of Oligocene to Miocene FFFs (see Itinerary 2).

Leave Baixas on the D18, then the D18a, which follows a Pliocene dry valley containing siliciclastic pebbles conveyed to it by an ancestral Agly River. You will reach the modern Agly channel at Cases de Pène. Take the D117 to Espira. The road follows the axis of an E–W syncline containing a core of Albian black shale. Cross the Agly River at Espira bridge on the D18. Here, at the edge of the Roussillon Basin, you can observe along the left bank and towards the hills to the north (spot elevation: 101 m) a limestone breccia formation resting unconformably on the Albian shale. The breccia has been assigned an Oligocene age, but it could equally be an occurrence of the coarse clastic unit of early Miocene age also observed elsewhere in the Roussillon Basin (e.g., 'Brèche de Thuir', see Itinerary 2). Descending to the Agly riverbed by the first gully downstream of the bridge will reveal an exposure, in the left bank, of coastal shelly Pliocene sediment resting unconformably on the older breccia. As documented by old photographs of the bridge ruins, the Agly channel has undergone substantial changes in the last 80 years: the thick pebble mass still functioned as an aggradational braided system at the time of the 1940 flood, but it has since been stripped away; as a result, channel incision is 5 m deep for several kilometres.

Berggren et al. 1995 modified		FOSSIL MICROMAMMALIAN LOCALITIES		MN Zones
Time (Ma)	Stages	Southern France Roussillon *Fissure fillings	Reference localities in Europe	
5	Pliocene			
6	MIOCENE Upper — Messinian			
7		Castelnou 3*	La Alberca	MN 13
8	Tortonian			
9		Castelnou 1*	Los Mansuetos	MN 12
		Lo Fournas 16-M*	Mollon	MN 11
10		Lo Fournas 6*		
		Lo Fournas 7*	Montredon	MN 10
11		Castelnou 1b*	Jujurieux 1-2 (11.6-11.8 Ma) La Grive L3 *	MN 9
12	Serravallian			
13		Lo Fournas 5*	La Grive M *	MN 8/7
		Lo Fournas 3*		
14		Lo Fournas 2* Cases de Penes*		
	Langhian	Lo Fournas 10*	Anwil	
15		Castelnou 6* Lo Fournas 8*	Sansan (15 Ma)	MN 6
		Blanquatère 3* Baixas*	Vieux-Collonges*	
16		Blanquatère 1*	La Romieu	MN 5
		Ste Catherine 8* Castelnou 2*, 9*		
17	Burdigalian	Lo Fournas 1* Ste Catherine 4* to 7*		MN 4
		Ste Catherine 9*	Beaulieu (17.5 Ma)	
18		Baixas 202b & c*		MN 3
19		Ste Catherine 2*, 3*		
20		Espira du Conflent Serre de Vergès*	Estrepouy	MN 2
21			Laugnac	
22	Aquitanian	Baixas 231*	Bouzigues*	
23				MN 1
24			Coderet*	MN 0 / MP 30
25	OLIGOCENE Upper — Chattian	Mas de las Fons* Castelnou 10*, 13*	Rikenbach	MP 29
26				

Fig. 6.10 An updated biochronological chart for the Neogene of southern France based on a synthesis of micromammalian taxonomy (after Gunnell et al. 2009, modified)

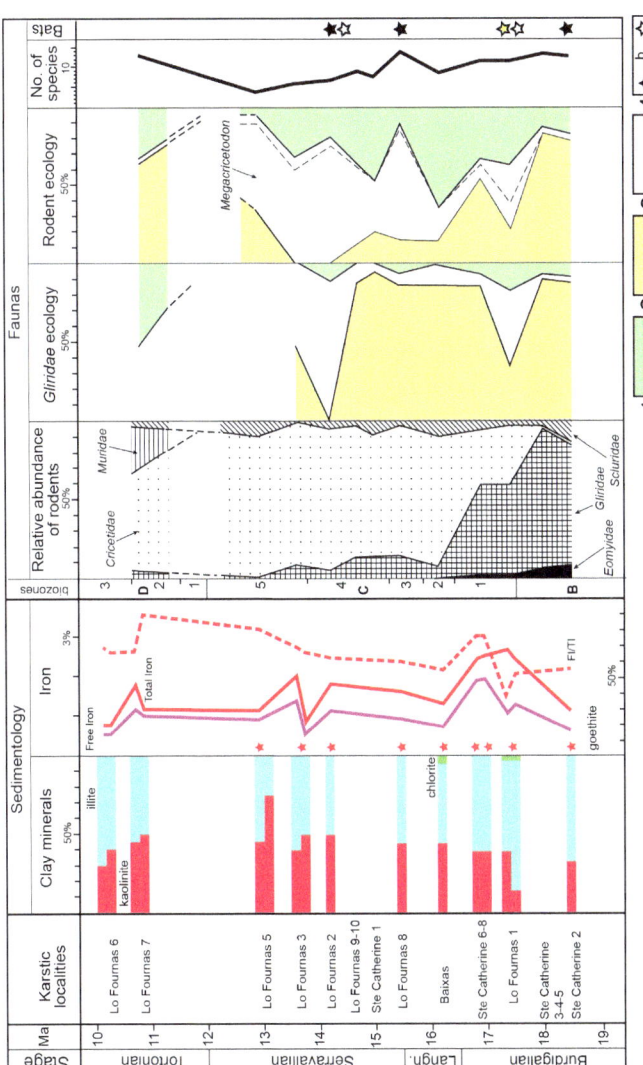

Fig. 6.11 **Environmental change during the Miocene based on micromammalian taxa** (after Calvet 1996, and Aguilar and Michaux 1990). Key to symbols—**1**: humid woodland habitats. **2**: dry open grassland mosaics. **3**: ubiquitous taxa or unknown habitat preference (*Megacricetodon*). **4**: *Chiroptera* (i.e., bats); **4a**: megabats such as fruit bats, suggesting a humid forest environment (black star); **4b**: tomb bat genus (*Thaphozous*) and other specimens of the sac-winged and sheath-tailed bat family (*Emballonurinae*), indicative of warm winters and wooded grassland (yellow star); **4c**: *Asellia* genus, a bat indicative of arid and steppe-like landscapes (white star). This synthesis updated by Aguilar et al. (1999)

Continue along the D18, D12 and D5. These roads follow the Crest de Rivesaltes, a vast gravel fan surface (Roboul alluvial fan) which grades to T2 — a Middle to Late Pleistocene Agly terrace (possibly coeval with MIS 6). Should you wish to skip **Stops 3** *to* **5**, *aim for Salses directly on the D5b.*

6.2   Opoul–Périllos plateau

The D5 follows to its left a line of hills known as Murta–Mas de la Xica. These consist of Neogene sedimentary rocks which have buried a small anticline where the Urgonian limestone core crops out through flanking Albian shale units. The anticline was buried by a sequence of coarse conglomerate of Oligocene to Miocene age, itself deformed and unconformably overlain by coarse-textured Pliocene red beds containing subangular gravel-sized debris. The anticlinal structure has been breached by a water gap—the Roboul gorge—probably as a result of initial drainage superimposition on the widespread, onlapping sediment fill of Pliocene age. The D5 rises to the Opoul Plateau, where a vestige of pediment P1 cross-cuts tight folds in an Albian–Aptian shale, sandstone and limestone sequence. Stop at Moli Nou, facing the village of Opoul.

Stop 3. Opoul plateau and its karst polje
The Opoul polje is countersunk 50 m into the tread of P1. It occupies a fault-angle basin which offsets the Miocene surface vertically (Figs. 6.13 and 6.15A). The floor of this polje cross-cuts the Aptian limestone strata (Urgonian facies) as well as interlayers of sandstone and shale, and the western scarp displays an apron of small rock pediments capped by rubified slope breccia. These features indicate a contribution from fluvial and runoff processes in addition to the water-table-driven rock solution typical of polje environments. Periodic flooding is restricted to the SW of the karstic depression. A discontinuous trail of small sandstone pebbles can be traced across a limestone bench situated on the eastern side of the polje, where the main sinkhole occurs (Fig. 6.15B; the sinkhole, or ponor, is named Aven des Amandiers, on the far side of Agouille du Barrant, 42°52′0.2″N, 02°53′10.4″E; reachable in 20 min return on foot, heading east, from **Stop 3**).

Drive through the village to join up with the D9, then aim for Périllos along the small road that rises up the fault scarp in thick and massive Neocomian–Barremian–Lower Aptian limestones. Once on the flat, parking is available in the intermediate outcrop of Lower Aptian (Bedoulian) shale, which contains urchin and oyster fossils. The

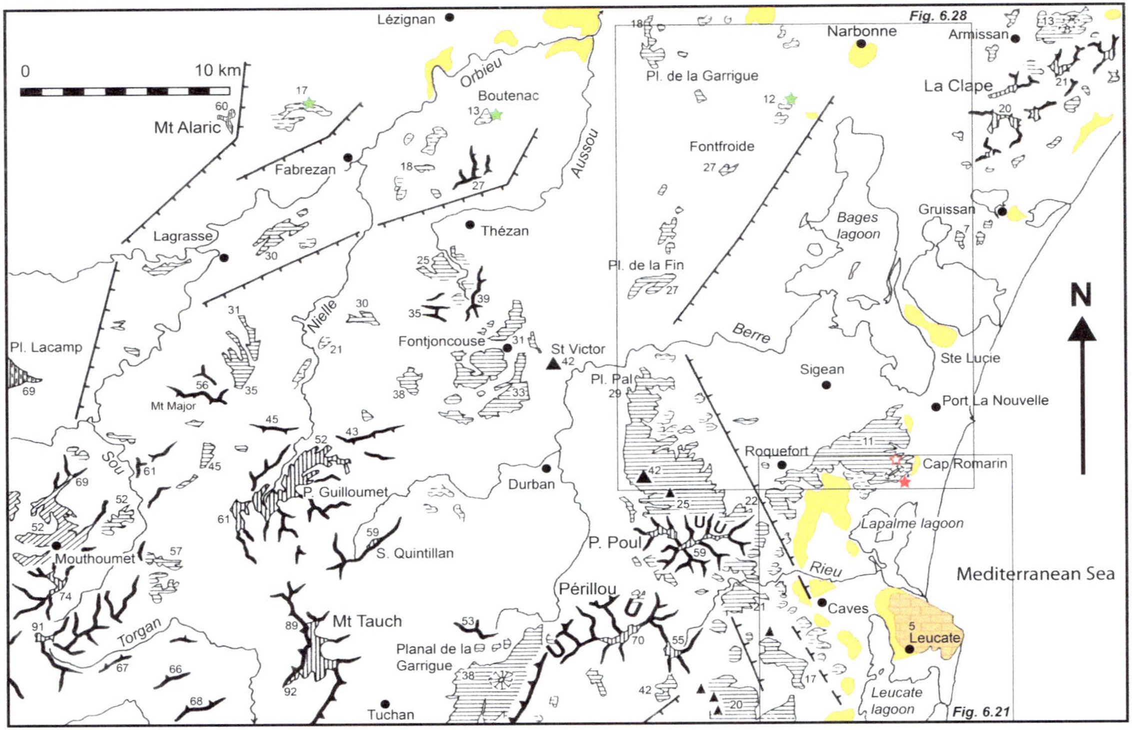

Fig. 6.12 Geomorphological map highlighting the Neogene erosion surfaces of the Corbières (northern sector) (after Calvet 1996, modified). Key to symbols and ornaments is provided with Fig. 6.13

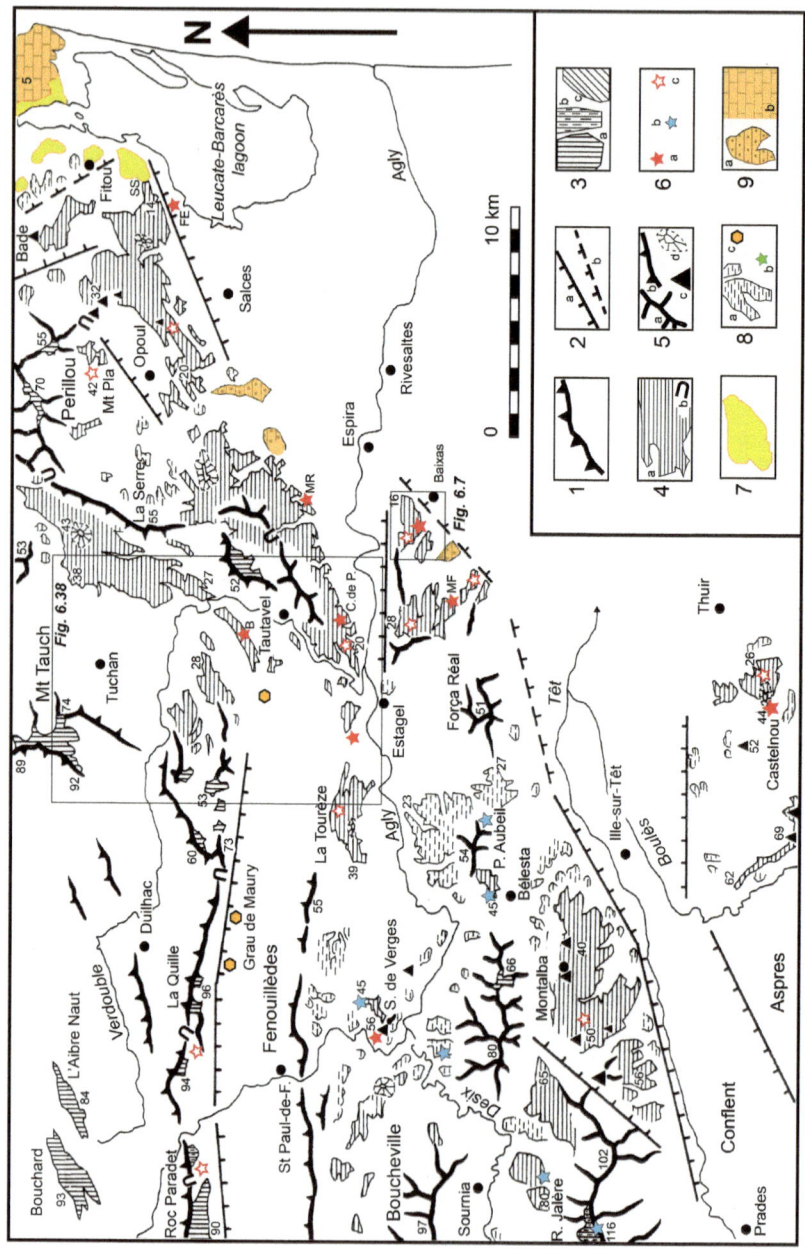

◄**Fig. 6.13 Geomorphological map highlighting the Neogene erosion surfaces of the Corbières (southern sector) and Roussillon Basin (northern edge)** (after Calvet 1996, modified). Altitudes in decametres. Key to symbols and ornaments—**1**: hogback scarp, Paleogene thrust front. **2**: Neogene tectonics; **2a**: normal fault and fault scarp; **2b**: flexure or blind fault. **3**: nature and origin of pre- or early Neogene planation surfaces; **3a**: summit surface (S); **3b**: elevated topography of uncertain status (summit surface, or tilted Neogene pediment P1); **3c**: possible exhumed sub-Cretaceous surface (Mouthoumet massif), locally regraded by components of P1. **4**: Miocene planation surface; **4a**: main Miocene pediment (P1); **4b**: embayments and dry valleys affecting the residual islands of older (generation S) topography. **5**: residual landforms among the Miocene landscape components; **5a**: ridge-and-valley topography in uniform bedrock; **5b**: structurally controlled residual relief rising above P1; **5c**: inselberg rising out of P1; **5d**: low-relief, convexo-concave hill rising out of P1. **6**: surficial deposits and age constraints on P1; **6a**: Miocene micromammalian fossil sites in limestone karst; **6b**: apatite fission-track and (U–Th)/He age data; **6c**: residual siliciclastic alluvium on vestiges of pediment P1. **7**: sandstone-textured marine molasse (mid-Miocene) grading to P1. **8**: late Neogene landforms; **8a**: pediments P2 and P3; **8b**: vestiges of siliciclastic alluvium on the pediments; **8c**: older travertine associated with levels P2 and P3. **9**: Pliocene deposits; **9a**: rubified debris fans (apex grades to P2); **9b**: Leucate marl and travertine

►**Fig. 6.14 Palaeogeography of the eastern Pyrenees during mid-Miocene time** (after Calvet 1996, modified). Key to symbols and ornaments—**1**: Carbonate lakes of Serravallian to Tortonian (i.e., Vallesian) age (provided by the Montredon mammalian fauna). **2**: marine Miocene beds; **2a**: marine domain; **2b**: coastline (pebble beach bars with clasts bearing bivalve-generated boreholes). **3**: continental Miocene beds. **4**: mid-Miocene pediment P1. **5**: residual topography rising above P1 (relative relief of residual landforms given in decametres). **6**: other indices of landscape evolution; **6a**: drainage floodways across the Miocene pediment (reconstruction based on vestiges of exogenous sand and gravel sourced by the Hercynian Mouthoumet massif and the Axial Zone); **6b**: boreholes reporting continental Miocene rocks beneath the younger fill sequences of the Roussillon Basin; **6c**: boreholes reporting marine Miocene deposits. The eastern wells of Elne and Canet document a thick continental sequence situated beneath the mid-Miocene marine beds, but the western well log of Ponteilla only reports continental Miocene deposits (examined in Itinerary 2)

flat butte at Château d'Opoul consists of the topmost plinth of Urgonian limestone (Lower Gargasian).

Stop 4. Château d'Opoul

Excellent view of the Opoul karst polje (Fig. 6.15C), the hangingwall tract of P1, the coastline, and the Roussillon Basin. The 360° panorama is even more complete from the castle itself (1 h walk return). The upthrown portion of Miocene

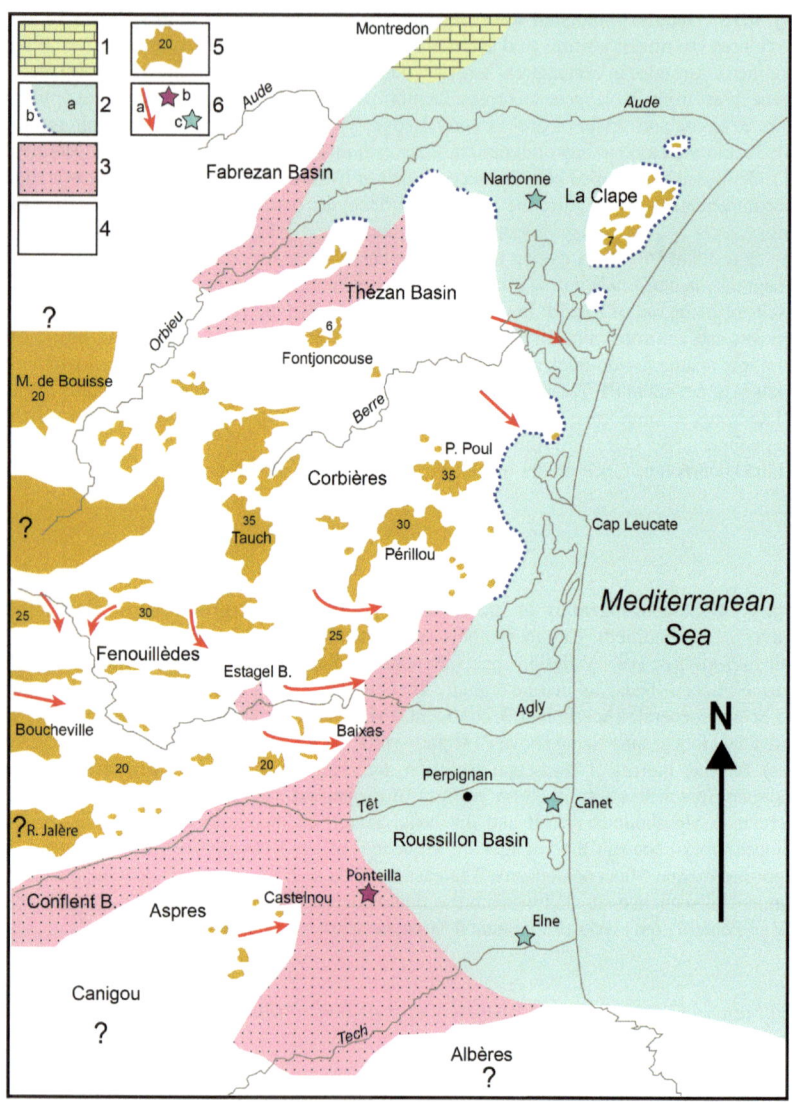

Fig. 6.15 Opoul karst polje. **A** Polje floor viewed from the ponor; in background, note the fault scarp, which has raised P1 to an altitude of 400 m. **B** Aven des Amandiers (the main ponor). **C** View of the polje from the scenic castle footpath; surface P1 in massive limestone extending towards Serre du Scorpion; Leucate–Le Barcarès lagoon and beach barrier in background

surface P1 has been tilted from 280 m in the SW (Pas de l'Escale, see description of return route) up to 400 m to the NE (Mt. Pla), and its tread is intensely degraded. From the castle you can distinguish residual monadnocks rising above it, e.g., Tautavel massif (Serrat de la Narède) in the west (520 m) and Montolier de Perellos to the north (709 m)—both capped by vestiges of summit surface, S. West of the castle, note the structural landforms involving a belt of outfacing hogback scarps in Urgonian limestone surrounding the synclinal basin of Mas Farines, which has been overdeepened by erosion of its outcrops of Albian shale.

Continue to the ghost village of Périllos. You will be driving across an ancient karst polje incised by the upper Roboul River and its tributaries (Fig. 6.16A). The area displays very ancient cave levels (Grand Barrenc, 42°53′41.6″N, 02°52′01.9″E; La Cauna). To reach Cauna Cave, park at elevation spot 297 m (Cortal de la Bosiga) and ascend the track to the right of the trunk road (1 h return trip on foot).

Fig. 6.16 Périllos karstic landscape. A The Périllos fossil polje, a late Neogene landform perforating the tread of P1; note in the foreground the Grand Barrenc (a sinkhole descending to −55 m). **B** Caune de Périllos, a cave containing stalagmitic pillars ~10 m tall

Stop 5. Périllos, Cauna Cave

The elevated cave entrance (42°54′12.4″N, 02°51′36.5″E, 370 m) lies slightly below the tread of Miocene pediment P1, where it connects to the residual hill of Montolier de Perellos. The cave is entered easily by a trail from the east. The ceiling of the larger inner cavity (Fig. 6.16B) has collapsed and is bounded to the north by a south-vergent reverse fault plane. The size of the cave implies a major underground flow system. It remains impossible to establish its flow direction, source, or outlet because it has been intersected by the modern topographic surface. The cave displays huge stalagmite pillars, currently corroding. This ancient karstic system could have formed at the same time as surface P1, or soon after.

To join up with the main itinerary, return to Opoul and follow the D5–D5b to reach Salses. Before the railway crossing, two streets striking east–west display marine deposits between +13 and +27 m containing a rich faunal assemblage of Ostrea lamellosa *and* Chlamys glabra *(42°50′10″N, 02°54′51.7″E). This level is almost certainly pre-Tyrrhenian. It has been ascribed an early Pleistocene age by* Ambert (1994), *but a Pliocene age by* Bourcart (1947). *Gastrochaenolites can be traced at a constant altitude for 2 km, from north of the fort to SW of the motorway. The Château de Salses is an impressive fortification constructed on the border with France between 1496 and 1502 by the Catholic Monarchs Isabella of Castille and Ferdinand of Aragon, and is worth a visit. Proceed northwards on the Narbonne road (D900) for 4 km and stop on your left at the large parking area of Font Estramar (or Estremera)* (Fig. 6.17A).

Stop 6. Font Estramar

Font Estramar is the main karstic spring of the southern Corbières (mean discharge: 1.8 m³/s, peak flood discharge: 25 m³/s). Together with neighbouring Font Dame (mean discharge: 0.8 m³/s, peak flood discharge: 6.5 m³/s), this Vauclusian spring collects the infiltrated rainwater of the limestone plateaus as well as the abstracted discharge (~60%) of the Agly and Verdouble rivers as they flow through the Aptian–Barremian (Urgonian) limestone outcrops of the Corbières nappe. The 300 km² catchment area extends westward to the Fenouillèdes (see Itinerary 5), forming an estimated reservoir of 70 M m³. Subterranean flow runs deep (−400 m), generating steady water temperatures of 18–20 °C. Contamination by lagoon water explains its brackish character. The exit pool (Fig. 6.17B) leads downward to a labyrinthine succession of subhorizontal passages extending at 15, 35, and 55 m below sea level. The passages are interconnected by a system of wide vertical wells, the deepest dive into one of these having reached −286 m in December 2019. This extremely deep underground karst system, the second

Fig. 6.17   **Font Estramar**. **A** Location of observation points (satellite imagery provided by Google Earth); **B** The Vauclusian spring

deepest in France after Fontaine de Vaucluse (–308 m), was emplaced at the time of the Messinian Salinity Crisis, i.e., when Mediterranean sea levels were much lower than ever before or since.

The NE–SW-striking scarp rising above the spring line has a long and complex history. The cliff face is an ancient feature of the landscape and was buried and exhumed on several occasions between the early Miocene and the Holocene (Calvet 1996). Its alternate states of burial or exposure have been controlled by the intensity of tectonic activity along the Têt Fault, which is a major structural feature in the region. Here, the Têt Fault is covered by recent alluvium but it runs along the base of the escarpment. The original fault scarp, which formed during early Neogene time, has been entirely buried by the Miocene marine sequences (see **Stop 7**). Rock structures in the footwall have themselves been truncated by Miocene pediment P1 along the entire coastal tract of the Corbières. The dramatic Messinian sea-level fall, including possible fault-slip activity, promoted exhumation of the scarp towards the end of the Miocene. After the Messinian Salinity Crisis, the Zanclean transgression subsequently buried at least the lower portion of the escarpment, as demonstrated by a limestone fissure at +30 m, NW of the karstic spring, in which the bone remains of terrestrial rodents (as well as a fish tooth) dating to the earliest Pliocene (Aguilar et al. 1991) have been preserved (Fig. 6.17A). Evidence of renewed scarp exhumation during the early Pleistocene is supported by several sites containing micromammalian taxa at the base of the scarp (Calvet 1996).

Gastrochaenolites and beach-sand levels define Middle Pleistocene sea-stands at +5, +10 and +15 m above present-day sea level. The +5 m strandline, displaying sand, marl, bivalve-bored limestone pebbles, and oyster shells, can be observed in the car-park ditch on the south side of the motorway embankment (Fig. 6.18A; 42°51′30″N, 02°57′30″E). The +10 m strandline, displaying quartz-rich marine sand, is exposed to the north of the motorway and above the karst-spring pool (Fig. 6.18B; 42°51′32.8″N, 02°57′29″E). Both generations of deposit have been buried by a Quaternary debris cone, itself sealed by a 1-m-thick, massive calcrete duricrust (Fig. 6.19A, B). The +15 m level is exposed immediately to the west of the Vauclusian pool, resting directly on the Urgonian limestone (Fig. 6.18C; 42°51′33.2″N, 02°57′26.2″E). From top to base, the highly indurated strandline sequence includes a cross-bedded sand dune, itself resting on a beach-rock plinth containing quartz pebbles whose provenance indicates a south-to-north longshore drift conveying debris supplied by rivers descended from the Axial Zone.

Fig. 6.18 Pleistocene marine beds at Font Estramar. A Oyster beds at +5 m NGF; **B**
Lithophaga boreholes at +8–9 m NGF, buried by rubified vestiges of marine sand (NGF
refers to the French Ordnance datum); **C** Highly cemented beach level at +15 m NGF; basal
layer of coastal sandstone and conglomerate (contains rolled quartz pebbles and locally-
supplied Urgonian facies limestone) is overlain by cross-bedded dune sand

*Continue on the D900 for 2.3 km until you reach a parking area on your right. Take
the motorway underpass and climb the slope along the motorway trench in order to
reach Serre du Scorpion (round trip: 1.5 h). After having reached the flat hosting
the radio repeater, you may follow the dirt road all the way up the hill, skirting it
from the south (summit at 128 m). A return trip via the north side is possible.*

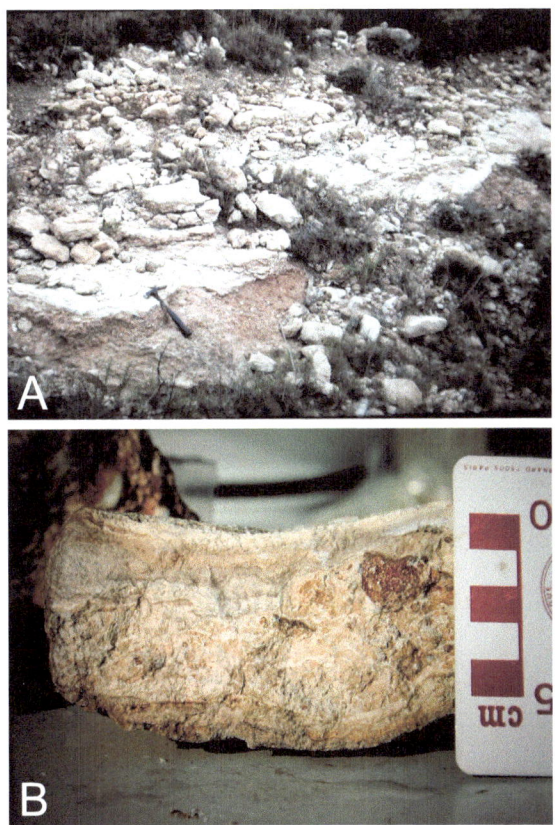

Fig. 6.19   **Indurated slope deposits of Font Estramar**, **A** The duricrust; **B** Close-up view of duricrust facies

Stop 7. Serre du Scorpion

At this key site, the Corbières erosion surface P1 grades directly to the mid-Miocene shoreline. This is France's most southerly marine Miocene outcrop (Magné 1978). Similar occurrences reappear in Catalonia near Barcelona, and in the Vallès graben. At Serre du Scorpion, the coastal Miocene deposits (Fig. 6.20A, B) rest on a marine abrasion platform at 50–60 m a.s.l., cut in the steeply dipping Urgonian limestone and riddled with bivalve boreholes (an exposure occurs ~40 m SW of the dirt track, along a sidetrack leading to a ruined building: 42°52′09.6″N, 02°58′48″E, 60 m). The deposits are 50–60 m thick and

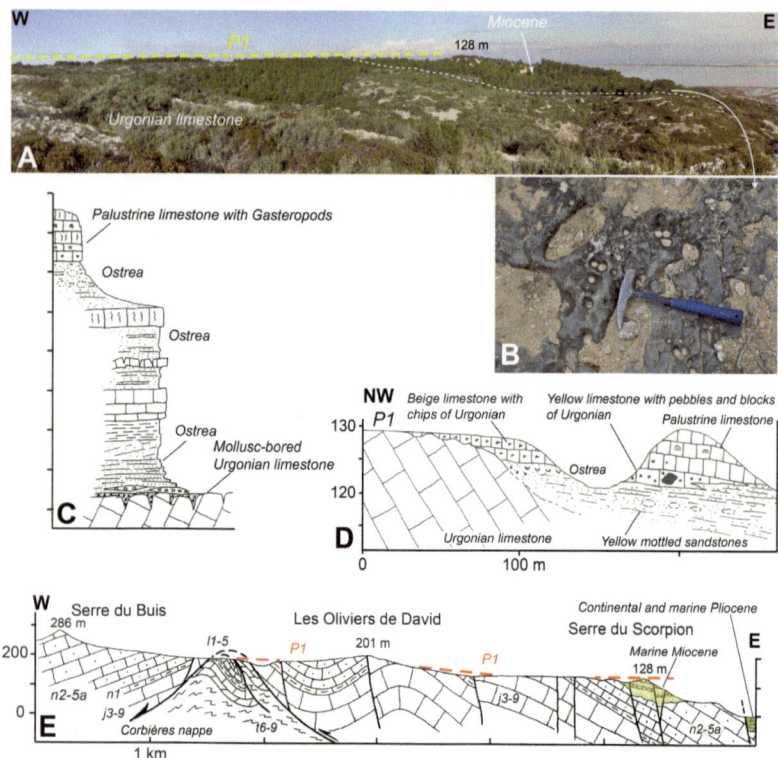

Fig. 6.20 Serre du Scorpion, where pediment P1 grades to the top of the marine Miocene sequence. **A** Panoramic view of the site from the SW; Leucate lagoon and beach barrier in background. **B** Shore platform in Urgonian limestone displaying bivalve borings at the base of the Miocene sequence. **C** Stratigraphy of the Miocene as observed at the NE hill face: alternating sandstone-textured molasse and yellow sand containing oyster shells, with grey shale and white limestone; the latter exhibits lacustrine and/or travertine features towards the top. **D** Close-up of the contact between the Miocene rocks and pediment P1; the palustrine limestone contains angular clasts and rounded pebbles of Mesozoic limestone from the hinterland, prograde over the palaeoscarp cut in Lower Cretaceous limestone, and grades to the tread of P1. **E** Cross-section through P1 at Serre du Scorpion: all thin-skinned Paleogene compressional structures have been bevelled by Miocene surface P1. Denudation of the updomed Paleogene nappe has formed a window in the stratigraphy, exposing the Triassic beds (C, D, E after Calvet 1996)

have buried the upper part of this ancient cliffline; they have themselves been tectonically deformed by a swarm of small normal faults, the most conspicuous among which strikes N10°E. The stratigraphy (Fig. 6.20C) consists of alternating beds of oyster-bearing sandstone and lacustrine to lagoonal limestone. The gastropod-rich limestone bed capping the butte and the tread of pediment P1 are in perfect alignment (Fig. 6.20A, D, E), and the butte caprock contains pebbles, boulders, and more angular clasts supplied by the pure-white Jurassic limestone of the immediate Corbières hinterland. This constitutes strong evidence that pediments belonging to generation P1 are quite probably mid-Miocene (age constraints provided at **Stop 8**).

The ascent to the butte summit (128 m) offers splendid views of Leucate–Le Barcarès coastal lagoon, with, in the foreground, some reedbed habitat taking advantage of the local karstic springwater; on the horizon, note the coastal lido with its seaside resorts. This continuous barrier beach was progressively constructed during and after the Flandrian sea-level rise (6500 yr BP), and a spit-accretion process promoted by (today mostly north-directed) longshore drift between coastal headlands is currently the favoured model (Brunel et al. 2014). About 2 km inland from the modern beach barrier, an older barrier attributed to the maximum Holocene highstand can be distinguished but becomes increasingly discontinuous towards the north.

Continue along the D900 (labelled D6009 as you enter the Aude département). At Fitou, you will enter the small Neogene extensional basin of Caves–Lapalme (Fig. 6.21). Turn right and follow the D627 towards the Leucate Plateau. Stop at the D627–D327 junction, near the large exposures of Miocene outcrop (south section: 42°54′37.7″N, 03°01′16″E; central section: 42°54′51″N, 03°01′17.5″E).

Stop 8. Leucate Plateau
Along the D627, and likewise in the gully parallel to the D327 and all the way to the lake shore, the Miocene deposits are exposed to direct observation (Fig. 6.22A, B). These outcrops correspond to the erstwhile marine base level of pediment P1 in the Corbières. The rock is a yellowish sandstone-textured molasse displaying beach pebble beds, date-mussel boreholes (*Lithophaga lithophaga*), and oyster-shell remains (Fig. 6.22D, E). A lateral facies transition to rubified deposits of a more clastic nature occurs in the section but these also exhibit littoral features. The facies is characteristically coastal but poorly constrained by age data. Bivalve fossils suggest Upper Burdigalian (*Pecten tournali*) and Langhian to Serravallian ages (*Ostrea crassissima*: Fig. 6.22D); foraminifera have suggested biozone N 8, i.e., Langhian (Magné 1978). Fossil rodent assemblages

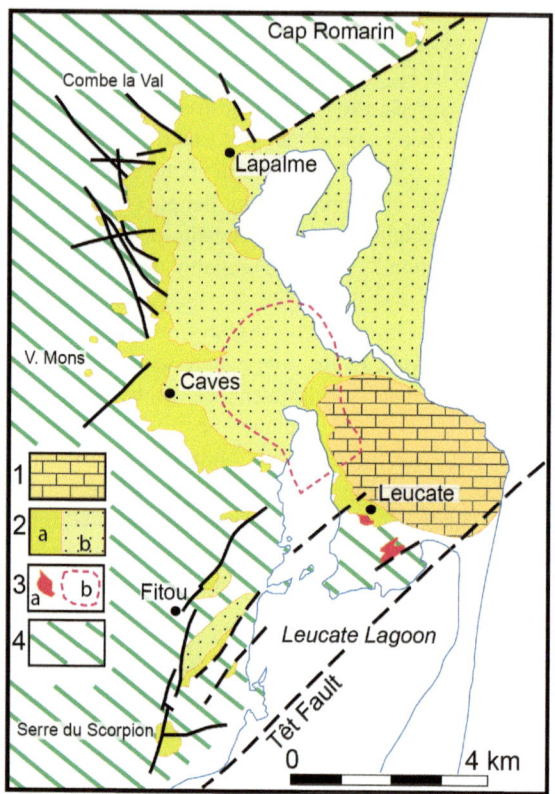

Fig. 6.21 **Lapalme–Leucate Neogene graben** (after Calvet 1996, modified). This tectonic basin contains three sequences, each separated by a major unconformity. Key to symbols and ornaments—**1**: Pliocene travertine and paralic shale (coeval with, or slightly younger than, the top of the Pliocene sequence in the Roussillon Basin). **2**: mid-Miocene marine and coastal outcrops (**2a**), at places covered by Quaternary deposits (**2b**). **3**: lacustrine limestone outcrops (**3a**), also encountered in boreholes (**3b**), of probable Oligocene to Aquitanian age. **4**: folded Mesozoic structures

contained in the coastal deposits have yielded the most precise age brackets so far (Aguilar 1979, 1980, 1981, 1982; see also revised biozones and updated tables in Gunnell et al. 2009; Aguilar et al. 2010; Hilgen et al. 2012). Among these, Port-la-Nouvelle is a littoral karstic cavity in Urgonian limestone, buried by the molasse and also containing foraminifera and shark-like teeth. The corresponding

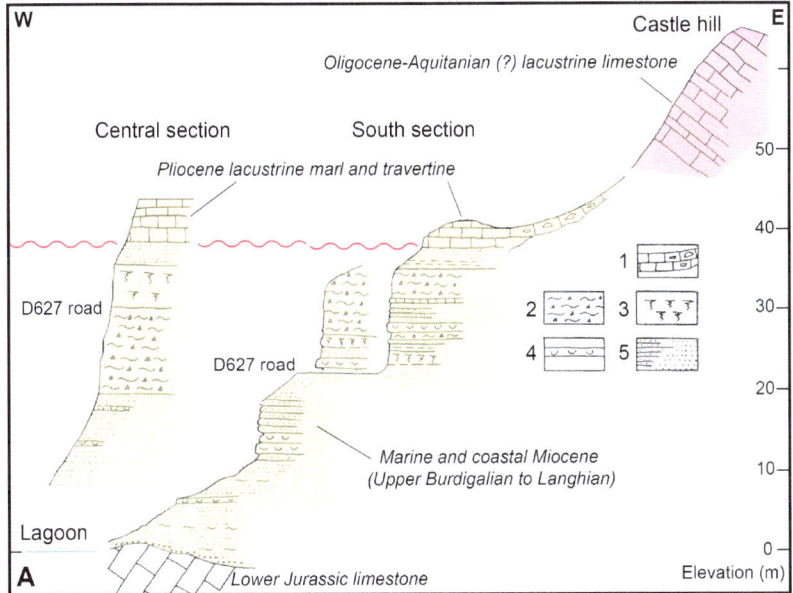

Fig. 6.22   Miocene outcrops at Leucate. A Cross-sections along the western plateau edge (after Calvet 1996, modified). **1**: Pliocene travertine or lacustrine limestone, with breccia associated with the travertine. **2**: rubified Miocene facies containing fragments of Oligocene to Aquitanian lacustrine limestone; coastal origin testified by *Ostrea* and pebbles displaying bivalve borings; merges laterally to 4 and 5. **3**: variegated paleosol ascribed to terrestrial conditions during the Messinian eustatic lowstand. **4**: bioherm containing *Ostrea*, *Pecten*, etc. **5**: sandy yellow marine molasse (Miocene). **B** South section; yellow molassic facies transitioning laterally to red littoral facies. **C** Central section, Messinian paleosol buried beneath Pliocene travertine. **D** *Ostrea crassissima* in situ. **E** Gastrochaeonolites (i.e., *Lithophaga* bore holes); D and E come from the upper portions of the Leucate outcrop, along the village footpath immediately south of section B)

biozone age (MN 4) captures the late Burdigalien sea-level rise (~16.4–17.2 Ma). Within the molasse, the assemblages of Leucate butte 1 (i.e., south section), similar to those at Luc-sur-Orbieu further NW, have yielded ages of ~13.4–15 Ma (MN 6)—i.e., Upper Langhian to Lower Serravallian. These Miocene rocks contain angular clasts of white limestone that could only have been supplied locally by the lacustrine limestone beds of possible Oligocene age at the top of the hill (where the Leucate fort was constructed). The relief is thus likely inherited from a reef structure buried by the mid-Miocene sea, in the midst of a tectonic basin

Fig. 6.22   (continued)

(Lapalme Basin, Fig. 6.21) similar in age and history to its Sigean–Narbonne counterpart (see Box 6.2).

Another exposed section 500 m to the north (central section, Fig. 6.22C) exhibits a thick red paleosol containing whitish streaks and waterlogging features. This paleosol occurs in the top layer of the marine Miocene sequence and is overlain by the Pliocene lake deposits capping the Leucate Plateau. The paleosol is thus interpreted as a legacy of the Messinian Salinity Crisis, i.e., a time when the marine Miocene deposits were temporarily exposed to terrestrial conditions.

The fort at Leucate (30 min return walk) offers a panoramic view of some key geomorphic features of the Corbières orocline and Roussillon Basin (Fig. 6.23A). Towards the south, note the Leucate lagoon and its modern barrier beach, and the plains of Roussillon in the middle ground, overlooked from east to west on the horizon by the Albères horst (1256 m; see Itinerary 2) and by Mt. Canigou (2785 m; see Itineraries 4 and 6). Towards the SW, W and NW, the nappe sequences of the eastern Corbières are truncated by Miocene erosion surface P1, above which residual massifs such as Pied du Poul and Périllos clearly stand out (Figs. 6.12 and 6.13; see **Stops 3, 4, 5**). Towards the NE, Leucate Plateau is a

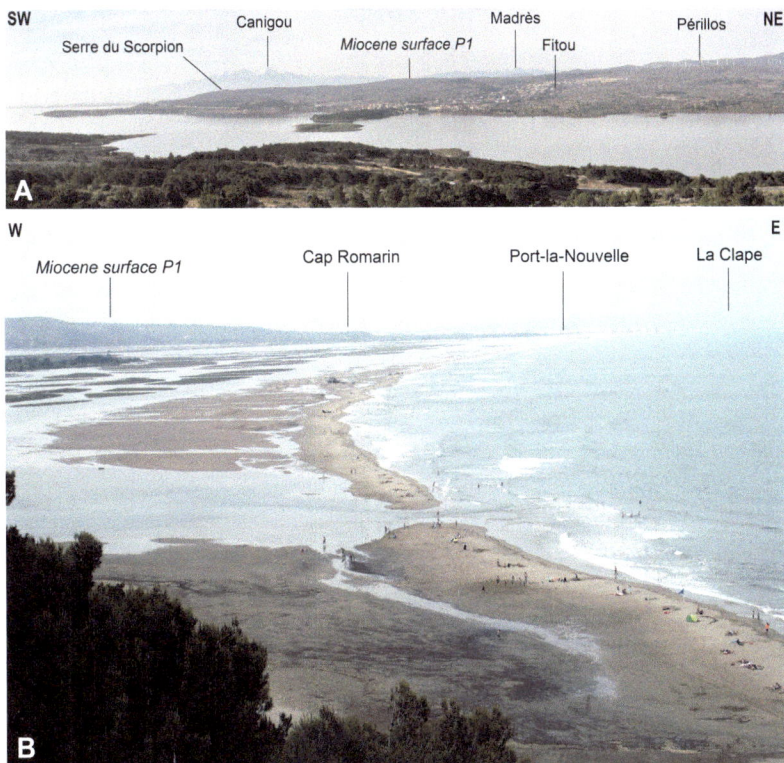

Fig. 6.23 Panoramic views from Leucate Plateau. A Looking west from the castle: the lagoon, Corbières, and footwall uplands of the Axial Zone. **B** Looking north from Cap des Frères: beach barrier from La Franquie to Port-la-Nouvelle after a heavy spring storm; to the NW: Cap Romarin plateau and its abandoned Holocene cliff face

structural surface formed by karst-corroded lacustrine shaly limestone and traver-tine beds (Fig. 6.24A, B). The latter contain numerous fossil plant impressions such as reed stalks. Their early to mid-Pliocene age has been established on the basis of fossil rodent taxa contained in the sediment (Aguilar 1977; Aguilar and Michaux 1977). The bright white colour of the limestone has given this former island its name (leukos: white, in Greek), and it was used as a landmark by Greek navigators. Good sections in the Pliocene limestone can be inspected along the cliffs of Cap Leucate, which are accessible via the road to the lighthouse. Here

Fig. 6.24 **Pliocene travertine and marl at Leucate.** **A** General view from Cap des Frères to Cap Leucate; note undulations at the surface of the travertine beds. **B** Aspects of the travertine facies (exposure at Leucate beach cliff face)

you may park and walk (1 h round trip) to Cap des Frères and appreciate the long sweep of the beach barrier towards La Nouvelle (Fig. 6.23B; see Box 6.3). En route, you may notice gentle tectonic warps and undulations in the lacustrine limestone beds (Fig. 6.24A).

Return to the Narbonne road (D6009) and follow it northwards. The Rieux allu-vial fan (Late Pleistocene) covers the Miocene units, which are otherwise exposed on the western border of the basin and occur as isolated patches further west (Fig. 6.21). These outliers are sometimes bounded by small normal faults and rest unconformably on a gently inclined topographic ramp that cross-cuts Juras-sic and Urgonian limestone strata. They could represent a vestige of range-top erosion surface, S — here tilted seaward and buried during the mid-Miocene high-stand (Fig. 6.25A). Marine Miocene deposits also underlie the stony buttes at Lapalme (the geological sheet of Leucate mistakenly reports Quaternary alluvial gravels, but these in fact correspond to a Miocene coastal pebble bed associated

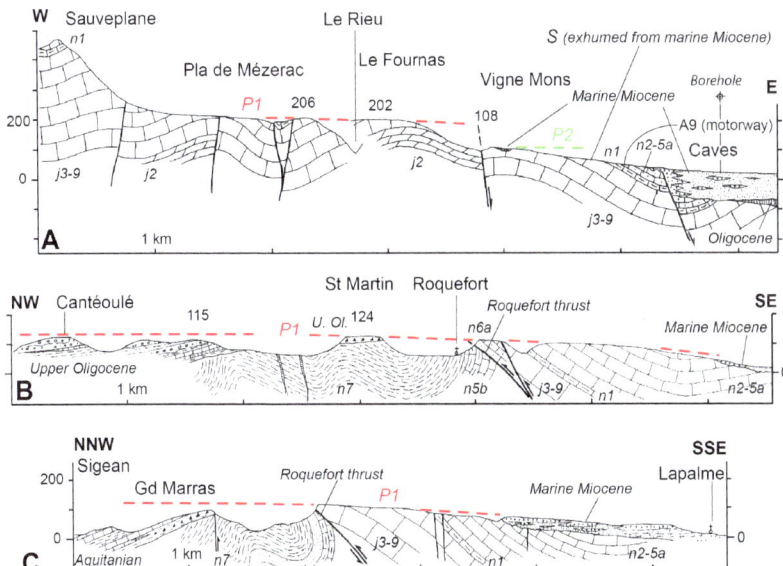

Fig. 6.25   Three cross-sections of Miocene pediment P1 through the Corbières (after Cal-vet 1996, modified). **A** Western edge of Lapalme Basin: concave connecting slope between pediment P1 and the residual Pied du Poul limestone massif (Sauveplane); note gentle tec-tonic deformation of the Miocene beds (Caves Fault) and extent of the Miocene sea-level rise (oyster-rich bioherm and marine shales present à Vigne Mons); range-top surface, S, which is downwarped and buried beneath the Neogene fill of Lapalme Basin, appears exhumed from Miocene molasse overburden on the basin's edge. **B, C** Northern edge of Lapalme Basin: erosion surface P1 bevels the Oligocene and Aquitanian beds of the Sigean–Narbonne Basin and the Roquefort Thrust, and grades to the top of the marine Miocene sequence near the coast to the SE

with populations of Ostrea crassissima*). The Miocene strandline grades exactly to the tread of surface P1 along the 68 m elevation contour at the plateau edge around Roquefort-des-Corbières–Cap Romarin (*Fig. 6.25C*), thus corroborating the context previously observed at Serre du Scorpion (***Stop 7***). In the north ditch where the road crosses the Paleogene Roquefort Thrust, note the erosional bevel on the Jurassic and Cretaceous limestone sequence (*Fig. 6.25B, C*). Drive on to the village of Roquefort-des-Corbières.*

Stop 9. Roquefort-des-Corbières

Good panoramic view from the butte summit at Saint-Martin's chapel. Opportunities are provided here to observe cross-cutting relations of pediment P1 with the Oligocene and Miocene stratigraphy. The erosion surface bevels the massive Jurassic limestone extensively to the NW and also to the south of the Roquefort Thrust, which is an offshoot of the Corbières orocline (Figs. 6.25B, C, 6.26). The tread of this surface has embayed, but not fully consumed, the upstanding residual butte at Pied du Poul in the west (Fig. 6.27).

Erosion surface P1 also bevels the Upper Oligocene fault breccia (28–23 Ma) on the interfluves and buttes around Roquefort and at St Martin's chapel. The breccia contains huge limestone boulders supplied from a local southerly source area. This site demonstrates that the erosion surface is post-Oligocene because it has destroyed the relief which was supplying this proximal breccia formation. P1 also cuts across the Aquitanian basal breccia (23–20.4 Ma) at Grand Marras (Fig. 6.25C), suggesting that P1 is also post-Aquitanian. In contrast, on the southern edge of Roquefort plateau, the erosion surface grades to the top of the coastal marine Miocene deposits of Lapalme Basin. This geometry confirms further that pediment P1 graded to this stable marine Miocene base level and had reached completion by the end of the mid-Miocene highstand (~12 Ma). The erosional episode resulting in P1, which has locally stripped away all of the Cretaceous outcrops, thus lasted 10 million years at most. This implies that there was a regional lull in base-level change conducive to generating relief-attenuating planar landforms rather than relief-accentuating structural landforms. This chronological inference is consistent with thermochronology results independently obtained by Milesi et al. (2020) along the Têt Fault in the Axial Zone (see Itinerary 6), which also records a relative lull in tectonic activity lasting ~10 Ma during the Miocene, and in an area where pediment benches affiliated to generation P1 also occur.

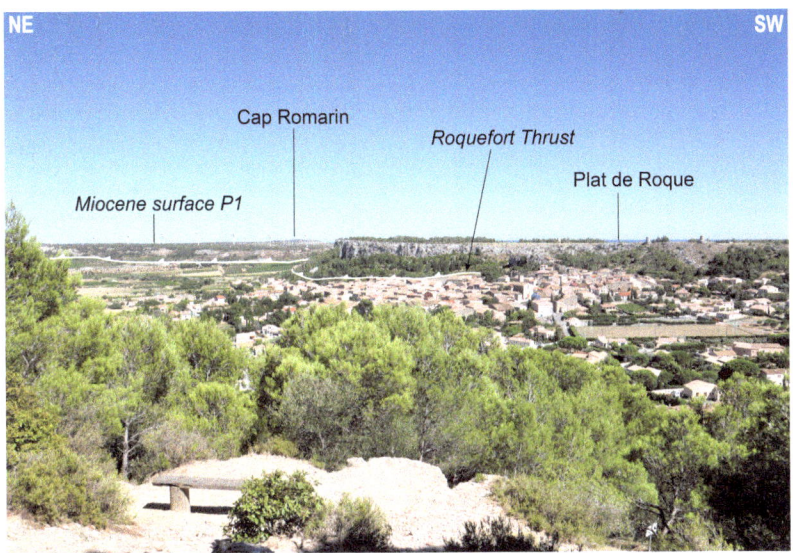

Fig. 6.26 Roquefort-des-Corbières. Viewed here towards the SE from Butte de Saint-Martin. Pediment P1 cross-cuts the crag in Aptian limestone and forms the treads of the plateaus fitted with ancient windmills and modern wind turbines; Oligocene breccia in foreground (see Fig. 6.25B)

Optional trailwalk to Roquefort village (2 to 3 hrs return)

To the south, the Combe de la Val is a dry valley displaying entrenched meanders. It previously functioned as an outflow for the Roquefort catchment towards Lapalme Basin but was more recently captured by the Rieu, a stream that flows north of the thrust through the Albian black shales. Follow the Sentier Cathare hiking trail (GR Sentier Cathare), initially a road then a series of tracks leading to the slope break where the two embayments of surface P1 (locally named Pla des Courbines and Pla de la Lauze) connect with Pied du Poul (or Pé del Poul), the conspicuous residual massif of the area. From here (plan for another 2 to 3 hrs, round trip), range-top surface, S, can be reached at Pla de la Serre (496 m) and at Pé del Poul (596 m). From these vantage points you will capture excellent views of vestiges of generation P1 to the north (Fig. 6.27), with the high ranges of the Pyrenees in the background to the SW and the Massif Central uplands to the north.

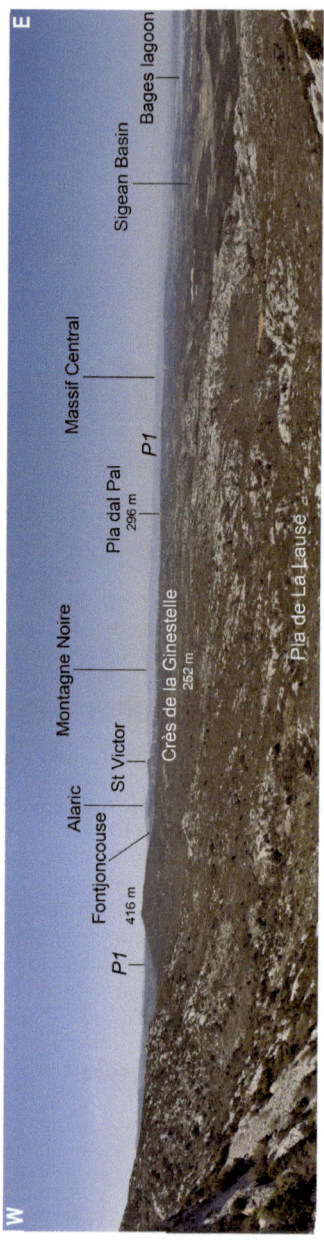

Fig. 6.27 **Where Miocene surface P1 meets an upstanding residual: Pied du Poul and Pla de la Lause embayment.** The wide-angle view reveals the extensive presence of pediment P1 in the eastern Corbières, from Sigean Basin to Fontjoncouse; local residual landforms: Saint Victor, and elevation spot 416 m. On the horizon: Massif Central and Montagne Noire

6.3   Sigean–Narbonne Basin

Box 6.2 The Sigean–Narbonne Basin

The total thickness of the Narbonne fill sequence (Fig. 6.28) is ~1 km, and it forms two biochronologically dated units (Aguilar 1977; Aguilar and Michaux 1977): the first is Oligocene (MP 28, ~26 Ma), and the other starts during the late Oligocene (MP 30, ~24 Ma at Doul) but consists mostly of an Aquitanian biozone (two MN 1 micromammalian assemblages among the lower beds). In each of the chronostratigraphic units, it appears that the basin was receiving slope breccia, small matrix-supported (red clay) debris cones, well-rounded alluvial gravels comprising quartzite from as far away as the Mouthoumet massif (to the west), but mostly shale and quite thick, white lacustrine limestone deposits—all suggesting somewhat low-energy topographic systems in the Corbières hinterland at the time (Calvet 1996). The Aquitanian gypsum deposits are indicative of a paralic environment (Rosset 1964), and thus of proximity to a coastline.

As most classic rift basins, this Oligocene to Aquitanian graben was eventually flooded and overtopped by a more widespread marine transgression. This occurred during the mid-Miocene (see **Stop 8**), promoting the deposition of a relatively thin mud- and sandstone sequence (i.e., molasse) that rests unconformably over the tectonically deformed earlier sequences and progressed landward onto the folded Mesozoic cover rocks of the Corbières orocline and La Clape outlier (see Box 6.4).

The continental and marine deposits of these sequences have been tilted and faulted by Neogene tectonics, although the marine sequence less intensely so than the continental, and the Aquitanian strata likewise less than their Oligocene counterparts. Differential erosion within the basin fill sequences has sculpted homoclinal scarps out of the lacustrine limestone, shale and marine molasse stratigraphy (Fig. 6.28).

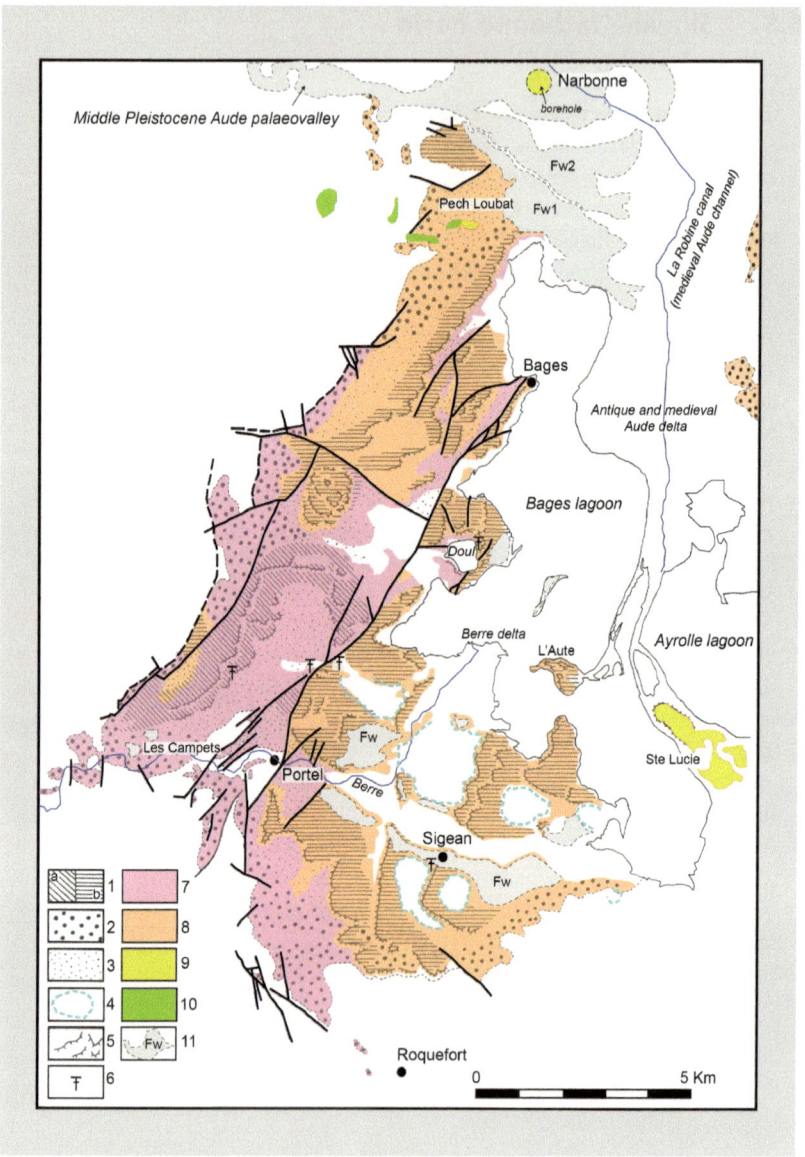

◄**Fig. 6.28** **The Oligocene to Neogene Sigean–Narbonne Basin** (after Calvet 1996, modified and updated). Key to symbols and ornaments—**1**: palustrine and lacustrine limestones and marls of Oligocene (oblique hatch) and Aquitanian (horizontal hatch) age. **2**: near-field breccia and far-field conglomerate transitioning laterally to lacustrine sequences of Oligocene and Aquitanian age. **3**: shale and argillite comprising sandstone and pebble beds. **4**: Quaternary aeolian deflation pans. **5**: homoclinal scarps in the lacustrine limestone beds. **6**: Oligocene and Aquitanian micromammalian fossil deposits. **7**: Oligocene outcrops. **8**: Aquitanian outcrops. **9**: marine and littoral sandstone, and gravel-textured molasse with limestone (Burdigalian to Langhian), also encountered in boreholes beneath Narbonne. **10**: quartz-rich Pliocene lag gravels conveyed from the west (former Aude channel) and capping the highest elevations in the local landscape. **11**: Middle Pleistocene alluvial deposits (Berre River terraces); dry valley (formerly the Aude) and succession of inset alluvial fans at Quatourze

From Roquefort you can reach Sigean by the small road that runs parallel to the A75 on its west side, then by the D205. At Col de l'Agrède, notice the Oligocene to Aquitanian beds resting unconformably on the Albian black shales, marls and yellow sandstone. The road then proceeds through white lacustrine limestones, with a good outcrop along the D205 before reaching Sigean (Fig. 6.29). A number of circular aeolian deflation pans, 0.1 to 1 km in diameter, can be observed at the surface of the friable limestone bedrock. These occur close to the gravel terraces of the Berre River, and were formed during the colder periods of the Quaternary, including the Late Pleistocene. The pans around Sigean (Sainte-Croix, Etang Boyé) are particularly representative of this population of landforms, which are otherwise fairly widespread across the molasse outcrops of Languedoc and Roussillon (Ambert and Clauzon 1992; Ambert 1994; Carozza et al. 2017). Sigean sits on the middle terrace of the Berre, a stream that used to strike east (you cross its former channel on the D6009) and reached the Bages–Sigean lagoon closer to Port-la-Nouvelle. During the Holocene, however, the Berre was diverted to the north, probably by avulsion into aeolian pans and lakes in that area. Today it feeds an actively prograding delta near the Réserve Africaine *(a zoo).*

Close to the Réserve Africaine, *the road crosses the N40°E Portel Fault, a normal fault which separates the basin into two compartments. Here, leave the D6009 and turn onto the D105. At Peyriac-de-Mer, circular Lake Doul is nestled in the core of a diapiric dome, its existence coinciding with the exposure of Keuper evaporites along the fault line. The road follows the shores of the Bages–Sigean lagoon, and* via *Bages bypasses a 2 × 3 km faulted anticlinorium. The Aquitanian limestone forming the*

Fig. 6.29 Cuesta in Aquitanian limestone at Sigean. Foreground: Sainte-Croix aeolian deflation pan

anticline overlies (perhaps unconformably) imprecisely-dated (perhaps Oligocene) ochre marl beds as well as a conglomerate unit, which can be observed to the SW and along the western edge of the basin. The road joins up with the D6009 trunk road at Montplaisir after having cut through the entire Aquitanian sequence (10° dip to the NW). It consists of limestone, marl and clay, including red and ochre clays cross-cut by gravel channels containing weakly rolled carbonate clasts supplied by local outcrops.

Stop 10. Narbonne–Croix Sud (business park)

Stratigraphic sections at Pech Loubat (a site previously named Terrier de Coudonne on older maps), which can be reached from behind the warehouses on the wooded southern hillside (42°09′12″N, 02°58′32.2″E), expose the upper portions of the Aquitanian stratigraphy (Fig. 6.30). The ochre and red silty claystone contains a few channels filled with sandstone and quartz gravel, whose provenance is probably a mix of Campanian sandstone and conglomerate beds of the Bizanet–Montserret Basin (these lie 8 km to the west). These units overlie

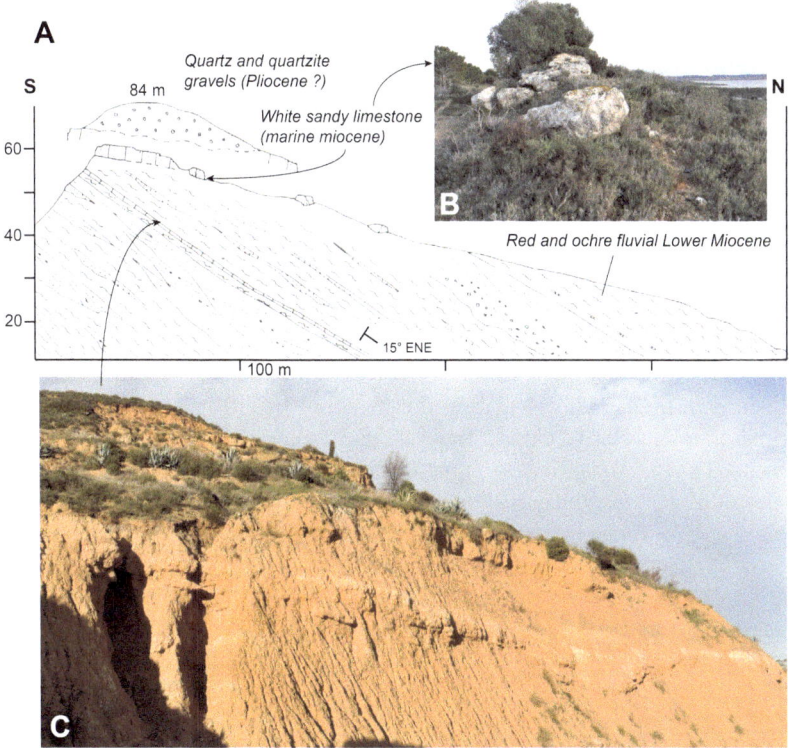

Fig. 6.30 Neogene deposits at Narbonne–Croix Sud. A Synthetic stratigraphic profile of the Neogene at eastern edge of Pech Loubat. Outcrops are accessible from the south face of the hill (walk up between the warehouses to the summit: 84 m); rip-rap now covers the north part of the section. **B** Mid-Miocene (marine): slab of white calcarenite. **C** Early Miocene (continental): section at SE of hill; conglomerate- and sandstone-filled palaeochannels embedded in ochre to red silt levels; dips variable, but predominantly N to NE; early indicator fossils (*Ostrea* towards the top) document the first Miocene transgression

the more locally-sourced Aquitanian conglomerate and ochre siltstone formation exposed in some road embankments of the D205 near the motorway. Dips vary from 25°NNE to 15°ENE, suggesting a possible unconformity between these late Aquitanian or early Burdigalian beds and the NW-dipping Aquitanian limestone outcrops along the motorway. Oyster shells reported from the uppermost beds

(Lespinasse et al. 1982) may document the first Miocene transgression. These beds could thus be Upper Aquitanian or Lower Burdigalian. A discontinuous white slab of marine shelly limestone containing *Crassostrea gryphoïdes* and capping the red beds, with a slight dip to the NNE and a low-angle unconformity, documents the second (mid-)Miocene transgression. A thick quartz-gravel layer of presumed Pliocene age highlights a ravinement surface cutting into the underlying Miocene sequence and caps all of the surrounding buttes up to elevations of 110–120 m. Given that mid-Miocene marine deposits have been detected at less than 70 m below sea-level beneath the city of Narbonne (see Box 6.4), it can be inferred that post-Miocene tectonic deformation (illustrated by the stratigraphic dips to the NNE) has occurred. The anomalously high elevation of the Pliocene alluvial outcrop near the coast also supports the inference.

From Pech Loubat, looking east below the business park, you overlook two Middle Pleistocene alluvial units vertically offset by~10 m from one another. Based on inventories of clast provenance, these alluvial fans suggest they were generated by the Aude River at a time when it still flowed through the abandoned valley of Montrodon (today hosting the road to Carcassonne and the railway line). This dry valley forms a deep-cut canyon (~100 m) in the eastern Corbières nappe (Fig. 6.28). Diversion of the Aude to the north is believed to have occurred during the Middle Pleistocene (Riss) or a little later, perhaps as a result of neotectonics, but avulsion was also promoted by the occurrence of the large aeolian pans (Ambert 1994).

Box 6.3 Holocene, Pleistocene and Miocene strandlines around Port-La-Nouvelle

Accessible from Sigean on the D6139, La Nouvelle (the place name refers to its maritime history) is a nineteenth century port constructed on an ancient breach through the barrier beach after the new Canal de Jonction connecting the Canal du Midi to the sea via the Robine waterway was inaugurated in 1767. Follow the D709 southward to reach Cap Romarin–Cap del Roc.

Despite poor exposures today amid the wave-cut Jurassic limestone and its cover of Miocene lumachella limestone, past observations in the area have reported (here and around the Bages–Sigean lagoon) a Tyrrhenian Stage strandline system (i.e., broadly middle to late Pleistocene) at ca.

+6 m NGF, displaying indurated shore and dune deposits, and another level at +3 m NGF (see Barrière 1966, and synthesis in Ambert 1994). An Electron Spin Resonance (ESR) age of 128 ± 15 ka has been obtained for the beach deposits forming the base of the cave deposit stratigraphy in Ramandils Cave (ca. +2 m NGF), which contains an important sequence of Mousterian artefacts (Rusch et al. 2019). A radiocarbon-dated, very well-preserved Holocene wave-cut notch can be followed continuously along the base of the scarp to the SW. Dated core samples have also provided constraints on beach-bar accretion between 4000 yr BP and the present, revealing a distinct acceleration after 2000 yr BP (Aloisi et al. 1978).

From La Nouvelle, visiting the Domaine de Sainte Lucie (owned and managed by the Conservatoire du Littoral, with signposted trails) affords access to outcrops of mid-Miocene marine deposits. Here the carbonate-rich Burdigalian molasse contains scallop shells (Pecten Tournali) and appears to rest conformably on lacustrine Aquitanian beds. The top of the molasse unit consists of a very coarse-textured conglomerate (cobbles up to 50 cm in diameter) containing a mix of rolled limestone debris and oyster shells. This terminal phase testifies to a powerful influx of fluvial sediment from the west. The landscape to the east displays expanses of salt pans on the modern barrier beach, with the modern cut through the barrier to the lagoon at Vieille Nouvelle (Figs. 6.28 and 6.31).

Box 6.4 Narbonne–Armissan–La Clape
Visiting Narbonne and its monuments, such as the cathedral, its Roman ruins and its museums, is also an opportunity to focus on the Holocene and historical evolution of its coastal lagoon environment. Crossing the coastal plain on the D168 should also allow you to explore the Clape massif and the surroundings of Armissan.

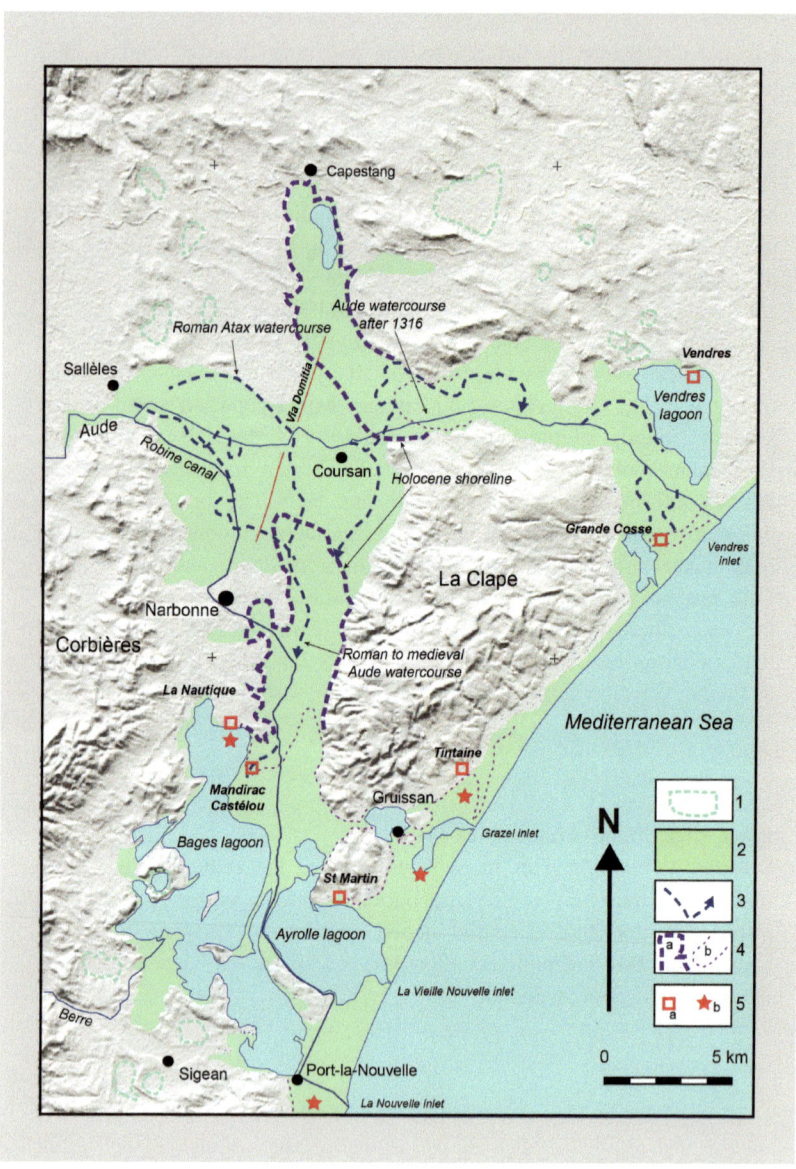

◀**Fig. 6.31**   **Gulf of Narbonne during Holocene and historical times.** Key to symbols and ornaments—**1**: aeolian pans. **2**: maximum extent of the Holocene continental and marine deposits. **3**: historical paleochannels of the Aude River (Atax: its Latin name). **4**: palaeoshorelines; **4a**: maximum extension of the Holocene transgression; **4b**: likely position of the Roman coastline (note port installations at Mandirac-Castélou, i.e., at the river mouth). **5**: antique archaeological sites; **5a**: known or suspected ports; **5b**: shipwrecks beneath the modern beach barrier (after Ambert 1993, 2000, 2011; Falguéra et al. 2000; Faïsse et al. 2018; Salel et al. 2019)

1. Narbonne and the Narbonnais Gulf

Founded in 118 BCE, Narbonne was the first Roman colony in Gaul. The city stands on a low hill (13 m a.s.l.)—effectively a gravel fan generated by the Aude River in Middle Pleistocene time—and is immediately surrounded by the marshy wetlands of the Narbonnais Gulf (Fig. 6.31). During the Holocene highstand, La Clape possibly became an island, with the sea reaching as far as Capestang (testified at this location by borehole cores through mudflat deposits; Ambert 1993). More recent work suggests that La Clape was connected to the mainland by a Pleistocene alluvial terrace deposit currently buried beneath Holocene deposits, and was thus a peninsula rather than an island (Ambert 2011; Salel et al. 2019). The chronology of depression filling by deltaic sediments during and since Antiquity, however, is still poorly constrained. Palaeoenvironmental indicators in more recently obtained cores in the Bages wetland area document a marine gulf as late as 2000–2500 yr BP, followed by an open lagoon environment. Closure by a beach barrier occurred later but its precise chronology is undocumented (Dolez et al. 2015; Faïsse et al. 2018). The northern gulf, from Vendres to Capestang, silted up progressively during Antiquity and then much faster after the fourteenth century (Salel et al. 2019).

Narbonne was the main port of the Languedoc coastline from Antiquity to the eighteenth century (Larguier 1999). During Antiquity, it was accessible to small and medium cargo ships through the mouth of the Aude River into the Bages–Sigean lagoon, and via natural gaps through the beach barrier. The Robine waterway, now a canal, is interpreted as an abandoned channel of the Aude; however, its anomalous strike through Narbonne's Middle Pleistocene alluvial butte rather suggests man-made channelisation during Antiquity (Ambert 2000), with the natural channel of the Aude

River skirting round the north side of the topographic obstacle. The 1316 flood, which caused a definitive avulsion of the Aude to its current position north of La Clape, threatened to deprive the port of Narbonne of its fluvial lifeline. Flow towards the port via the abandoned former channel was artificially maintained by constructing a diversion dam on the Aude at Sallèles. Farther downstream, wetland dessication as a result of sediment influx accelerated during the Little Ice Age, further compounded by accretion of the beach barrier offshore. Geoarchaeological investigations have mostly focused on locating the Roman port and on the Medieval and post-Medieval avulsions of the Aude. Ambert (2000) initially imagined that the ancient city stood directly on the shores of the lagoon, but more recent work has shown that the alluvial plain was already aggrading and the Aude was building its delta in the Bages lagoon. The Roman port and its jetties (100 to 500 CE) have been found there, at the former Aude rivermouth. The bay at the time was still open to the sea, and displayed a discontinuous beach barrier that was still absent, for example, at La Clape (Falguera et al. 2000; Faïsse et al. 2018).

2. Massif de la Clape

The Montagne (or Massif) de La Clape is an outlier of the Corbières orocline rising from the coastal plain, and thus a Pyrenean anticline later uplifted as a small horst within the extensional Sigean–Narbonne Basin. On the western face of La Clape, the unconformable Oligocene to Miocene sequence has undergone tilting and attains dips of ~20°, but it has not been precisely dated. Its base, which consists of lignite-bearing marls followed up-sequence by lacustrine limestone, is probably Upper Oligocene (Lespinasse et al. 1982) and contains the rich botanical assemblages of Armissan (with distinct subtropical affinities). These units are covered by red conglomerate beds of plausible Aquitanian age, their clast content consisting almost exclusively (98%) of Albian green sandstone from the outer envelopes of the anticline. This evidence indicates that the Urgonian core of the anticline was not yet exposed at the time. A geomorphological corollary is that the erosion surface cross-cutting the Clape limestone outcrop is post-Aquitanian (Ambert 1994). To the north and NE of the massif, marine sands and limestone beds encasing the edge of the anticline in Urgonian limestone are Aquitanian or Lower Burdigalian, here marking the westernmost encroachment by those marine environments onto the continental margin. The east and SE edge of the massif is partly buried by mid-Miocene molasse, but given the small outlier also discovered at Les Bugatelles it appears the marine molasse reached deeper into the core of La Clape. The low dip angle of the Cretaceous strata makes cross-cutting relations with the topography difficult to ascertain, particularly for surface S, which has been assumed to occur on the summits of a few residual hills. Matters are more straightforward in the case of pediment P1, which grades to the top of the marine Miocene sequence. Around Armissan, it cross-cuts Cretaceous strata as well as the Oligo-Aquitanian sequence. The marine vestige at Les Bugatelles suggests that the Miocene sea may have covered the entire massif and contributed to generate a wave-cut platform.

From Narbonne, follow the D168, which at first cross the Holocene plain and then cuts through the Oligocene and Aquitanian conglomerates, followed by the limestones. After the junction with the D68, you begin to drive across P1, here ornamented by a number of residual hills.

Stop 11. Panorama of massif de la Clape (junction between the D168 and D68)
From east to west, ca. 130 m a.s.l., the pediment cross-cuts the Barremian–Bedoulian (i.e., Lower Urgonian) sequence, the Bedoulian marl, the Upper Urgonian (Gargasian) limestone razorback at Plan du Roy, and lastly the Oligocene–Miocene sequence at Plan d'Izard (Fig. 6.32). The itinerary can be extended as far as Narbonne-Plage, with a return route via Gruissan and the southern edge of La Clape (see also Box 6.4, and Fig. 6.31); or via Saint-Pierre and the northern side of La Clape, visiting along the way the Oeil Doux sinkhole — here a veritable cenote providing a vertical window through the bedrock to the underground karst watertable. The quickest return route is via the D28 to Armissan.

Returning to Narbonne and the D6009, you are soon back in the Sigean Basin. Aim for Portel-des-Corbières via the D611a. While driving through the village as far as the bridge, note the normal fault plane which separates the graben into two sub-basins (contact between the Jurassic limestone and, to the east, the Aquitanian carbonates and evaporites). You may also arrange to visit Terra Vinea, a wine cellar installation at Rocbère in a reclaimed underground gypsum quarry.
Proceed up the Berre River valley, where it cuts the Oligocene outcrops of the Sigean–Narbonne Basin. The beds correspond here to biozone MP 28 (two sites

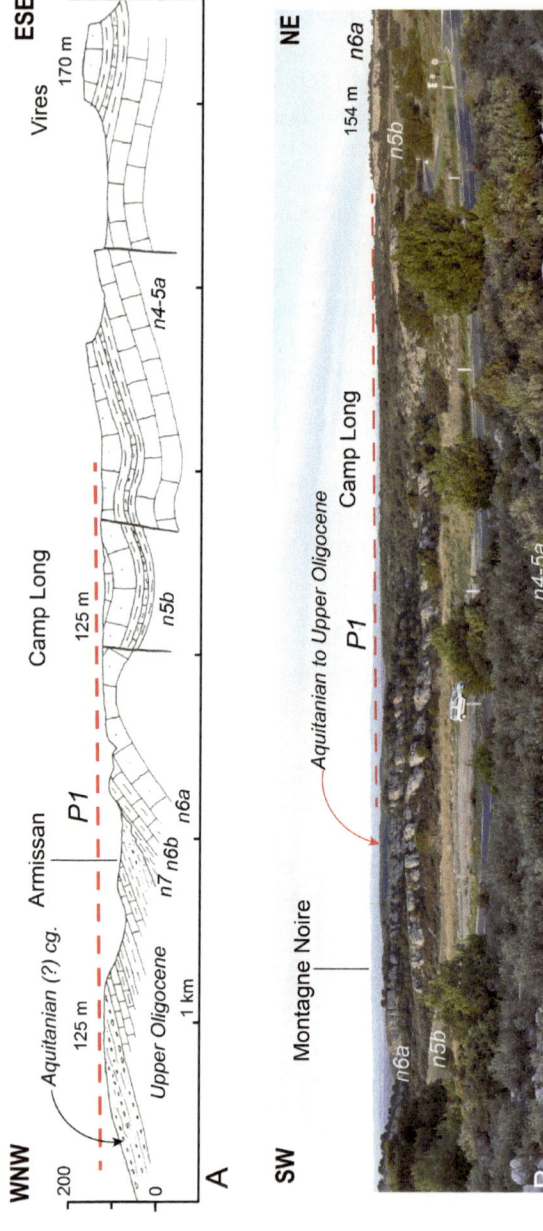

Fig. 6.32 Surface P1 on western flank of La Clape. A Geological section at Armissan. **B** View from junction between roads D168 and D68; pediment P1 cross-cuts a succession of NW-dipping Urgonian limestone beds, which are also locally replicated by fault offsets, as well as the Oligocene to Aquitanian limestone and conglomerate beds (in far distance)

*dating to ~26 Ma), with one age obtained at the base and the other in the middle
part of the sequence.*

Stop 12. Oligocene exposures at Les Campets
Several observation points are possible in the area. The road to Lastours provides
access to the base of the sequence, which rests on massive Upper Jurassic limestone units (e.g., at La Serre). These Jurassic beds display a conspicuous erosional
bevel which was subsequently buried ca. 26 Ma by the basal Oligocene marl and
lacustrine limestone strata. This is regionally one of the rare localities where we
come closest, i.e., with relative precision, to assigning a late Oligocene age to
range-top erosion surface, S. The marl and limestone unit switches up-sequence
to a pale yellow marl and sandstone unit displaying conglomerate-filled channels.
Towards the north, the sequence transitions to marl and limestone exclusively, displays travertine concretions (of palustrine origin given the plant stalk imprints)
capping the Canto Perdrix hogback ridge, and dips 20–30° NW towards the
basin's boundary fault. After Les Campets, you should find some good sections in
the matrix-supported red conglomerates (Fig. 6.33; 43°03′01.5″N, 02°52′45.2″E),
which in this proximal setting contain mainly grey limestone and Cretaceous
sandstone pebbles, all of a fairly small calibre. Exposures at the next bend in
the valley exhibit an even more proximal facies containing locally supplied and
poorly rounded cobbles, some over 1 m in length.

Fig. 6.33 Upper Oligocene beds in Sigean Basin. Exposure along the D611a at Les
Campets. Fluvial facies (gravel bedload channels), dips to the NW

Up to Ripaud, the road follows the Corbières nappe front (here its Taura lobe). This component of orocline was emplaced during the Eocene, and in this area it displays a number of thin, extremely imbricated duplex structures of Triassic, Lower Jurassic and Lower Cretaceous units. At the Ripaud crossroads, opt for the D611 going NW: here you will first drive across the thrust sole (exposure of Keuper evaporites on the right bank of the gully) before entering the Sub-Pyrenean Zone. The road follows the axis of the Donos anticline, an asymmetric N160°E fold breached by an anticlinal valley striking through its continental Campanian and Maastrichtian marl and sandstone core (noted 'Valdo-Fuvelian' and 'Begudo-Rognacian' on geological maps, borrowing here from a Provençal rather than Pyrenean stratigraphic reference frame). To the east, the outfacing hogback scarp in Upper Maastrichtian lacustrine limestone forms the edge of the synclinal valley of Surroque. At Montplaisir, take the D123 towards Fontjoncouse. The road bisects the western limb of the fold, successively exposing the thin and vertically upturned Upper Maastrichtian limestone and sandstone strata, the red Danian–Selandian siltstone and conglomerate formation, and lastly the thick Thanetian–Lower Ypresian lacustrine limestone unit. At the bridge you re-enter the Corbières nappe (here its Fontjoncouse lobe: a fairly undulating stratigraphy of Jurassic limestone and dolomite overlying a sole of Keuper evaporites). Follow the D323 to reach Fontjoncouse, then prepare for a 1 hr round trip on foot across the Devès Plateau (Fig. 6.12; see also Fig. 3.2, Part I). Note that reforestation of the plateau now obstructs views to the SE. A better panoramic view of the area is also gained from the residual hill of Saint Victor (2 hr round trip).

Stop 13. The Corbières erosion surface from Fontjoncouse
Walking across the Devès plateau provides an opportunity to appreciate the topographic smoothness of the regional erosion surface, which, at elevations of 320–370 m, consistently cross-cuts the stratigraphy of the North Pyrenean Zone (Corbières nappe) to the south and west at the same time as the upturned Thanetian limestone strata of the Sub-Pyrenean Zone towards the east and north. Once on the other side of the reforested area (43°02′09.1″N, 02°47′07″E), you will gain 360° views of the landscape (Fig. 6.34). The pyramidal peak at Saint Victor (420 m) to the SE is in Thanetian limestone. Further afield to the SE you will recognise Pied du Poul (596 m) and Montolier de Perellos (709 m), and towards the SW Mt. Tauch (920 m) and Pech Guilloumet–Plateau de Lacamp (633 m), all bearing vestiges of range-top surface, S.

Resume your itinerary on the D123 southward (signposts to Durban-Corbières). At Col de Carla, turn east onto the D40 at a junction where the road once again

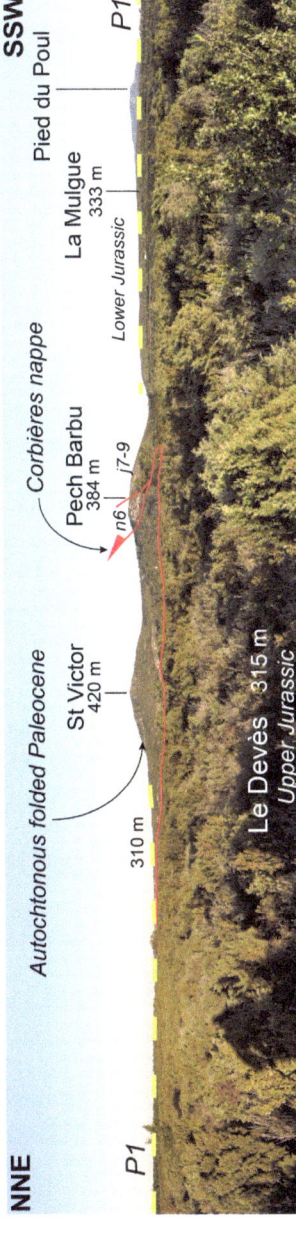

Fig. 6.34 Pediment P1 at Fontjoncouse. Viewed here from south of the Devès, looking ESE. The erosion surface cross-cuts the overriding Corbières nappe structures as well as the overridden terranes

cuts through the base of the Corbières nappe. Looking west, note the Albas razor-back ridge (Thanetian limestone) and the erosional depression in Maastrichtian and Paleocene marl and red conglomerates, which themselves rest unconformably on the thin Permo-Triassic cover of the Paleozoic basement of the Mouthoumet massif. Still farther west, the Maastrichtian sequence rests directly on the Mouthoumet basement. This regional unconformity is a legacy of the first major episode of Pyrenean tectonics during the late Cretaceous. To the south, the overturned stratigraphic duplexes of Lower Jurassic and Aptian–Santonian–Campanian units at Pinède de Durban have nourished a long-standing controversy as to their structure and origin (Durand-Delga and Charrière 2012), but they represent another legacy of this early tectonic stage. Essentially, this limestone, sandstone and marl sequence represents the Mouthoumet cover sequence, deformed and unconformably overlain by the Maastrichtian, and subsequently ploughed into by the base of the advancing Corbières nappe. The road follows the Triassic basal nappe unit up to Durban.

Stop 14. Catastrophic flash floods on the Berre
In 1999, and again in 2005, like many other catchments in the eastern Corbières the Berre valley suffered exceptionally intense Mediterranean-style flash floods. The 1999 floods were triggered by a rainstorm totalling 500 to 700 mm in just 30 h, generating networks of gullies on sloping vineyards (Fig. 6.35C, D), debris flows, shallow landslides, as well as substantial reactivation of the fluvial system such as channel scouring and bedload transport across the floodplains (Fig. 6.35A, B; Calvet 2001; Calvet and Lemartinel 2002). The villages of Durban, Cascastel and Villeneuve sustained severe structural damage. For example, torrential flow swept through the cooperative wine cellars at Cascastel, and a large cistern of muscat wine was transported 15 km and dumped at Les Campets (**Stop 12**). Overbank channel deposits up to 2 m thick, i.e., sufficient to bury vinestocks over dozens of hectares, are still observable at sites such as the large meander upstream of Durban (Fig. 6.35A).

Follow the D611 via Villeneuve and Col d'Extrême. This will take you through the Hercynian Mouthoumet massif, here consisting of low hills in Lower Ordovician pelitic outcrops, above which rise a few steeper-sided monadnocks in sandstone and conglomerate (e.g., Serre de Quintillan, 590 m). The road eventually reaches the Tuchan–Paziols Basin and its famous vineyards. This is an Oligocene half-graben (Calvet et al. 1991), but its master fault has been buried beneath Quaternary alluvium (Fig. 6.38). At Tuchan, aim for the small road climbing up to Mt. Tauch, which starts from behind the wine cooperative building.

Fig. 6.35 Catastrophic floods in the Corbières. **A** Sediment-choked meander upstream of Durban: the vineyards were buried under 1 to 2 m of gravel bedload as a result of the November 12–13, 1999 flood. **B** Petit Verdouble channel, upstream of Tuchan: young vines, replanted after the 1999 event, were once again buried beneath 1 m of pebble beds by the November 13, 2005 flood. **C** Gully near Quintillan: this landform appeared overnight during the November 12–13, 1999 flood. **D** Gullies in a young vineyard at Col d'Extrême (November 12–13, 1999)

Stop 15. Panoramic view of the eastern Corbières from Mt. Tauch

From Salavet summit flats (743 m), the panoramic view to the east overlooks the Mouthoumet massif and topographic surface P1, which cuts extensively across the Corbières limestone nappe. The vast Planal de la Garrigue (Fig. 6.36), which extends above Aguilar castle, slopes to the SW from 410 to 250 m and displays distinctive residual summits at Perellos, La Serra (576 m) and Monts de Tautavel (Serrat de la Narède, 520 m). Surface P1, to the south of Aguilar castle, cross-cuts the Upper Jurassic and Berriasian–Valanginian stratigraphy, as well as west-dipping (15–20°) Oligocene conglomerate beds (the latter are also vertically upturned by fault drag along the Aguilar Fault). This configuration documents a post-Oligocene age for P1, consistent with other occurrences discussed throughout this Itinerary. The Verdouble River, which drains the Tuchan–Paziols Basin, became superimposed on this low-gradient erosion surface and now joins the

Fig. 6.36 **Eastern Corbières from Mt. Tauch**. The Tuchan Basin has been overdeepened by erosion into its Oligocene clastic fill sequence. Note the extent and smoothness of the P1 erosion surface

Agly River via a succession of water gaps cutting through all the folds of the North-Pyrenean Zone.

At Tuchan, take the D611 followed by the D14 towards Padern. Stop at the Verdouble gorge.

Stop 16. Grau de Padern
This site provides insight into the period of extensional tectonics during Albian to Cenomanian time, which subsequently provided the accommodation space for flysch accumulation during the late Cretaceous. The Cenomanian unit consists of alternating strata of limestone and dark shale, and is lacerated by series of normal faults (downthrow to the N or NE) which produce offsets in the limestone ridge (Fig. 6.37). If you examine the faults in the road embankment (e.g., at elevation spot 195 m), or better in the bedrock channel of the Verdouble, you may note stylolites parallel to the stratigraphic planes, and orthogonal tensional cracks. These extensional structures have been buried beneath the North-Pyrenean Frontal Thrust (NPFT), which appears at the foot of the crags in Jurassic limestone on the right bank of the river.

The Verdouble gorge is a major channel knickpoint, ca. 40 m high, and the first in a series going upstream. Active coupling between channel and hillslope processes is highlighted by the presence of large Holocene and historical rockfalls by the bridge.

Drive back towards Paziols on the D14, then the D611. The road between Paziols and Col de la Bergerie d'Alcide offers a cross-section through the entire Oligocene sequence (Fig. 6.38), here dipping 15–20° NW and consisting of conglomerate beds (chiefly containing Cretaceous limestone pebbles) confined to palaeochannels cutting into grey, fine-textured flood-basin deposits. The sediment becomes finer up-sequence and towards the NW, and coarser and boulder-rich in the SE, thereby suggesting a supply area situated somewhere to the east or SE. While displaying conformable boundaries with the overlying units, the fluvial deposits at the base of the sequence consist of pebble beds almost entirely supplied from quartzite outcrops in the north (Serre de Quintillan). From such a sharp change in provenance and texture during the depositional history of this continuous basin-fill sequence, it can be inferred that the Mouthoumet massif had already lost its cover sequence and was already shedding quartzite from its basement during Oligocene time. Block tilting in the SE subsequently generated locally greater local relief, with erosion supplying debris to the basin and terminating its cycle in that area with the production of range-top surface, S. Ten million years later, pediment P1 developed across the entire area,

Fig. 6.37   Grau de Padern gorge (Verdouble River), looking west. Synsedimentary normal faults in the Cenomanian limestone. The Verdouble riverbed is strewn with boulders from a rockfall

indiscriminately cross-cutting the deformed Oligocene beds, the limestone nappe of the Corbières orocline, and the Paleozoic outcrops of the Mouthoumet massif (see also Itinerary 5).

After the col, you gain a good view of the Verdouble water gap cutting through the Devèze mushroom fold, with surface P1 cutting across all of the Urgonian limestone strata. The road then follows an outcrop of Albian black marl in syncline position before cutting the Devèze anticline through another short water gap. Drive past the hydrothermal spring at Fouradade below the road; just after the junction with the D59, also note the outcrop of travertine (presumed age: Pliocene) on the right side of the road (42°49′18.6″N, 02°42′39.3″E). Well-preserved plant impressions in other travertine deposits at the base of the Oligocene sequence can be observed at Roc de la Cest (42°49′03.1″N, 02°42′36.4″E; Fig. 6.38). Follow the D59 towards Tautavel along the Tautavel syncline—again consisting of Albian black marl and bounded to the south by the Vingrau Thrust.

Stop 17. Tautavel, a major Middle Pleistocene Palaeolithic site
The museum, scheduled to be under renovation at the time of writing, is located just above the wine cooperative. The exhibitions are devoted to the Lower Palae-olithic in general, and to the Tautavel excavations in particular. One highlight among the exhibits is a facial replica of *Homo erectus tautavelensis* (the origi-nal is kept in a safe; de Lumley 2015); other items include faunal remains and a large number of lithic artefacts. The cave entrance (la Cauna de l'Arago) lies 3 km upstream from the village at the mouth of the Verdouble gorge (Figs. 6.38 and 6.39A, B). Visits are possible at times of ongoing excavations (enquire at the museum). The cave stratigraphy is ~10 m thick and spans the 700–100 ka interval, i.e., several glacial–interglacial cycles of the Middle Pleistocene. The deposits include fine aeolian sand, and local gravel and clay deposits from the cave itself and from the plateau surface (Fig. 6.39C). The cave is thus the aban-doned karstic channel of a subterranean ancestor of the Verdouble, here recording one stage of canyon incision during Pliocene time.

Leave Tautavel towards Vingrau on the D9, driving through picturesque vineyards overlooked by the thrust-front limestone scarp. The road crosses the Vingrau Thrust below the pass. The col itself (Pas de l'Échelle, or Pas de l'Escale), is the vestige of a broad, shallow palaeovalley which formed at the time of pediment P1 (Fig. 6.40), and belongs to a population of pediment passes situated between the residual masses of Monts de Tautavel, La Serra, and Monts d'Espira (Fig. 6.13). Rare, scattered occurrences of quartz and quarzite pebbles testify to ancient floodways descending from the Mouthoumet massif, or from some outcrops of Upper Cretaceous detrital

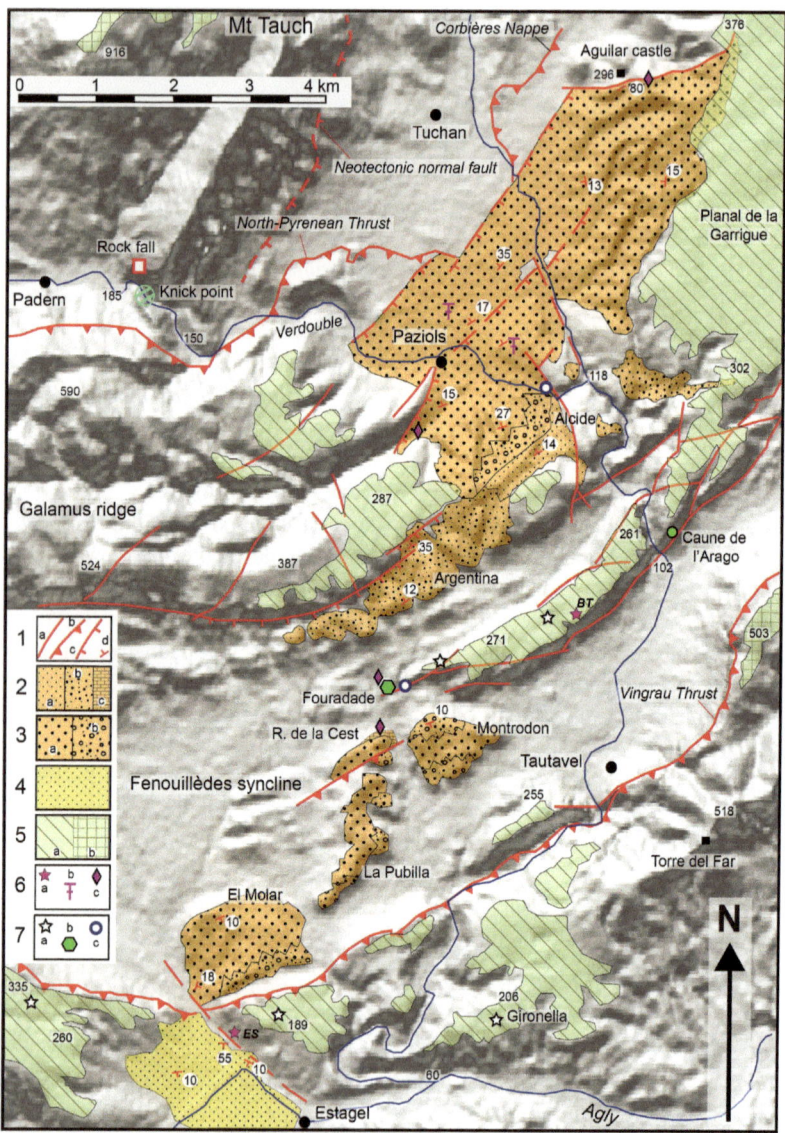

◄**Fig. 6.38   Oligocene to Neogene basins of Tuchan–Tautavel–Estagel** (after Calvet et al. 1991, redrawn and modified). Key to symbols and ornaments—**1**: tectonic features; **1a**: fault (dashed line: blind or hypothetical fault); **1b**: Paleogene thrust; **1c**: Oligocene to Neogene normal fault; **1d**: dip values in degrees. **2**: Oligocene basal units; **2a**: pebble facies (exclusively quartz); **2b**: mixed facies of quartz pebbles, (predominant) and limestone pebbles and blocks; **2c**: massive travertine with leaf impressions at Roc de la Cest. **3**: Oligocene upper units; **3a**: alternation of grey marl and conglomerate-filled palaeochannels (limestone clasts); **3b**: proximal facies displaying poorly rolled limestone cobbles. **4**: continental Miocene beds of Estagel Basin: variegated clays, limestone pebble beds conveyed from the north, siliciclastic sand and pebbles conveyed from the west, rubified limestone boundary-fault breccia; this sequence has buried a Lower Burdigalian micromammalian fossil deposit in the surface karst. **5**: erosion surfaces cross-cutting Paleogene fold structures; **5a**: main surface P1 cutting across tilted Oligocene basin-fill units and the boundary faults; **5b**: vestiges of summit surface S. **6**: chronological and palaeoenvironmental constraints; **6a**: Estagel FFF (Es) containing Lower Burdigalian rodent assemblages (MN 2), and Blanqueterre FFF (Bq) containing 2 Langhian rodent assemblages (MN 5 and MN 6); **6b**: Oligocene rodent fossil deposits in sedimentary beds; **6c**: site remarkable for its plant impressions. **7**: other features; **7a**: remains of quartz-rich alluvium resting on the tread of P1; **7b**: travertine of estimated Pliocene age (Fouradade); **7c**: hydrothermal springs. Note that the four outcrops of Tautavel Basin are undated, and that the Oligocene age is only assumed by analogy with the Tuchan–Paziols Basin. Other symbols highlight the prehistoric site of Caune de l'Arago, and the Padern water gap, knickpoint and rockfall. The unusual prominence of Mt. Tauch suggests it may be a neotectonic horst

rocks, and directed towards the Roussillon Basin and the mid-Miocene coastline (Fig. 6.14).

*The road descends the gentle gradient of pediment P1, here tilted towards the Roussillon. Around the 200 m elevation contour along this traverse it is possible to appreciate the countersunk population of younger pediments belonging to generation P2, such as around Baixas Plateau. These P2 landform units occur at elevations of ~150 m. A roadside exposure near Montpin–Mas Romani (42°49'16"N, 02°49'45"E) shows that the floor of local pediment P2 grades downward to the top of a very coarse-textured debris cone (Fig. 6.41) of probable Pliocene age (see beginning of the itinerary), which rests unconformably on a tectonically deformed clastic unit exposed in the left bank of the gully (Calvet 1996). The road soon reaches the Crest de Rivesaltes, previously visited after **Stop 2**. Follow the signs back to Perpignan.*

Fig. 6.39 Caune de l'Arago prehistoric site. **A** Mouth of the Verdouble canyon. **B** Bedrock channel of the Verdouble. **C** Caune de l'Arago, showing the archaeologically well-documented Middle Pleistocene stratigraphic section

Fig. 6.40 **View from Pas de l'Escale, looking west.** Note the treads of S and P1 and the topographic perfection of level P1; the Albian-cored syncline (black marl) was breached by fluvial erosion during Pliocene and Quaternary times

Fig. 6.41 View from Montpins, looking north. Local pediment P2 has cut a bench extending ~50 m below the tread of P1 and grades to the top of the near-horizontal red conglomerate bed sequence observable along the D12; the proximal facies displays metre-sized boulders but transitions rapidly eastward to beds of red clay and pebbles not exceeding 10–20 cm in length

References

Aguilar JP (1977) Données nouvelles sur l'âge des formations lacustres des bassins de Narbonne-Sigean et de Leucate (Aude) à l'aide des micromammifères. Geobios 10:643–645

Aguilar JP (1979) Principaux résultats biostratigraphiques de l'étude des rongeurs miocènes du Languedoc. C R Acad Sci Paris D 288:473–476

Aguilar JP (1980) Rongeurs du Miocène inférieur et moyen en Languedoc. Leur apport pour les corrélations marin-continental et la stratigraphie. Palaeovertebrata 9:155–203

Aguilar JP (1981) Évolution des rongeurs miocènes et paléogéographie de la Méditerranée occidentale. Ph.D. thesis, USTL Montpellier, 203 p

Aguilar JP (1982) Biozonation du Miocène d'Europe occidentale à l'aide des rongeurs et corrélations avec l'échelle stratigraphique marine. C R Acad Sci Paris II 294:49–54

Aguilar JP (2002) Les Sciuridés des gisements karstiques du Miocène inférieur à moyen du sud de la France : nouvelles espèces, phylogénie, paléoenvironnement. Geobios 35:375–394

Aguilar JP, Michaux J (1977) Remarques sur la stratigraphie des terrains tertiaires des bassins de Narbonne-Sigean et de Leucate (Aude). Geobios 10:647–649

Aguilar JP, Michaux J (1987) Essai d'estimation du pouvoir séparateur de la méthode biostratigraphique des lignées évolutives chez les rongeurs néogènes. Bull Soc Géol Fr 8:1113–1124

Aguilar JP, Michaux J (1990) A paleoenvironmental and paleoclimatic interpretation of a Miocene rodent faunal succession in Southern France. Critical evaluation of the use of rodents in paleoecology. Paléobiol Contin, Montpellier, XVI:311–327

Aguilar JP, Michaux J (1997) Les faunes karstiques néogènes du Sud de la France et la question de leur homogénéité chronologique. Actes du Congrès BiochroM'97, Biochronologie mammalienne du Cénozoïque en Europe et domaines reliés, Aguilar JP, Legendre S, Michaux J (eds) Mém Trav EPHE Inst Montpellier 21:33–38

Aguilar JP, Calvet M, Crochet JY, Legendre S, Michaux J, Sige B (1986a) Première occurence d'un mégachiroptère pteropodidé dans le Miocène moyen d'Europe (gisement de Lo Fournas 2, Pyrénées orientales, France). Paleovertebrata 16:173–184

Aguilar JP, Calvet M, Michaux J (1986b) Découvertes de faunes de micromammifères dans les Pyrénées-Orientales (France) de l'Oligocène supérieur au Miocène supérieur ; espèces nouvelles et réflexions sur l'étalonnage des échelles continentales et marines. C R Acad Sci Paris II 303:755–760

Aguilar JP, Escarguel G, Michaux J (1999) A succession of Miocene rodent assemblages from fissure fillings in southern France: palaeoenvironmental interpretation and comparison with Spain. Palaeogeogr Palaeoclimatol Palaeoecol 145:215–230

Aguilar JP, Lazzari V, Michaux J, Sabatier M, Calvet M (2007) Lo Fournas 16-M (Miocène supérieur) et Lo Fournas 16-P (Pliocène moyen), nouvelles localités karstiques à Baixas, Sud de la France): partie I- Description et implications géodynamiques. Géol Fr 1:55–62

Aguilar JP, Michaux J, Bachelet B, Calvet M, Faillat JP (1991) Les nouvelles faunes de rongeurs proches de la limite mio-pliocène Roussillon: implications biostratigraphiques et biogéographiques. Paleovertebrata 20:147–174

Aguilar JP, Michaux J, Aunay B, Calvet M, Lazzari V (2010) Compléments à l'étude des rongeurs (Cricetidae, Eomyidae, Sciuridae) du gisement karstique de Blanquatère 1 (Miocène moyen, Sud de la France). Geodiversitas 32:515–533

Aloisi JC, Monaco A, Planchais N, Thommeret J, Thommeret Y (1978) The Holocene transgression in the Golfe du Lion. Géogr Phys Quat 2:145–162

Ambert P (1993) Preuves géologiques de l'insularité du massif de la Clape (Aude) pendant la transgression flandrienne. C R Acad Sci Paris II 316:237–244

Ambert P (1994) L'évolution géomorphologique du Languedoc central depuis le Néogène (Grands Causses méridionaux-Piémont Languedocien). Document du BRGM 231, ed. BRGM, Orléans 210 p

Ambert P (2000) Narbonne antique et ses ports, géomorphologie et archéologie, certitudes et hypothèses. Rev Archéol Narbonnaise 33:295–307

Ambert P (2011) Potentiel et contraintes du cadre géologique de Narbonne pour l'aménagement de Narbonne Antique. In Sanchez C, Jézégou MP (eds) Zones portuaires et espaces littoraux de Narbonne et sa région dans l'Antiquité. Monogr Archéol Méditer 28:13–20

Ambert P, Clauzon G (1992) Morphogénèse éolienne en ambiance périglaciaire : les dépressions fermées du pourtour du Golfe du Lion (France méditerranéenne). Z Geomorph Suppl B 84:55–71

Bache F, Olivet JL, Gorini C, Aslanian D, Labails C, Rabineau M (2010) Evolution of rifted continental margins: the case of the Gulf of Lions (Western Mediterranean Basin). Earth Planetary Sci Lett 292:345–356

Barrière J (1966) Le rivage tyrrhénien de l'étang de Bages et de Sigean (Aude). Bull Assoc Fr Ét Quaternaire 3:251–283

Bourcart J (1947) Étude des sédiments pliocènes et quaternaires du Roussillon. Bull Serv Carte Géol Fr XLV:395–476

Brunel C, Certain R, Sabatier F, Robina N, Barusseau JP, Aleman N, Raynal O (2014) 20th century sediment budget trends on theWestern Gulf of Lions shoreface (France): an application of an integrated method for the study of sediment coastal reservoirs. Geomorphology 204:625–637

Calvet M (1992) Aplanissements sur calcaire et gîtes fossilifères karstiques. L'exemple des Corbières orientales. Tübinger Geogr Stud H 109:37–43

Calvet M (1996) Morphogenèse d'une montagne méditerranéenne : les Pyrénées orientales. Documents du BRGM, Orléans, 255, 1177 p

Calvet M (2001) La catastrophe exemplaire: premiers enseignements géomorphologiques de la crue de novembre 1999 dans les Corbières. In: "Au chevet d'une catastrophe", actes du colloque de Perpignan, 26–28 juin 2000 , 63–86, Presses Universitaires de Perpignan, coll. Etudes

Calvet M, Lemartinel B (2002) Précipitations exceptionnelles et crues éclair dans l'aire pyrénéo-méditerranéenne. Géomorph Rel Proc Environ 1:35–50

Calvet M, Aguilar JP, Crochet JY, Dubar M, Michaux J (1991) Première découverte de mammifères oligocènes et burdigaliens dans les bassins de Paziols-Tautavel-Estagel (Aude et Pyrénées-Orientales), implications géodynamiques. Géol Fr 1:33–44

Calvet M, Gunnell Y, Laumonier B (2021) Denudation history and palaeogeography of the Pyrenees and their peripheral basins: an 84-million-year geomorphological perspective. Earth-Sci Rev 215:103436

Canva A, Peyrefitte A, Thinon I, Couëffé R, Maillard A, Jolivet L, Lacquement F, Martelet G, Guennoc P (2020) The Catalan magnetic anomaly: its significance for the crustal structure of the Gulf of Lion passive margin and relationship to the Catalan transfer zone. Mar Pet Geol 113 https://doi.org/10.1016/j.marpetgeo.2019.104174

Carozza JM, Llubes M, Carozza L, Danu M, David M (2017) Les processus de formation et évolution des dépressions fermées du golfe du Lion au cours du Pléistocene supérieur et du tardiglaciaire. Quaternaire 28:323–336

De Lumley MA (2015) L'homme de Tautavel. Un Homo erectus européen évolué Homo erectus tautavelensis. L'Anthropologie 119:303–348

Delmas M, Calvet M, Gunnell Y, Voinchet P, Manel C, Braucher R, Tissoux H, Bahain JJ, Perrenoud C, Saos T, Aster Team (2018) Terrestrial [10]Be and electron spin resonance dating of fluvial terraces quantifies quaternary tectonic uplift gradients in the eastern pyrenees. Quat Sci Rev 193:188–211

Dolez L, Salel T, Bruneton H, Colpo G, Devillers B, Lefèvre D, Muller SD, Sanchez C (2015) Holocene palaeoenvironments of the Bages–Sigean lagoon (France). Geobios 48:297–308

Durand-Delga M, Charrière A (2012) Tectonique tangentielle fini-crétacée au front NE du massif de Mouthoumet (Pinède de Durban – Serre de Ginoufré, Corbières, Aude). Géol Fr 2:25–48

Faillat JP, Aguilar JP, Calvet M, Michaux J (1990) Les fissures à remplissages fossilifères néogènes du plateau de Baixas (Pyrénées orientales, France), témoins de la distension oligo-miocène. C R Acad Sci Paris II 311:205–212

Faïsse C, Mathé V, Bruniaux G, Labussière J, Cavero J, Jézégou MP, Lefèvre D, Sanchez C (2018) Palaeoenvironmental and archaeological records for the reconstruction of the ancient landscape of the Roman harbour of Narbonne (Aude, France). Quatern. Int. 463:124–139

Falguéra F, Falguéra JM, Guy M, Marsal A (2000) Narbonne: cadre naturel et ports à l'époque romaine. Méditerranée 94:15–24

Granado P, Urgeles R, Sàbat F, Albert-Villanueva E, Roca E, Muñoz JA, Mazzuca N, Gambini R (2016) Geodynamical framework and hydrocarbon plays of a salt giant: the NW Mediterranean Basin. Pet Geosci 22:309–321

Gunnell Y, Calvet M, Brichau S, Carter A, Aguilar JP, Zeyen H (2009) Low long-term erosion rates in high-energy mountain belts: insights from thermo- and biochronology in the Eastern Pyrenees. Earth Planet Sci Lett 278:208–218

Gunnell Y, Zeyen H, Calvet M (2008) Geophysical evidence of a missing lithospheric root beneath the Eastern Pyrenees: consequences for post-orogenic uplift and associated geomorphic signatures. Earth Planet Sci Lett 276:302–313

Hilgen FJ, Lourens LJ, Van Dam JA, with contributions by Beu AG, Boyes AE, Cooper RA, Krijgsman W, Ogg JG, Piller WE, Wilson DS (2012) The Neogene period. In: Gradstein et al. (eds) The Geologic Time Scale, Elsevier, chap. 29, pp 923–978

Jolivet L, Romagny A, Gorini C, Maillard A, Thinon I, Couëffé R, Ducoux M, Séranne M (2020) Fast dismantling of a mountain belt by mantle flow: late-orogenic evolution of pyrenees and liguro-provençal rifting. Tectonophysics 776(228312):15 p

Larguier G (1999) Le drap et le grain en Languedoc, Narbonne et Narbonnais, 1300–1789. Presses Univ. Perpignan, 3 t, 1368 p

Lespinasse P et al. (1982) Handbook, sheet Narbonne, Carte géologique de la France n°
    1061, BRGM, Orléans, 51 p
Magné J (1978) Études microstratigraphiques sur le Néogène de la Méditerranée nord-
    occidentale. t. 1: Les bassins néogènes catalans, 259 p; t. 2.: Le Néogène marin du
    Languedoc méditerranéen, 435 p, Édit. CNRS, Toulouse
Milesi G, Monié P, Münch P, Soliva R, Taillefer A, Bruguier O, Bellanger M, Bonno M, Mar-
    tin C (2020) Tracking geothermal anomalies along a crustal fault using (U–Th)/He apatite
    thermochronology and rare-earth element (REE) analyses: the example of the Têt fault
    (Pyrenees, France). Solid Earth 11:1747–1771
Rosset C (1964) Les formations du basin oligocène de Sigean-Portel et leur chronologie.
    C R Somm Soc Géol Fr, pp 415–417
Rusch L, Boulbès N, Lartigot-Campin AS, Pois V, Testu A, Bahain JJ, Falguères C, Shao Q,
    Moigne AM, Saos T, Boutié P (2019) Contexte paléoenvironnemental et chronologique
    des occupations néanderthaliennes de la grotte des Ramandils (Port-la-Nouvelle, Aude,
    France): apport des restes de grands mammifères. Quaternaire 30:151–165
Salel T, Bruneton H, Degeai JP, Mulot M, Lefèvre D (2019) Nouvelles données sur la
    dynamique des environnements fluvio-lagunaires de la basse vallée de l'Aude (France)
    au cours des sept derniers millénaires. Quaternaire 30:351–368
Séranne M (1999) The gulf of lion continental margin (NW Mediterranean) revisited by IBS:
    an overview. In Durand B, Jolivet L, Horvath F, Séranne M (eds) The mediterranean
    basins: tertiary extension within the alpine orogen. Geol Soc Lond Spec Publ 156:15–36
Sissingh W (2006) Syn-kinematic palaeogeographic evolution of the West European plat-
    form: correlation with alpine plate collision and foreland deformation. Neth J Geosci
    85:131–180
Zeck HP (1999) Alpine plate kinematics in the Western Mediterranean: a westward directed
    subduction regime followed by slab roll-back and slab detachment. In: Durand F, Jolivet
    L, Horvath M, Séranne M (eds) The mediterranean basins: tertiary extension within the
    alpine orogen. Geol Soc Lond Spec Publ 156:109–120

From Perpignan back to Perpignan (150 km, excluding optional excursions; 1 to 2 days depending on number of stops, optional loops, and walks). See Fig. 7.1.

Itinerary 2 focuses on the Neogene evolution of the Roussillon extensional basin. Its most characteristic outcrops are examined successively along the southern, western and northern boundaries of the basin. At Perpignan, the Natural History Museum has a display of Pliocene vertebrate fossils from the region (Philippe and Bourgat, 1985), although the majority of specimens are kept by the Natural History Museum in Paris and at the university of Lyon. A climb up to the Palais des Rois de Majorque will reward you with a wide view of the plains of Roussillon and the surrounding hills and mountains.

© Springer Nature Switzerland AG 2022   197
M. Calvet et al., *Geology and Landscapes of the Eastern Pyrenees*, GeoGuide,
https://doi.org/10.1007/978-3-030-84266-6_7

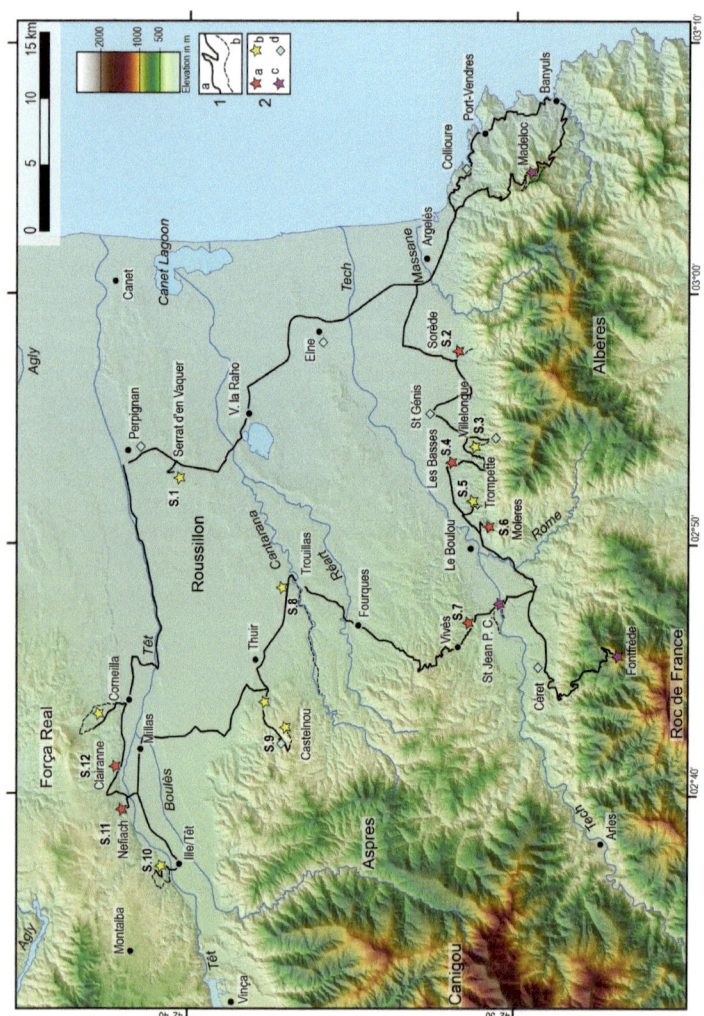

Fig. 7.1 Itinerary 2: route map and overview. Key to symbols—**1**: roadways, irrespective of road rank, size, or quality. **1b**: Trailwalks. **2**: stopping points (outcrops, exposures, landforms, or landscapes). **2a**: roadside stops. **2b**: roadside stops involving a short walk, or marking the start of a more major trailwalk. **2c**: subsidiary stops. **2d**: cultural stops. Labels S.1 to S.12 locate the sequence of roadside stops

Box 7.1 The Roussillon Basin, general geological background

The Roussillon Basin (Fig. 6.2) is a Neogene half-graben, an onshore component of the Western Mediterranean rift system (Nercessian et al. 2001; Mauffret et al. 2001; Gallart et al. 2001; Bache et al. 2010). At the surface, the master fault occurs on the south side of the basin (Tech and Albères faults, which present a succession of E–W and NE–SW segments), with secondary faults within the basin and along its northern edge. At crustal scale, however, the Roussillon Basin is controlled by a shallow-angle detachment fault dipping SE that can be traced to a depth of 11 km. Based on this crustal model, a fault such as the Tech Fault requires to be interpreted as an antithetic fault of secondary importance, and requires the presence of transfer faults between the grabens of the rift system. A major NW–SE, right-lateral transfer fault (Fig. 6.2) controls the offshore termination of the Albères massif (Mauffret et al. 2001).

The syn-rift deposits attributed to Oligocene and Aquitanian extensional tectonics are a collection of rubified breccia outcrops preserved on the edge of the Corbières (see Itinerary 1) and of the Aspres ('Thuir breccia'; see **Stop 9**), respectively. Most of the Miocene fill onshore is continental in origin and clastic in nature (lower 'Série Rouge' of the Céret Basin: see **Stops 6, 7, 8**; and Conflent Basin: see Itinerary 6). It has been also reported from the Ponteilla borehole (Fig. 7.2), with subsequent onlap by the marine mid-Miocene sequence (recognised in the Canet and Elne boreholes, Fig. 7.3). Offshore (Cravatte et al. 1974), the entire Miocene sequence is marine, with an offlapping trend during the Tortonian. The Tortonian beds were themselves severely eroded during the eustatic lowstand caused by the Messinian Salinity Crisis (Fig. 7.3B). The Messinian disconformity (ravinement surface) is well documented across the continental shelf (Gorini et al. 2005; Lofi et al. 2005) and around the edges of the Roussillon Basin, where displacement along Pliocene and post-Pliocene faults has nonetheless interfered with (and thus blurred) the stratigraphic and geomorphic effects of the sea-level fall (Fig. 7.2).

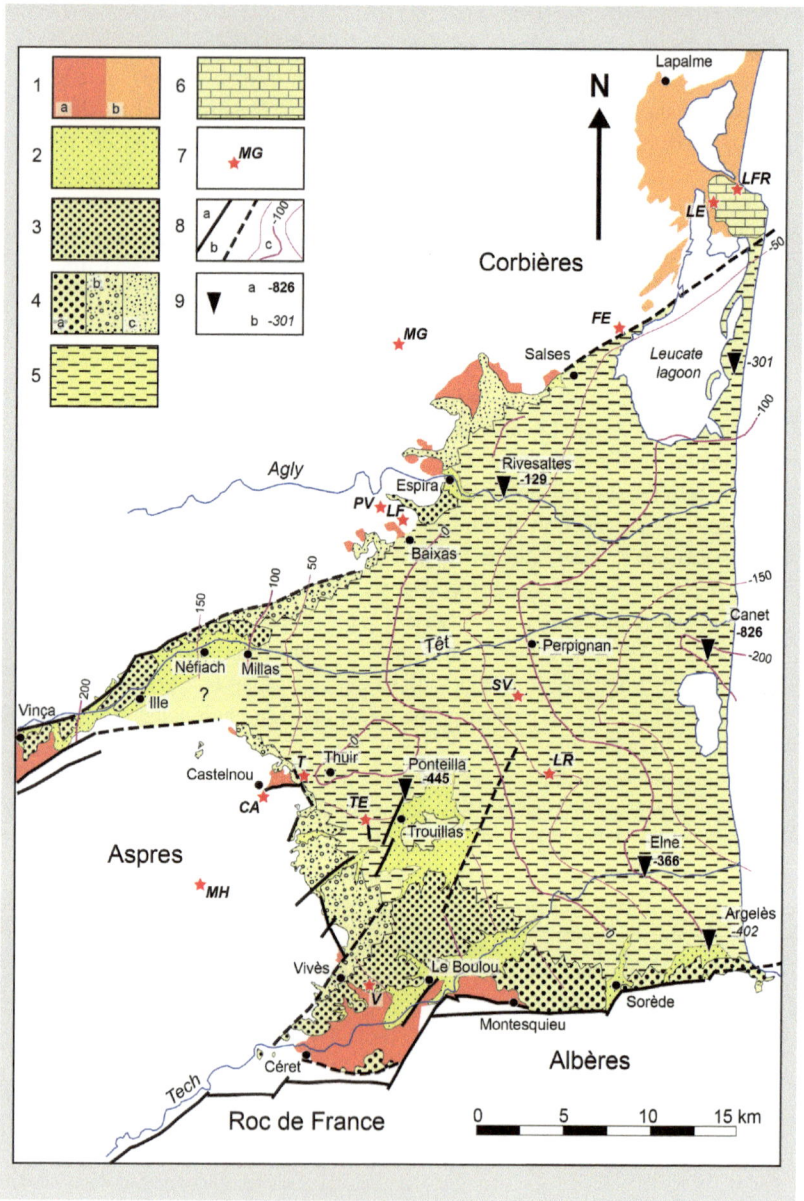

◄**Fig. 7.2 Map of the Roussillon, a Neogene sedimentary basin** (after Calvet, 1996, redrawn and modifed). Key to symbols and ornaments—**1**: Pre-Pliocene sedimentary sequences; **1a**: Oligocene to Miocene continental deposits, chiefly alluvial fans of the Conflent and Céret basins and Lower red limestone-breccia ('Série rouge inférieure', Aquitanian to Burdigalian); **1b**: Miocene marine depocentre. **2**: Lower Pliocene marine sequence, clay and sandy littoral or deltaic facies (Zanclean). **3**: Lower Pliocene alluvial fans of the rivers Têt and Tech. **4**: Lower Pliocene, locally sourced clastic formations; **4a**: Bouldery formation of the Albères piedmont zone; **4b**: Red beds with schist debris; **4c**: Upper red limestone-breccia. **5**: Lower Pliocene floodplain facies (bedload channels with flood-basin silts and post-depositional calcrete). **6**: Lower to mid-Pliocene marl and travertine. **7**: Pliocene microvertebrate fossil site. **8**: Stratigraphic and structural boundaries; **8a**: Exposed boundary fault; **8b**: Buried fault; **8c**: Depth contours of the Pliocene marine-to-continental sequence boundary (after Salvayre and Sola, 1982; Duvail et al. 2001, modified). **9**: Borehole; **9a**: Confirmed base of the Pliocene sequence; **9b**: Depth reached within the Pliocene sequence. **Pliocene microvertebrate sites**. Occurrences as fossil-filled fissures (FFFs)—MG: Mas Genegals, PV: Pla de la Ville, LF: Lo Fournas, MH: Mont Hélène, CA: Castelnou, FE: Font Estramar. Occurrences within the fill-sequence stratigraphy—LR: Villeneuve de la Raho, SV: Serrat d'en Vaquer–Perpignan, TE: Terrats–La Jasse, V: Vivès, T: Thuir, LFR: La Franquie, LE: Leucate.

The geometry of the Pliocene fill sequence is partly influenced by syntectonic displacements (Calvet 1996; Carozza 1998) but is also conditioned by the irregular palaeotopography that resulted from fluvial dissection driven by the Messinian sea-level fall. The resulting ravinement surface at places involves deeply incised and ramified valleys, which were abruptly filled and buried by the Zanclean parasequence during subsequent sea-level rise (Clauzon and Cravatte, 1985; Clauzon et al. 1987, 1990; Clauzon, 1990; Duvail et al. 2005; Gorini et al. 2005). The distribution of the bluiesh clay and deltaic sand deposits forming the base of the Zanclean beds indicates that the sea initially filled the basin to its edges, but that the marine sequence was soon covered by a prograding sequence of sediments supplied by the hinterland. In their proximal part, these exhibit the facies of torrential debris fans, whereas more distal portions involve silt-dominated flood-basin facies with calcrete layers and a network of mixed load shoestring channels (Serrat d'en Vaquer, **Stop 1**). Onshore, the Zanclean sequence is the only legacy of a short-lived marine environment in the Roussillon Basin. It has been dated at its top and base by vertebrate fossil assemblages (Fig. 7.4), including by the Perpignan faunal assemblage (**Stop 1**). This sediment package is quite thick (826 m at Canet, 366 m at

Elne and 402 m at Argelès, Figs. 7.2, 7.3). It consists of a series of six prograding systems, and exhibits a basinward shift east of the coastline to a Piacenzian (3.6–2.588 Ma) clastic wedge. The latter further connects to a progradational and subsequently aggradational parasequence of Gelasian (2.588–1.806 Ma) to Pleistocene age. The switch to aggradation indicates renewed margin subsidence (Fig. 7.3B; Duvail et al. 2005; Lofi et al. 2003). A hinge zone operating close to the current coastline has controlled the spatial distribution of tectonic subsidence within the Roussillon Basin since at least Gelasian time. It has accordingly promoted topographic dissection of the hinterland and the formation of 5 major generations of alluvial terraces in the landscape (Calvet 1996; Delmas et al. 2018) (see Itinerary 1, Itinerary 6, and the closing section of this itinerary).

►**Fig. 7.3** **Sections across the Roussillon Basin interpreted from seismic profiles. A** N–S, onshore (after Duvail et al. 2000, Calvet et al. 2015). Note compression-related arching of the cover sequence beneath the Tech and north of the Têt. Arrows locate incision maxima in the ravinement surface caused by the late Messinian sea-level fall. Incision of the sequence boundary is shallow, suggesting that the deep canyons, which are emblematic elsewhere of the Messinian Salinity Crisis, are not a prominent feature of this region. **B** E–W correlation between on- and offshore sequences (after Duvail et al. 2005). Tectonic deformation of the syn-rift sequence is intense (listric faults, tilted graben stratigraphy) but more subdued in the post-rift Pliocene sequence (some deformation also apparent in section A). **C** Location of cross-sections and exploration wells

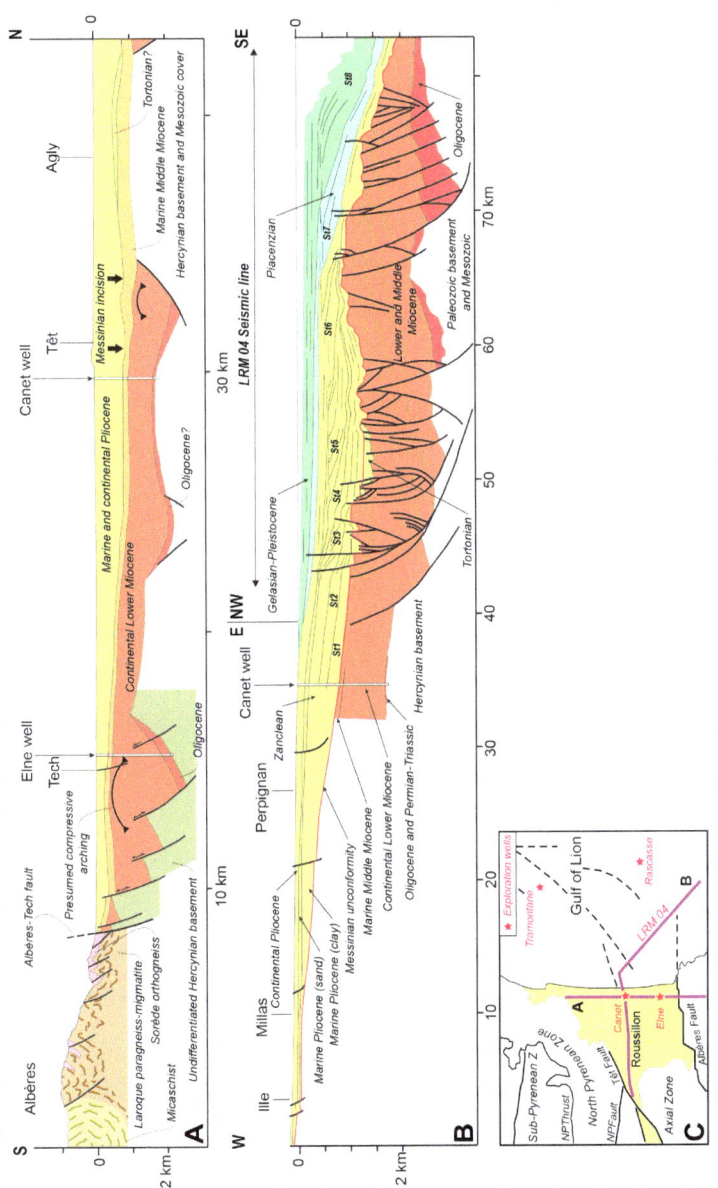

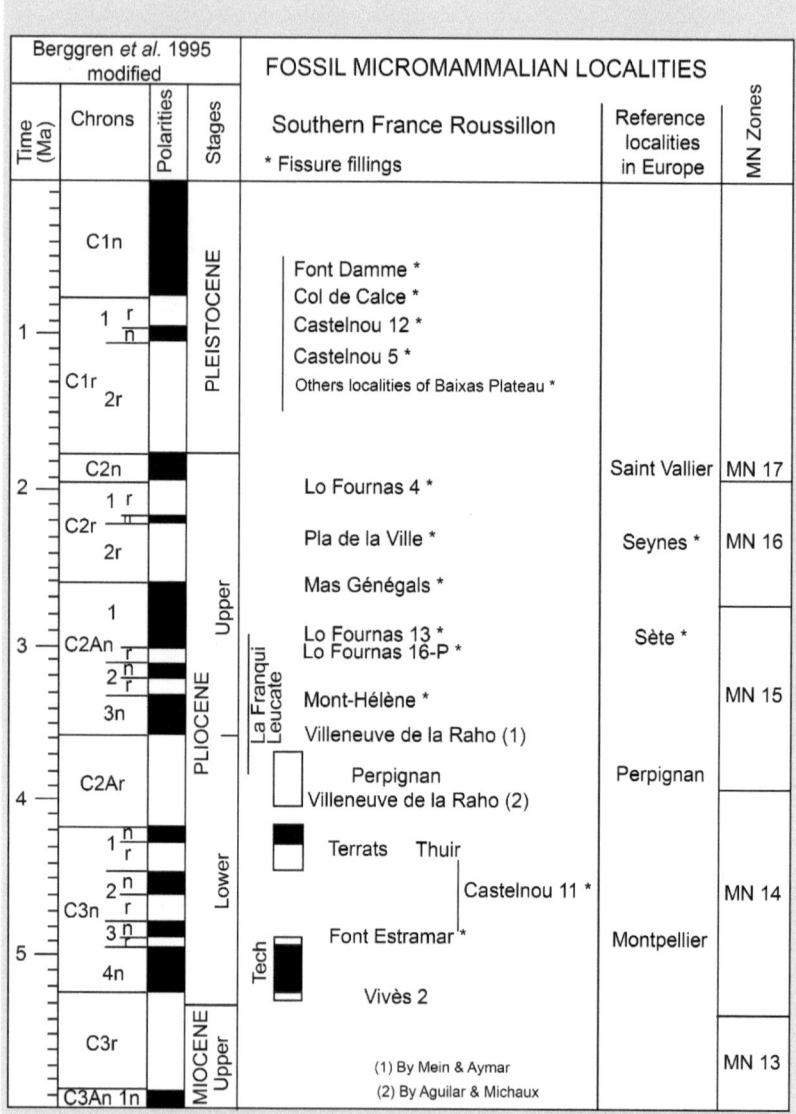

Fig. 7.4 Pliocene chronostratigraphy of the Roussillon based on micromammalian index fossils (after Aguilar, unpublished)

Drive out of Perpignan on the road signposted to Spain. At the fourth roundabout in the 'Zone commerciale Sud', exit on your right on the D900. You should pass under the fourteenth century Arcades aqueduct, still in operation and hosting the Canal de Perpignan (see also Itinerary 6). At the following roundabout, follow the signs to the left to Serrat d'en Vaquer, a fort constructed in 1885 and now converted to a municipal park. Access is on foot.

Stop 1. Serrat d'en Vaquer, type locality for the Ruscinian period

The top of the fort offers a 360° panoramic view of the Roussillon. The palaeontological site was discovered during fort construction but is no longer visible. In addition to a few other sites around the city (e.g., some brick quarries around Perpignan), it yielded fossils of Pliocene biota in sufficient diversity and abundance for the Ruscinian to be elected to the chronostratigraphic rank of 'age' in the European Land Mammal biozone chronology (ELMA). The Ruscinian (biozones MN 14 and MN 15, 5.3–3.5 Ma) is the continental equivalent of the Zanclean, or Lower Pliocene. The most characteristic species are represented by a primate, *Dolichopithecus ruscinonensis* (Fig. 7.5C); by various carnivores: *Machairodus cultridens, Viverra Pepratxi, Vulpes Donnezani, Hyoena arvernensis*; by bears, a beaver, various rodents; *Mastodon arvernensis, Rhinoceros leptorhinus, Tapirus arvernensis, Hipparion crassum, Gazella Borbonica*; various *Cervidae*; birds; and by tortoises, including a giant species with a shell > 1 m in diameter: *Testudo perpiniana* (Fig. 7.5B), which indicates an absence of cold winters in the region during the Pliocene (Depéret 1885, 1890).

Exposures of the fluvial deposits can be observed below the fort on its south side, in the embankments of the shopping centre, and west of the hill below elevation spot 74 m (indicated on topographic maps) (Fig. 7.5A). The alluvium is predominantly silt and contains calcrete levels, with meandering sand-filled channels containing occasional gravel clasts. Three small normal faults (1 m throws) offset the alluvial beds near the entrance to the shopping-centre car park, but the exposures are difficult to inspect at close quarters (Fig. 7.6).

Return to the D900, and leave it using the slip road after the shopping centre by taking the D91 towards Villeneuve-de-la-Raho. This road will lead you through low, vineyard-covered hills consisting of Pliocene sediment and capped by vestiges of older Quaternary alluvial terrace deposits. It then strikes across terrace T3 (Fw on geological maps) of the Réart River before crossing it. The river channel is usually dry but prone to catastrophic flash floods, e.g., from 0 to 1000 m³/s in just 3 h on Sept. 26, 1992; this flood occurred in response to a rainstorm which generated 329 mm of precipitation in 15 h, including a peak of 189 mm in 3 h. The road then skirts around the edge of the circular (Pleistocene) deflation pan of La Raho, now

Fig. 7.5   The Ruscinian stratotype at Serrat d'en Vaquer. **A** Outcrops of Pliocene fluvial deposits (floodplain and channel-fill). **B** Shell and breast plate of *Testudo Perpiniana*, a large land tortoise unearthed near the gateway to Serrat d'en Vaquer fort (Muséum National d'Histoire Naturelle, Paris). **C** Face and skull of the adult male primate, *Dolichopithecus ruscinensis,* discovered at Serrat d'en Vaquer (Museum National d'Histoire Naturelle, Paris), here a reproduction of the sketch made by C. Depéret in 1890

converted to a water reservoir. After driving through Villeneuve, take the D914, which crosses the low Holocene alluvial plains of the Réart and Tech rivers (T0, i.e., Fz) and passes Elne, ancient capital of the Roussillon and Medieval bishop's see (eleventh century Romanesque cathedral and 12–thirteenth century cloister) perched on a hill carved out of Pliocene continental beds. Drive to Argelès, then follow the D618 and D11 to Saint-André and Sorède.

Fig. 7.6 Normal faults at Serrat d'en Vaquer. Note how they offset the calcrete-encrusted floodplain deposits as well as, to the right, a sand- and gravel-filled palaeochannel. Site location: disused track cut in the embankment overlooking the shopping-centre parking area, below the water tower (location: 42°40′14″N, 2°52′59″E). Fault characteristics indicate fault strike direction, fault dip and direction, and fault striation rake

Optional excursion: exploring the Côte Vermeille (40 km)

From Argelès, and all the way to Le Racou, the expressway crosses the last outcrops of continental Pliocene boulder beds before entering the Hercynian basement of the Albères. The basement rocks here consist of Ediacaran (Canaveilles Group) and Lower Cambrian (Jujols Group) metasedimentary series, which form a monotonous succession of schist intruded by pegmatite and microgranite veins around the migmatite core of the Albères (Laumonier, in Calvet et al. 2015). Drive up the small D86 to Madeloc tower, which stands on a sandstone ridge. Along the way you will catch views of the terraced vineyard of Banyuls, which have been organised into a sophisticated contributary network of paved stormwater evacuation channels lined with stone walls (Fig. 7.7A, B). The rocky coast is a succession of drowned valleys forming narrow bays, but several segments are controlled by NW–SE Neogene faults. One of these controls the straight coastline between Le Racou and Cap Béar, and further offshore puts Pliocene deposits and the basement in abrupt juxtaposition. A two-hour loop can be walked from Casernes du Centre to Madeloc Tower via Col de Taillefer and the ridgetop footpath, then back along the road. From the summit you get a far-reaching view of the Albères extending from Pic Sallfort (981 m) to Maçana tower. In a succession progressing from cordierite–andalusite micaschist, to sillimanite micaschist, to migmatitic anatexite, metamorphic grade increases westward. The Albères still harbours the Couloumates beech forest (600–900 m), a relict, cool-climate enclave at the heart of the Mediterranean biome. The Medieval towers are vestiges of a network of signalling towers designed to ensure coastal surveillance against pirates. The D86 descends to Banyuls (famous wines and related attractions); you can then return to Argelès via the D914 through Port-Vendres and Collioure.

Fig. 7.7   Eastern Albères and Banyuls vineyards. A Panoramic view from Tour de Made-loc road. Interfluves are in schist; in distance: the Roussillon Basin and its sandy beaches. **B** Traditional vineyard features; the network of rainwater collector channels is named *Peu de gall* (rooster's foot, in Catalan)

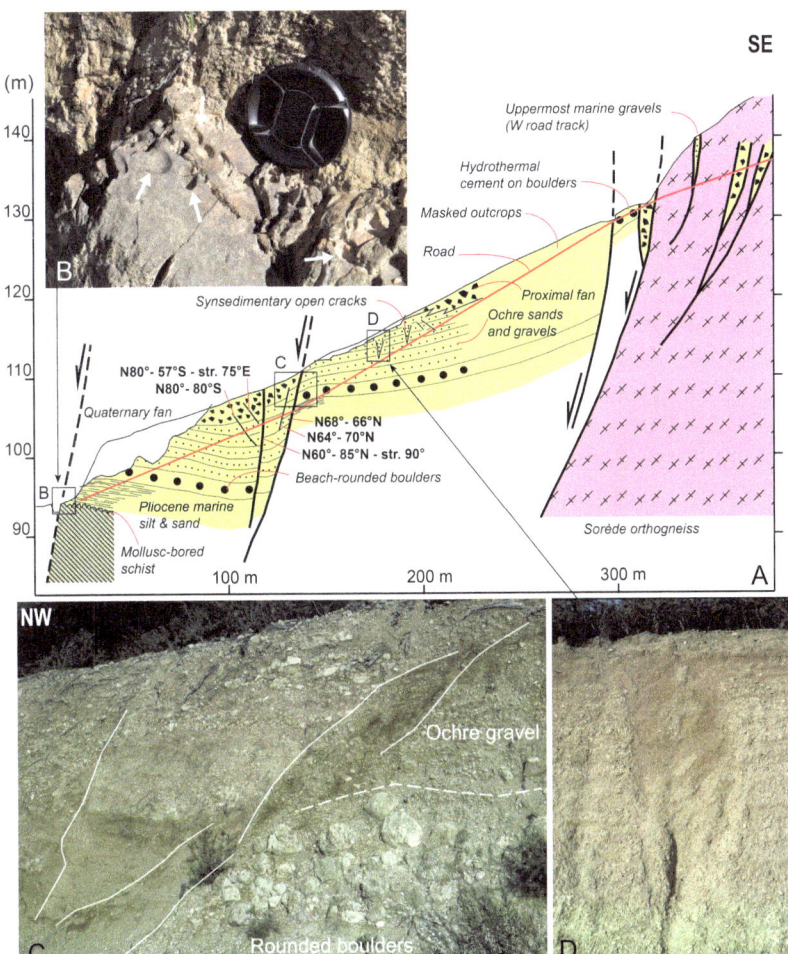

Fig. 7.8 Faulted Pliocene sequence at Sorède, along the track to Ultrera. **A** Geological section (see text for details); fault characteristics indicate fault strike direction, fault dip and direction, and fault striation rake. **B** *Lithophaga* holes bored into in the schist, covered here by marine sandy silt (location: 42°31′34″N, 2°57′36″E). **C** Detail of main fault plane (42°31′32″N, 2°57′39″E). **D** Detail of synsedimentary extensional cracks in ochre beach-gravel beds

Stop 2. Sorède: faulted boundary between the Albères basement and the Pliocene series

The villages of Saint-André and Sorède are located on Pleistocene debris fans with their apices at the Albères mountain front. They form two weakly offset treads T1 (noted Jy on geological maps) and T2 (Jx), both deeply incised by modern stream channels. At Sorède, cross the bridge and turn right in the old village. Look out for the trail leading up to Notre-Dame d'Ultrera. You can park your vehicle near the last few houses at the foot of the hill.

The Ultrera trail embankment exposure is 380 m long, somewhat degraded but its features still distinguishable (Fig. 7.8). Marine Pliocene beds (blueish sandy clay) occur at the base but are not well exposed. However, as you enter the footpath leading past the back of the houses on the left side of the street, you may observe some cobbles of schist preserved in situ and bearing date-mussel holes (Fig. 7.8B). Above the blue clay beds, a coastal deposit ~1 m thick and containing large, well-rounded pebbles and boulders (including intensely rolled quartz clasts) is covered by several metres of alternating ochre beach sands and gravel beds. The sand and gravel beds display synsedimentary open cracks (Fig. 7.8D). The basal deposit (coarser gravel) is upthrown by a fault (Fig. 7.8C), and upthrown further still at the top end of the trail (Mas Cordovès), where another outcrop can be observed opposite a small oratory. Here, the debris are cemented by silica, probably from hydrothermal fluids. Continuing onwards up the hill, the embankments of both tracks reveal several tectonic slivers of basement bedrock (gneiss) alternating with Pliocene breccia and beach quartz pebbles. The cumulative throw of these normal faults is ~50 m.

Cultural stop: St-Genis-des-Fontaines. This abbey was founded in the eighth century after Pepin the Short and Charlemagne recaptured the north of the califate of Cordoba. The white marble lintel is one of the oldest Romanesque sculptures in Europe. An inscription ascribes it to the 24th year of the reign of Robert II the Pious, king of France (996–1031 CE). The late Romanesque cloister (thirteenth century) was sold after the French Revolution; one part of it is preserved at the Philadelphia Museum of Art, and the other has been recently reconstructed on site.

From Sorède, the road crosses alluvial fans T1 and T2, with vestiges of generation T3 (small hills at Laroque) rising above their treads. The fan heads occur at the base of the fault-controlled Albères mountain front, where triangular faceted spurs increase in tallness towards the west. Despite this clear pattern in the scarp morphology there has so far been no geological evidence of faulting of the middle and late Quaternary fan deposits themselves. Aim for Villelongue-dels-Monts, and at the village follow for 1.3 km the track leading to the priory of Santa Maria del Vilar. Parking space is available at the start of the forest trail (42°30'52"N, 02°54'09"E).

Stop 3. Pliocene alluvial fan at Puig Janer (30 min round trip, on foot)
The forest trail crosses the stream before rising westward. You may observe a pegmatite and leucogranite intrusion in the migmatite before passing a few metres of breccia containing pegmatite clasts and reaching the Pliocene formation itself, which contains boulders of granite, gneiss and migmatite. It can then be followed all the way up to Puig Janer. Its boundary with the basement is fault-related but not clearly identifiable without trenching; a small N70°E normal fault plane has nonetheless been observed. The Pliocene alluvial fan sequence here is 80–100 m thick, contains boulders several metres in diameter, and is the typical product of an active fault scarp. The summit of Puig Janer offers a good perspective view of the alignment of triangular faceted spurs at the mountain front. Each individual facet is 100 to 150 m high, with evidence of more degraded facets further up the slopes, each time forming a staircase of erosional benches and/or truncated vertices. The landform assemblage suggests multiple stages of fault slip and erosion, with tectonic activity clearly occurring during the Pliocene, but perhaps also during the early Pleistocene in the case of the lowermost, better preserved spur facets (Figs. 7.9, 7.10 and 7.11C).

From Villelongue, follow the D11 to Montesquieu. As you leave Puig Janer, note the large boulders on its western slopes. The road then follows the boundary fault. Its position is underlined by a wide, grey-coloured belt of uncemented crushed rock (an outcrop of it can be inspected on the left-hand side of the road, just before entering the village of Montesquieu). In the village, take a first right (Chemin du Roy). Continue as far as elevation spot 90 m, an area marked on maps as Les Basses (42°31'53"N, 02°53'04"E).

All of the Neogene sediment beds present in the Roussillon Basin can be inspected in the area situated between Montesquieu and Céret around Le Boulou (**Stops 4, 5, 6, 7**), thus allowing their links with the Albères horst to be examined (Figs. 7.9, 7.10A). The Lower Miocene consist here of a thick clastic sequence known as 'Série Rouge inférieure'. The Moulas sandstone (see **Stop 6**) is a marine facies of uncertain Miocene or late Oligocene age. Incised valley fills ascribable to fluvial incision during the Messinian Salinity Crisis are well preserved. The Pliocene outcrops exhibit terrigenous facies ranging from debris-flood bedload deposits forming topset beds, to sandy deltaic foresets, to blueish argillaceous marine bottomset beds.

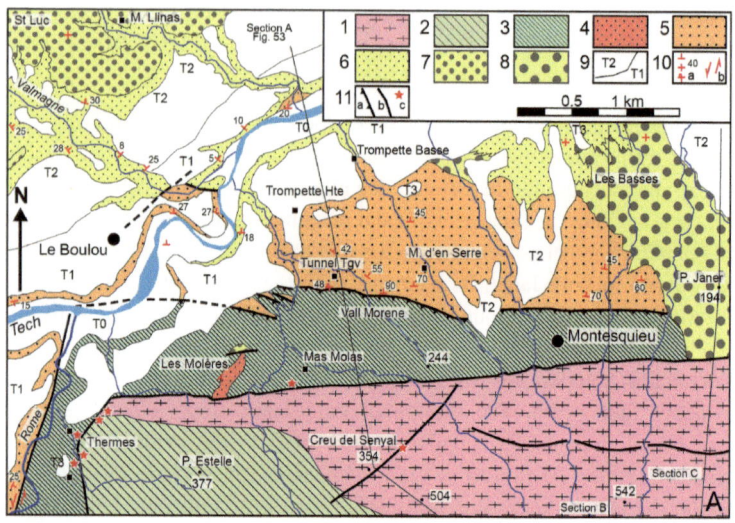

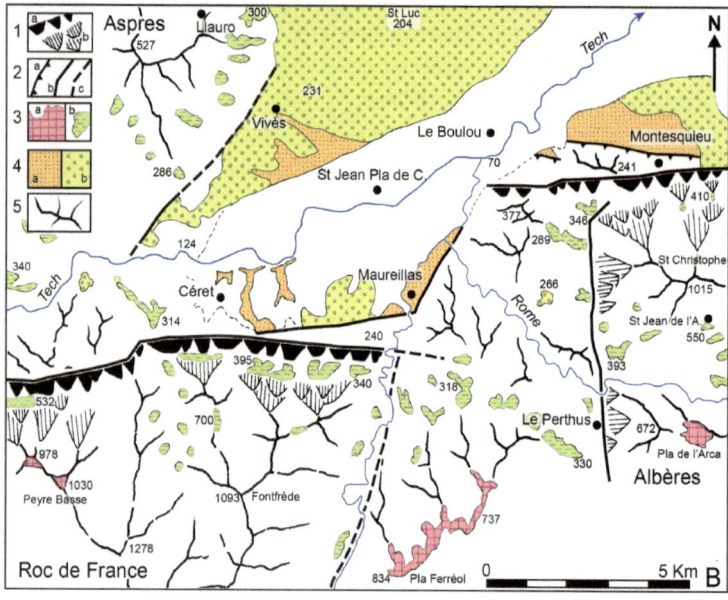

◄**Fig. 7.9 Geology and geomorphology of the Albères piedmont.** A Geology of the
north Albères piedmont (after Calvet, 1996). Key to symbols and ornaments—Hercynian
basement: **1**: Paragneiss and migmatite. **2**: Micaschist (Lower Canaveilles Fm.). **3**: Epimeta-
morphic schist (Jujols Group). **4**: Marine Moulas sandstone (Oligocene or early Miocene).
5: Miocene 'Série rouge', including the Trompette boulder-bed facies. **6**: Marine and deltaic
Pliocene formations (gravel, sand, blueish clays). **7**: Far-field continental Pliocene formations
(ancient alluvial fan of river Tech). **8**: Near-field (transverse) continental Pliocene formation
of Puig Janer (transverse boulder fan). **9**: Quaternary alluvium and terraces (chronological indices compatible with generations T0 to T3). **10**: Rock deformation features; **10a**: dip
(degrees), horizontal beds, overturned beds (dips in the Pliocene deltaic beds may be primarily depositional); **10b**: left-lateral offset in pebble beds of the 'Série rouge' (Rome Fault). **11**:
Tectonic features; **11a**: Miocene reverse fault; **11b**: Neogene normal faults and other fractures; **11c**: Thermal springs associated with the Neogene faults. B Geomorphological map
of the Albères fault-controlled range front (after Calvet, 1999). **1**: Faceted spurs, **1a**: well
preserved, and **1b**: degraded. **2**: Fault characteristics; **2a**: Neogene reverse fault, **2b**: Normal fault, **2c**: probable or blind fault. **3**: Pediment surfaces; **3a**: Vestiges of mid-Miocene
pediment (P1); **3b**: Vestiges of later Neogene erosional benches (pediment P2). **4**: Neogene
clastic sequences; **4a**: Miocene continental sequence (Série rouge); **4b**: Pliocene debris-
cone deposits containing large boulders and basal marine Pliocene beds. **5**: Ridge-and-ravine
topography in dissected basement outcrops

Stop 4. Chemin des Basses
From here you can appreciate the faceted fault scarp of the Albères and the
hilly terrain underlain by the Pliocene boulder beds of Puig Janer. Given that the
Pliocene formation is faulted along the entire length of the Albères range front,
it is likely that the base of the faceted spurs (100–150 m of relief) is Pleistocene
or late Pliocene in age. The topographic level of mid-Pliocene pediments in the
Albères (generation P2) forms hanging benches above the faceted spurs (Figs. 7.9,
7.11), and these pediment vestiges have undergone clear vertical offset further to
the west near Céret. Substantial dip-slip movement also occurred during the early
Pliocene given the ~100 m thickness of boulder-rich deposits at Puig Janer. As
a result, vertical uplift along this range front has likely totalled several hundreds
of metres during the last 5 Ma (Calvet 1996, 1999).

From elevation spot 90 m, the roadside section reveals a formation (equivalent
to the Puig Janer Formation) containing large rounded boulders of Laroque gneiss
(migmatites) and other granitoid rock. These overlie and prograde over Pliocene
ochre-coloured coastal gravels, identical to those observed at Sorède, and greyish-
blue submarine deltaic sand deposits (Figs. 7.9A and 7.11B).

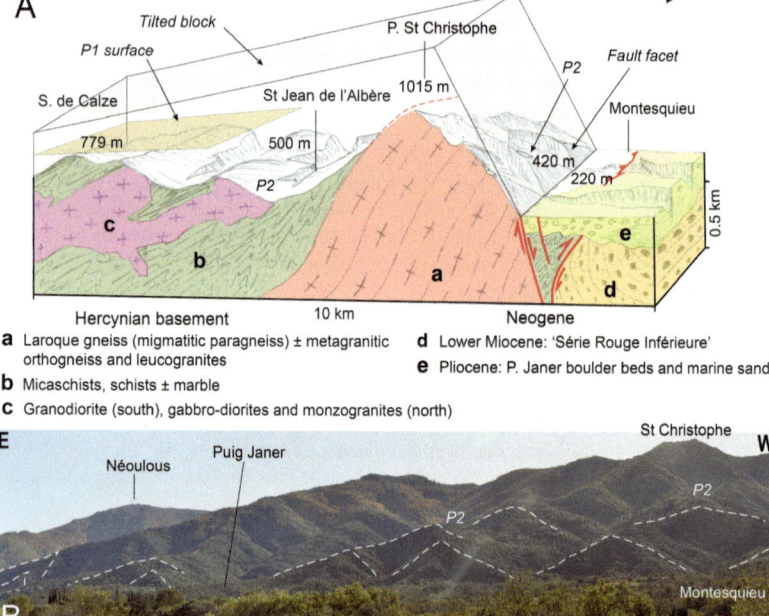

Fig. 7.10   The Albères massif, a Neogene demi-horst. A Schematic block-diagram near Montesquieu. Note possible existence of pediment P1 on the south side of the range, and size of P2 embayments in micaschist and gabbro-diorite outcrops around Saint-Jean and Saint-Martin-de-l'Albère. **B** Panoramic view of the fault scarp and its faceted spurs near Villelongue and Montesquieu

At the base of the exposure at Les Basses, turn left and follow the country lanes through Mas Santraille, Trompette Basse and Trompette Haute, where at one point you take an underpass beneath the high-speed rail track (TGV) linking Paris to Barcelona. Having passed the outcrop of Miocene deposits on the north side of the hill, turn into the TGV tunnel service road, where you may park your vehicle in the open space near a helipad. From here you can walk up the forest track (20 min round trip) until you find an exposure of the reverse-faulted contact with the metamorphic basement.

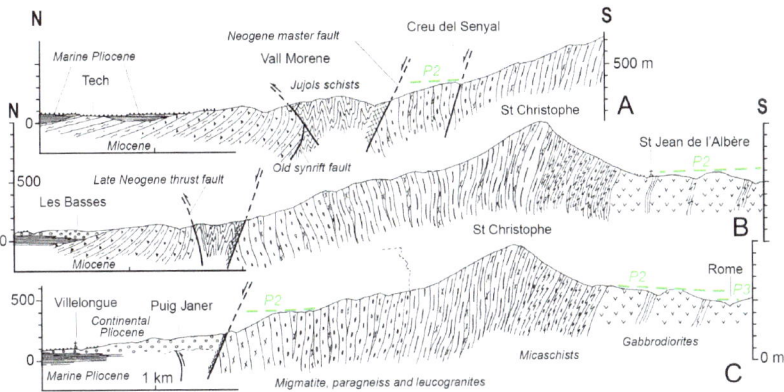

Fig. 7.11 **Geological sections through the north Albères range front** (after Calvet 1996). Sections located in Fig. 7.9A

Stop 5. Trompette and TGV tunnel exposures

This section exposes the Miocene 'Série Rouge inférieure' (Figs. 7.9A and 7.11A). At Mas Trompette Haute (42°31'36"N, 02°1'13"E), the 'Série Rouge' (Fig. 7.12A, B) is coarse-textured, with at its top some intraformational boulders of granodiorite and migmatitic gneiss several metres in diameter. Dips of 30° towards the north progressively steepen to 90° in Balmourène gully (Fig. 7.11A). Total thickness here is 650 m. The petrographic composition of the pebble assemblages changes rapidly near the base of the formation, a testimony to the intensity of denudation cutting through the diverse outcrops of the Albères metamorphic dome. Clasts from the deeper migmatitic gneiss (Laroque gneiss) only begin to appear towards the top of the sequence, with schist, granitoids, and some marble and diorite forming the entire basal assemblage. Given the absence of offlapping contacts in the stratigraphic architecture, it can be inferred that tectonic deformation of the 'Série Rouge' occurred mostly after deposition of the entire sequence had been completed (this inference is based on the absence of progressive unconformities, i.e., angular syntectonic unconformities resulting from rotational offlap caused by accelerated uplift of the hinterland). On the surface, the 'Série Rouge' appears to be overthrust by the metamorphic basement (Fig. 7.12C), but this could just result from minor reverse motion affecting the older normal fault plane. The compressional structure is nonetheless evident from crushed alluvial pebbles throughout the entire exposure. The episode of convergent deformation appears to

◀**Fig. 7.12**   The Miocene 'Série rouge' and its tectonic deformation features in the Albères foothills: **A** TGV embankment trench at tunnel entrance (east side). Note N to NNE dips, conformable bed sequence, and numerous fractures. **B** 'Série rouge' at Mas Trompette Haute, here displaying lateral inputs from the Albères. **C** Reverse-fault contact between the basement (schist) and the 'Série rouge' in Vall Morene ravine (location: 42°31′14″N, 2°51′28″E). Fault characteristics indicate fault strike direction, fault dip and direction, and fault striation rake

be pre-Pliocene because the reverse fault between Montesquieu and Villelongue is buried by the basal unconformity of the Puig Janer Formation (Figs. 7.9A and 7.11C; **Stop 3**).

Box 7.2 General description of the Miocene 'Série Rouge'

The aggregate thickness of this continental series (Fig. 7.13), which underlies the Pliocene marine beds, was unknown until recently. Moreover, scattered exposures were often confused with other continental red-bed facies of Pliocene age, which also occur in the Roussillon Basin. Bourcart (1947) was the first to emphasise the unique and chronostratigraphically uniform character of these deposits, and proposed to classify them as Rhodanian—a stratigraphic equivalent of the Pontian (i.e., Upper Miocene or Messinian, in modern nomenclature). They signal a major 'mountain-building' event (Bourcart 1947, p. 417). Magné (1978) also recognised the uniformity of this series. He correlated it with the early Miocene continental sequences detected in the deep boreholes of Canet, Elne and Ponteilla, and with the early Burdigalian sequences of the Conflent Basin, which have been dated using microvertebrate fossil assemblages (see Itinerary 6). Geometric correlations with the chronostratigraphy of offshore exploration wells are also robust (Cravatte et al. 1974; Fig. 7.14). In the Céret Basin (Fig. 7.13A), it has been possible to infer from the steady south-easterly dip of these Miocene beds that the 'Série Rouge inférieure' is at least 2 km thick, and that it consists of very coarse clastic levels at its base (Vivès beds) and top (Lo Regatiu alluvial beds, which were laid down by an ancestor of the Tech; and Maureillas and Trompette beds, which consist of tributary fan deposits originating from the Albères). The middle sequence is mainly sand- and clay-textured (upper Vivès beds and

Mas Marty beds). Provenance analysis of clasts among the debris delivered by this early Neogene ancestor to the Tech River provides evidence of intense denudation in the Canigou massif. The Vivès and Mas Marty beds contain an almost exclusive assortment of schist, marble, micaschist, Triassic sandstone, and granitoid pebbles, clearly all supplied by the outermost metamorphic envelopes of the Canigou dome; clasts supplied by the augengneiss, which forms the deeper core of Mt. Canigou, only occur towards the top of the 'Série Rouge', notably in the Regatiu beds (Calvet 1986, 1996; Donzeau-Wiazemski et al. 2010). Given that the main erosion surface in the Canigou massif cross-cuts the gneiss outcrops and fabrics at Pla Guillem and belongs to generation P1 (see Itinerary 4), it can be inferred that the lull in base-level change, which provided suitable conditions for pediment P1 to form, occurred after the phase of intense denudation that supplied the 'Série Rouge'.

▶**Fig. 7.13 Stratigraphy of the Céret Basin and its Miocene 'Série rouge'** (after Calvet, 1996). **A** General section showing the Pliocene sequence resting over the tectonically deformed Miocene *Série rouge* (regional unconformity). **B** Mas Forcade exposure, showing angular unconformity between the Pliocene marine and Miocene sequences. Key to symbols and ornaments—**1**: Terrace T1. **2**: Bedding in blue sandy clay (marine, Pliocene). **3**: Ochre gravel beds containing lignite debris. **4**: Blueish fine sand (base of marine Pliocene sequence). **5**: Lower 'Série rouge', Lo Regatiú-type facies comprising rolled gneiss pebbles. **C** Schematic lithostratigraphy of the Miocene 'Série rouge' (after Calvet 1996). **1**: Channels filled with angular debris descended from the Albères. **2**: Cobble-filled channels containing abundant rounded clasts of gneiss. **3**: Cobble-filled channels containing rare gneiss clasts; some wider channels are debris-flow deposits rich in angular clasts (Mas Marty beds). **4**: Presence of intraformational, metre-sized boulders (granite, schist, marble, Triassic sandstone). **5**: Flood-basin facies (red sandy clay beds, grey arkose, and sporadic pebble layers). **6**: White marly limestone; thin, greyish brown palustrine beds. **7**: Observational hiatus. The widespread occurrence of white calcrete layers and nodules filling intraformational cracks is an important diagnostic feature of the Miocene 'Série rouge'. The younger Pliocene continental sequence, which exhibits similar sedimentary facies at exposure level and can, therefore, lead to misleading stratigraphic interpretations, is entirely decalcified. Granite and gneiss pebbles are also intensely weathered in the Miocene beds (mineral grains crumble to a powdery texture), whereas their counterparts in beds of Pliocene age rather have the consistency of grus

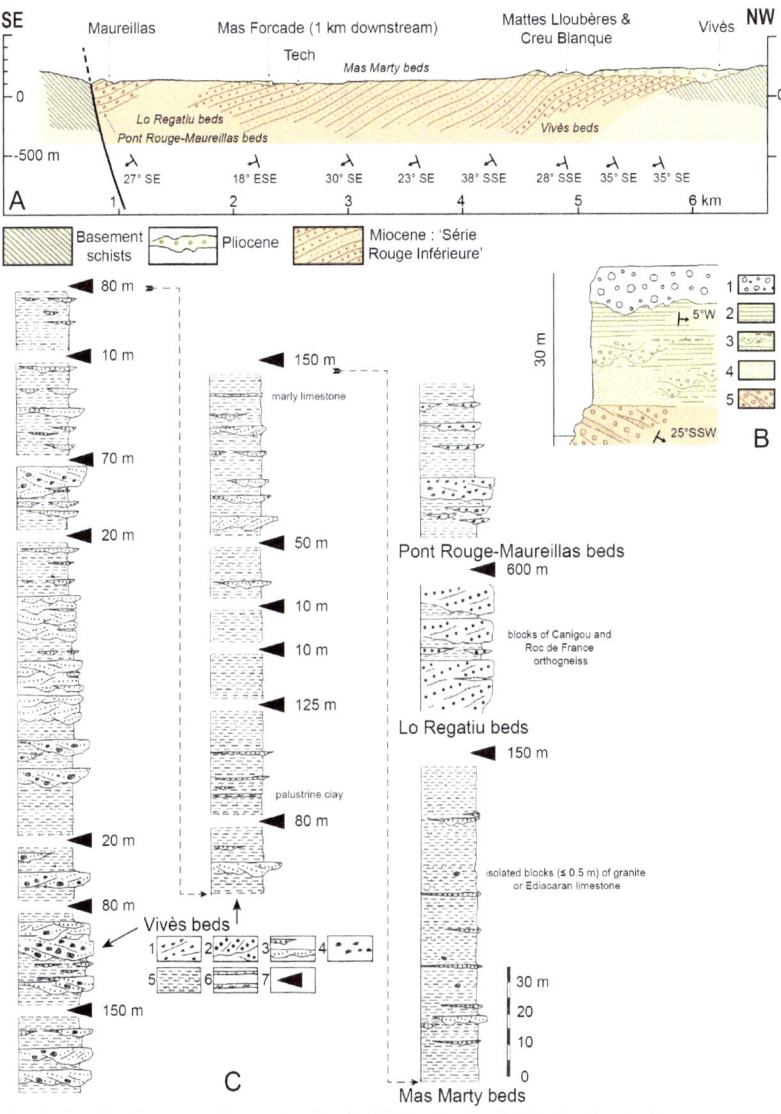

◄**Fig. 7.14 Facies features of the Miocene 'Série rouge'. A** Angular unconformity at Mas Forcade (location: 42°30′46″N, 2°48′40″E). **B** Sections at Mas
Marty (42°30′28″N, 2°46′40″E). **C** Vivès beds, mid-sequence facies (42°31′23″N,
2°46′49″E)

*Return to the road running parallel to the D618 and drive through Lo Naret housing
estate. At the eastern entrance to the estate, or behind the casino building of the spa
centre, drive up any of the two tracks leading to Moleres stone quarry. The quarry
(which was visible from **Stop 5**) is situated at the vertex of the lowermost faceted
spur, below elevation spot 242 m.*

Stop 6. Les Moleres quarry (meaning 'millstone'; often 'Moulas' in the geological literature)

The quarry face exhibits greenish grey to ochre arkose and sandstone (with
intraformational hummocky bedforms and recognisable tidal bundles) resting
unconformably on the Jujols Group (Lower Cambrian) schists (Fig. 7.15). The
green sandstones contain silicified woody debris and tree-trunk moulds. The clastic beds transition southward to deltaic foreset beds consisting of locally-sourced
conglomerates. The boundary (normal) fault clearly offsets the sandstones and
conglomerates (Fig. 7.9A). These deposits were originally believed to be of continental origin and were assigned a late Eocene age (Depéret, 1912), but more
recent revision of their fish fossil content (which includes perch) has validated
instead a marine origin and a mid-Miocene to Tortonian age (Gaudant 1999).
In summary, the palaeontological criteria are currently too discrepant to be of
any practical use in assigning a precise age to the sandstone unit. A Serravallian
to Tortonian age seems implausibly recent given the absence of this sandstone
formation, despite its mechanical resistance to erosion, from the stratigraphy of
the Boulou depocentre (just a few hundred metres from the Moulas quarry exposure). Given that the Pliocene sequence rests directly on the early Miocene 'Série
Rouge', it seems implausible to advocate that sandstones of Tortonian age were
entirely eroded away before loading by Pliocene sediments began, or alternatively
that the absence of sandstone in the Boulou depocentre is just a local stratigraphic
lacuna. In conclusion, a late Oligocene to earliest Miocene age is a more likely
option for the Moulas sandstone. This preferred interpretation concurs with evidence that the sandstone formation rests directly on the metamorphic basement
and should thus be stratigraphically older than the 'Série Rouge'. The evidence
also concurs with the presence of a similar sandstone formation containing plant
impressions at the base of the Elne borehole.

Fig. 7.15 **Les Molères quarry: Oligocene to Aquitanian base of the Neogene sediment fill of the Roussillon Basin**. **A** Main exposed section, showing green bedded sandstone and deltaic gravels descended from the Albères. **B** Close-up of tidal facies

Take the D900, then the D618 towards Maureillas and Céret. The road crosses the treads of fans T1 and T2, fed during the Pleistocene by the Roc de France massif. Note the fault scarp overlooking Céret. Like the north face of the Albères, this scarp is also staircase of faceted spurs indicating successive episodes of active faulting. The lowermost facets are very steep and display a trapezoidal morphology (Fig. 7.9B). Their vertices (elevation: ~400 m) are topped by a very smooth erosional bench compatible with P2, the regionally widespread generation of late Pliocene pediments.

At Céret, you may wish to visit the Museum of modern art, with paintings by Picasso, Chagall, Miro, Matisse… At the back of the town, a small road (the D13f) will take you up to Pic de Fontfrède (1093 m). The road rises up the faceted spurs, where vestiges of a pediment are well preserved in the weathered granite and gabbro outcrops around 700 m. The top offers good views of the Roussillon Basin.

Turning back to Céret, continue to Maureillas and turn down the D13 to Saint-Jean-Pla-de-Corts. Once at the bridge over the Tech, you can access outcrops of the 'Série Rouge'. The exposures along the left bank, upstream of the camping ground at Mas Marty, are remote (Fig. 7.14B), *but an equivalent exposure can reached by the roadside to the north of the campsite entrance (La Ribera de les Aygues); other outcrops—depending nonetheless on water levels—are easier to reach at Lo Regatiu, by following the right bank up river* (Fig. 7.13). *Down river, below Mas Forcade, a vertical exposure* (Figs. 7.13B, 7.14A) *shows a clear angular unconformity between the steeply-dipping 'Série Rouge' (southerly dips) and overlying marine Pliocene beds, but the section is difficult to get reach for close-up views.*

At Saint-Jean, take route D13 towards Vivès. The 'Série Rouge', displaying continuously steep dips to the SE, is exposed at several places along the way as well as in the stream bed, but access is limited (better exposures occur in the river bend upstream of where the Mata Lloberes ravine joins the Tech, 42°31'23"N, 2°46'45"E). As you enter the valley, two quarries—one to the NE and the other to the SW—provide exposures of the Pliocene sequences. The SW quarry (Mata Lloberes) can still be visited, although exposures today are no longer as good as they used to.

Stop 7. Pliocene beds at Vivès
Bedding in the Pliocene sequence is almost horizontal and forms an angular unconformity over the underlying 'Série Rouge' (Fig. 7.13A). The access road to Creu Blanque quarry, to the NE, cuts through the 'Série Rouge' in the gully, but quarry-face degradation rules out any clear views of the stratigraphy. The fresh exposures used to show the boundary between the greyish-blue submarine deltaic sands and the terrestrial debris-flood sequence produced by an ancestor to the Tech. This high-energy deposit displays a number of the large gneiss boulders that otherwise form much of the entire hill itself (Fig. 7.16A). At the top of the deltaic sequence, a black lignite bed documenting emergence of the continental margin was covered by palustrine marl. The marl beds have yielded an assemblage of terrestrial mammalian fossils typical of the base of the Pliocene (Vivès 2, Fig. 7.4). The Mata Lloberes quarry, still open and accessible to the SW (Fig. 7.16), contains blueish-grey exposures of lacustrine or paralic sands and clays (A) that contain antelope fossils. They are covered by the ochre debris-flood sequence (B).

Take the D13 towards Llauro. After Vivès, some Pliocene debris-flood deposits are exposed in road cuts, distinguishable by the platy schist debris in a bright red matrix. You then enter the schist outcrops of the Aspres massif, which are classified as Cambrian (upper Jujols Group). At Llauro, take the D615 towards Fourques. The

Fig. 7.16 Pliocene basal beds at Vivès (Mata Lloberes quarry). The Pliocene Tech fan is still well preserved (**A**), but the palustrine clays (**B**) are now only exposed in the upper quarry (location: 42°31′12″N, 2°46′37″E), and currently extracted for the brickworks at Saint-Jean-Pla-de-Corts

road exits the schist and strikes across the Pliocene fan, which was fed exclusively with schist from the Aspres massif; its bright red matrix is conspicuous among the badlands of the Forêt domaniale du Réart (Réart National Forest). Around Fourques you can observe outcrops of distal continental facies of beige and ochre argillaceous siltstone containing calcrete duricrust levels. At Fourques, take the D23 towards Trouillas. Along the way, the road cuts across the deflation pan of Estany Alt, which is eroded out of a marine sand outcrop surrounded by the duricrust-reinforced siltstone. As you enter Trouillas, turn left down to the ford, where the Cantarane River valley offers a series of good exposures (1.5 h round trip as far as La Jaça).

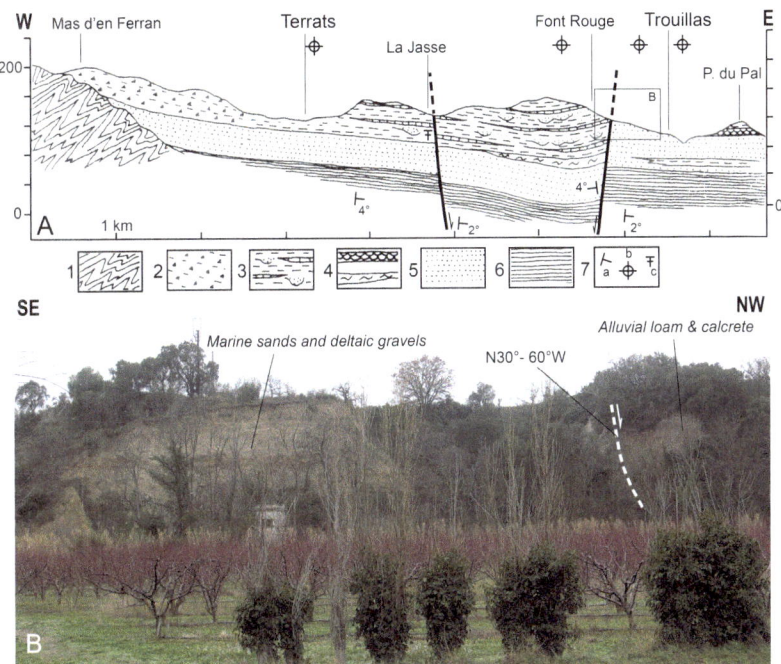

Fig. 7.17 **Pliocene exposures in Cantarana valley. A** Cross-section from the basin's edge to Trouillas. **1**: Aspres basement (schist); **2**: Proximal alluvial deposits descended from the Aspres massif (bright red clays and silts, with quartz and schist gravel beds); **3**: Distal facies of the continental Pliocene sequence (ochre to red flood-basin silts, calcrete, and shoestring sands); **4**: Oyster-rich marine level at Trouillas, with dark-blue clays reported from the borehole situated upstream (correlations by Magné, 1978); **5**: Deltaic to marine Pliocene facies (grey arkose sand containing gravel beds and granite and gneiss pebbles); **6**: Marine Pliocene (blueish sandy clay, here only identified from borehole cores); **7a**: dip (degrees); **7b**: Groundwater exploration borehole; **7c**: Fossil rodent remains (Terrats–La Jasse). **B** Site at Trouillas, showing fault location (location: 42°36′32″N, 2°48′09″E)

Stop 8. Tectonic deformations of the Pliocene sequence at La Cantarane

La Cantarane is a small canyon (Fig. 7.17) cut in grey sand layers capped by alternating sand and pebble beds, the latter containing gneiss, granitoïds, schist, and rare clasts of Permo-Triassic sandstone. These assemblages indicate supply from an ancestor to the Tech River. The sequence is coastal deltaic, and at an elevation of 80–90 m donwstream of the village (at the stadium and at the foot of Puig del Pal) it exhibits a mudstone level containing oyster beds (*Crassostrea*

Fig. 7.18   Castelnou Plateau and its fossil-filled fissures, or FFFs (after Calvet 1996, modi- ▶ fied). **A** Chronology and characteristics of the micromammalian deposits; note that CA5 and CA12 fall within MN17. Topographic position of FFFs (key to symbols)—**1**: on slopes of a residual hill rising above P1; **2**: on shallow valley side incising P1. Assemblage matrix: **3**: Exogenous quartz-rich sand and gravel deposits. Assemblage occurrence: **4**: within scattered clumps of sediment on the slope surface; **5**: in a thin tegument of in situ sediment. **6**: in a tectonic fissure; **7**: in a corroded clint or shaft in the limestone pavement. CA: Castelnou; MH: Mont Hélène. **B** E–W section through the Aspres massif. Tilted vestiges of Neogene erosion surfaces; along the edge of the Pliocene basin, range-top surface S and range-flank pediments P1 and P2 converge and become indistinguishable. **C** Detail of cross-section 17.8B showing where S and/or P1 cross-cut the Devonian limestone beds on Castelnou Plateau, with location of FFF sites

(Saccostrea) virleti). The muds and sands also contain Foraminifera, including rare plankton varieties (Magné, 1978). Further upstream, about 400 m from the ford and near the pumping station, the marine sands come abruptly into contact with the ochre, red and beige continental siltstone. The contact (Fig. 7.17B) is a N30°E, west-facing normal fault with a throw of several tens of metres (Salvayre and Sola, 1975). The siltstone formation is dislocated by a number of other more minor faults such as at Font Rouge and at La Jaça (spelling often given as 'La Jasse') still further upstream. Riprap unfortunately hides these outcrops from view, including around the micromammalian-fossil-rich lignite bed at Terrats–La Jasse (Michaux 1976). These faults do not appear to offset local Quaternary terrace levels T2 and T3, nor the older Quaternary alluvial deposits capping the surrounding hills. It is possible to extend the inspection of this cross-section by driving upstream as far as Terrats, where you can observe the marine sand and pebble beds re-emerge from beneath the continental siltstone and the more proximal detrital facies of the sequence (note gentle eastward dips), and rest directly on the Cambrian schist (Calvet 1996; Donzeau-Wiazemsky et al. 2010). The Cantarana River, a perennial stream in the schist, becomes a losing stream after reaching the Pliocene sands, i.e., at the stratigraphic unconformity, and in doing so contributes to recharge the deeper aquifers of the Roussillon Basin.

Between Trouillas and Thuir, the road first strikes across the tread of Cantarana terrace T3, then descends to level T2 at Thuir. At the base of the riser between the two levels, the wetland known as La Prade de Thuir was originally a deflation hollow. To the left of the road, the higher Aspres hills consist of intensely dissected Paleozoic schist. The syncline cores contain outcrops of massive Devonian limestone, and this harder rock has preserved occurrences of either surface S or pediment P1, with small vestiges of these occurring at increasing elevations from 250 m on the Thuir

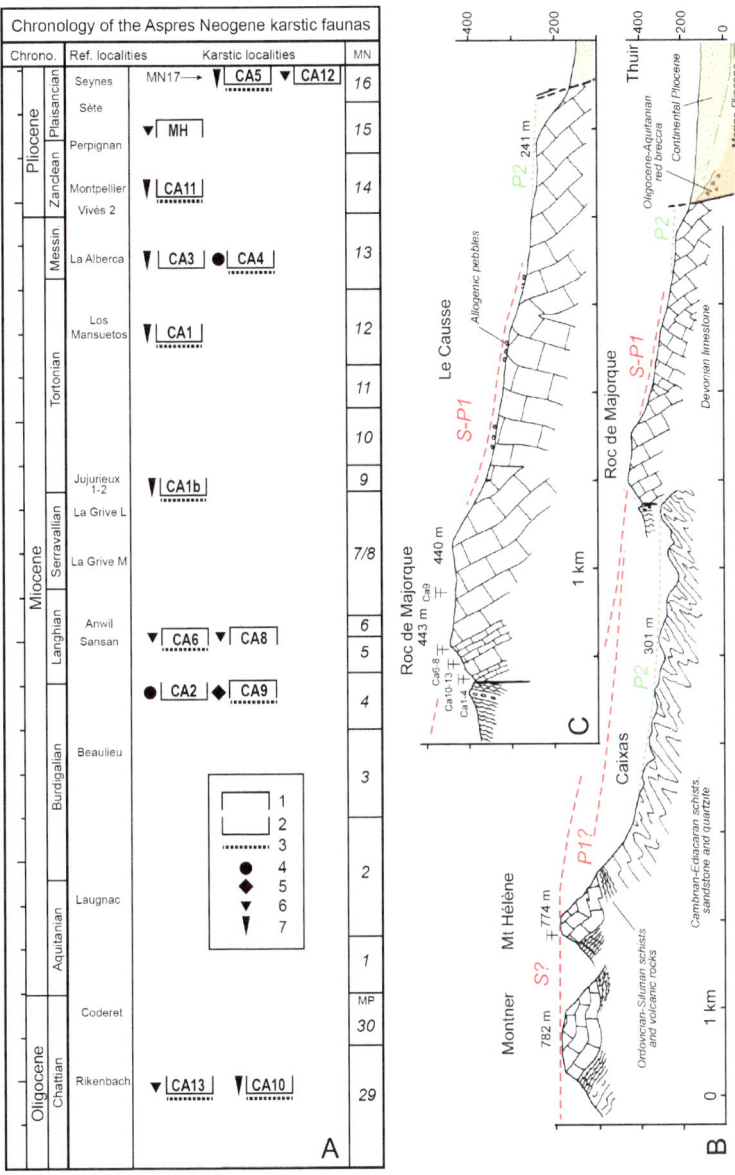

Fig. 7.19 **Thuir breccia, base of the Neogene Roussillon basin fill**. Note the synsedimentary fault boundary between the red 'breccia' (actually the proximal facies of a boulder bed) and the Devonian limestone. Cobbles and boulders are often striated, crushed, and occasionally split. Site location: Correc del Causse (location: 42°37′29″N, 2°43′43″E; 'correc', in Catalan: ravine)

plateau to 750 m at Mt. Hélène (Fig. 7.18B, C). The *Neogene fossil-filled fissures (FFFs) of Castelnou occur on the Thuir plateau itself* (Fig. 7.18A). *At Thuir, take the D48 towards Castelnou and consider the following stops.*

Stop 9. Thuir Plateau

- *The Thuir breccia.* The D48 embankment cut after the hospital (spot elevation: 152 m) shows the distal marl facies of the Pliocene. The micromammalian fossil deposit of Thuir (Fig. 7.4) was discovered in this formation, near the shooting range. The marl passes up-sequence to a thin band of proximal limestone pebble beds in a red matrix which have undergone patchy cementation and rest unconformably around the 200 m elevation contour on the Thuir breccia, which presents a much more massive facies and distinctive red and

Fig. 7.20 **Fossil-filled fissures in the Castelnou karst**. **A** West face of Roc de Majorca. **B** Site CA1, small solution pits containing ochre sediment (Tortonian, MN 12). **C** Site CA10, large pocket containing allochtonous pebbles (quartz visible) (Chattian, MP 29)

orange tones (Fig. 7.19). The road cuts through this breccia almost all the way to the pass. This 'Brèche de Thuir' is actually a stratified debris-flood deposit made up of thick beds containing rolled cobbles of Devonian limestone, diverse varieties of schist, and quartzite and quartz, all supplied by outcrops in the Aspres massif. The beds dip to the east and exhibit intraformational palaeostress markers such as pitted cobbles. The contact with the Devonian bedrock to the south is a N100°E normal fault. The fault plane can be observed in the Ravin du Causse (Fig. 7.19) and its small quarry. The upturned breccia formation has been clearly cross-cut by pediment P1, i.e., by the limestone surface of the Thuir plateau (Causse de Thuir). Based on this

Fig. 7.21 **Pliocene beds and their tectonic deformation features at Les Orgues d'Ille** ▶ (after Calvet 1996). **A** Simplified geological map. Key to symbols and ornaments—**1**: Millas–Quérigut granite (~300 Ma). **2**: Marine Pliocene (grey deltaic sand, littoral facies containing rolled pebbles, and cliff-face rockfall debris). **3**: Continental Pliocene (ochre channel-bed deposits of the ancestral Têt River). **4**: Continental Pliocene (short-range sedimentary input from the footwall upland: arkose with intraformational boulders and pebbles). **5**: Quaternary alluvial deposit T5. **6**. Tectonic features; **6a**: Exposed fault plane; **6b**: Buried fault plane; **6c**: Viewpoint indicator, a good vantage point for inspecting the faults. **B, C** Pebbles offset by a south-facing reverse fault (location: 42°40′52″N, 2°36′43.5″E). **D, E** Pebbles offset by left-lateral (D) and right-lateral (E) faulting (42°40′56.5″N, 2°36′48″E). **F** Faults at the viewpoint indicator (42°41′02″E, 2°36′56.5″E). **G, H** Fault plane exposed in the roadside ditch, Pliocene over basement (42°40′56.5″N, 2°36′48″E). Fault characteristics indicate fault strike direction, fault dip and direction, and fault striation rake

criterion, the Thuir breccia has been assigned a latest Oligocene or Aquitanian age; it thus dates from a time when tectonics was beginning to transform the continental margin along active extensional faults such as this one.

- *The Neogene fossil-filled fissures of Castelnou*. You can reach this site via the D48 at the back of the medieval village of Castelnou, then up the small road leading to Col de la Creu. From the col, walk to the narrow ridge in Devonian limestone of the Roc de Majorca. The fossil-bearing sites are situated discontinuously along the sharp ridge between elevations of 390 and 444 m. Whether of tectonic or shallow karstic origin, the fissures are often small (Fig. 7.20), rarely more than 1 m deep and usually less that 40 cm wide. Several FFFs contain allochtonous lag deposits such as quartz sand and small quartz and schist pebbles, documenting the transport of thin gravel trains across the land surface from the west and from the Aspres massif. The large fossil age range (Fig. 7.18A) within a shallow reach (max. ~50 m) below the topographic surface implies a very stable land surface in this area, with index fossils spanning late Oligocene to Pliocene biozones. it is thus likely that along this south-western border of the Roussillon Basin, as likewise along its northern border near Baixas (see Itinerary 1), erosion surfaces S, P1 and P2 were tangential to one another at very low angles or offsets, thus forming slowly evolving, polygenetic land surfaces such as the Causse de Thuir. The substantial tilt and fluvial dissection of the Thuir plateau are thus geologically recent, post-Pliocene features.

From Thuir to Millas, route D612 follows the contact between the Pleistocene alluvial fans of the Aspres and terrace T2 of the Têt River, which here hosts irrigated orchards and market gardens. The road descends a 15 m terrace riser to T1, which

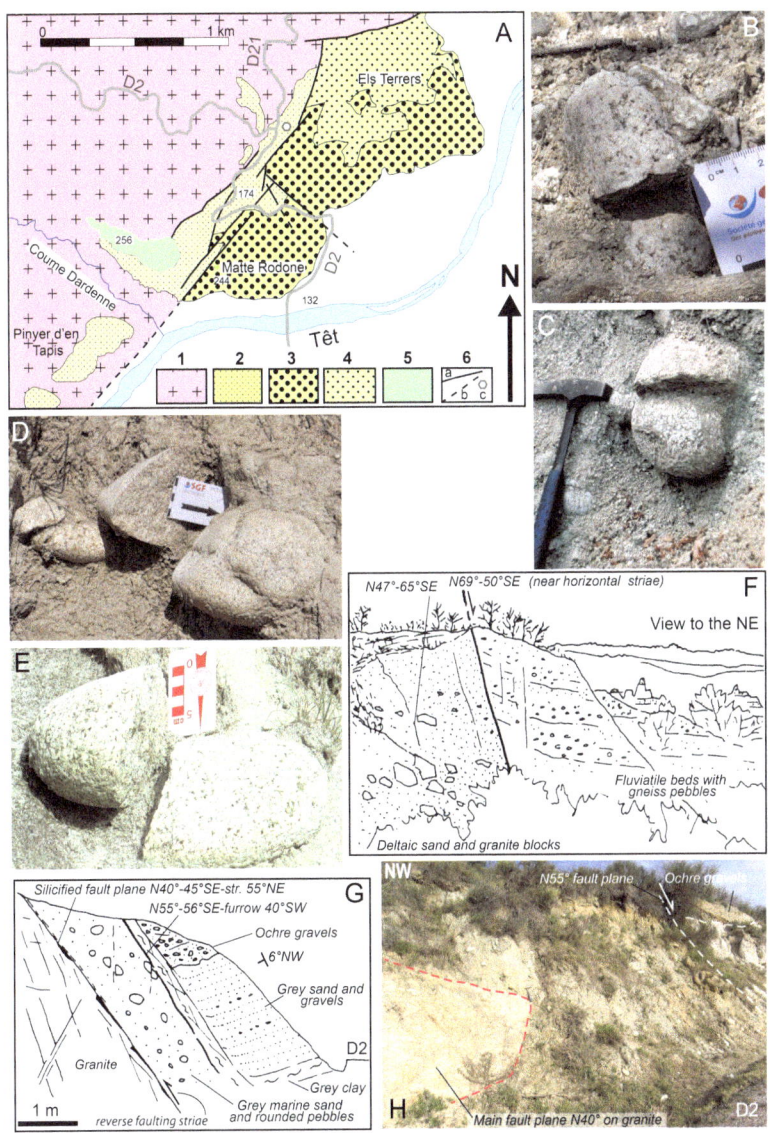

Fig. 7.22   The Orgues d'Ille rock pillars (west site). A Panoramic view from the south: ▶ note the two facies, i.e., marine and continental, and their tectonic deformation features. **B** Panoramic view from the summit (la Reixa, on terrace T5): Pliocene high-energy fluvial facies capped by early Quaternary deposit T5; the faults here are hidden from view. **C** Panoramic view from the north along the D2; the two facies, fluvial and deltaic, are separated by a fault array which vertically offsets the tread of terrace T5, suggesting tectonic activity during the Quaternary

extends all the way to Ille-sur-Têt. At Ille, cross the bridge to join the left bank of the Têt. Note that the active channel has narrowed substantially in the last 70 years, whereas the braidplain generated by the 1940 millenial flood was at places up to 500 m wide.

Stop 10. Ille-sur-Têt–Les Orgues

This is a spectacular protected geoheritage site (Fig. 7.21a) where badlands and rock pillars reveal the stratigraphy and Quaternary tectonic deformation of the Pliocene clastic sequence. The rock pillars are evocative of organ pipes, hence 'Les Orgues'.

You may begin by inspecting the western end of the Orgues, first along their southern face. For this it is recommended to park at the picnic area situated on your left at the bridgehead. To the SW of the picnic area, beyond the Coma d'Ardena gully, a footpath should lead you to the outcrops at Pinyer d'en Tapis. On the opposite side of the gully, another footpath also leads to the top of the preserved fill sequence at La Reixa (30 min round trip). Next you may inspect the north face and sections along route D2, where the fill sequence abuts the boundary fault and granitic footwall of Montalba plateau. The stop terminates at the roadside viewpoint indicator, from which you also get a panoramic view of the Roussillon Basin. A full visit up and down the footpaths, via la Reixa and the D2, should take about 2.5 h.

- *Pliocene marine deposits.* The basal deposits (Fig. 7.22A, C) rest directly on the granite basement at Pinyer d'en Tapis. The irregular nonconformity— a ravinement surface—exhibits gullies and washouts and correlates with the regionally extensive Messinian stratigraphic unconformity. The light grey debris are entirely of local provenance and consist of well-rolled granite pebbles and cobbles capped by texturally uniform arkose beds. The facies is coastal but includes sandy deltaic foreset beds. These are clearly visible both on the north face of the protected site and in the D2 roadside quarry. Chaotic masses of angular clasts, some of them huge, correspond to Pliocene rockfall

Fig. 7.23 **Facies characteristics of the Pliocene continental deposits at Orgues d'Ille. A,** ▶
B Fluvial deposits of the ancestral Têt River; alternating ochre-coloured, cobble-rich beds,
and lighter-coloured, gravelly sand beds; they form laterally extensive but thin units displaying erosional boundaries at their base, typical of braided channel systems. **C** Aspects
of the fluvial sequence; it displays 9 coarse-bedload depositional units within the 100-m-
thick exposure. **D** Eastern portion of Les Orgues: arkose containing locally supplied granite
boulders (access from the tourist trail)

or sturzstrom debris as a result of cliff collapse. Other boulders are granite
spheroids sourced by weathering fronts that were being stripped at the time
by denudation in the Montalba footwall upland.

- *Pliocene continental deposits.* In contrast to the deposits preserve in the northern part of the site, the deposits to the south of the Orgues correspond to a
  high-energy fluvial sequence emplaced by an ancestor of the Têt (Figs. 7.22
  and 7.23A–C). They contain a large proportion of augengneiss pebbles. The
  conglomeratic facies and its internal fabrics are typical of highly unstable
  braided channel bars, with channel fills several metres thick. Some boulders
  are up to 1 m in diameter. Overall, the sequence is a deep ochre colour, with
  lighter-coloured bands of sand and well rolled fine gravel. This alluvial formation is 100 m thick. A borehole at Ille reports 130 m of Pliocene deposits,
  but it was impossible to identify the continental / marine sequence boundary.
- *Early Quaternary alluvial terrace.* Vestiges of an ancient alluvial deposit cap
  the Pliocene Orgues sequence. Relics can be observed at the Matte Rodone
  rock pillar (spot elevation: 244 m), with another at la Reixa (spot elevation:
  256 m) (Fig. 7.22B). The distinctive feature of this formation compared to the
  underlying Pliocene beds is the horizontal bedding and its content in coarse
  debris, with quartz boulders exceeding 2 m in diameter. This terrace has been
  named T5 (Fu on geological maps) and ESR dating has provided for it, much
  further upstream (see Itinerary 6), an age of 1.1 Ma (Delmas et al. 2018).
- *A major tectonic boundary* involving two main faults separates the deltaic from
  the fluvial deposits at Les Orgues (Figs. 7.21A and 7.22A, C). The rake of lineations on the N40°E-striking fault-plane exposures is 20°NE, which indicates
  a normal to left-lateral strike-slip motion. This fault array can be observed
  parallel to the D2, just below the roadside viewpoint signpost (Fig. 7.21F).
  The roadside ditch cross-cuts the outermost boundary fault in two places, and
  the Pliocene coastal facies and granitic basement, as a result, find themselves
  abruptly juxtaposed (Fig. 7.21G, H). Here, the fault plane exhibits 50°NE lineations and therefore normal to left-lateral displacement. A large number of
  sheared cobbles also indicate a right or left-lateral shear sense (Fig. 7.21D, E),

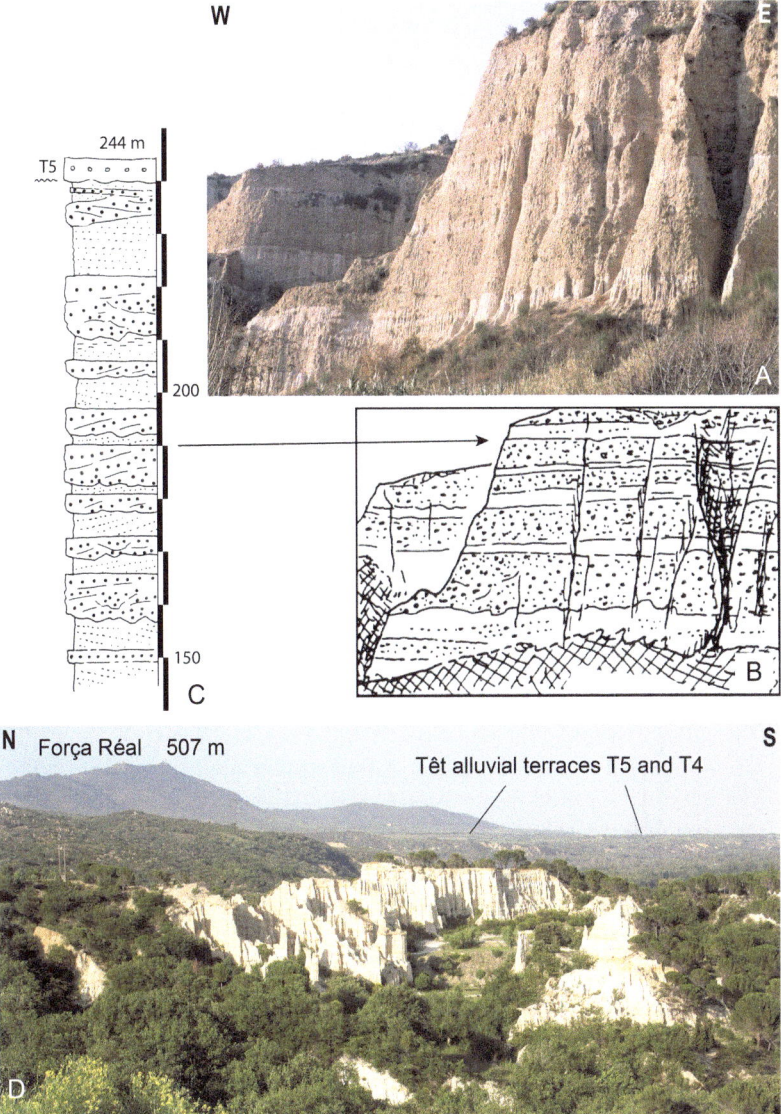

and even occasionally reverse slip motion (Fig. 7.21B, C). Overall, evidence suggests that these faults were active during and after the Pliocene period, under stress regimes that appear to have varied from transtentional to transpressional. It is unlikely that fault movements were exclusively Quaternary as this would imply a vertical Quaternary throw in excess of 200 m. If this were the case, vestiges of the 200 m-thick topset beds should occur at much higher elevations in the landscape, and would probably show an onlapping stratigraphy over the edge of the footwall upland (Montalba plateau, to the north). No such deposits, however, have been reported from the adjacent plateau surface or slopes. The more realistic scenario involves synsedimentary fault slip during aggradation of the thick fluvial sequence, which explains why the fill sequence remained confined to the Roussillon hangingwall and did not overfill and bury the Montalba footwall. Post-Pliocene fault reactivation probably generated the strike-slip fault lineations and the 7° dip to the NW of the ochre conglomerate beds (Fig. 7.22A, C). The angular unconformity between this Pliocene fill sequence and the overlying 'Fu' (i.e., T5) alluvium nonetheless indicates that the stratigraphic backtilt occurred prior to the deposition of T5. Subsequent fault reactivation is nonetheless also plausible because a ~10 m vertical offset affecting the base of 'Fu' has been observed (Fig. 7.22A, C).

- **Eastern portion of Les Orgues.** This is visible from the roadside vantage point. It consists of the ochre conglomerate facies on its southern side; but, nearer the boundary fault to the north, we note a lateral shift in facies to whitish arkose (Fig. 7.23D) containing poorly sorted and poorly rounded granite pebbles. This suggests debris-flow deposits resulting from denudation in the granitic catchments of the footwall upland. A tour involving an entrance fee allows you to visit this part of the geoheritage site.

Return to Ille over the bridge and turn down the D916 as far as Néfiach. In the village, take a left turn down rue Gabriel Péri, and through a narrow maze of streets you will find your way to a causeway across the Têt River. Here you may observe the deltaic facies of the Pliocene, with east-dipping sand and clay beds. At this small knickpoint in the river channel, the Pliocene bedrock is exposed on the left bank of the river (just east of the causeway) mostly as a result of over 50 years of unregulated in-channel alluvium mining (most of the bedload from the 1940 flood has been removed). After the crossing, drive another 700 m upstream (to your left) to the sand quarry known as Venta Farina.

Stop 11. Section through the Pliocene Gilbert delta at Néfiach

The quarry offers a spectacular exposure of the internal structure of a Gilbert delta (Fig. 7.24). Blueish, silty-sandy bottomsets can be seen at the base of the quarry face and upstream of the hill, where they also contain abundant seashells (including scallop shells). Ochre-coloured sand and gravel foresets otherwise make up the bulk of the outcrop, while the topset beds mostly document a supply of schist debris from the north by a tributary of the Têt. A little farther east, near the start of the road to Bélesta, Astre (1937) also reported nummulites in the Pliocene sediment, presumably supplied by Ypresian to Lutetian outcrops in the vicinity. To the west of Venta Farina hill, the Ravin des Teuleries is a gully where the Pliocene marine beds display a N45°E strike-slip fault plane.

Return to the junction near the causeway and turn left onto the small road that

winds around the hill named Poc Garbell. This road cuts through sub-Pliocene clastic deposits containing schist debris in a brown matrix. The beds have undergone intense tectonic deformation and are overlain by marine beds containing shell levels, followed by deltaic sands, then by continental beds, and lastly capped by a Quaternary alluvial deposit mapped as unit 'Fu'. The track descends eastward towards the Ravin des Clairannes, which offers a series of good exposures of Pliocene and sub-Pliocene units.

Stop 12. Clairannes ravine

Exposures of marine Pliocene bioclastic beds with 20° dips to the south occur on the right side of the ravine in its downstream section (Fig. 7.25B). Those exposures, however, are not easily accessible. In a bend in the ravine further

Fig. 7.24 Pliocene Gilbert-delta sequence at Venta Farina (Néfiach), viewed from the east. Gravel input, mostly schist, entered via a Pliocene tributary of the Têt arriving from the N to NW. The bottomsets are mostly hidden from view, but consist of clay- and shell-rich, blueish-coloured beds

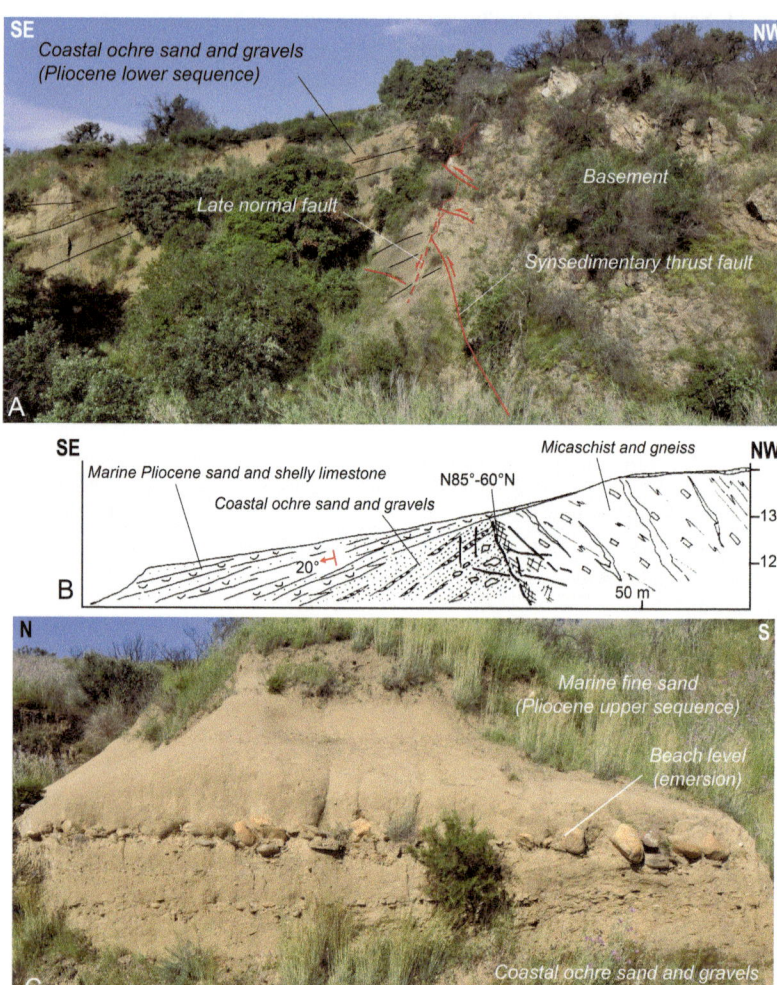

Fig. 7.25 Tectonically deformed exposure of the basal Pliocene at the Ravin des Clairanes. A General view of the fault line from the left bank (location: 42°42′13″N, 2°40′57″E); **B** Schematic cross-section of the right side of the ravine wall; **C**: Pliocene sequence on the upstream part of the left wall (42°42′18″N, 2°41′01″E), showing two distinct sequences separated by an erosional disconformity (beach level; see text for details)

upstream, the sub-Pliocene sequence is likewise deformed and comes into contact with the basement along a N85°E reverse fault, which suggests a brief period of convergent tectonics. The best views are gained when looking across from the left bank (Fig. 7.25A, B). Farther upstream and on the left bank of the gully, bedding in the entire Neogene sequence is perfectly horizontal. In this exposure, the sub-Pliocene sequence consists of large, oblique sand and gravel lenses cross-cut by a thin shell-rich marl layer; the overlying Pliocene is marine and consists of grey fine sand and bioclastic (lumachelle) beds. The boundary between the two stratigraphic units is a coarse, clast-supported beach gravel (Fig. 7.25C).

The small road continues and joins up with the D612 at the bridge to Millas. From there, follow the D614 to Corneilla-la-Rivière. At Corneilla, you can drive or walk (2 h round trip) up the rue de Força Réal, which winds through vineyards towards the Mas de la Garriga, NW of the village. This excursion provides a full introduction to the strath sequence of older alluvial terraces of the Têt (Figs. 7.26 and 7.27). From Corneilla, which stands on terrace T1, the road rises to T3 (Clot de Rico), then to the extensive tread of T4 (Plane d'en Burgat, Fig. 7.27A), and lastly to T5. Both this uppermost Quaternary level and T4 rest on the Pliocene beds, their proximal facies recognisable by the abundant flakes of schist in a red matrix (Fig. 7.27A). The stairway ends above T5 with pediments cut in the bedrock (schist) of Força Réal, a hill visited in Itinerary 5 (warning: inexplicably, terrace labelling on the geological map of Rivesaltes departs from convention by noting these alluvial levels as Fxb, Fxa, and Fw). Return to Perpignan either following the left bank of the Têt, or along the right bank via the causeway back across the river at Pézilla–Sant Feliu d'Avall. This second option takes you across lowermost terrace T0, which was locally dated to the late Middle Ages (see Fig. 6.4B, Itinerary 1).

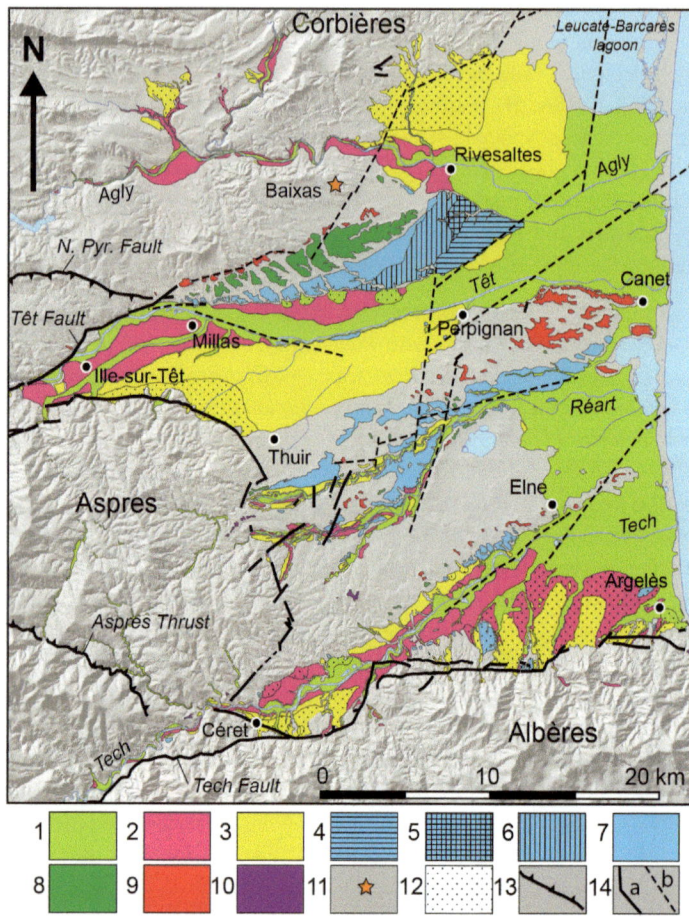

Fig. 7.26 Quaternary terraces of the Roussillon Basin (after Calvet 1996, and Delmas et al. 2018, modified). Key to symbols and ornaments—**1**: T0, Holocene to present (coeval colluvium and shore deposits not mapped). **2**: T1, Late Pleistocene. **3**: T2 (ESR age: 174 ± 44 ka). **4**: T3a. **5**: T3b (ESR age: 374 ± 47 ka). **6**: T3c. **7**: T3d, and other undifferentiated T3-generation deposits. **8**: T4. **9**: T5 (ESR age: 1099 ± 179 ka). **10**: T5 or latest Pliocene. **11**: Fossil-rich fissure sites in the limestone karst containing alluvium and bone assemblages dating back to the late Pliocene and Gelasian (2–3 Ma) (see Itinerary 1). **12**: Debris cones grading to the alluvial terrace deposits. **13**: Paleogene fault, possibly reactivated during the Neogene (the Aspres thrust was inverted, operating as a normal fault facing NE). **14**: Neogene faults; **14a**: Fault with evidence of Neogene activity; **14b**: Buried fault with indirect evidence of Neogene activity

Fig. 7.27 Oldest terrace levels of the Têt above Corneilla-la-Rivière. **A** Level T4 at La Plana d'en Burgat; note vestige of T5 and a degraded occurrence of pediment P2 on Paleozoic schist in the background. **B** Exposure of T4, left wall of Ravin de Campells; here the Quaternary deposits have cut and filled the underlying Pliocene beds, with their distinctive red facies containing gravel-sized clasts of schist locally supplied as lateral debris cones. The terrace tread is a bright-red chromic Luvisol, highly eluviated and desaturated

References

Astre G (1937) Nummulites remaniés dans le Pliocène de Néfiach en Rousillon. Somm Soc Géol Fr, 347–351

Bache F, Olivet JL, Gorini C, Aslanian D, Labails C, Rabineau M (2010) Evolution of rifted continental margins: the case of the Gulf of lions (Western Mediterranean Basin). Earth Planet Sci Lett 292:345–356

Bourcart J (1947) Étude des sédiments pliocènes et quaternaires du Roussillon. Bull Ser Carte géol de Fr, 218, XLV, 1945, 395–476

Calvet M (1986) La stratigraphie du Néogène du Roussillon et le problème des séries détritiques de bordure. Essai de mise au point. Géol Fr 2:205–220

Calvet M (1996) Morphogenèse d'une montagne méditerranéenne : les Pyrénées orientales. Documents du BRGM, Orléans 255:1177 p

Calvet M (1999) Régime des contraintes et volumes de relief dans l'Est des Pyrénées. Géomorph: Rel, Proc, Environ 3:253–278

Carozza JM (1998) Évolution des systèmes géomorphologiques en contexte orogénique : l'exemple des bassins d'alimentation du Roussillon. Approche morphotectonique. Thèse de doctorat, Université de Toulouse 2:398

Clauzon G (1990) Restitution de l'évolution géodynamique néogène du bassin du Roussillon et de l'unité adjacente des Corbières d'après les données écostratigraphiques et paléogéographiques. Paléobiol Contin XVII:125–155

Clauzon G, Cravatte J (1985) Révision chronostratigraphique de la série marine pliocène traversée par le sondage Canet 1 (Pyrénées-Orientales): apport à la connaissance du Néogène du Roussillon. C R Acad Sc Paris II 301:1351–1354

Clauzon G, Aguilar JP, Michaux J (1987) Le bassin pliocène du Roussillon (Pyrénées-Orientales, France): exemple d'évolution géodynamique d'une ria méditerranéenne consécutive à la crise de salinité messinienne. C R Acad Sc Paris II 304:585–590

Clauzon G, Suc JP, Aguilar JP, Ambert P, Capetta H, Cravatte J, Drivaliari A, Domenech R, Dubar M, Leroy S, Martinell J, Michaux J, Roiron P, Rubino JL, Savoye B, Vernet JL (1990) Pliocene geodynamic and climatic evolutions in the French Mediterranean region. In: Iberian Neogene Basins, Paleontologia i Evolució, Mem. Especial n° 2, Sabadell, p 131–186

Cravatte J, Dufaure JF, Prim M, Rouaix S (1974) Les forages du Golfe du Lion. Stratigraphie, sédimentologie. Notes et Mémoires du C.F.P., 11:209–274

Delmas M, Calvet M, Gunnell Y, Voinchet P, Manel C, Braucher R, Tissoux H, Bahain JJ, Perrenoud C, Saos T, Aster Team (2018) Terrestrial [10]Be and electron spin resonance dating of fluvial terraces quantifies quat tectonic uplift gradients in the eastern pyrenees. Quat Sci Rev 193:188–211

Depéret C (1885) Description géologique du bassin tertiaire du Roussillon. Ann Sc Géol 17, 1–136. Thèse, Paris. Description des vertébrés fossiles du terrain pliocène du Roussillon. Ann Sc Géol 17:137–268

Depéret C (1890) Les animaux pliocènes du Roussillon. Mém Soc Géol Fr, Paléontologie, Mém 3:195

Depéret C (1912) Sur le grès éocène de Moulas, près le Boulou (Pyrénées-Orientales). C. Somm Soc Géol Fr, 21–22

Donzeau-Wiazemski M, Laumonier B, Guitard G, Autran A, Llac F, Baudin T, Calvet M (2010) Carte géologique de la France au 1:50,000, sheet 1096 Céret. Handbook by Laumonier B, Calvet M, Donzeau-Wiazemski M, Barbey P, Marignac C, Lambert A, Lenoble JL, avec la collaboration de Autran A., Cocherie A., Baudin T., Llac F., 2015, BRGM édit., Orléans, 164 p

Duvail C, Gorini C, Lofi J, Le Strat P, Clauzon G, Dos Reis T (2005) Correlation between onshore and offshore pliocene-quaternary systems tracks below the Roussillon basin (Eastern Pyrenees, France). Mar Pet Geol 22:747–756

Duvail C, Le Strat P, Alabouvette B, Perrin J, Seranne M (2000) Évolution géodynamique du bassin du Roussillon : analyse des profils sismiques calibrés par les sondages profonds de Elne 1 et de Canet 1. Rapport n° GTR/BRGM/1200–137, Montpellier, 23 p

Duvail C, Le Strat P, Bourgine B (2001) Atlas géologique des formations plio-quaternaires de la plaine du Roussillon (Pyrénées-Orientales). Rapport BRGM, BRGM/RP-51197—FR

Gallart J, Diaz J, Nercessian A, Mauffret A, Dos Reis T (2001) The eastern end of the pyrenees: seismic features at the transition to the NW Mediterranean. Geophys Res Lett 28:2277–2280

Gaudant J (1999) Présence du genre Lates Cuvier et Valenciennes dans les grès de Moulas (Pyrénées-Orientales). Géol Fr 4:67–75

Gorini C, Lofi J, Duvail C, Dos Reis T, Guennoc P, Le Strat P, Mauffret A (2005) The late messinian salinity crisis and late miocene tectonism: interaction and consequences on the physiography and post-rift evolution of the gulf of lions margin. Mar Pet Geol 22:695–712

Lofi J, Gorini C, Berné S, Clauzon G, Dos Reis T, Ryan WBF, Steckler MS (2005) Erosional processes and paleo-environmental changes in the Western Gulf of Lions (SW France) during the Messinian Salinity Crisis. Mar Geol 217:1–30

Lofi J, Rabineau M, Gorini C, Berné S, Clauzon G, De Clarens P, Dos Reis T, Mountain GS, Ryan WBF, Steckler MS, Fouchet C (2003) Plio-Quaternary prograding clinoform wedges of the western Gulf of Lion continental margin (NW Mediterranean) after the Messinian salinity crisis. Mar Geol 198:289–317

Magné J (1978) Études microstratigraphiques sur le Néogène de la Méditerranée nord-occidentale. t. 1: Les bassins néogènes catalans, 259 p; t. 2: Le Néogène marin du Languedoc méditerranéen, 435 p, Édit. CNRS, Toulouse

Mauffret A, Durand de Grossouvre B, Dos Reis AT, Gorini C, Nercessian A (2001) Structural geometry in the eastern pyrenees and western gulf of lion (Western Mediterranean). J Struct Geol 23:1701–1726

Michaux J (1976) Découverte d'une faune de petits mammifères dans le Pliocène continental de la vallée de la Cantarrane (Roussillon) ; ses conséquences stratigraphiques. Bull Soc Géol Fr, XVIII, 165–170

Nercessian A, Mauffret A, Dos Reis AT, Vidal R, Gallart J, Diaz J (2001) Deep reflection seismic images of the crustal thinning in the eastern Pyrenees and western Gulf of Lion. J Geodyn 31:211–225

Philippe M, Bourgat R (1985) La collection de vertébrés pliocènes du Muséum de Perpignan. Bull Mensuel Soc Linn Lyon 54:146–160

Salvayre H, Sola C (1975) Observations sur la stratigraphie et la néotectonique du Pliocène de la vallée de la Cantarrane (Pyrénées-Orientales). Bull Soc Géol Fr XVII:1121–1125

Salvayre H, Sola C (1982) Conceptions modernes de la structure géologique du réservoir multistrates des nappes captives du Roussillon en rapport avec leur étude hydrodynamique. Actes du 106ème Congrès national société savantes, Perpignan 1981, Sciences III:195–234

From Perpignan to Perpignan (95 km; 1 day). See Fig. 8.1.

Itinerary 3 mostly focuses on the geologically recent evolution of the alluvial lowlands of Roussillon. The event chronology is underpinned by archaeological vestiges, principally Roman antiquity and Romanesque chapels and churches which cover ~2000 years of sediment dynamics across the alluvial floodplains of the Tech, Réart, Têt and Agly rivers.

Leave Perpignan via the D617a towards Canet, and turn left at the shopping centre towards Château-Roussillon.

Stop 1. Ruscino, capital of Roussillon during Antiquity

The site of Ruscino was an Iberian Iron Age oppidum occupied by the Sordones. It later became a Roman colony. Ruscino stands on a spur of the uppermost terrace T5, and is cut off from the surrounding plains of the lower Têt and the Via Domitia by deep ravines (good panoramic views to the NW). A small museum and some outdoor archaeological remains show the foundations of living quarters to the east of the road, and a forum to its west (public access is unpredictable). The highlight of the site, discovered by the late R. Marichal, archaeologist of the city of Perpignan, is the evidence of building structures deformed by seismic disturbance: stone-lined gutters warped and fractured by left-lateral strike-slip fault displacement, and forum pillars also twisted by anticlockwise torque motion (Fig. 8.2). The seismic event responsible for this structural damage may have precipitated the ruin and abandonment of the upper part of the site towards the end of the first century CE. Up to 6 m of colluvium containing archaeological artefacts from Late Antiquity have filled the ravine to the east, and probably buried a stretch of the Via Domitia passing through this area (Marichal and Rébé

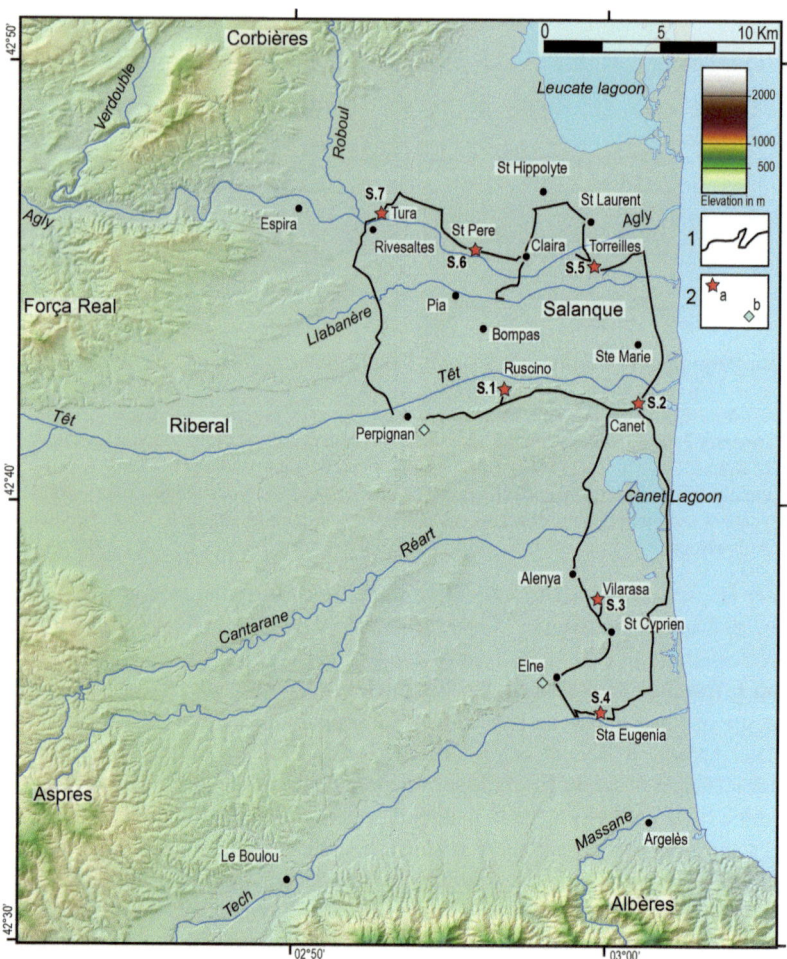

Fig. 8.1   Itinerary 3: route map and overview. Key to symbols—**1**: Roads and trails, irre-spective of road rank, size, or quality. **2**: Stopping points (outcrops, exposures, landforms, or landscapes). **2a**: Roadside stops. **2b**: Cultural stops. Labels S.1 to S.7 locate the sequence of roadside stops

Fig. 8.2 Seismic damage on buildings at the site of Ruscino. A left-lateral offset in gutter. **B** pivoted forum pillars

2003). Ruscino appears nonetheless to have been reoccupied during the early Middle Ages at the time of the Wisigoths, then briefly by Arabian settlers. It was decreed capital of the county of Roussillon by emperor Charlemagne, but was eventually transferred slightly westward to the site where Perpignan now stands (and has since vastly expanded). For further details on the geohistory of the area, see Calvet and Lacquement (2019).

Return to the D617a, which as far as Canet follows the tread terrace T5 along the Têt River ('terrasse de Cabestany'). This alluvial deposit has been dated (see Itinerary 6) at ~1.1 Ma (Delmas et al. 2018). The older generations of Roussillon terrace deposits T3 to T5 contains quartz ventifacts and highly patinated artefacts ascribed to the region's Pebble Culture (Collina-Girard 1976). A large number of open-air sites containing handaxe-culture assemblages have been studied in terrace deposits T5 to T2. Based on comparisons with artefacts excavated and dated at Caune de l'Arago, a protected cave site near Tautavel (see Itinerary 1), this early

*Palaeolithic technology using quartz cobbles as primary raw material dates back to 530–350 ka, and onward possibly as recently as 150 ka (*Martzluff 2006*). These clues thus provide an indirect age bracket for the terrace sequence. At Canet, turn left towards Le Barcarès (D81) and stop immediately after the first traffic roundabout.*

Stop 2. Early Pleistocene terrace T5 at Canet

The ditch on the east side of the road provides a good exposure of the alluvial deposits, which here rest directly on the sand and gravel beds of the Pliocene sequence. Alluvial sediments belonging to generation T5 are deeply weathered: highly friable, powdery gneiss and granite pebbles in a rubified, clay-rich matrix. The topmost 1 m of the section is made up exclusively of a residual assemblage of highly patinated quartz pebbles displaying red and ochre tones, and some ventifacts. A lens of red sandy matrix was sampled for Electron Spin Resonance dating, but the age was deemed inconclusive probably because of sample pollution by small decomposed granite pebbles visually undetectable at the time of sampling (most consistent replicate ESR age: 1058 ± 498 ka; Delmas et al. 2018).

Return to the roundabout and drive back towards Perpignan for 1 km before turning left at the junction with the D11 towards Saint-Nazaire. The road skirts along Canet lagoon marshes, at the base of the Pliocene hills capped by T5 (buttes at Cabestany to the west, and at L'Esparrou and Saint-Nazaire to the east). Drive through Alenya, then follow the D22 towards Saint-Cyprien. After 2 km from Alenya village centre, turn left into a country lane towards Saint-Estève-de-Vilarasa.

Stop 3. Vilarasa

The restored Romanesque church was consecrated in 1150 CE and served a parish village now destroyed. The building rises out of a hollow 1.5 m deep (Fig. 8.3) resulting from the excavation and restoration recently undertaken. It was buried by alluvium during the Little Ice Age (LIA) by an avulsing arm of the Tech River, which until then had taken a course north of the hill at Elne (Carozza et al. 2012, 2013). During the great flood of 1940, an arm of the Tech temporarily reactivated this palaeochannel. Alluvial aggradation has nonetheless not been ubiquitous since the fourteenth century across the lowlands between here and Canet. Between Saint-Nazaire, Alenya and Saint-Cyprien, quite a number of Roman-age and even Neolithic sites (Kotarba et al. 2007) have thus been uncovered by shallow ploughing rather than by deep excavation, indicating a degree of geomorphological stability of these lowlands at some distance from the floodways of the Tech and Réart streams (Fig. 8.4).

Fig. 8.3 Romanesque chapel of Sant Esteve de Vilarasa (Saint-Étienne-de-Villerase). Note thickness of alluvial deposits from the Little Ice Age

Get back on the D22 towards Saint-Cyprien, then turn off on the D40 towards Elne. Drive through the town and make sure you come out on the D612 while avoiding the bypass towards Argelès. Follow the road parallel to the bypass as far as the Tech along a country lane that goes eastward and passes under the Tech bridge. At the first junction after the bridge, turn right on the track leading to Mas Batlle. Aim to stop at the chapel of Santa Eugènia in the midst of fruit-growing polythene tunnels, ~1.2 km east of the Elne bypass.

Stop 4. Santa Eugènia de Tresmals
A 12th century chapel buried by historical flood deposits. Overview of Neolithic through to Medieval archaeological markers of the Holocene alluvial history in the Roussillon Basin.

- *Features and event chronology of the present-day alluvial plain.* This entire area was flooded in October 1940, the hill at Elne forming an island for several days. The Tech river burst its levees at several locations and distributed its sediment load extensively over cultivated fields (Fig. 8.4). Coarser sand

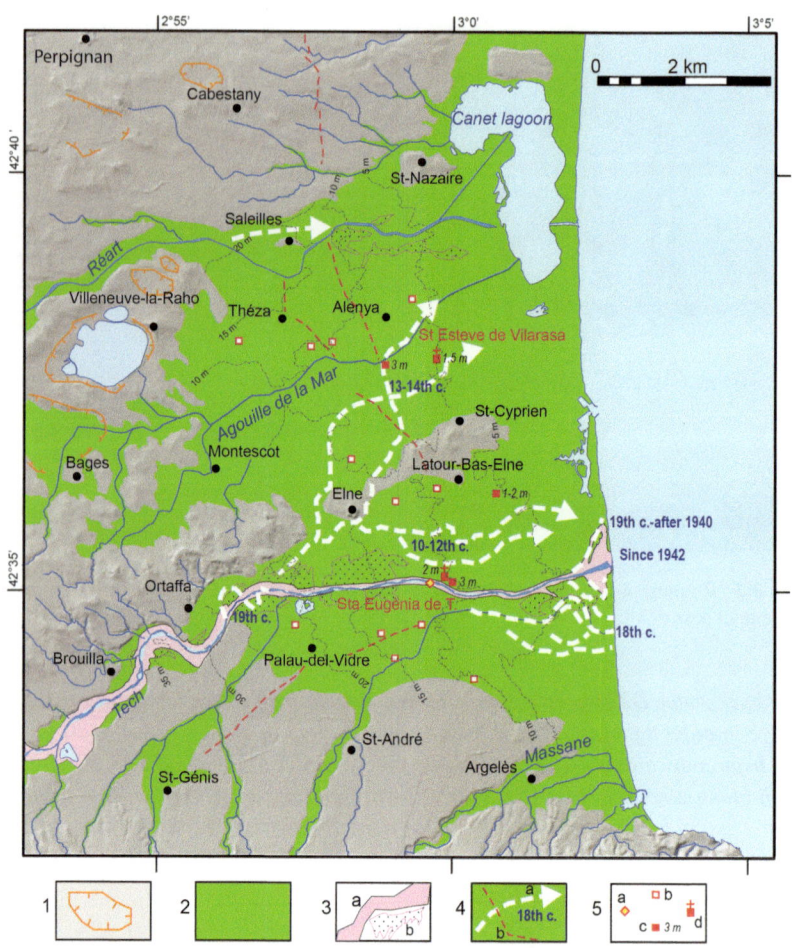

Fig. 8.4 The lower alluvial plains of the Tech and Réart. Key to symbols and ornaments—**1**: Pleistocene aeolian deflation pans in outcrops of continental Pliocene. **2**: Holocene
to recent alluvial fill. **3**: Deposits from the October 1940 catastrophic flood, mapped from
aerial photographs (1942 flight by the Institut Géographique National); **3a**: Tech channel,
and gravel fill; **3b**: Sand and silt of the floodplain. **4.** Linear features; **4a**: Fluvial palaeochannels, with indications of their age and morphology based on various maps, aerial photographs
and archaeological constraints. **4b**: Roman roads (Via Domitia and 'Chemin de Charlemagne'). **5**: Archaeological sites; **5a**: Cardial site (early Neolithic) in the Tech channel;
5b: Gallo-Roman site in the subsurface (depth unspecified but < 1 m and within depth
range of ploughshares); **5c**: Buried Gallo-Roman site, depth indicated; **5d**: partially buried
Romanesque church (burial depth indicated). Compiled from Calvet et al. (2002), Kotarba
et al. (2007), Carozza et al. (2012)

and gravel was typically deposited closer to the overtopped channel banks, with silt and finer sand transported further afield (mean thickness 15–20 cm, with accumulations locally reaching 2 m; Jacob 1997). Because of the steep longitudinal profile of the floodplain (0.3%), the spillways and splay channels cut gullies 0.1 to 1 m deep and formed scour pools up to 4 m deep. The 1940 flood was exceptional (Pardé 1941), with an estimated 200- to 300-year return period (Bénech 1993). Similar or perhaps more intense events have been reported in historical accounts for all rivers in Roussillon: e.g., in 878, 1264 (or 1261: probable destruction of the Roman bridge at Céret), 1421 and 1553 (destruction of Boulou bridge), 1632, 1763 (a major catastrophe in the upper Tech catchment, see Itinerary 4) (Bénech 1993); with perhaps less intense or widely impacting flooding events in 1766, 1772, 1777, 1833, 1842, 1892, and 1907, all particularly well documented for the Tech (Desailly, 1992). The Tech carries the largest sediment load among Roussillon rivers. During the 4-day 1940 flooding event, when peak discharges reached 3500 m^3/s, 10 to 15 Mt of sediment reached the sea, at least 25 Mt reached Céret, and overall more than 40 Mt of mud and gravel were moved by that river (Pardé 1941). A large proportion of this was bedload, with an estimated 415,000 m^3 (~650,000 t) conveyed past Amélie-les-Bains; by comparison, estimates for a 100-year flood are closer to 116,000 m^3 (Capolini and Lefort 1993).

- *Holocene stratigraphy.* The Holocene depositional history is poorly known and apparently complicated. Sediment thicknesses reach 20–30 m near the coast. Over a 1 km distance downstream of the bridge over the D914, the Tech has incised an anthropogenic deposit and exposed grey silty sands and dark muds resting at 5 m NGF (the Ordnance datum in France is abbreviated as NGF: Nivellement Général de la France) on a fluvial sequence containing pebble-filled palaeochannels (possibly generation T1/Fy, or early Holocene). The site is accessible from the right bank of the Tech along a track imme-diately downstream after the bridge. This Neolithic deposit has been dated to 6800–7000 yr BP based on impressed ceramic ware typical of the Cardial Culture (Martzluff et al., 1994–1995), and it pre-dates the Holocene Optimum eustatic high. Given that the Ancient Roman living floor at the chapel of Santa Eugènia occurs at a depth corresponding to 8–8.5 m NGF, the 5000 years of alluvial aggradation, which add up to just 3 m of sediment, imply that the sequence must have been interrupted by several erosional hiatuses (Fig. 8.5) (Calvet et al. 2002). At the chapel, a drill core has struck the top of the Pliocene sequence at 0.5 m NGF. According to the core sequence, the Pliocene is thus capped by 0.5 m of Pleistocene gravels, followed by another 3–4 m of palustrine deposits that share the same characteristics as the site containing

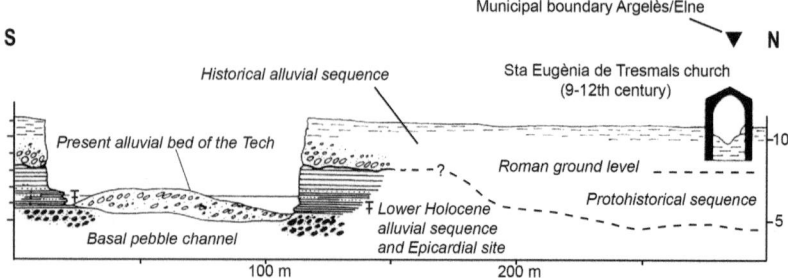

Fig. 8.5 Stratigraphic sections in the lower floodplain of the Tech near Santa Eugènia de Tresmals (Sainte-Eugénie-de-Tresmal). The Holocene fill consists of three successive generations of cut-and-fill alluvial deposits: early Holocene, Protohistoric, and historic to present. After Calvet et al. (2002), modified

vestiges of Cardial ceramic. Lastly, these beds have been cut by gravel-filled channels supplied by an ancestral Tech. These can be inspected in the river-bank section. Although undated, they probably document an event during the second half of the Holocene, and are overlain at a depth of 3 m by the Roman archaeological level (Carozza et al. 2012; Carozza 2012).

- *Historical evolution of the lower Tech alluvial plain.* The historical evolution of the lower Tech and Réart river floodplains has been described as a flood-dominated regime (FDR, *sensu* Hickin 1983) linked to the onset of the LIA, which began around 1303–1307 CE in this region according to climate chronicles available in temperate continental Europe (Carozza et al. 2012). However, the 878, 1264, 1907, 1940 and 1999 floods occurred outside the LIA interval and may be related instead to episodes of rapid regional climate switches documented by the Western Mediterranean Oscillation Index (Martin-Vide and Lopez-Bustins 2006).

Achaeological excavations at Santa Eugènia (Figure 8.6) document a net increase in sedimentation rates after construction of the chapel, rising from 1 to 3 mm/yr (Carozza et al. 2012). The building was buried by 2 m of sediment, and at least 5 or 6 flood events have been detected from the sediment stratigraphy in the vicinity of the construction, and up to 11 inside it. Burial would have been rapid after the 1421 flood, which relocated the Tech channel to a position close to the chapel after centuries of it flowing north of Elne (until at least 1261–1264). The village of Santa Eugènia de Tresmals was progressively deserted during the 15th century, i.e., during the coolest period of the LIA. This phenomenon of village desertion during that time was common

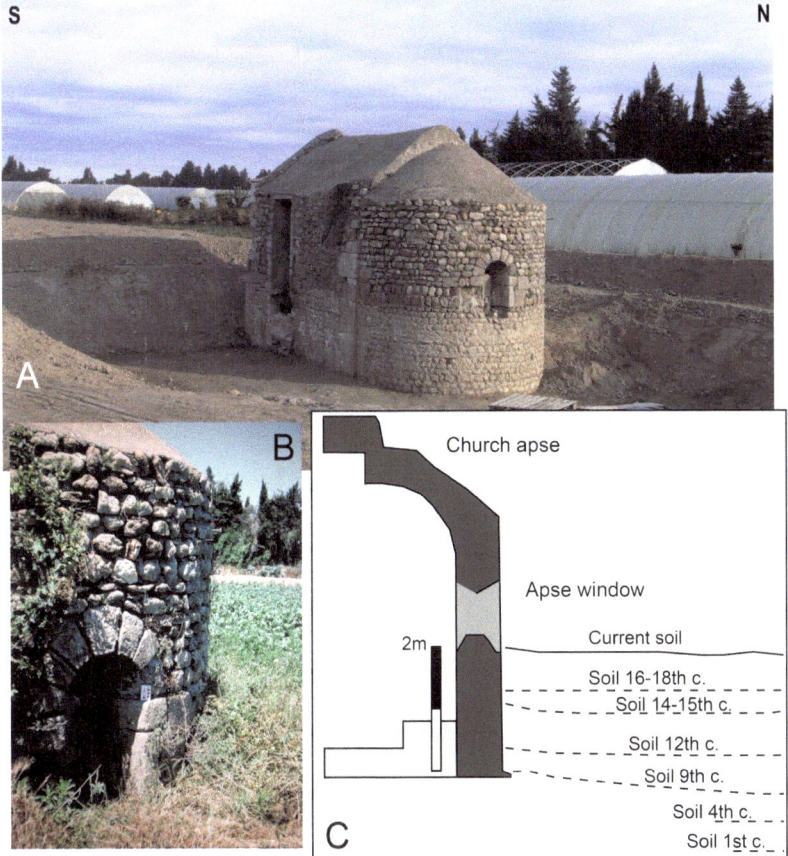

Fig. 8.6 **Archaeological site of Santa Eugènia de Tresmals. A** The chapel after its excavation; note lighter tones of the masonry previously buried in the flood deposits. **B** The apse prior to its excavation. **C** Stratigraphy adjacent to the apse; note substantial acceleration in aggradation rate (rising from 1 to 3 mm/yr) among the six alluvial sequences after chapel construction, i.e., between the tenth and sixteenth centuries (after Carozza et al. 2012, redrawn)

throughout Roussillon. Its link with flood frequency and floodplain aggrada-tion, however, clearly needs to be considered on a case-by-case basis because climatic parameters were not the only cause behind economic or demographic instability (Puig 2009; Carozza et al. 2009).

Follow the country lanes back eastward until you reach the D8. Drive on to Saint-Cyprien Plage and follow the beach barrier towards Canet, where the road is separated from the sea by a low line of sand dunes. Canet lagoon is a protected site managed under the stewardship of the Conservatoire du Littoral. *Having lost ~50% of its open-water area since 1750 as a result of infrequent connections with the sea through the outlet channel, of a predominance of freshwater input, and of rapid deltaic sedimentation by the Réart River (which terminates in the lagoon), its natu-ral cycle of sediment filling is nearing completion. As you drive past the reconstituted fishing village, note the traditional reed huts whose shape, according to archaeolo-gists, has probably remained unchanged since the Bronze Age. As you join up with the traffic roundabout of Stop 2, turn into the D81 towards Torreilles. The road crosses the lower Têt plain and Bourdigou channel, which was an ancient estuary shared by the Têt and Agly rivers at various times of their wandering histories. Go all the way to Torreilles on the D11e.*

Stop 5. Torreilles: 3000 years of avulsions by the Agly River

The low coastal plain shared by the Têt and Agly is known as La Salanque, a name referring to the salinity of its soils. The Holocene and younger history of this plain has until now not been much studied in its southern part (Têt flood-plain), but is better documented in the north (Agly floodplain; Buscail-Martin 1978; Serrat 2000; Calvet et al. 2002; Carozza et al. 2013; Fig. 8.7). Exposures of the stratigraphy in this flat landscape are scarce, and interpreting its sedi-mentary history therefore requires some faith in the evidence obtained by past excavations that are no longer open to inspection.

A first stop at the eastern entrance to Torreilles, on the bank of the Bourdigou channel (Fig. 8.8A), is a good vantage point for reconstructing the deltaic fill sequence of the ancestral lagoon of Leucate–Barcarès which, in the context of the Holocene sea-level maximum ca. 6000 BP, originally extended much further to the south than today. Palustrine muds were documented in all of the drill cores, which generally encountered the top of the Pliocene sequence at a depth of 15 to 20 m, sometimes more. The Bourdigou channel is about 3 m deep, and shelly lagoon muds have been exposed in trenches at 3 to 6 m depths north of this point. The base of the mud beds has produced a radiocarbon age of 3000 cal yr BCE; the top of the mud sequence, where palustrine sedimentation documents

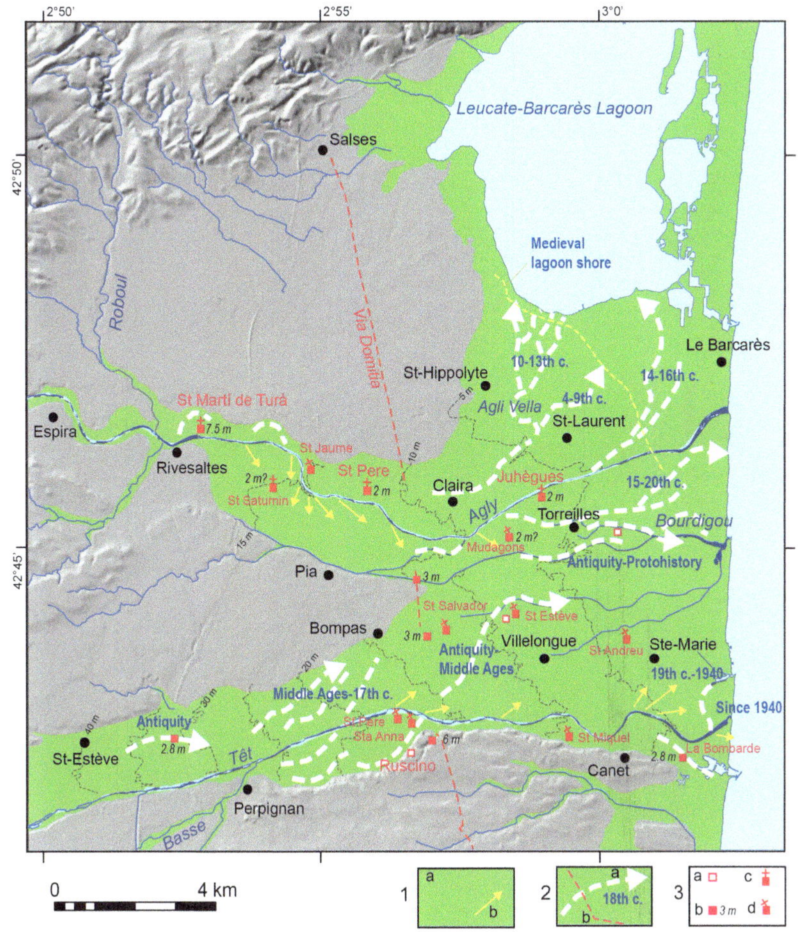

Fig. 8.7 The Salanque: Holocene lower plains of the Têt and Agly. Key to symbols and ornaments—**1**: Plain dynamics; **1a**: Holocene to modern alluvial fill. **1b**: Levee breaches and sandy-silt-filled crevasse splays from the Oct. 1940 flood (from IGN aerial photographs, 1942). **2**: Linear features; **2a**: Fluvial palaeochannels, with indications about their age and morphology based on various archives and archaeological age constraints. **2b**: Known trace of the Via Domitia. **3**: Archaeological sites; **3a**: Roman or Protohistorical site at the surface or in the subsurface. **3b**: Buried Roman site (depth below alluvium or colluvium specified). **3c**: Buried Romanesque church (burial depth indicated). **3d**: Abandoned or destroyed Medieval locality (burial depth unknown). After Calvet et al. (2002) and Carozza et al. (2013), modified

Fig. 8.8 Over 2500 years of Agly avulsions around the village of Torreilles. A Aerial view showing the main deltaic lobes (after Carozza et al. 2013, and Google-Earth-supported satellite imagery). White dashes: palaeochannel distinguishable on aerial photographs. **B** Abandoned Medieval village of Mudagons; vestiges of destroyed Romanesque church of Saint-Sébastien. **C** Notre-Dame-de-Juhègues; the nave aisle on the south side has undergone at least 2 m of burial; these Romanesque archaeological structures are best appreciated from inside the building

evidence of emersion and an influx of floodplain sand deposits, contains late Bronze Age artefacts. Overlying this stratigraphy comes a cut-and-fill floodplain sequence consisting of sand, silt and gravel with paleosol interlayers containing Roman artefacts. Further up, at 1 m depth, artefacts are Medieval (14th–sixteenth century). On the right bank of the Bourdigou channel, the archaeological site of Les Parrudes yielded surface artefacts ranging from the first century BCE to the fifth century CE (Kotarba et al. 2007). Such a surface assemblage spanning more than 500 years suggests a long period of stability around this living floor, unimpacted by encroaching lobes of the Agly immediately to the north and southwest (Fig. 8.8A). Slightly further to the SW, the record of fluvial dynamics was indeed more continuous, with three inset alluvial sequences each presenting channel–levee–flood-basin facies assemblages. Based on their respective artefact assemblages, they have been dated to the Protohistorical, Roman, and late-Medieval–early Modern periods.

The site of Juhègues, at the north exit of Torreilles, has been standing by the banks of the Agly River only since the river's last avulsion in the fourteenth century (Fig. 8.8A). The church architecture, which was renovated in the seventeenth century, is highly composite. Its base is Romanesque, and a nave aisle on its south side reveals the top of a window and door arch currently at ground level (Fig. 8.8C). Roughly 100 m SW of the church, the apex of a partially buried Medieval motte still rises 3 m above the floor of the plain. To the west, near the area named Mudagons, the only remaining vestige of the church of Saint-Sébastien is a leaning apse at the corner of a field (Fig. 8.8B). This entire area corresponds to the most recent of the Agly's deltaic lobes, which began to spread across this area during the fifteenth century in conjunction with the LIA.

Cross the Agly and continue on the D11 through the town of St-Laurent-de-la-Salanque, then onto the D83 expressway. You will get a good view of the northern expanses of the Salanque plain from the bridge. Trenching for the expressway has allowed several kilometres of stratigraphy to be observed and radiocarbon- or archaeologically dated beneath the modern alluvium. At the base of the sequence, to the east of the viewing point, lagoon deposits dating back to the Roman period (1st to third century CE) are covered by coarse fluvial sands from the eighth century. These represent the Agly's deltaic lobe during Late Antiquity. It strikes SW–NE. The third unit, further to the west, is also a sandy lobe. It dates back to the Middle Ages and strikes north. It sweeps past the village of Saint-Hippolyte and is highlighted by the Aglí Vella palaeochannel (noted thus on the IGN 'Top 25' map), which ceased to be active after 1280 (Figs. 8.7, and Fig. 8.8A) (Carozza et al. 2013).

Take the D83 westward for 1.3 km and aim for Claira on the D41. As you cross the Agly and Bassa de Pia (an artificial channel), engineering works on the Bassa some 325 m east of the bridge have revealed vestiges of the Via Domitia and its successive causeways across small local streams, all of this beneath 3 m of alluvial silt and channel gravel fills (Fig. 8.9). Unfortunately, these exposures are no longer visible. Return to Claira, and at the village centre turn left towards the signposted chapel of Saint-Pierre, then left after the water tower, and continue another 2 km from the village.

Stop 6. Chapel of Saint-Pierre-del-Vilar (Sant Pere del Vilar)
This site is located at the apex of the Agly River's Holocene alluvial fan. The width of the lowermost terrace, T0, is at most 2.5 km between the unpaired risers of T2 to the north (Crest de Rivesaltes) and T3 to the south (Pia terrace; see Itinerary 1). As with many other Medieval sites of this kind, nothing is left apart from the ruined church which, here as elsewhere, has been restored in different styles over time and still attracts pilgrims today. The road embankments expose sand and gravel channel deposits containing debris from the destroyed wall still visible in the ditch; the channel was cut in overbank silts and is itself capped by 1.5 m of floodbasin silts. The top of the apse window frame is flush with the ground, and the upper arch of a fourteenth century doorway juts out 1 m above ground (Fig. 8.9B). As at the previous sites encountered in this itinerary, the burial depth of the chapel is ~2 m; evidence here, as at other buildings elsewhere, of a raised doorway suggests belated attempts at adaptation in response to initial flooding and partial burial. On the main façade you can observe attempts at raising the nave by 2–2.5 m in the context of renovation, probably at some time during the seventeenth century (Fig. 8.9C). About 1 km further upstream, and closer to the Agly channel, the village and church of Sant Jaume de la Ribera were totally wiped off the map (only the place name has been preserved); in contrast, Saint-Saturnin des Hortolanes (alias Notre-Dame-de-la-Salut) stands on the right bank and was entirely rebuilt in the seventeenth century using the original building materials and reproducing the original Romanesque style. It is open to visitors, but the twelfth century floor has never been identified.

Follow the country lane going west from Sant Pere del Vilar and proceed through the shopping centre, aiming for the D900 and the link road to the A75. At the round-about before the motorway tollgate, branch off onto the D12, then at the following roundabout take the D5 towards Rivesaltes. At the third roundabout thereafter, turn left onto a road taking you to Saint Martin de Tura, near a ford across the Agly River.

Fig. 8.9 **Historical alluvial aggradation around Claira. A** Excavation of the Via Domitia (El Bogariu Alt) where it intersects the Bassa de Pia channel. Note the road paving, and two causeway structures for crossing a small local stream (probably the Llabanère). Other associated vestiges testify to continued use of this Roman road through the Middle Ages. The road rests on, and is also covered by, blueish-grey palustrine silts, then by at least 2 m of ochre to beige, slightly sandy floodplain silts. A quartz-rich gravel bed at the surface, containing mostly reworked older deposits from alluvial terraces in the vicinity, highlights a recently active channel of the Llabanère (base indicated by white dashes). **B** Church of Sant Pere del Vilar (Saint-Pierre-del-Vilar); the modern chapel stands on Romanesque walls. **C** Church façade, showing raised brickwork above the eleventh century stone-and-mortar structure. **D** Detail of the apse; the window is flush with the ground, implying at least 2 m of sediment

Stop 7. Saint Martin de Tura (Sant Martí de Turá)

The site of this ancient hamlet is the loop of a cut-off meander on the Agly River. Access is restricted because the site is on private property, but looking over the perimeter wall you can see the steeple of the seventeenth century reconstruction of the church. Against the east wall you will also see the departure of the Claira canal, which today is at the bottom of a 5-m-deep trench shored up by stone walls. Such was not the case at the time of its construction, thus suggesting subsequent work to extend its useful life in the face of sedimentation on the plain. Given that all of the irrigation canals in Roussillon were constructed (or reconstructed) during or before the fifteenth century, the preservation of the water intake sluice on the Claira canal at this location provides a means of quantifying historical sediment aggradation on the plain. Note the benchmarks in the sluicegate wall showing floodwater levels attained at the time by the Agly River, all remarkably similar to the levels observed in 1940 and 1999 (Fig. 8.10).

Tura church is where post-twelfth century alluvial aggradation in Roussillon has been the most spectacular (Alessandri 1990; Serrat 2000; Calvet et al. 2002). The original paved floor (perhaps 7th–8th century) lies 7.4 m below ground. It has not been archaeologically excavated because it lies below the Agly water line, but it is estimated to date back to Late Antiquity or to the construction of the initial Romanesque or pre-Romanesque church. A 13th–fourteenth century sill plate, apparently of the same age as the southern gate (restored in a post-Romanesque style), lies 1.9 m above the original floor. A series of undated arches standing on the eighth century floor show signs of having been raised at least once. The Romanesque nave is filled with laminated silts up to its pointed barrel vault but no analysis of palaeoflood chrononology has as yet been carried out. The deposits can nonetheless be estimated to date back to the LIA, i.e., between the thirteenth and seventeenth centuries, because the foundations of the chapel rebuilt during the seventeenth century rest on top of the Romanesque vault (Fig. 8.10A), and they have since then only been weakly affected by overbank sediment aggradation (20–30 cm of silt deposited by the October 1940 flood).

The protracted history of fluvial aggradation on the Agly plain has become a threat to the town of Rivesaltes. The old town was initially built on the tread of terrace T1/Fy at a place where it merges with T0/Fz, but it is now chronically threated by the bankfull and flood regimes of the river. The substantial height of the upper terrace riser, which gave its name to the town (*Ripas altas*, ca. tenth century CE), has now been diminished by the active historical aggradation on the plain below, thus now exposing Rivesaltes to flood hazards from which it was originally protected. As a result of centuries of historical backfill across the floor

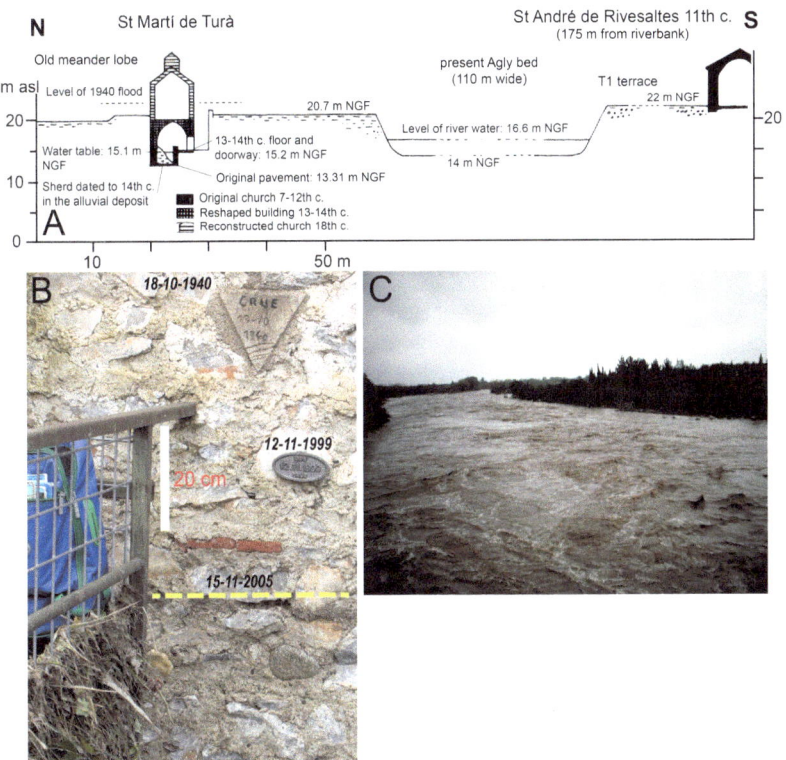

Fig. 8.10 Rivesaltes and Sant Martí de Turà (Saint-Martin-de-Tura). A Section through the Agly floodplain and its archaeological vestiges. On the right bank (higher embankment: *ripas altas*), you can still see exposures of late Pleistocene terrace T1, on which the town is built. On the left bank, alluvial aggradation since the tenth century attains thicknesses of 7.5 m and has entirely buried the earliest church building, of Romanesque or pre-Romanesque age. NGF refers to the French Ordnance datum (Nivellement Général de la France, established in 1969 by reference to the Marseille tide gauge). **B** Marked flood-level records on the walls of Domaine de Tura, above the sluice of the Claira canal. The water marks for 1940 and 1999 (an event of lesser magnitude) are close; this is probably explained by the construction, between those two dates, of the adjacent ford causeway, which raised the waterline above natural levels. **C** The Agly River on the falling limb of its flood-wave, morning of November 13, 1999, viewed downstream from the bridge at Espira. At this time, discharge was still 1000 m^3/s; the flood peak, around midnight, had reached 2000 m^3/s. In October 1940, peak flow was also estimated at ~2000 m^3/s but it lasted for three days; changes to the landscape have been too intense since 1940 to apply rigorous comparisons based on fixed landmarks or human witness accounts

of T0, the tread of T1 stands only ~2 m above the left bank of the river's modern floodplain.

Return to Perpignan via the D5, D614 and D117, driving across the treads of the Têt middle terraces (T3d, c, b) around La Llabanère, near the airport.

References

Alessandri P (1990) L'église Saint-Martin de Rivesaltes. Archéo du Midi Médiéval 8–9:174–177

Bénech C (1993) Estimation des périodes de retour de 'l'aiguat' d'octobre 1940 dans quelques vallées des Pyrénées–Orientales. In : "L"Aiguat del 40", actes du congrès de Vernet-les-Bains, Generalitat de Catalunya Edit., pp 297–313

Buscail-Martin R (1978) Evolution holocène et actuelle des conditions de sédimentation dans le milieu lagunaire de Salses-Leucate. PhD thesis (unpubl.), Université Toulouse 3 Paul Sabatier, Perpignan, 228 p

Calvet M, Lacquement G (2019) Perpignan, la Têt et la Basse. In: Carcaud N, Arnaud-Fassetta G, Evin C (eds) Les rivières en ville—aujourd'hui, hier, demain. Edit. CNRS, Paris, pp 194–205

Calvet M, Serrat P, Lemartinel B, Marichal R (2002) Les cours d'eau des Pyrénées orientales depuis 15 000 ans : état des connaissances et perspectives de recherches. In: Bravard JP, Magny M (eds) Les fleuves ont une histoire, paléoenvironnements des rivières et des lacs français depuis 15 000 ans. Errance, Paris, pp 279–294

Carozza JM (2012) Les sociétés du passé face aux crises environnementales : une approche géoarchéologique et géomorphologique. Mémoire d'Habilitation à Diriger des Recherches, Université Paris 7-Diderot, vol 1, 266 p

Carozza JM, Puig C, Odiot T, Galop D, Passarrius O, Valette P (2013) La basse plaine du Roussillon au cours du dernier millénaire : évolution paléogéographique et impact sur le peuplement médiéval et moderne d'une crise morphogénique sans précédent (13e-16e s.). Quaternaire, 24:141–151

Carozza J-M, Odiot T, Valette P, Puig C, Pequinot C, Alessandri P, Passarius O (2009) Réponse des bassin-versants du Roussillon entre le XIIe et le XIXe siècle: un impact du Petit Age Glaciaire ? Archéol Midi Médiéval 27:207–215

Carozza J-M, Puig C, Odiot T, Passarrius O, Valette P (2012) Lower Mediterranean Plain accelerated evolution during Little Ice Age: geoarchaeological insight in the Tech basin (Roussillon, Gulf of Lion, Western Mediterranean). Quat Int 266:94–104

Collina-Girard J (1976) Les alluvions fluviatiles des fleuves côtiers dans le Roussillon. In: De Lumley H (ed) La préhistoire française, vol 1. Édit. CNRS, Paris, pp 78–82

Capolini J, Lefort P (1993) Le risque d'inondation à Arles-sur-Tech et à Amélie-les-Bains pour une crue de type octobre 1940 et une crue centennale. In: l'Aiguat del 40, actes du congrès de Vernet-les-Bains, Generalitat de Catalunya, pp 285–294

Delmas M, Calvet M, Gunnell Y, Voinchet P, Manel C, Braucher R, Tissoux H, Bahain JJ, Perrenoud C, Saos T, Aster Team (2018) Terrestrial [10]Be and electron spin resonance dating of fluvial terraces quantifies quaternary tectonic uplift gradients in the eastern pyrenees. Quat Sci Rev 193:188–211

Desailly B (1992) Le temps des aiguats. In Broc N, Brunet M, Caucanas S, Desailly B, Vigneau J-P (eds), De l'eau et des hommes en terre catalane. Trabucaire, Perpignan, pp 167–217

Hickin EJ (1983) River channel changes: retrospect and prospect. Spec Publ Int Ass Sedim 6:61–83

Jacob N (1997) La crue d'octobre 1940 dans la basse vallée du tech (Roussillon) d'après les dossiers des sinistrés. Ann Géog 596:414–424

Kotarba J, Castellvi G, Mazière F (2007) Carte archéologique de la Gaule, les Pyrénées-Orientales. Académie des Inscriptions et Belles-Lettres, Paris, 712 p

Marichal R, Rébé I (2003) Les origines de Ruscino. (Château-Roussillon, Perpignan, Pyrénées-Orientales) du Néolithique au premier âge du Fer. Lattes, Edition de l'Association pour le Développement de l'Archéologie en Languedoc-Roussillon, 2003. Coll. Monographies d'Archéologie Méditerranéenne (MAM), 16:298 p

Martzluff M, Passarius O, Vignaud A, Donès C (1994–1995) Nouvelles données sur le Néolithique ancien du Roussillon. Études Roussillonnaises, XIII:7–16

Martzluff M (2006) Entre pebble culture, bifaces et érosion, le « Tautavélien » des terrasses quaternaires en Roussillon. Bull Assoc Archéo Pyr-Or 21:89–112

Martin-Vide J, Lopez-Bustins JA (2006) The western Mediterranean oscillation and rainfall in the Iberian peninsula. Int J Climat 26:2245–2255

Pardé M (1941) La formidable crue d'octobre 1940 dans les Pyrénées-Orientales. Rev Géogr Pyr Sud-Ouest, XII: 237–279

Puig C (2009) Les prémices du Petit Age Glaciaire en Roussillon à travers le prisme des sources écrites. Archéol Midi Médiéval 27:191–205

Serrat P (2000) Genèse et dynamique d'un système fluvial méditerranéen: le bassin de l'Agly (France). PhD thesis (unpubl.), Université de Perpignan, 653 p

Perpignan to Perpignan (191 km excluding optional extras; 1 day, 2 days including walks). See Fig. 9.1.

Itinerary 4 takes you up the Tech valley, with opportunities to observe Hercynian and Alpine structures as well as geomorphological features on the southern flank of the Axial Zone (e.g., two generations of pediment surfaces, P1 and P2). An even more striking feature is the geomorphological record of high-intensity meteorological events from the past years and centuries. The Vallespir lies in the humid enclave of Catalonia, i.e., in a small, semi-enclosed area sheltered from westerly weather systems by the mountain range but open to overheated and humidity-laden air masses from the adjacent Mediterranean, which penetrate via the low-lying Empordà graben and get funnelled into the Tech valley where they promote thunderstorm clouds. Thus, although autumn storms periodically generate major hydrological extremes as elsewhere in Roussillon, unlike adjacent areas summers are also rarely dry. These paroxysms have left scars in the landscape as well as in the collective memory of residents (Pardé 1941; Soutadé 1980, 1993, 2010).

Leave Perpignan following signposts to Spain, then take the D900 towards Le Boulou followed by the D115 towards Céret through an area previously described in Itinerary 2.

© Springer Nature Switzerland AG 2022
M. Calvet et al., *Geology and Landscapes of the Eastern Pyrenees*, GeoGuide,
https://doi.org/10.1007/978-3-030-84266-6_9

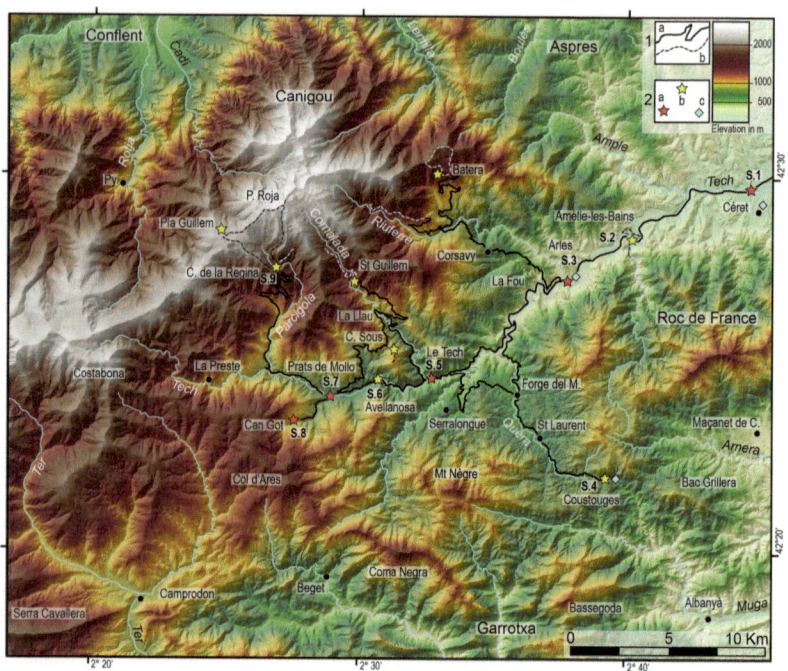

Fig. 9.1 Itinerary 4: route map and overview. Key to symbols—**1**: roads and trails. **1a**: roadways, irrespective of road rank, size, or quality. **1b**: trailwalks. **2**: stopping points (outcrops, exposures, landforms, or landscapes). **2a**: roadside stops. **2b**: roadside stops involving a short walk, or marking the start of a more major trailwalk. **2c**: cultural stops. Labels S.1 to S.9 locate the sequence of roadside stops

Stop 1. Devil's Bridge at Céret

The Pont du Diable (Fig. 9.2) is a Medieval structure situated downstream of the two modern bridges. Its single arch spanning the Tech is 31 m high. It was constructed between 1321 and 1341, after a flood which destroyed the older 'Roman bridge' (thus named today, and vestiges of which are preserved upstream). The flood could have been the 1264 event, which throughout Catalonia prompted thereafter the construction of 'overcalibrated' bridges (i.e., overengineered for their time; Pardé 1941; Bénech 1993). On October 17–19, 1940, water levels under this bridge reached heights of 5.94 m. The peak discharges of 3900–4275 m^3/s initially estimated by the Inspector General of the Ponts et Chaussées, B. Quesnel, were later revised by M. Pardé (1941) to 3500 m^3/s.

Fig. 9.2 The three bridges at Céret, viewed from downstream. Medieval bridge in foreground. Dashed line: water level in October 1940

A retrospective critical analysis of that evidence by Lalanne-Berdouticq (1993), however, suggests that stream velocities were probably overestimated and that 2200–3000 m^3/s would be a more accurate range.

The D115 will take you to Amélie-les-Bains along the right bank of the river, across the treads of alluvial terraces T1 and T2. The narrow valley is cut in the Lower Cambrian Évol schists (Jujols Group) and in the limestone-banded schist of the Ediacaran Canaveilles Group. Vestiges of pediment P2 form erosional benches in those rocks on both sides of the river at elevations of 300 m and above (see Itinerary 2). On the left (Saint-Paul) and right banks (spot elevations: 293 and 301 m) there is an opportunity to observe the vestiges of an ancestral Pliocene river course, strewn with gneiss pebbles and boulders and filling a cut in the underlying bedrock—itself probably generated by the Messinian Salinity Crisis. Topographic benches on the left bank display alluvial deposits consisting mostly of schist. Labelled as 'Fu-p' on the geological map, they correspond either to the youngest stratigraphic vestiges of the Pliocene sequence, or to uppermost Quaternary terrace T5. At Améølie, the sites of interest are best inspected on foot (allow 1 to 2 hrs).

Stop 2. Amélie-les-Bains

The Amélie syncline, an Alpine structure overprinted by Neogene tectonics. The Tech forms an entrenched meander around the only outcrop of Mesozoic cover rocks of the Axial Zone in the eastern and central Pyrenees (Fig. 9.3A), here compressed into an asymmetrical syncline with its NE flank overturned and over-ridden to the SW by the Aspres Thrust. This major Paleogene Alpine structure has displaced the Aspres unit (a component of the Upper Thrust units) over the Canigou–Vallespir unit (Middle Thrust units of the Axial Zone; see Figs. 1.2 and 1.3D). The sole of the Aspres unit can be inspected around Palalda, where it consists of a thick mass of Ediacaran limestone, brecciated by crushing and heavily enriched in iron mineralisations ('brèches du Mas Manès', at El Mener). The syncline is truncated immediately south of Amélie by the Neogene Tech Fault, which exposes outcrops of Roc de France gneiss belonging to the lower Alpine thrust units (Lower Thrust sheets). Thermal springs along the normal fault (water at 47 to 67 °C) have been utilised since Roman times and the modern spa installations have restored two Roman-style rooms (still in use). A trail down the gorge starting from behind the spa buildings will take you to Mondony canyon, where the fault plane exposes outcrops of deep migmatitic gneiss.

For a closer look at the Mesozoic outcrops, a good area is the downstream end of town near the ancient 'Roman bridge', which was swept away by the 1940 flood and has been replaced by a new footbridge. In the riverbed immedi-ately upstream of the footbridge, you may observe the unconformity (Fig. 9.3B) between the granite-veined micaschists of the Hercynian basement and the bur-gundy tones of the Bundsandstein sandstones, conglomerates and pelites. The steeply dipping beds (50–60° to the NE) are surmounted on the right bank and downstream of the footbridge by grey Muschelkalk limestone and dolomite, also forming a bedrock scarp above the adjacent road (Fig. 9.3C). These sequences underlie the spur extending to the village of Montbolo, where they form thin imbricate thrusts associated with gypsum beneath the main thrust (observable from the D53). Along the right bank of the river, the Triassic beds are uncon-formably overlain by Upper Cretaceous marine beds. These are not within sight from the footbridge site but can be reached via tracks heading north above the Super Amélie housing estate or leading to Can Tormenta valley; sections typically expose clays and sandstones of probable Santonian age, followed by Campanian sandstone and rudist-bearing limestone, and then by a thick, oyster-rich paralic sequence of Maastrichtian marl, sandstone and sandy limestone. This

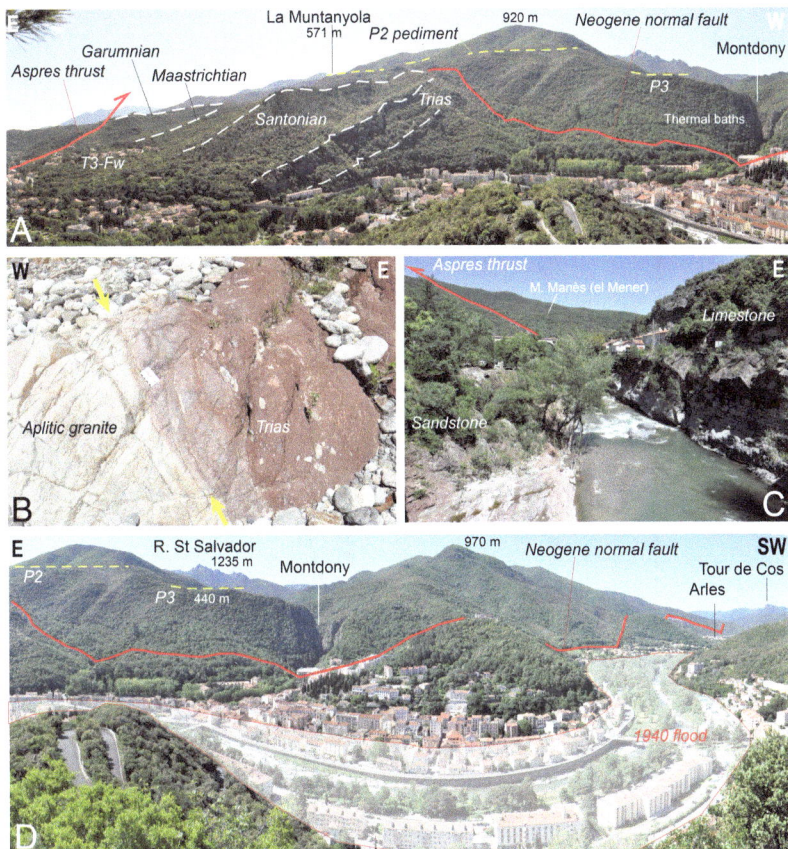

Fig. 9.3 Amélie-les-Bains sedimentary basin. A Wide-angle view from the calvary looking downstream. The Paleogene syncline is overridden by the Aspres Thrust; note the close
succession of hogback scarps in the Triassic, Santonian, Maastrichtian and Garumnian beds
(see also Fig. 9.5). The Neogene Tech Fault scarp has evolved into triangular faceted spurs;
the base of the facets has been exhumed, and the offset of pediment P2, which also cuts
across hogback ridges in the syncline, strongly suggests a geologically recent vertical tectonic displacement of ca. 100 m. **B** Unconformity of basal Triassic sandstone over Hercynian
basement in the Tech riverbed. **C** Triassic series, here viewed from the footbridge ('Pont
Romain'). **D** Wide-angle view from the calvary, looking upstream; note continuation towards
the SW of the Tech Fault scarp and pediment bench P3 above Montdony gorge. The highlighted October 1940 flood zone, with flow velocities exceeding >8 m/s, was 230 m wide

entire Mesozoic sequence is itself covered by the 'Garumnian' (i.e., continental Maastrichtian to Paleocene) detrital sequence relating to the erosion of the rising mountain range and consisting of conglomerates, sandstones, red clays, lacustrine limestone, and *Microcodium*-bearing marls.

Legacies of the 1940 flood. The entire lower part of the town was under water in 1940 (Fig. 9.3D), including the D115 below the town centre (23 deaths through-out the municipality). The high street just above was at the edge of the water line and of its gravel deposits. Over the footbridge, you will find a flood channel west of the Triassic sandstone ridge; it has been landscaped into public gardens. Heading downstream along the trail through the sandstone outcrop you will pass a metre-sized gneiss boulder abandoned by the Tech in 1940. Just upstream of the railway bridge, a multistorey residential building was entirely destroyed by the flood and the railway embankment was swept away, leaving the railtrack hanging over a precipice. Note that a new building (Résidence l'Oratori) has been rebuilt exactly at the site of its destroyed predecessor.

Return in an upstream direction by walking up Avenue du Vieux Pont. Here, all the buildings to your left had water up to their roofs, under mean flow veloc-ities of >8 m/s. Photographic archives have shown the Villa Beauséjour in those circumstances (second house down from the bridge by the Casino, whose orig-inally double-sloping roof has been reprofiled as a result of removing the third floor of the building). At the bridge head, walk up the steps joining with the D53 (road to Montbolo) and follow that road as far as the calvary. The bends in the road cut through the Triassic series, which forms imbricates and duplications in relation to subsidiary tectonic displacements along the Aspres Thrust. From the calvary you get a good general view of the Amélie syncline and the Roc de France normal fault (Fig. 9.3A, D). From here you are also overlooking the 'Petite Provence' suburb, which has been entirely built over since the 1940 flood and initially was the site of the old railway station (Fig. 9.4A, C). If the station were still here today, its location would be approximately in the middle of the modern channel. After the flood, this area was a vast expanse of coarse bedload (Fig. 9.4B), on average 250 m wide, 8–10 m thick, and entirely deposited during that three-day flood event. Flood defence works since the flood have consisted in clearing the alluvium from the active channel, blasting the bedrock reach near the bridge, and erecting artificial levees. The stock of alluvium farther upstream is nonetheless still substantial and could repeat the 1940 disaster at Amélie in just a few hours on a future occasion, emphasizing the fact that flood disasters are not just a matter of peak water discharge but also of in-channel sediment aggradation raising the water line.

Fig. 9.4 **Effects of the 1940 flood upstream of Amélie, going back in time from A to C. A** Present state (2020). **B** Alluvial aggradation immediately after the 1940 flood (Terra Nostra collection). **C** Situation at the beginning of the twentieth century, i.e., before the 1940 flood, here around the railway station (postcard). Buildings A to D (labelled in yellow) in the three pictures provide bearings

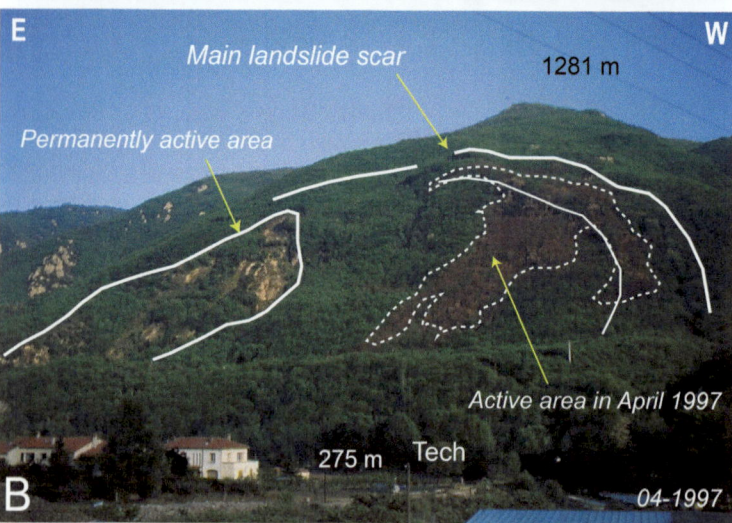

Fig. 9.5 **Large, composite rock slope failure at Arles-sur-Tech**. The valley side coincides with the Neogene Tech Fault scarp and its crush zone. **A** In May 1993. **B** In April 1997. Illustration B shows the same area as in A, but slope failure during an exceptionally wet spring in 1997 caused a major disturbance to the root system and entailed massive vegetation die-off (note brown tree foliage)

Continue towards Arles along the D115, which follows the right bank. By the bridge before Arles, note the large quartz dyke injected in the leucogneiss. This feature marks the trace of the major Canigou Thrust, here tilted vertically, which separates the Canigou nappe (outcrops at Amélie) from the Vallespir nappe (outcrops near Arles; see section in Fig. 9.5B). A brief stop as you enter Arles from the north (parking is possible at the crossroads signposted 'Bonabosc–chemin du cimetière') allows you to get a view of the 1940 flood deposits on the opposite bank: the terrace tread is now a stand of poplar and willow, but the 8 m, post-1940 vertical cut by the river is a stark reminder of the thickness of the deposit emplaced in just one three-day event. Drive through Arles (gothic cloister, Romanesque churches, including a ninth century abbey) and stop before the bridge over the Riuferrer, at the western exit of town.

Stop 3. Arles-sur-Tech
The town stands at the junction between the Tech and the Riuferrer, and was built on the Riuferrer debris cone. The hourglass-shaped watershed of the Riuferrer (steep catchment, gorge, and debris fan), typical of many high-energy alpine valley sides, descends straight from Mt. Canigou (elevation drop from 2500 to 280 m in 15 km), with 1940 discharge statistics as follows: peak flows at 700 m^3/s (specific discharge: 14.7 $m^3/s/km^2$), and 644,000 m^3 of sediment delivered. The Tech, as a result, was pushed away by the Riuferrer and its bedload, thus saving the town from the main flood—albeit with the help of its residents, who spent their first night erecting and maintaining a sandbag dam on the high street, aptly named 'Boulevard du Riuferrer'. Destruction, as a result, was reduced, mostly impacting the low-lying industrial area (textile and biscuit factories). Urbanisation has now engulfed the trail of debris from both the Tech and the Riuferrer, but such a catastrophic event could be repeated, particularly on the high slopes to the south of Arles, which coincide with the Tech Fault. You can see there a 400-m-high rockfall scar and a 50–60-m-high Pleistocene fallen mass at its base (downward from Mas de la Guàrdia). Because of its multiple signs of ongoing instability (active gullying on the NE edge of the unstable slope mass, scars beneath the land cover entailing distinct springtime wilting of the forest canopy in 1997, Fig. 9.6), the slope is being instrumentally monitored.

Optional excursion with trailwalks: valley of the Riuferrer, Batère iron mines (26 km return trip)
At Arles, take the D43 towards Corsavy. The road rises rapidly above the valley, and you will soon see to the west the 'Gorges de la Fou', a metre-wide but 200-m-deep fissure in the Lower Cambrian dolomitic limestones (Valcebollère Formation, Jujols Group), which here

Fig. 9.6   Upper catchment area of the Riuferrer. A Extensive gully systems above the tree line, here seen in 1995 from Tres Vents ridgetop. These gullies were reactivated in 1940 and still active in 2020. **B** Riverbed at Léca soon after the 1940 disaster (image code ET1)

are exceptionally thick (can be reached via the D115; paying visit provides access to foot-bridges and walkways). From Corsavy, the road winds across low-gradient pediments ca. 700–800 m, cut in deeply weathered granite outcrops and corresponding to generation P2. The Riuferrer ('iron river') bears that name because of the numerous iron forges on its banks, supplied by the iron ore deposits at Batère. Below the village of Léca, the road crosses the Riuferrer at a spot where it remains hard to imagine its state after the 1940 storms, when the swollen stream tore through the flanking meadows, cutting a strip 50 to 75 m wide and dumping here an unsorted mass of debris several metres thick (Fig. 9.6B). Vast expanses of planted protection forest occur further up in the catchment, and deep gully scars on slopes above the forest belt (Fig. 9.6A) can be reached via the dirt track linking Léca to El Faig. The main road eventually reaches the mines at Batère, at 1400 m a.s.l. Iron was mined here from the second century BCE (numerous slag heaps in the valleys surrounding the massif: Aspres, Arles...), and was the last in Pyrenean history to close down (1987, and definitively shut down in 1999). The ore consists of siderite, oxidised at the surface into hematite and goethite, and forms either manto deposits or cross-cutting veins; the bodies are intimately associated with the carbonate and dolomitic rocks at the base of the Canaveilles Formation, which here dips almost vertically between the Batère diorites to the east and the orthogneisses of Mt. Canigou. The abrupt topographic rise to the summit area (ca. 2400 m) from the ridgetops around Batère at barely 1800 m suggest displacement along a N–S fault during Neogene time (Fig. 9.7A). Mining in the early days was conducted from open pits at the surface, and later down adits at multiple levels striking through the entire massif and joining up with the sister mines at La Pinosa, on the north face of Puig de l'Estella. Ore evacuation was ensured by cableway gondolas. The road passes the most recent mining sites at 1100 and 1280 m (tailings visible below the road), and you can reach the more ancient open pits (90 min walk there and back, Fig. 9.7C) from the parking area at Coll de la Descarga via the trail leading to Coll de la Cirera. Weather permitting, a 3 h return walk to Puig de l'Estella and Tour de Batère guarantees very broad panoramic views over to the Albères and across the Vallespir (Fig. 9.7B).

Follow the D115 from Arles towards Prats-de-Mollo. At Pas del Llop, you will be leaving the Vallespir nappe / thrust unit and entering the Saint-Laurent-de-Cerdans thrust unit (Figs. 9.8B and 9.9B). Turn left towards Saint-Laurent-de-Cerdans on the D3, which leads you across large, flat-floored granite depressions hanging above the Tech V-cut and lined with a thick mantle of grus (Fig. 9.8). Despite minor retouching by Quaternary processes, these pediment landscapes of P2 and P3 affinity are late Neogene. The only outcrops having resisted intense weathering consist of less susceptible pink granite, monzogranite, leucocratic alkaline granite and miarolitic granite, all of which coincide with the eye-catching bornhardts at Tours de Cabrenç, Mont Nègre, Serrat del Cogull, and Coustouges. Pass through Saint-Laurent, then drive onward to Coustouges and stop by the Retirada monument at the town's eastern exit, at a junction between the road to the Spanish border and the road to Can d'Amunt.

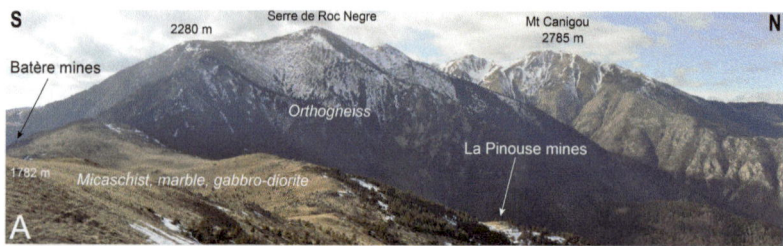

Fig. 9.7 Iron mines at Batère. A Mt. Canigou from Puig de l'Estella (1782 m). The rock dome in Paleozoic orthogneiss (plutonic protolith: 470 ± 10 Ma) rises abruptly from the envelope of Ediacaran marble and schist, which itself was intruded by the Hercynian batholith of Batère (granites and gabbrodiorites, 310–300 Ma). **B** panorama of the Vallespir from Puig de l'Estella. Note the Neogene fault scarp (Tech Fault), which forms the boundary fault of the Albères and Roc de France massifs farther east (Roussillon Basin visible in far distance). Note also the large relict pediments (generation P2), particularly extensive among the weathered granites around Corsavy. **C** opencast iron mine (elevation: 1700 m), on the path to Coll de la Cirera. The Ediacaran dolomitic limestone was metamorphosed into Cipollino marble, and dips steeply to the east (red arrow). The limestone hosts a still active groundwater karst network, intersected by some of the mineshafts and adits

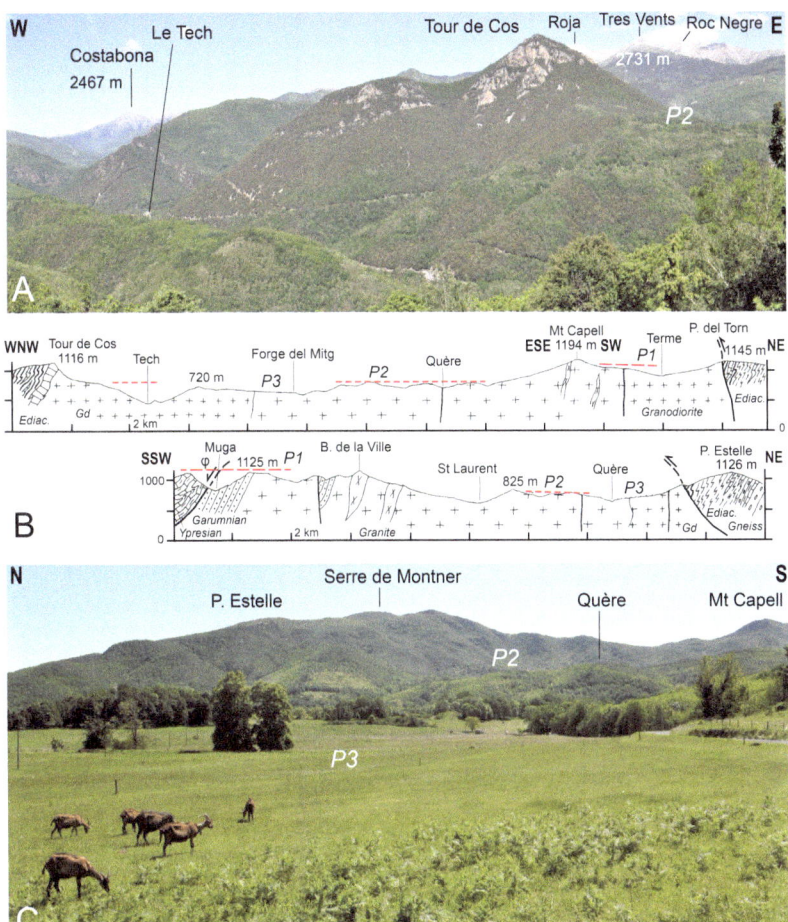

Fig. 9.8 Granite basin at Saint-Laurent-de-Cerdans and its late Neogene erosion surfaces. A North edge of the basin: Tech gorge surrounded by P2 pediment benches and overlooked by summits in Ediacaran limestone (Tour de Cos); background: gneiss massifs of Canigou and Carança (here seen from the track above Coll du Sagué). **B** Two perpendicular geological sections through the Saint-Laurent basin. **C** Erosion surfaces P2 and P3 around Forge del Mitg, as seen from the D64

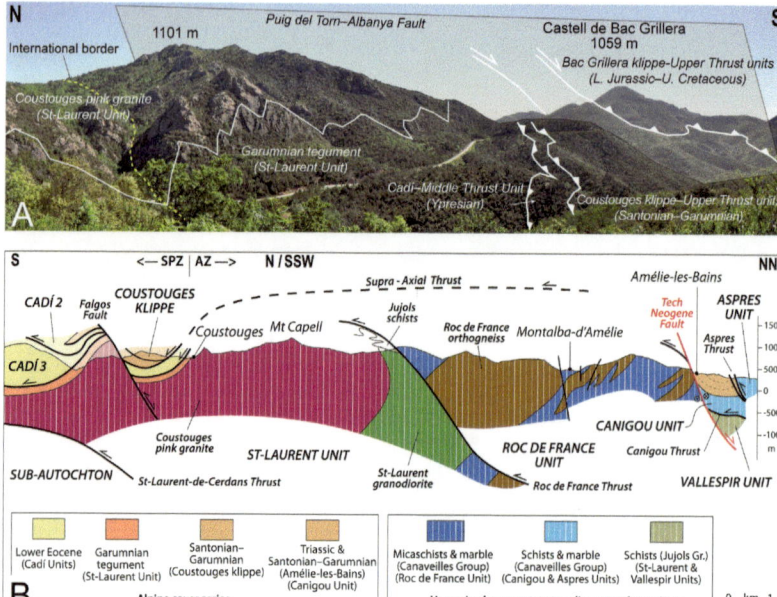

Fig. 9.9 **Alpine structures along the southern edge of the Axial Zone. A** Panoramic view of the frontal ramp anticline, here seen from the first bend in the road to Spain. The pink granites and their tegument of purplish-red Garumnian sandstone plunge beneath the overriding thrust sheets. The uppermost nappe forms the isolated massif of Bac Grillera. **B** Geological cross-section of Alpine structures in Vallespir. These structures are overprinted on the older Hercynian structures, particularly on the orthogneiss-cored, late-Variscan metamorphic domes of Canigou and Roc de France. This complex pile of Alpine thrust units includes, from top to bottom: (i) Bac Grillera (no outcrops) and Supra-Axial units (Upper TS; TS: thrust sheet); (ii) Coustouges–Aspres units (Upper/Middle TS); (iii) Cadí–Canigou–Vallespir nappe (main Middle TS); (iv) Roc de France Unit (Lower TS); (v) Saint-Laurent-de Cerdans Unit (lowermost TS). The Falgos Fault is a normal, partly inverted, polyphase Alpine fault, duplicating the Axial Zone–SPZ monocline boundary. To the north, the tectonic pile is cross-cut by the Tech Fault—a Neogene normal fault recording a throw of at least 1500 m. The nappe stack in this landscape was emplaced between ~47 and ~37 Ma

Stop 4. Coustouges and the southern flank of the Axial Zone

This area provides insights into the very complicated Alpine structures of this portion of the Pyrenean orogen. Here, the miarolitic pink granite is covered by a thin layer of pelites and purplish-red sandstone, upturned (southerly dips of 70°) and preserved as a succession of flatirons at the base of the escarpment. These rocks are of Garumnian age, and best viewed from the first bend in the road to the Spanish border (Fig. 9.9A). The porch of the Romanesque church at Coustouges is sculpted out of the local sandstone, and its walls are made of pink granite. The unconformity was passed along the road leading to Coustouges (42°22′10″N', 2°38′32″E), and a closer inspection of that site would show the sandstone resting on a weathered (rubefaction to a depth of 3 m) exposure of underlying granite. The Paleogene tectonic structure (Fig. 9.9B) is a frontal ramp anticline overlying a thrust (at depth) over which the Axial Zone and the Sub-Pyrenean Zone (i.e., the entire Pyrenean orogenic wedge at this longitude; see Part I, Figs. 1.2 and 1.3D therein) were jointly displaced over the southern foreland area. This major structure can be followed westward along the base of Mont Nègre and Tours de Cabrenç, as well as eastward towards Massanet de Cabrenys, Boadella dam, and as far as Darnius. In this area, the Axial Zone also displays its deepest structural units, i.e., Saint-Laurent to the west, and Roc de France to the east.

The village of Coustouges stands exactly on the tectonic boundary formed by the base of the Cadí cover nappe, which consists exclusively of marine marls and limestones of Ilerdian-Cuisian age (Sagnari Formation). This nappe is preserved in a half syncline bounded to the south by a normal fault (Falgos Fault). Initially a legacy of the Eocene foreland flexure, and thus a normal fault, it was later inverted (Fig. 9.9). The nappe can be seen in the embankment of the path leading up from the Retirada monument to the village church. Over the Cadí nappe lies the Coustouges klippe complex, apparently forming a sequence similar to that previously observed at Amélie, but in the present case inverted—i.e., with the Garumnian below and the marine Upper Cretaceous beds on top. The stratigraphy can be inspected between the monument and the first bend in the road to Spain. This upper unit broadens in the Bac Grillera massif (1059 m), which can be seen on the other side of the valley to the SE, where it actually becomes a stack involving two nappes. The lower unit is a continuation of the Coustouges nappe, whereas the stratigraphic sequence of the upper unit, which has furthermore been duplicated by the prevailing thrust tectonics, offers a more complete series from the basal Keuper sandstone to Liassic limestones, followed by Upper Cretaceous sandstones, marls and limestones. The Upper Thrust units (Coustouges and Bac Grillera klippen), were displaced over the Axial Zone by a Supra-Axial Thrust; the home area of these nappes was located somewhere above the Axial Zone (see Part I and Calvet et al. 2021) and, in the case of the upper unit: north of the Amélie Basin, where the Lower Jurassic (Lias) is absent (Laumonier 2015).

Further to the SW, the elevated limestone crags of La Garrotxa closing the horizon are made up of the lower Eocene series of the Cadí unit, here forming complicated and highly fractured duplexes where Hercynian rock outcrops

Fig. 9.10 Rainfall intensity distribution during the mid-October 1940 hydrometeoro- ▶ logical disaster. A Cumulative rainfall for October 16 to 20, 1940. Raingauge stations (black circles) and rainfall totals (in mm) are indicated on both maps. As with the 1999 storm, the impact of the event was regional, extending from the Ter River in the south to the Montagne Noire in the north. However, unlike the 1999 event (epicentre over the low-lying Corbières, and also briefer), the epicentre struck the elevated mountain front of Mt. Canigou under a regime of saturated and highly unstable warm air entering Vallespir from the south and SE. The magnitude of destruction in 1940 is thus ascribable to the magnitude and duration of airmass obstruction by the high massifs of the Axial Zone. The meteorological archives are incomplete and rainfall distribution over those four historic days is imprecisely known (successive syntheses of the existing data in Pardé 1941; Bénech 1993; Boutin and Pascual 1993; Vigneau 1993; Llasat 1993; Soutadé 1993). **B** Spatial distribution of peak rainfall on October 17. The trajectory of the southeasterly airmass is perceptible from the isohyets on this map: the hot air spilled over into the Tech valley through the Coustouges saddle, then followed the Saint-Laurent basin before being funnelled up into Comalada valley (after Boutin and Pascual 1993, redrawn and modified)

form the summits of basement-cored anticlines of the Garrotxa. Together with the Canigou nappe of the Axial Zone, these nappes form the Middle Thrust units.

From Coustouges it is possible to return to Perpignan via Massanet de Cabrenys, Darnius, La Junquera, and Col du Perthus (74 km), i.e., by following the southern border of the Axial Zone. The itinerary otherwise continues by returning towards Saint-Laurent and turning left onto the D64 at Farga del Mitg. The road takes you across a well-preserved pediment surface (generations P2/P3) between 650 and 700 m, cut in deeply weathered biotite and hornblende granodiorite (Fig. 9.8B, C). Just before leaving the plateau (clear views can be gained from Col du Saguer, further up the forest road), note to the north the topographic mass of Mt. Canigou, with, before it, Torre de Cos (1116 m), a peak in Ediacaran limestone—itself also an upstanding residual bornhardt on a vestige of P2 at its base (800–750 m), on the left side of the Tech (Fig. 9.8A, B). The road descends into the deeply cut valley of a tributary to the Tech, affording glimpses of P2-generation benches around Serralongue. You will join the Tech valley via the D44. Aim to stop at Le Tech.

Stop 5. Le Tech village

The village is situated at the junction between the Tech and the Comalada (spelling on old maps: Coumelade), and was at the epicentre of the hydrometeorological disaster of 1940 (Fig. 9.10A, B). The church, which was situated on the bedrock spur at the confluence and surrounded by a few houses, and the school, just by the Comalada above the bridge, were swept away with the teachers and their entire families (13 reported dead or missing in the municipality).

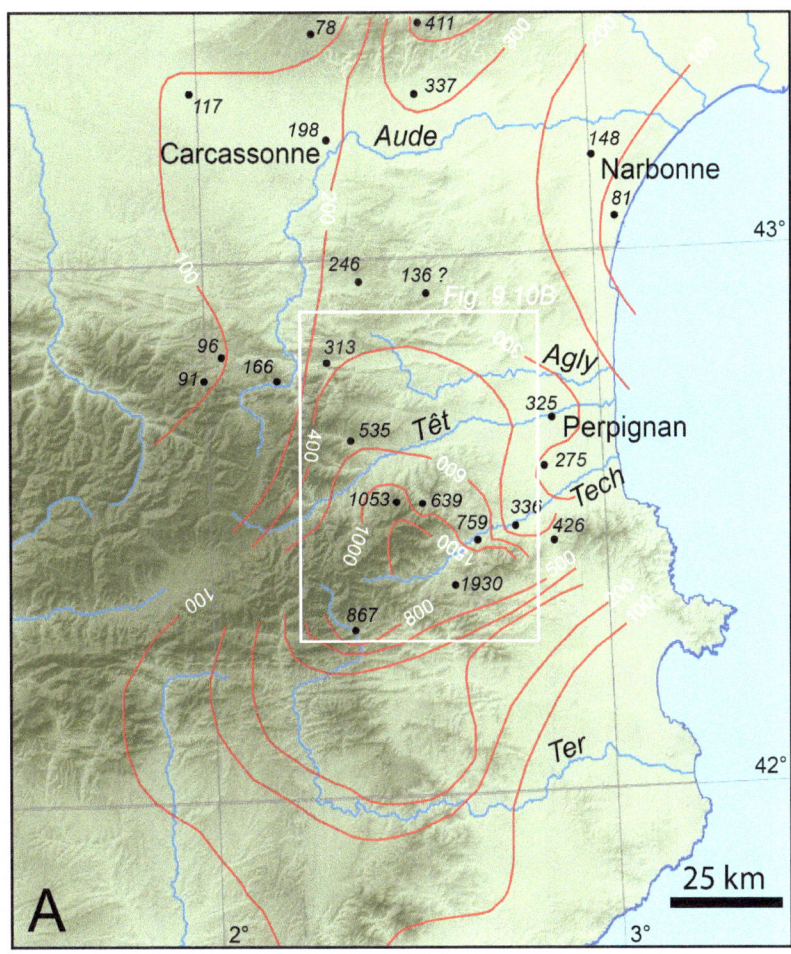

The Comalada is an alpine torrent quite similar to the Riuferrer (peak discharge in 1940: 800 m³/s; specific discharge: 33 m³/s/km²). Throughout this reach of the Tech valley, alluvial aggradation attained, and at places exceeded, 10 m.

The D115 after Le Tech was entirely destroyed in 1940. At first, the road was temporarily re-established in the riverbed with Bailey bridges; its current tread and the two bridges date back to 1962. Still 80 years after the event, erosion scars

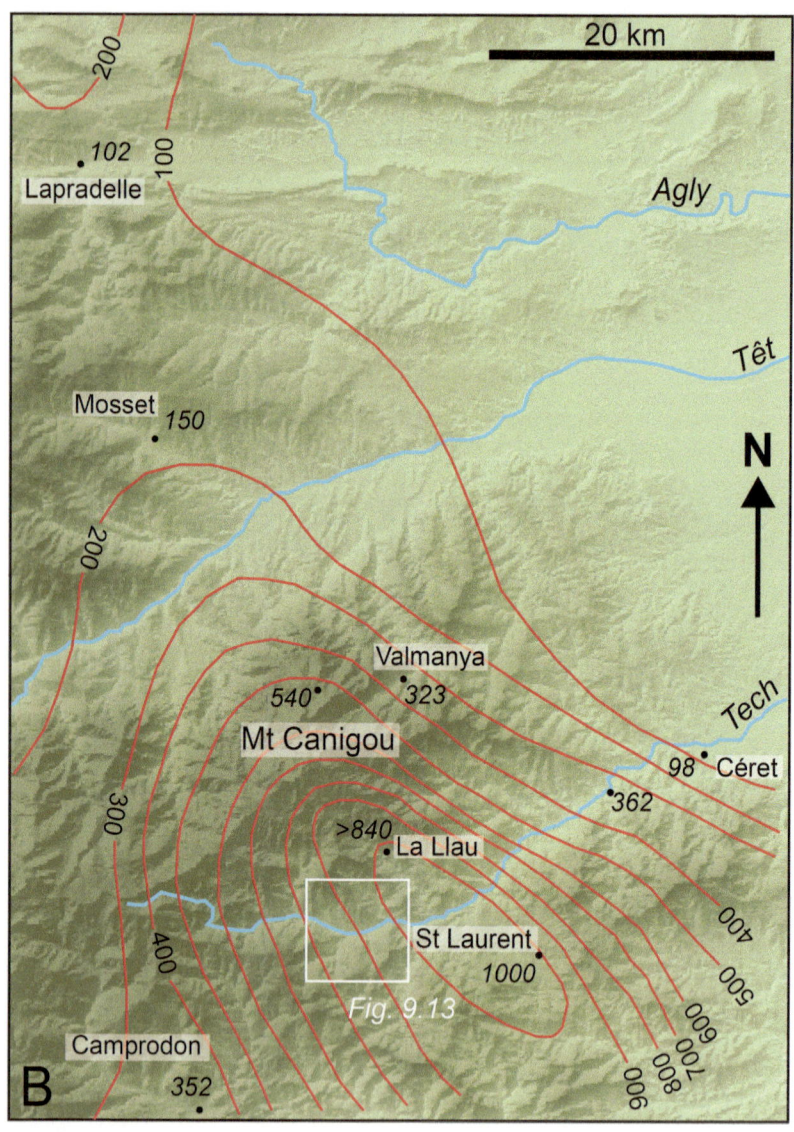

Fig. 9.10 (continued)

in Avellanosa gorge (alternate spelling: Baillanouse) are spectacular. Stop at the
parking area as you exit a small tunnel, just before the second bridge.

Stop 6. L'Avellanosa

On the valley side to the right (Canaveilles schist outcrops of Puig Cabrers, volcano-sedimentary protolith), the scars of a 500-m-high, 800-m-long, and 250-m-wide landslide are still conspicuous. Two display boards can be consulted at the parking area and at the viewing point (Fig. 9.11A). The embankment in the roadbend is actually the landslide toe, which has upthrust by ~50 m some rolled gneiss cobbles from the riverbed, various debris from buildings destroyed during the storm, and tarmac slabs from the old road surface. The landslide crown is still visible, as likewise the flat bench formed by the top of the slipped mass on the left bank (Fig. 9.11C). The history of this landslide, however, goes back further than the 1940 storm, which appears to have mostly displaced a mass of bedded slope debris with a silty matrix (periglacial *grèzes*). These talus deposits formed an apron over the underlying concave bedrock slope (Fig. 9.11B). Pardé (1941) suggested a displaced volume of 6–7 M m^3, but the surface area impacted in 1940 was ~0.11 km^2, with a zone of depletion 20–30 m deep—thus yielding a more likely displaced volume of 2.5–3 M m^3 in 1940. The flat area on the left bank (Fig. 9.11A, B) is thus more probably the top of an older slipped mass, perhaps of Pleistocene age, and consisting mainly of a chaotic jumble of weathered schist blocks and boulders. This is thus, effectively, the original rock slope failure (RSF). Photographs from before 1940 already showed the gullied slope intersecting the flat area (Fig. 9.11B); the entrance to the penstock tunnel for the Tech hydroelectric plant, constructed in 1930–36, was also visible and was not displaced by the 1940 event (Soutadé 2010). Despite the composite and ancient history of the Avellanosa valley side, the 1940 landslide was still sufficiently substantial to generate a 50-m-high dam, thus impounding the river and forming a lake and a 21-m-thick deltaic sequence within it (sediment jointly supplied by the Tech and its tributary, the Figuera). Luckily, rather than a catastrophic outburst the lake drained out over an interval of several days through a breach in the debris dam. Follow the path down to the river before the bridge to gain a different view of the landslide (Fig. 9.11B). A cut in the first loop of the trail exposes 2 m of flood debris containing moderately weathered gneiss boulders resting on bedded deltaic sands (Fig. 9.12). Those units themselves overlie an 8-m-thick torrential debris-flood sequence floored by a bedrock strath. Those deposits probably document an older natural dam associated with a Pleistocene episode of slope failure.

Fig. 9.11 The Avellanosa composite landslide. A View from the D115. **B** View from upstream at the water intake dam. **C** View from the summit of Puig Colom

Fig. 9.11 (continued)

Upstream of the bridge, near Pollangarda farm, a daylighting schist rockfall impacted the right bank of the Tech in 1940, its toe reaching the road. M. Pardé estimated an accumulation of 1 M m³, but the main scarp is only ~230 m high for a surface area of 45,000 m² and an unknown thickness which, however, did not exceed 10–20 m. The failure furthermore occurred in a pre-existing hollow, thus making that 1940 mass failure somewhat smaller than initially estimated (perhaps between 0.5 and 1 M m³).

Drive on to Prats-de-Mollo and park in the main square. The aim is to examine the Tech riverbed from the road over the bridge to Spain.

Stop 7. Prats-de-Mollo

Alluvial backfill of the Tech valley caused by the Avellanosa and Pollangarda debris dams reached all the way to Prats-de-Mollo, where it affected the area with the factories, the current swimming pool and small sports stadium over a width of 150 m, widening to 250 m farther downstream (Figs. 9.13 and 9.14A). The Tech upstream buried the Medieval bridge, which was exhumed by recent reincision of the river into its own alluvial fill: you can observe the ruins of an arch on each bank exactly below the modern footbridge (Fig. 9.13B). The main

Fig. 9.12 Traces of a Pleistocene alluvial dam at L'Avellanosa. Exposure showing the deltaic deposits (with dip value and direction: 10°S) capped by a high-energy fluvial sequence

bridge was unscathed by the flood and stood above the alluvial fill by ~10 m. The modern street, parking lot, stadium, and a few houses, however, were constructed on the 1940 depositional surface.

Drive up the road to Spain, which follows the Canidell valley. The steep catchment of this high-energy stream, which broke the region's specific discharge records in 1940 (42.3 m³/s/km², for a peak discharge of 550 m³/s), underwent extreme gullying in 1940 (Fig. 9.13), although the scars are now obscured by the dense forest canopy following sustained efforts by the RTM service (Restauration des Terrains en Montagne) to restore protective land cover. An aerial photograph from 1942 shows the extent of the damage quite clearly, with the grey debris from the tributary darkening the lighter alluvium of the Tech over a distance of >600 m downstream of the stream junction. About 1 km after the bridge, turn right onto the small road to El Xatard and park at the forest cabin of Can Got. From there, follow the forest trail on foot for ~300 m.

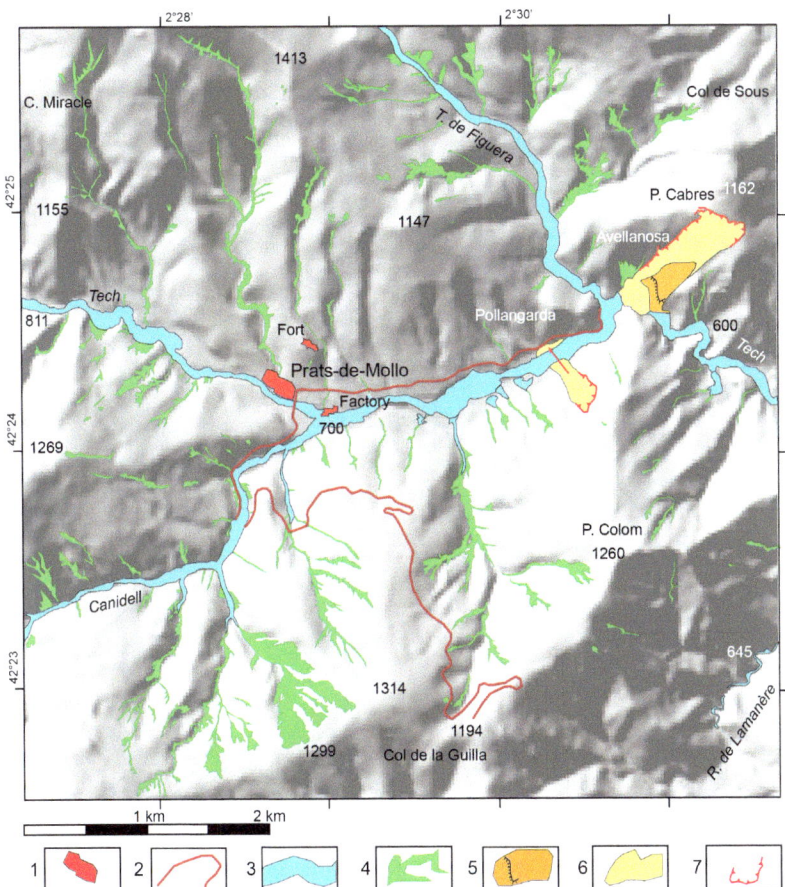

Fig. 9.13 **Geomorphological map of erosional and depositional landforms generated by the 1940 floods around Prats-de-Mollo.** Key to symbols and ornaments—**1**: older buildings. **2**: main roads in 1940 (the road to Coll d'Ares, still under construction at the time, only reached up to Coll de la Guilla). **3**: longitudinal alluvial deposits and lateral debris cones; the alluvial backfill is 10 m thick at Prats, rising to 21 m upstream of Avellanosa landslide. **4**: gullies, debris-flow scars, and landslips. **5**: Pleistocene landslide and bench-like top of the displaced mass. **6**: 1940 landslide. **7**: upslope scars of the various 1940 landslides. Map produced from an analysis of the 1942 and 1953 aerial photographs. Elevations in metres

Fig. 9.14   Flood impacts around Prats-de-Mollo. A View downstream from the bridge on the Tech. Vestiges of the 1940 alluvial aggradation, which has since undergone 10 m of reincision; destroyed textile factories in background. **B** Remains of the old Pont d'Espagne, at the time entirely choked by alluvium (note rolled cobbles on the left bank) but subsequently exhumed by fluvial reincision

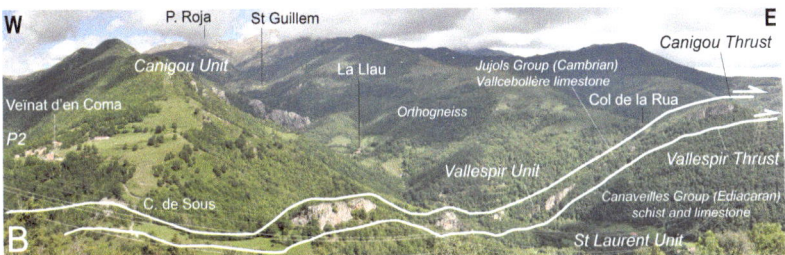

Fig. 9.15 The Canigou Thrust near Prats-de-Mollo. A Thrust plane near Can Got. **B** Panoramic view from Puig Cabrers (spot elevation: 1124 m, above Col de Sous). Between the major Canigou and Vallespir thrusts (both Alpine features), the Vallespir thrust sheet here is reduced to little more than a strip of Valcebollère limestone; in the background: Canigou orthogneiss; middle ground: leptynite outcrop cut by Comalada gorge. See also Fig. 9.11C

Stop 8. Can Got: Alpine Canigou Thrust

About 10 m above the trail, note a well-preserved fault plane beneath a steep crag of Ediacaran limestone (Canaveilles Formation; Fig. 9.15A). This is the Canigou thrust plane, where the Canaveilles Formation (Canigou unit) overrides the Jujols schists (Vallespir unit) out of which the Prats-de-Mollo basin was eroded. The two units themselves override the Lower Thrust units of Saint-Laurent-de-Cerdans and Roc de France along the Vallespir Thrust, previously encountered (**Stop 2** and **Stop 4**; Fig. 9.9B; Laumonier 2015; Laumonier et al. 2015).

Return to Prats-de-Mollo and follow the D115a towards La Preste. After 3.5 km turn right into the Parcigoule (Parcigola) valley. The track (in poor condition) will

*lead you up a valley which was completely devastated in 1940 (peak discharge: 600 m^3/s; specific discharge: 20 $m^3/s/km^2$) and definitively deserted thereafter. Today the flood deposits are overgrown by riparian forest, and the gullies and debris-flow scars on the slopes have been covered by spontaneous regrowth, and most of all by RTM reforestation programmes. Farmsteads were definitively abandoned soon after the flood, and farmers expropriated for tree planting. Landscapes of the Vallespir changed radically in the aftermath of the 1940 floods: only 13% of forest in 1827, at the time of maximum montane economic activity (mining, pastoralism), but 60% today. The geomorphological scars of that period can only be appreciated today in areas above the timberline (**Stop 9**).*

It is important to emphasise that sites inspected along Itinerary 4 were impacted by many more extreme events prior to October 1940. Quite precise records exist for the 1763 disaster, including this same valley. The energy of the floods could have exceeded 1940 levels, and the deluge occured almost on the same dates, i.e., October 16 and 17, and thus on Saint Gaudérique Day—a ninth century local patron saint and protector against storms, floods, as well as extreme droughts. Civil engineer François de Lescure produced a very detailed report on the 1763 events following a visit to the Vallespir in November 1763 (Archive Départementale-AD66, 1 C 1078) (Dessailly, 1992). His descriptions of the geomorphological transformations of the landscape, whether on slopes (gullying, rockfalls) or in riverbeds (alluvial aggradation, bank erosion), are strikingly similar to the records for 1940.

At the crossroads before Can Pitot, turn onto the track to your left towards Cal Cabus. It rises in a series of hairpin bends to Col de la Regina (1760 m) and at several points cuts through unconsolidated fault breccia along the Neogene 'Parcigoule–Col du Miracle' normal fault, with a downthrow to the W.

Stop 9. Col de la Regina, Les Estables (sheperd's refuge hut)

Following the track eastward above the refuge you will approach the vast, gully-scarred slopes of Serra Vernet–Comall Escur (Fig. 9.16A, B). The vernacular name for such gullies in the region is *xalade* (pronounced 'shalad'). They are very widespread above the treeline on Mt. Canigou, particularly on its Vallespir side. The channels are cut in sand- and gravel-rich periglacial slope deposits as well as the in-situ grus of the underlying weathered gneiss. The gullies are generated by pore saturation of the regolith, where hydrostatic pressure suddenly causes a state change and triggers a debris flow. The process was observed and described by woodcutters trapped in the deluge at these altitudes in 1940. 'Xalades' only form during such extreme hydrometeorological events (e.g., 1763, 1940), how-ever. Modal rainstorms essentially perpetuate them and widen them upstream by progressive encroachment into the alpine meadows and gorse moorland. Another

Fig. 9.16 Upper Parcigoule valley. A Col de la Régine–Les Estables. Note thickness and nondescript morphology of the Late Pleistocene moraine deposits, clearly emplaced by a low-energy but highly loaded glacier. The 1940 gullies ('xalades') make up at least 30% of the landscape; even 80 years on, their scars are unhealed. **B** Detail of the Comall Escur mountainside in 2012. Note the chronic instability of these slopes despite the intensity of soil conservation engineering. **C** El Saiol (Sayol) dam on the Parcigoule (1200 m). Another dam occurs further upstream (1550 m). Both dams were constructed on bedrock knickpoints and were aimed at arresting the advance of headward stream incision. They became, however, choked with sediment within just 2 years, thus also providing at El Saiol a measure of modern erosion rates in this 11.36 km^2 catchment: 3300 m^3 in 1961, i.e., an equivalent erosion depth of 0.29 mm for that year; and 7165 m^3 in 1962— i.e., 0.63 mm, on that occasion during a year marked by a fairly major rainfall episode (November 3–11) involving 24-h cumulative downpours of 365.6 mm and 170 mm on November 4th (Soutadé 1969, 1980)

variety of shorter, deeper gullies, displaying broad, semicircular heads, is more typical of headward erosion occurring in the thick Würmian moraine deposits of the alpine zone, such as below the Estables refuge.

Historical archives suggest that this population of gullies was generated only for the first time in 1763, when the environment crossed a tipping point after centuries of land-cover change peaking in the mid-eighteenth century with forest clear-cuts to fuel the forges, with huge flocks of transhumant sheep congregating on the alpine meadows of Mt. Canigou during the summer months, and when agricultural was pushed to its uppermost limits close to the 2000 m elevation contour. Under such conditions of strain on the environment during the Little Ice Age, a single extreme climatic event would have been sufficient to overshoot a number of critical limits, driving the metastable montane environment into a lasting state of geomorphic instability, which engineers even today struggle to control with gabions and check dams (Soutadé 1969, 1980). It is nonetheless a possibility that a similar cycle was set in motion by similar magnitudes of strain on the landscape in the more distant past, for example at the time of mining and montane pastoralism peaks during the Bronze Age, Roman times, or the Middle Ages. Allée and Denèfle (1989), for example, have documented overgrown 'xalades' in the headwaters of the Tech, whose debris cones contained (uncalibrated) radiocarbon-dated pollen assemblages dating to 2800 yr BP. On the north side of Mt. Canigou, the minimum Medieval age for a major debris-flow deposit in the valley of Mantet was inferred from the remains of a charcoal kiln established on one of the levees, with a radiocarbon age of 1450–1635 cal CE (in Laumonier et al., 2015; see Itinerary 6).

Optional trailwalks over Mt. Canigou

1. Pla Guillem, an uplifted vestige of P1 at 2300 m a.s.l. (3 to 4 hrs round trip)
From Col de la Regina, follow the well-marked trail GR T83 up the rounded ridgetop towards the NW. The trail passes between two SE-facing armchair cirques, in an area where no glacial cirques occur on north-facing slopes. This paradox highlights the importance of wind-blown snow accumulation driven by northwesterly winds sweeping across the extensive Pla Guillem snowfield, which could act as a snow reservoir in a context where the Late Pleistocene equilibrium line altitude (ELA) in this massif stood around 2050 m. The low-gradient summit surfaces (generation P1 in the present case) extend around 2300 m, and Pla Guillem, in augengneiss, is the most iconic (Fig. 9.17A). Two apatite fission-track (AFT) ages of 33.3 and 26.5 Ma indicate that these erosion surfaces are late- to post-orogenic landforms (Gunnell et al. 2009), which have been preserved for a long period of geological time, unscathed by cycles of gullying and fluvial incision. The Parcigoule–Mariailles Fault strikes across Coll de Bocacers. Its dip-slip movements have raised the summit area of Mt. Canigou by an additional 350–400 m above the tread of Pla Guillem. The Pla Guillem plateau surface is padded by at least 10 m of in-situ weathered gneiss, which can be appreciated at the En Felip gully head, below the refuge. You may wish to return via Coll de Bocacers and the Estables valley, which displays a succession of small recessional moraines (unmarked path: just follow the topography).

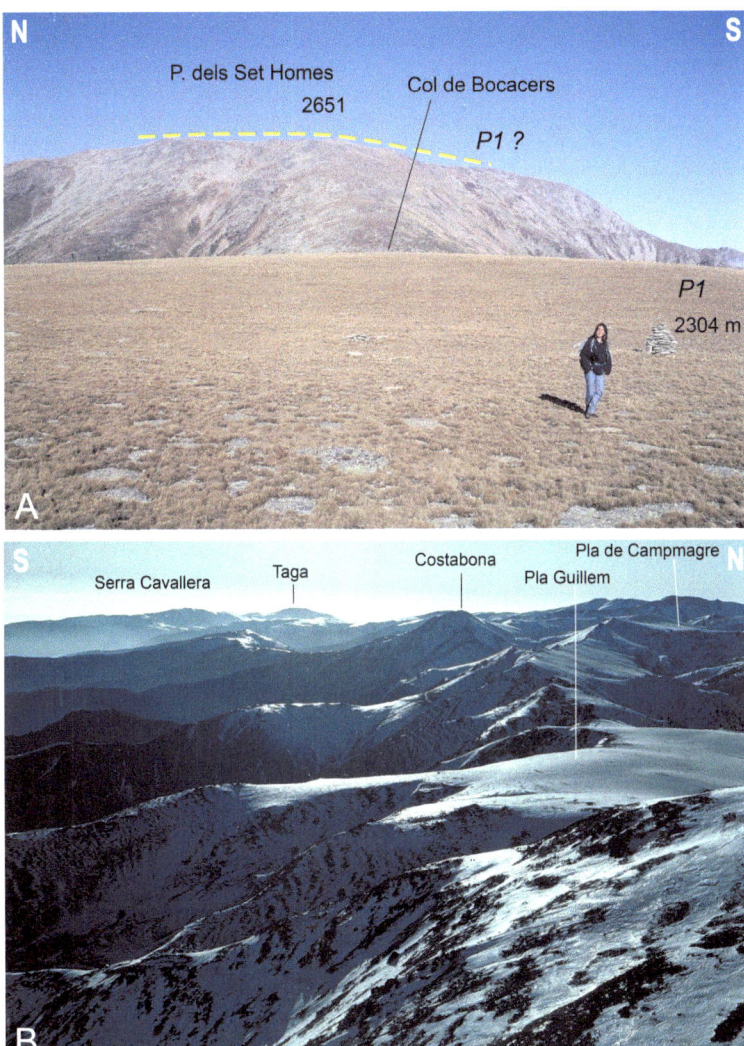

Fig. 9.17 **Pla Guillem, a remarkable landform among the population of summit erosion surfaces of the Pyrenees. A** View of Pla Guillem and Pic des Sept Hommes, looking east. Here, the topography bevels orthogneiss. In the background, Mt. Canigou is an upthrown block uplifted by the Neogene fault striking through Coll de Boucacers. **B** General view of the multiple relict erosion surfaces ('plas') from Pic des Sept Hommes, looking SW. The area extends westward for 14 km (Pla Guillem, Pla de Campmagre, Pla de Coma Ermada). Massifs of Carança (W) and Ripollès (SW) in the distance

2. Pic Roja and Pla Guillem (5–6 hrs round trip, Unmarked path)
From Les Estables, where the Late Pleistocene terminal moraine occurs nearby at ~1600 m, walk up the Coma de Bacivers, a glacial valley containing recessional moraines at 1900 m and 2100 m, and displaying rock glaciers higher up. The valley is by no means a U-shape, and the cirque walls are gashed by a large number of 'xalades'. The estimated Late Pleistocene ELA is 2150 m, but the supply of windblown snow was much more limited that in the environment described in Trailwalk no. 1 because of limited plateau-like surfaces at the top. At 2724 m, Pic Roja is a very small vestige of an erosion surface, which continues to the west (Pic des Sept Hommes, 2651 m); it is probably the same generation as P1 at Pla Guillem, but upthrown during the late Neogene by the Parcigoule–Mariailles Fault. From the summit you will get excellent views of the glacial cirques of the Cady and Pic Canigou to the north, and over the extensive vestiges of P1 to the west from Pla Guillem to the Carança massif (Fig. 9.17B), on the horizon. To the NW, the panoramic view of the Conflent Basin and, beyond it, the Carlit massif, is also stunning (see Itineraries 6 and 7). To the south, you will see the outer Pyrenean fold belts (Garrotxa) and the Catalan Ranges. Two conspicuous features of the landscape, distinguishable on the horizon on a clear day, are the regular outline of Montseny, which is a tilted block of Hercynian basement displaying vestiges of an exhumed sub-Triassic peneplain; and the finely serrated outline of Montserrat, a series of rock pillars carved out of highly cemented, synorogenic Eocene conglomerate beds. You may return via Coll de Bocacers, Pla Guillem, and the ridgetop ascended at the beginning of Trailwalk no. 1.

Return journey options

- *Additional full day: from Prats-de-Mollo to Perpignan via the Pyrenean mountain front in Catalonia (not detailed here). Recommended route over Coll d'Ares, onward to Camprodon, Ripoll, Vallfogona (Pyrenean folds in the marine Eocene sequence), Olot, Castellfullit de la Roca (normal faults of the Transverse Ranges, Pleistocene to recent volcanism around Olot), Besalu (early Pleistocene lakes, neotectonic features; see Part I, Figure 3.5 therein), Figueras (Neogene sequences of Empordà).*
- *Via the Tech valley: as you exit Prats, take the D74 towards La Llau. At Col de Sous, between the road and the col itself to the south, the Alpine Canigou and Vallespir thrust planes are almost indistinguishable, with the Vallespir unit consisting merely of a thin sliver of limestone (Valcebollère Formation) dipping NE towards Cingle del Coll de Sous and Roc de la Rua. A good view of these structures can be gained from the track situated 500 m to the SE of the col, or even better at elevation spot 1124 m, which can be reached along a trail starting from the powerline pylon (Fig. 9.15B). A walk up the track (1 h return trip) to the top of Puig Cabrès leads you to the crown of the Avellanosa landslide. The thrusts are made conspicuous in the landscape by a limestone scarp with*

*beds dipping 50°NW (Valcebollère Formation, Jujols Group), overridden from the north by Canigou orthogneiss and overriding to the south the Canaveilles Group micaschists, which belong to the Lower Thrust unit of Saint-Laurent-de-Cerdans (*Figs. 9.11C and 9.15B*). The road descends from the col into the Comalada valley before reaching La Llau (also spelled L'Allau). This hamlet is best known for its French Meteorological Office raingauge, which famously recorded 840 mm in 24 h on Octobre 17, 1940—the official European record to this date (*Fig. 9.10*). The engineer in charge of the weather station let the raingauge overflow four times before it was eventually damaged beyond use around 7.30 p.m. and replaced by a bucket. The official value is thus a low estimate, easily on a par with monsoon downpours in South Asia. On the same day, the schoolmaster at Saint-Laurent-de-Cerdans, G. Julia, conscientiously recorded 1000 mm over the same 24 h interval, and 1930 mm in total over the 5 days of deluge (16–20 October); out of that total, 1700 mm fell over the 3 days of peak downpour (17–18–19 October) (*Boutin and Pascual 1993; Soutadé 1993; Bénech 1993). From La Llau and over Coll de la Rua, a badly maintained road will take you to the Romanesque chapel of Saint Guillem (1300 m), from where it is possible to walk up to the spectacularly gullied mountain side of the upper Comalada (*Fig. 9.18*) around Jasse des Troncasses (1730 m) (2.5 hr return trip). From La Llau, the road winds down to the hydroelectric power plant, which initially was situated on a bridge in the middle of the stream; its presence in the active channel in 1940, however, generated a major debris jam, which eventually broke and triggered the devastating floodwave all the way to the Tech and the small village of Benat (Baynat). Along the right bank below the plant, note the well-preserved vestige of 1940 aggradational fill, benath the road, with boulders several metres in diameter. The road cuts through the limestone scarp of the Canigou thrust front, and thereafter you join up with the D115 at Le Tech. From there the journey back to Perpignan follows the same route as the outbound journey.*

Fig. 9.18 Upper Comalada valley, viewed from Tres Vents ridgetop in 1995. Note thickness of the moraine deposit. The 1940 gullies ('xalades') still look fresh

References

Allée P, Denèfle M (1989) La Coma del Tech. Un exemple de ravinement protohistorique dans les Pyrénées orientales. Bull Assoc Géogr Fr 66:57–72

Bénech C (1993) Estimation des périodes de retour de 'l'aiguat' d'octobre 1940 dans quelques vallées des Pyrénées–Orientales. In : l'aiguat del 40, actes du congrès de Vernet-les-Bains, Generalitat de Catalunya Edit., pp 297–313

Boutin A, Pascual M (1993) L'aiguat d'octobre 1940. In : "L'Aiguat del 40", actes du congrès de Vernet-les-Bains, Generalitat de Catalunya Edit., pp 67–76

Calvet M, Gunnell Y, Laumonier B (2021) Denudation history and palaeogeography of the Pyrenees and their peripheral basins: an 84-million-year geomorphological perspective. Earth-Sci Rev 512:103436

Gunnell Y, Calvet M, Brichau S, Carter A, Aguilar JP, Zeyen H (2009) Low long-term erosion rates in high-energy mountain belts: insights from thermo- and biochronology in the Eastern Pyrenees. Earth Planet Sci Lett 278:208–218

Lalanne-Berdouticq G (1993) Aspects méthodologiques de la reconstitution des écoulements des grandes crues catastrophiques. In : "L'Aiguat del 40", actes du congrès de Vernet-les-Bains, Generalitat de Catalunya Edit., pp 229–264

Laumonier B (2015) Les Pyrénées alpines sud-orientales (France, Espagne) – essai de synthèse. Rev Géol Pyrén, 2:44. http://www.geologie-despyrenees.com/

Laumonier B, Le Bayon B, Calvet M (2015) Handbook to the Carte géologique de la France (1:50,000 scale), sheet Prats-de-Mollo La-Preste (1099). Geological map by Laumonier B et al., Orléans, BRGM, 189 p

Llasat MC (1993) Les inondations de 1940 en Catalogne espagnole. Les inondations semblables des cinquante années suivantes. In: "L'Aiguat del 40", Actes du congrès de Vernet-les-Bains, Generalitat de Catalunya, pp 137–146

Pardé M (1941) La formidable crue d'octobre 1940 dans les Pyrénées-Orientales. Rev Géogr Pyr Sud-Ouest XII:237–279

Soutadé G (1969) Un milieu sub-alpin de glyptogénèse. Les ravins de Comall Escur, versant sud du Massif du Canigou (Pyrénées-Orientales). Rev Géogr Pyrén Sud-Ouest 40:353–370

Soutadé G (1980) Modelés et dynamiques actuelles des versants supraforestiers des Pyrénées orientales. Imprimerie Coopérative du Sud-Ouest, Albi, 452 p

Soutadé G (1993) Les inondations d'octobre 1940 dans les Pyrénées-Orientales. Conseil Général, Direction des Archives départementales, Perpignan, 351 p

Soutadé G (2010) Quand la terre s'est ouverte en Roussillon, l'Aiguat – octobre 1940. Publications de l'Olivier, Perpignan, 171 p

Vigneau JP (1993) Un épisode pluvieux méditerranéen parmi d'autres ? Enquête sur les précipitations d'octobre 1940 dans les Pyrénées-Orientales. In: "L'Aiguat del 40", Actes du congrès de Vernet-les-Bains, Generalitat de Catalunya, pp 77–86

Perpignan back to Perpignan (322 km; 387 km including optional excursions: 2 to 3 days in total, 4 including hikes) See Fig. 10.1. Itinerary 5 begins arbitrarily from Perpignan, but given it can only be fully completed in 3 days or more, its modular nature would justify planning alternative base camps and/or stopovers.

The tour offers a discovery of Alpine geological structures in the North-Pyrenean Zone and clues to the area's geomorphological evolution during the Neogene and Quaternary. Itinerary 5 crosses the Internal Metamorphic Zone (IMZ) on several occasions, and likewise the folded and faulted Mesozoic cover sequence and the Axial Zone's satellite massifs, which are northern outposts of Paleozoic basement sheared off the Axial Zone along north-vergent thrusts. You will be able to inspect some major tectonic discontinuities of the Alpine crustal fabric, e.g., the North-Pyrenean Fault (NPF) and North-Pyrenean Frontal Thrust (NPFT), and to make a traverse through the Sub-Pyrenean fold belt (Western Corbières, Plantaurel and Petites Pyrénées) all the way to the synorogenic Paleogene conglomerate sequences (Palassou series). The latter record uplift and erosion at the time of mountain building, and underwent tectonic deformation in the retro-foreland basin. From a geomorphological perspective, the itinerary showcases the westward continuation of the erosion surfaces previously presented and age-bracketed in the eastern Corbières (Itinerary 1), and offers clues for explaining links between those different generations of erosion surfaces, their respective base levels, and the tectonic regimes that initially drove and subsequently interrupted their completion. Their respective ages and past connections with the Paleogene conglomerates of the Aquitaine foreland are discussed, and their presence in elevated parts of the high range is inventorised. In any given area, three distinct populations of low-gradient landforms can be distinguished: generation S, always a summit surface; generation P1, usually (though not systematically) encountered as mountain-flank pediments rather than as summit surfaces, vestiges of which nonetheless tend to become scarce among the massive limestone outcrops of the North-Pyrenean Zone compared to the extant populations encountered in the Corbières farther east (Itinerary 1); and generation P2, for which inventories of relative age and altitudinal position provide partial constraints on late Neogene and Quaternary palaeoaltimetry, crustal

M. Calvet et al., *Geology and Landscapes of the Eastern Pyrenees*, GeoGuide,
https://doi.org/10.1007/978-3-030-84266-6_10

deformation, and surface uplift. Groundwater karst landforms in the thick Jurassic and Lower Cretaceous carbonate cover sequences, which throughout the area are widespread and conspicuous, also provide important clues to landscape evolution and palaeoaltimetry.

Leave Perpignan on the N116 and take an exit at Millas, following thereafter the D612 signposted to Estagel. After the bridge over the Têt, note the roadside section in sandy foreset beds of the marine Pliocene fill sequence the Roussillon Basin. The road then rises up to the tread of terrace T3 (Les Planes), with residual hills to the right bearing vestiges of T4 (L'Arbocera) and T5 (Mas Ferriol, below which the Pliocene basin's boundary fault is also documented). The footwall (Agly massif) consists of Proterozoic to Paleozoic schists (Col de la Bataille and Força Réal series). At Col de la Bataille, drive up the road signposted to Força Réal chapel,

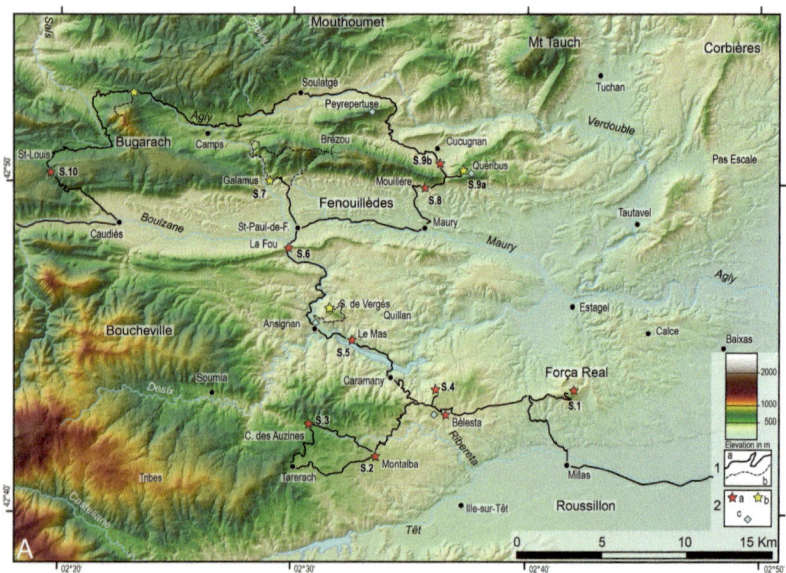

Fig. 10.1 Itinerary 5: route map and overview. A Eastern region. **B** Western region. Key to symbols—**1**: Roads and trails. **1a**: Roadways, irrespective of road rank, size, or quality. **1b**: Trailwalks. **2**: Stopping points (outcrops, exposures, landforms, or landscapes). **2a**: Roadside stops. **2b**: Roadside stops involving a short walk, or marking the start of a more major trailwalk. **2c**: Cultural stops. Labels S.1 to S.18 locate the sequence of roadside stops

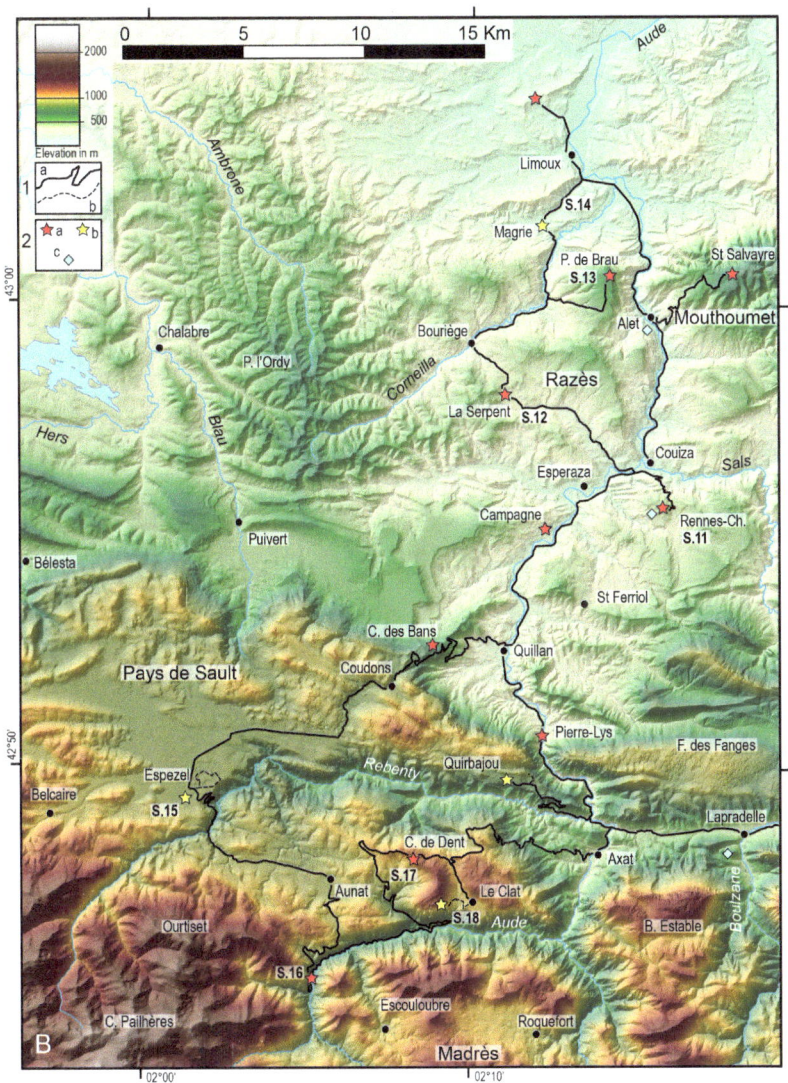

Fig. 10.1 (continued)

which, on clear days, provides the region's most accessible and most rewarding
360° panoramic view.

10.1   Agly massif

Stop 1. Força Réal: Paleozoic rocks, landscapes and landforms of the Agly massif

A transect from Bélesta to Força Réal showcases the entire lithostratigraphic sequence of the Hercynian basement, from 'catazonal' gneiss and migmatite (Caramany and Bélesta gneisses) in the west, to the fossil-rich Upper Paleozoic beds in the east (Caradoc puddingstones; Ashgillian brachiopod- and corallite-bearing schists; see also Itinerary 6, Fig. 11.27G; Silurian schists, sandstones and carbonates; Devonian limestones and dolomites). The sequence displays steady eastward dips throughout (Fonteilles et al. 1993). The core of the Agly massif consists of the 1500-m-thick homoclinal micaschist cover of the Col de la Bataille and Força Réal series, which rest on the gneiss (in map view, the contact is highlighted by a marble bed). The protolith consisted of laminated turbidites, which are correlatable to the Jujols Group and are of Cambrian age *senso latu*. The singular feature of this sequence is the rapid downward transition (in just 1.5 km) from the 'epizone' to the top of the 'catazone' through all of the classic metamorphic isograds, i.e., the chlorite, biotite, cordierite, andalusite, and sillimanite–K-spar zones, respectively. The anomalously high geothermal gradient (~100 °C/km) required to achieve such a rapid succession has been attributed to rapid crustal thinning towards the end of the Variscan orogeny, i.e., ca. 300 Ma.

Força Réal is an inselberg standing on the tread of pediment P1, which rises gently (Figs. 6.13, 6.14) from Baixas (150–200 m) and Calce (250–270 m) plateaus in the east to Bélesta and Montalba plateaus in the west (450 m; see forthcoming **Stop 2**). The greenish, banded schists (mudstone protolith) are quite massive, and thus relatively insensitive to chemical weathering. Erosional benches belonging to generation P2 form an apron around the hill at elevations of 240–250 m. Such benches are conspicuous on the south (Mas de la Garrigue), north, and most of all west flanks (vineyard-covered land surface at Col de la Bataille around Caladroy, 260–290 m), thus 50–70 m above the level of early Quaternary terrace T5.

The view from the chapel can be complemented by the view from the telecommunications tower (Fig. 10.2). Towards the south (Fig. 10.2A), note the terrace sequence of the Têt (Itinerary 2, Figs. 7.26 and 7.27), the wider Roussillon Basin and, further on the horizon, the Albères–Roc de France and Aspres horsts,

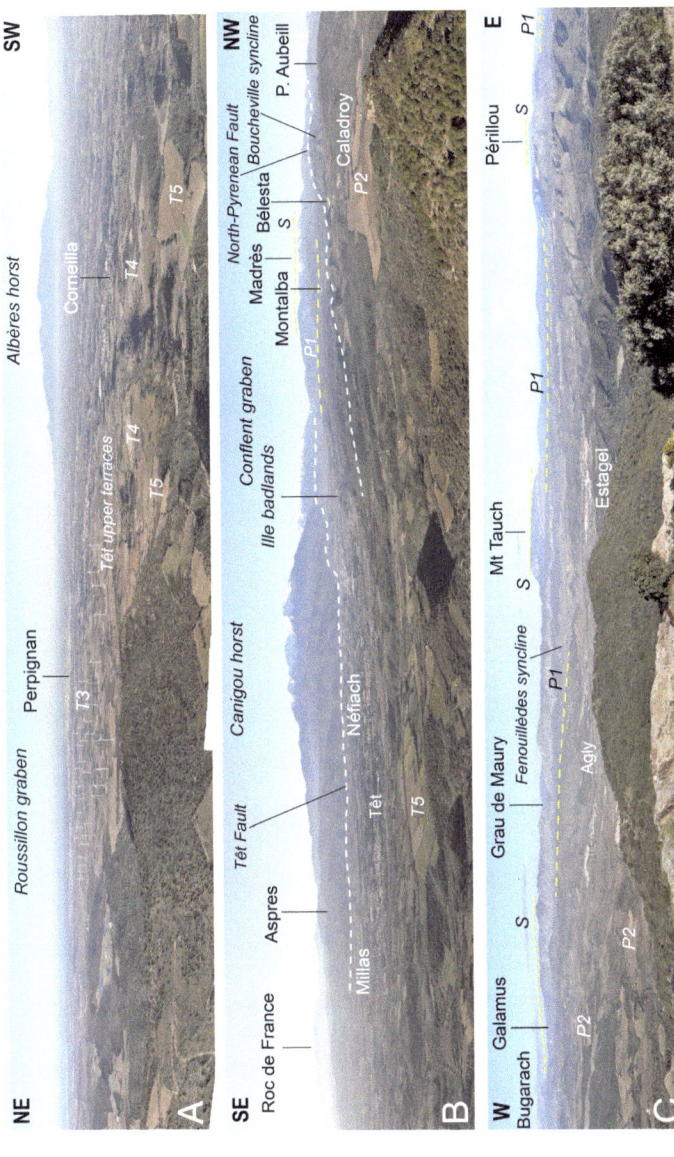

Fig. 10.2   View of the landscape from Força Réal. A Towards the Roussillon Basin. **B** Towards the Conflent Basin and elevated massifs of the Axial Zone. **C** Towards the North-Pyrenean Zone and Fenouillèdes

whose skyline gradients sloping eastward correspond to occurrences of S and/or P1 (often undifferentiated; Fig. 7.18B, C; see also Itinerary 2). In the west (Fig. 10.2B), the Neogene Conflent Basin, flanked to the south and north by the Canigou horst and the tilted Madrès footwall upland, respectively, are capped by vestiges of S or P1 (see Itineraries 4 and 6). Closer, note Montalba Plateau (P1) and the local vestige of P2 around Caladroy. The NW to NE quadrant (Fig. 10.2C, and Itinerary 1, Fig. 6.13) is the area of the North-Pyrenean Zone where folds are highlighted by the razorback and hogback ridges of white Urgonian limestone. These structural landforms typically rise above the topography of the Agly basement, where the gneiss and granite are comparatively more susceptible to subaerial weathering and systematically form topographic depressions in the regional landscape. These low-lying erosion surfaces and benches flanking the limestone ridges generally correspond to pediment generation P2 (Pliocene) and host the best vineyards of the area. The east and NE quadrant is dominated by extensive vestiges of surface P1, cross-cutting massive limestone structures and previously highlighted as part of Itinerary 1 (Fig. 6.13 and 6.14). Note the population of residual monadnocks in limestone (Galamus ridge, Mt. Tauch, Tour de Tautavel, Périllou) or schist (eastern Mouthoumet massif). The Oligocene to Miocene basins of Paziols–Tuchan and Estagel (Itinerary 1, Fig. 6.38) are also distinguishable in the distance.

Descend back to Col de la Bataille and follow the D38 towards Bélesta. You will be driving across the tread of P2 at Caladroy, after which the road climbs up onto degraded vestiges of P1 as you skirt around the residual inselberg of Pic Aubeil (Bélesta gneiss). The road also follows to its south a series of razorback hills in Mesozoic metamorphic limestone (Pic Haut, Sarrat del Bouix), sculpted out of upturned beds belonging to the southern limb of the Boucheville syncline (structurally, this is part of the 'Internal Metamorphic Zone', or IMZ; the HT–LP metamorphic event occurred at the time of Cretaceous extensional tectonics ca. 90–95 Ma, i.e., prior to the existence of the Pyrenees). These beds are crushed against the eastern termination of the North-Pyrenean Fault (NPF). At Bélesta, between houses near the southern exit of the village (42°42′56.6″N, 02°36′24″E), an outcrop of the NPF can be inspected. Perhaps atypically, the NPF dips to the north, and the NPF forms a boundary between the Mesozoic marble and the weakly metamorphosed Cambrian/Ordovician schist (Fig. 10.3A). At the castle, a small museum of prehistory displays vestiges of the local Neolithic necropolis known as Caune de Bélesta. The large cave where the prehistoric remains were found lies to the west of the village and is accessible (390 m). This cavity, with galleries up to 10 m wide, is disproportionately large compared to its minuscule drainage catchment area, and thus is very likely quite ancient—potentially coeval with, or slightly younger than, the development of surface P1, a time when the cave system still harnessed waters from the former Agly watershed and directed them straight to the Roussillon Basin through the marble ridge at Bélesta.

At Col de Bélesta, follow the D21 towards Caramany, then the D17 towards Montalba-le-Château. Along the way, the road once again cuts through the Boucheville syncline, here overturned towards the north. The syncline is flanked by two steeply upturned marble ridges, its core consisting of Albian black shales metamorphosed into massive hornfels. You then cut across the NPF, and finally enter the Montalba granite, which is part of the Axial Zone. At Montalba, continue on the D17 towards Tarerach and stop by the roadside where possible.

Stop 2. Montalba plateau

The granite is an outcrop of the vast Millas–Quérigut batholith of Hercynian age. Its structure is complex (Messaoudi et al. 1993). The dominant texture here is

Fig. 10.3 Aspects of the North-Pyrenean Fault. A Exposure at the south entrance into Bélesta. **B** View to the west from Col des Auzines. Differential erosion has hollowed out the Axial Zone granites and topographically emboldened the Mesozoic rocks of the Boucheville syncline, which contains a core of black hornfels (Internal Metamorphic Zone). **C** North-Pyrenean Fault at Sournia, here seen from the plateau around Roc Jalère. The granite (gr) is an outcrop of the Millas–Quérigut massif; marble (m) is Mesozoic, hornfels (h) Albian

porphyritic (K-feldspar phenocrysts), grading to the south to a non-porphyritic variety containing enclaves of basic rock in addition to numerous veins of white granite, aplites, albitites, and schist septa. The albitites, which are widespread in the Agly massif and quarried, have been dated (110–90 Ma) and ascribed to metasomatic reactions as a result of hydrothermal circulation at the time of Cretaceous extensional tectonics (Fallourd et al. 2014).

Montalba plateau is a vestige of surface P1 (Fig. 10.4A), with scattered insel-bergs rising above it in the west at elevations of 460–500 m. The mantle of grus has been partly stripped away, generating flat-floored etch basins, presumably under conditions of a temporarily steady base level. Today, these topographic depressions contain wet meadows sustained by a permanent water table, which are used as pasture and form ponds during the rainy season (numerous wells also harness the underground water). These depressions are often closed, countersunk

Fig. 10.4 Miocene erosion surface P1 across granites of the Millas–Quérigut massif. **A** Montalba plateau, looking west; surface P1, here forming a population of etch basins in the granite, continues to the west, tilted and faulted, across Roc Jalère plateau. **B** Roc Jalère plateau and its koppies and granite tors

by a few metres into the tread of the etch basin floor as a result of aeolian deflation during the drier and cooler periods of the Pleistocene. They contain quartz
ventifacts coated in ochre or red weathering rinds, including dreikanters similar
to those encountered on the upper and middle terrace treads of the Roussillon
(see Part I, Fig. 3.6C therein, and Itinerary 2) and in the Cerdagne Basin (see
Itinerary 7). The large, circular etch basin of Tarerach, which today is a wine-
growing enclave, lies a few dozen metres below the tread of P1, and could belong
to generation P2 (Pliocene age based on constraints from other sites).

The continuation westward of these P1- or P2-generation pediments is unclear.
The escarpment above Tarerach to the west continues along the edge of Conflent
Basin (see Itinerary 6) and was probably generated by a N40°E neotectonic fault
(Lagasquie 1984), but its profile here appears substantially degraded by weath-
ering and militates against very recent tectonic activity (Figs. 6.13, 10.4B and
10.12).

*At Tarerach, take the D13 towards Col des Auzines. The road ascends the escarpment
and overlooks Montalba plateau. Just before the col on the left, a small track leads
up to the footwall summit area (2 km return trip), which is an upthrown occurrence
of P1 in granite (Séquières plateau). This upper plateau has retained sometimes
quite large upstanding residual hills (650–700 m), and the tread of pediment P1
rises progressively to >1000 m in the west (Roc Jalère–Col de Tribes) (Fig. 10.4B).*

Stop 3. Col des Auzines and the North-Pyrenean Fault
The nature of the NPF (its geometry and origin) remains controversial, but it
is supposed to mark the boundary between the Iberian and European plates. Its
presence in the landscape is made conspicuous by the white marble 'flatirons'
along the overturned southern limb of the Boucheville syncline, particularly the
flatiron to the west of Sournia, which also hosts a quarry (Fig. 10.3B, C). North
of the col, follow the GR trail for ~900 m until it intersects the NPF and a series
of tight folds generating three outcrops of marble in succession (La Trufère). The
topography here is inverted with respect to the Paleogene tectonic structure; this
means that, despite forming the tectonically uplifted footwall compartment, the
Montalba granite forms a long topographic furrow periodically widening to larger
basins such as at Trévillach in the east, and Sournia and Rabouillet in the west
(Fig. 10.3B), because the granite is first and foremost vulnerable to chemical
weathering. The Desix, a tributary of the Agly, follows the furrow and thus the
strike of the NPF, but eventually makes a right-angle elbow to the north and cuts
a gorge through the long hornfels ridge of the Boucheville syncline. The river
presumably flowed initially across the tread of P1, and the elbow of capture was

facilitated by tectonic tilting to the north along a hinge zone striking through the summit of Roc Jalère and Col de Tribes.

Return to Col de Bélesta along the NPF, driving through Trévillach and the D2 as far as Montalba, then back along the D17. At Col de Bélesta, take the small road that rises eastward to Moulin de Bélesta, spot elevation: 448 m. From there, walk over to the small knoll at the plateau edge to see the view. You will also find a dolmen with its preserved circular tumulus ~100 m to the SE of the road.

Stop 4. Moulin de Bélesta, panoramic view of the Agly massif
This small plateau ~450 m a.s.l. is a component of P1, here well preserved on Bélesta paragneiss. To the east, a convex hill (Pic Aubeill) is carved out of somewhat more massive, less susceptible paragneiss and forms a residual inselberg ca. 100 m high. The Bélesta and Caramany gneisses are interpreted today as equivalent to the schists of the Canaveilles Group, which display here high-grade metamorphic features typical of the catazone (Laumonier 2008).

To the NNE, surface P1 also tops the plateau of La Tourèze (Fig. 6.13), where it cross-cuts the folded Jurassic to Neocomian limestone sequence, and farther still the ridge above Galamus gorge. To the SW, P1 on Montalba plateau appears very clearly, and likewise to the west the dark hornfels forming the core of the Boucheville syncline. Throughout this area, differential erosion has completely inverted the relief, with synclinal structures underpinning the highest topography (Figs. 10.5A, B and 10.6). The Hercynian basement of the entire Agly massif, which comprises the Lesquerde granites, catazonal Caramany gneisses, and catazonal Ansignan dark charnockite, is topographically low-lying compared to the limestone, marble and hornfels structures of the Mesozoic cover sequence. The silicate rocks are deeply weathered, and their outcrops host all the wine-growing land use. The many depressions, including the benches on either side of the Agly River, everywhere correspond to pediments and straths labelled regionally as generation P2 (Figs. 6.13 and 10.12).

The age of surfaces S and P1 has been documented by a number of thermochronological cooling ages on crystalline rocks, as well as by mammalian fossil assemblages preserved in limestone karren (Fig. 10.6A, B) (Gunnell et al. 2009). Limestone plateau surfaces contain innumerable pockets filled with mostly Miocene micromammalian bone and teeth remains (refer to Itinerary 1, Figs. 6.13 and 10.12) and one of these, of early Burdigalian age, was discovered at the top of Serre de Verges (Meurisse et al. 1969). These very shallow fossil deposits post-date the formation of S, and perhaps even of P1, their preservation thus also testifying to long-term immunity of these land surfaces from erosion. Apatite

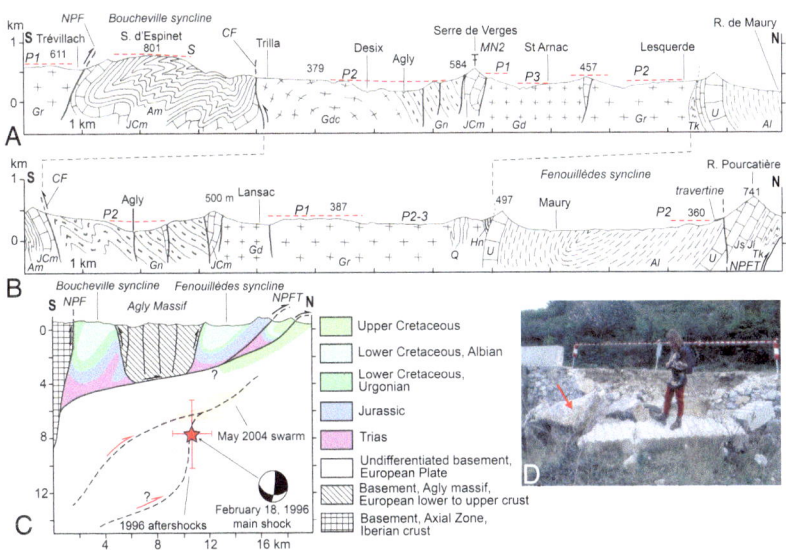

Fig. 10.5 Structure of the North-Pyrenean Zone in the Fenouillèdes area. A, B Geological and geomorphological cross-sections (after Calvet 1996, redrawn). All elevations in metres. The sections show how differential erosion has systematically placed basement outcrops in low-lying positions; in contrast, also note the counter-intuitive geomorphological position of the synclines: the Fenouillèdes forms a topographic furrow in its Albian marls, whereas the Boucheville syncline forms upstanding relief because the same marl protolith has undergone metamorphic transformations to hornfels (a hard rock, resistant to erosion). Key to petrography and stratigraphy—*Gr*, granite; *Gd*, granodiorite; *Gdc*, charnockitic granodiorite; *Gn*, gneiss; *Q*, quartz dike; *Hn*, Hercynian hornfels; *Tk*, Triassic (Keuper); *Ji*, Lower Jurassic; *Js*, Upper Jurassic; *U*, Urgonian (Lower Cretaceous) limestone; *Al*, Albian marl; *JCm*, Jurassic–Cretaceous marble (Alpine metamorphism); *Am*, Albian hornfels (Alpine metamorphism). **C** Crustal structure of the North-Pyrenean Zone. Focal depth of the 1996 earthquake and location of the aftershocks are indicated, along with the depth of the 2004 seismic swarm (seismotectonic data after Sylvander et al. 2007, redrawn). **D** Effects of the 1996 earthquake: fallen boulders and road damage at the north end of La Fou gorge

fission-track and (U–Th)/He ages obtained from Agly gneiss, Agly granite, and Axial Zone granite outcrops along a transect from Pic Aubeill to the top of the Madrès massif document rapid exhumation during the Paleogene (Gunnell et al. 2009), also more recently confirmed by apatite (U–Th)/He exhumation ages (Ternois et al. 2019). The peak of intense rock denudation clearly pre-dates the completion of the low-gradient surfaces, which is typically documented, on

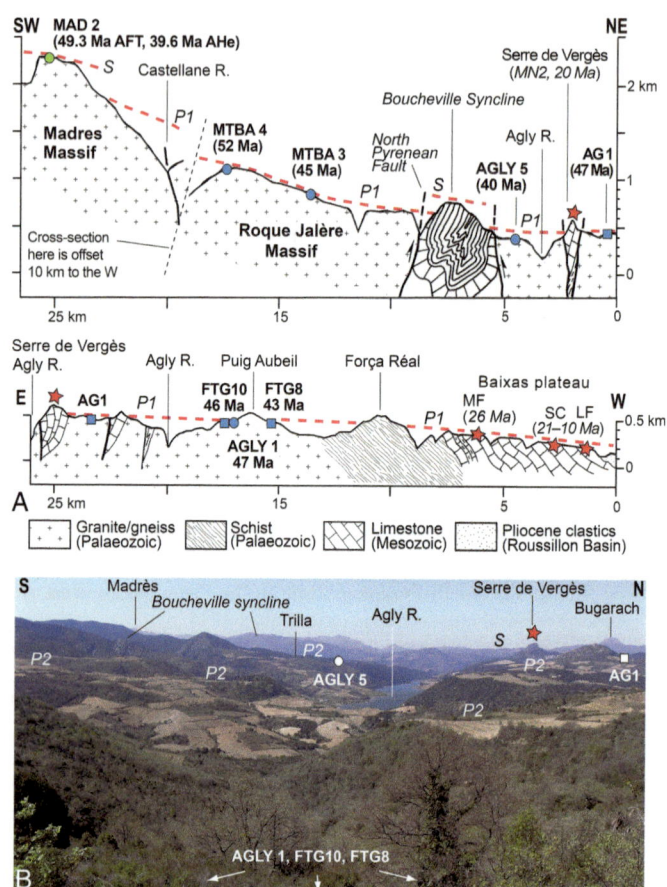

Fig. 10.6 **Age constraints on erosion surfaces in the Pyrenees based on two independent methods: thermochronology and biochronology.** **A** Schematic cross-sections from the Fenouillèdes to the Madrès. All elevations in metres. Red stars: sites hosting micromammalian fossils which have been correlated with well-established biozones (MF: Mas de las Fonts; SC: Sainte-Catherine; LF: Lo Fournas); blue squares: apatite fission-track (AFT) samples of Yelland (1991), with their names and apparent ages; blue and green circles: AFT and combined AFT and (U–Th)/He (or AHe) samples, respectively, of Gunnell et al. (2009), with their names and apparent ages. **B** View of the Agly valley from Moulin de Bélesta. Note the extent of partial erosion surface P2 (Pliocene) hanging above the Quaternary valleys (after Gunnell et al. 2009, redrawn)

time–temperature graphs, by the termination of rapid cooling and prevalence of a temperature plateau compatible with shallow crustal depths (Figs. 10.6A, 6.13 and 11.15; Calvet et al. 2015a).

10.2 Fenouillèdes

Continue to Caramany on the D21. The road descends to the treads of P2 (Fig. 10.6B), with probably several short-lived generations forming a staircase of intermediate bench levels. The most conspicuous exemplars of P2 in the area are a pediment sloping away from the structural ridges of the Boucheville syncline, and erosional benches at elevations of 320–310 m on both sides of the Agly valley. The benches are padded by yellowish to red, and at places silt- and clay-rich, grus. To the west of the village of Caramany, the Holocene tufa cascade containing leaf impressions is worth a stop. It was generated by a hydrothermal karstic spring (constant temperature of 18.5 °C) that comes out of the Boucheville marble outcrop along the Caramany Fault.

The road passes over the Agly dam, then follows the D9 along the left bank of the reservoir. At several places it crosses a band of red breccia consisting of Mesozoic limestone and Keuper gypsum crushed between upthrust slabs of gneiss. The band of tectonic breccia is parallel to a much larger band of breccia which forms steeply upturned imbricates of bedrock such as at Tour de Lansac and Serre de Verges. For a closer inspection, stop below Roquo Roujo, before Le Mas.

Stop 5. Quaternary neotectonics at Le Mas?

A possible neotectonic feature the neotetconic feature can be seen in the road embankment (42°45′11.5″N, 02°32′38″E). A partially cemented and entirely rubified talus deposit appears to be offset by a reverse fault where crushed gneiss overrides the Quaternary slope accumulation (Fig. 10.7). At the base of the exposure, a fault plane striking N145°E, dipping 55°NE, and displaying vertical striations documents the strain parameters at the site. The presence of Triassic gypsum beneath this valley segment could justify the sensitivity of this area to neotectonic stress fields (Philip et al. 1992; Calvet 1996, 1999). Geomorphic indices of Quaternary tectonics are also suggested by the hydrological consequences of a shutter-ridge effect cutting obliquely through the valley with, (i) upstream, anomalously large volumes of recent alluvial deposition around Le Mas; and (ii) an occurrence of rapids through a bedrock channel where the Agly intersects the fault (Fig. 10.7A; this is only visible when the reservoir is empty, usually in autumn and winter). A re-examination of this exposure in late 2021

Fig. 10.7 Possible evidence of Quaternary neotectonics at Le Mas (municipal boundary between Ansignan and Caramany). **A** General view from right bank of the Agly. **B** Roadside exposure in 1993. The late Middle Pleistocene scree deposits have been supplied by the Roquo Roujo limestone outcrop, and are quite strongly cemented. **C** Fracturing through the cemented scree deposit and through an individual clast. **D** Same roadside exposure in 2020

did not, however, fully confirm this indication of Quaternary neotectonics. Further clarifications are required.

The D9 crosses the Agly, then the Desix (charnockite outcrop). The D9b leading to the village of Trilla offers good views of an occurrence of pediment P2 (Fig. 10.8B). At Ansignan, a Roman acqueduct (third century CE, reconstructed in the ninth century) is still functional. From the left bank, a track suitable for vehicles provides access to Saint-Arnac plateau, then to Serre de Verges on foot (1.5 hr round trip), from which you get very good views of the Agly massif and its landscape pattern of vineyards on silicate rocks surrounded by structural landforms in limestone or marble (Fig. 10.8A, B). After Ansignan, the D619 cuts through the Mesozoic ridge

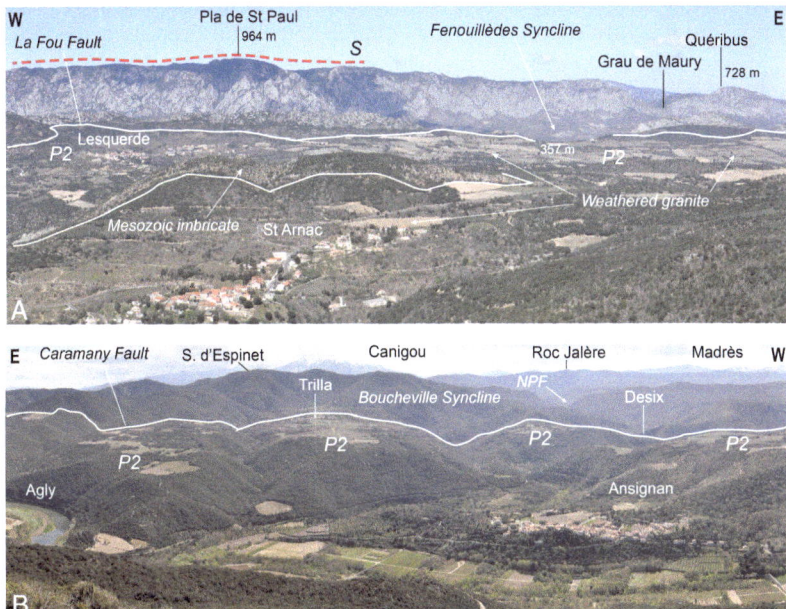

Fig. 10.8 **Views of the landscape from the summit of Serre de Verges.** **A** Towards the north; pediments P2 and P3 in weathered granite (land use: vineyards) surrounded by structural landforms in Mesozoic limestone; summit surface, S, has been preserved on the north edge of the Fenouillèdes syncline. **B** Towards the south; pediment P2 at Trilla slopes away from the rounded hornfels hills of the Boucheville syncline (e.g., Sarrat d'Espinets); background: major footwall uplands of the Axial Zone (Canigou, Madrès). All altitudes in metres

of Serre de Verges (structurally a horse, probably generated by transpressional tectonic inversion of a Cretaceous graben), where an outcrop (similar to an outcrop near Caramany) displays tectonic breccia traditionally ascribed to the Eocene, but more recently reattributed to the episode of Cretaceous extension (Motus et al. 2020). After Saint-Arnac water mill, the road cuts through another horse of Mesozoic limestone before entering a narrow anticline cored by Lesquerdes granite, here intensely crushed and weathered.

Stop 6. Clue de la Fou

Fou gorge is cut in massive Valanginian to Aptian limestones (Urgonian facies) and breaches the southern overturned limb of the Fenouillèdes syncline (Fig. 10.5A, B). The bedrock channel through the gorge is narrow and abundantly potholed. The contact with the granites in this area is a vertical fault, with in-situ weathered granite daylighting through the talus deposits at the junction between the D619 and D7. Several thermal springs (21–27 °C) are reported along the gorge, and this site was the epicentre of the last recorded earthquake in the eastern Pyrenees (M_L 5.2, left-lateral strike-slip motion on a E–W-striking vertical fault plane). It occurred on Feb. 11, 1996, and caused a rockfall with large boulders damaging the road at the upstream end of the gorge (Fig. 10.5D). The earthquake, however, does not seem linked to the aforementioned fault separating the granite from the limestones, which connects with the NPFT at a depth of 6 km beneath your feet in this area. The focal depth of the 1996 earthquake was situated instead at a depth of 9 km and is interpreted rather as a manifestation of blind thrusting in the European Plate's lower crust (Fig. 10.5C; Sylvander et al. 2007).

The course of the Agly River is riddled with drainage anomalies similar to Fou gorge: the Agly rises in the Sub-Pyrenean Zone, ignores the broad, shale-cored Fenouillèdes syncline and, instead, follows a tortuous line through gorges in massive limestone such as La Fou, then crosses the entire crystalline basement outcrop of the Agly massif before reaching a marl-cored syncline in its lower course and eventually joining the Roussillon Basin. Such occurrences of anomalous drainage could result from a case of drainage superimposition on erosion surface P1, or perhaps even on surface S, with subsequent (post-Miocene) tectonic deformation of the land surface guiding the river's path in piecemeal fashion (Figs. 6.14 and 10.12). This theme is expanded through examples and evidence at forthcoming stops further west along the mountain front.

Drive through Saint-Paul-de-Fenouillet and enter Galamus gorge on the D7.

Stop 7. Gorges de Galamus

Depending on traffic conditions, try either the first or the second parking area in preparation for a choice between two trailwalks—both of which will afford you sweeping views of the Fenouillèdes syncline, with massifs of the high range looming in the background: Canigou, Carança, Madrès (Fig. 10.9).

Optional trailwalks through Galamus gorge

1. **Roadside walk** (2 hrs, round trip)
The gorge forms a natural cross-section through the entire Mesozoic fold sequence, from the Albian black shales (observable from the road before the first parking area) to the Keuper gypsum beds at the NPFT sole near Moulin de Cubières (water mill). The different out-crops are not always very clearly distinguishable because the rock is massive and intensely deformed, but inspection is best achieved on foot given the parking constraints and nar-rowness of the road. The main rock mass (where the road has been cut laterally into the overhanging gorge wall) consists of reef limestones and breccia (Upper Jurassic). These highly resistant beds have blocked the headward migration of a major knickpoint on the Agly, analogous to the knickpoint already encountered on the Verdouble River at Padern (see Itinerary 1). The relatively bedload-starved bedrock channel is only mildly eroded, thus allowing tufa accumulations on the upstream side of potholes. At the point where the road joins the river, note the subvertical and refolded bed of Middle Jurassic limestone, clearly outlining the geological structure (Fig. 10.9B). The Lower Jurassic shale (Pliensbachian, Toarcian) is relatively inconspicuous but the upstream part of the gorge cuts through lime-stones of the lower Lias (Hettangian, Sinemurian). The Triassic gypsum is likewise not easy to distinguish at Moulin de Cubières, and you need to walk further up the valley to encounter the Upper Cretaceous (Albian to Santonian) marl and sandstone, which is highly deformed into a succession of thin imbricates beneath the thrust sole.

2. **Pech d'Auroux: Structure of the North-Pyrenean Mountain Front, Erosion Surfaces, Dry Valleys** (4 to 5 hrs round trip)
This walk follows the GRP 'Tour des Fenouillèdes', then the GR 367—both well-marked and signposted hiking trails. Start from elevation spot 323 m on the D7. The trail first cuts through the Albian marls and dark sandstones. At Col de la Corbassa, note the faulted bound-ary between those Albian strata and the Aptian nodular, then massive, limestone. This normal fault, with the downthrown block to the south, was probably reactivated during the Neo-gene because its en-echelon segments can be traced eastward all the way along the Galamus razorback ridge, with hydrothermal springs along it. This fault has also offset the Oligocene Paziols gravel beds (Figs. 6.13, 6.40).
Along this geological traverse, the Urgonian carbonate series contains lacunae, but splits fur-ther west into several reef bodies separated from one another by marl beds. The path skirts around the first reef unit, which consist of middle Aptian (Gargasian) to Valanginian white limestones containing index fossils such as rudists and orbitolinids. Around the 486 m ele-vation contour, this upturned limestone ridge displays Neogene karstic cavities containing rolled quartz pebbles supplied from a northern source area. A lacuna in the stratigraphy (gap

Fig. 10.9 **Landscapes of the Fenouillèdes syncline.** **A** View from the trail between Prugnanes and Campeau; note systematic ridgetop bevels, here corresponding to paleoplain S, more uplifted in the west than the east. **B** View to the west of the NPFT from eastern edge of Galamus gorge; note staircase of landforms belonging to generations S, P1, and P2, and anomalous altitude of Pic de Bugarach (potentially a neotectonic uplift). In **A** and **B**, note obvious neotectonic offset in the suite of landforms between the two sides of the Fenouillèdes syncline

Fig. 10.10 Intra-Aptian bauxite ore deposit. Exposure situated above trail GR36, at the exit of the Brézou dry valley

between the Upper Barremian and Lower Aptian) displays a bed of red to green pisolithic bauxite (Fig. 10.10), suggesting emersion and subaerial exposure of this area to humid tropical climates at the time. The bauxite has not been mined commercially, but crosscuts and reddish spoils above the path (after elevation spot 486 m, and others further west) indicate that this option was once considered.

The Valanginian to Upper Berriasian marls and platy limestone beds are not well exposed, but the trail soon cuts past the second mass of reef limestone, which is very thick and upholds most of the mountain range: white, massive limestone, dolomite and breccia of Berriasian and Upper Jurassic age. Here, the Galamus ridge has been breached by the (now dry) valley of Pla de Brézou, which displays a box-shaped cross-profile with a 200-m-wide flat floor lying at an elevation of 660 m in its upstream part (Fig. 10.11B), i.e., 280 m below extant vestiges of range-top palaeosurface, S, which occurs on the surrounding ridgetops. Scarce but important rolled pebbles of quartzite, vein quartz, and sandstone, similar in kind to those observed in the cavities mentioned earlier, document an ancient river entering this area from the Mouthoumet massif and Cretaceous outcrops of the Sub-Pyrenean Zone to the north. This abandoned palaeovalley could be coeval with P1, or slightly younger. The lowermost stratigraphic components of the Mesozoic sequence are poorly exposed but the basal red marl and gypsum (Keuper) along the NPFT can be reached directly by walking another 600 m along the trail towards Duilhac until you reach a spring.

At the col, follow the path going west and make the ascent to the narrow ridgetop of Pech d'Auroux. Midslope, ca. 800 m, you will gain an excellent view of the NPFT (Fig. 10.11A), which consists of two successive tectonic units: to the south, the main thrust, which has transported the North-Pyrenean Zone over the Sub-Pyrenean Zone; and to the north, the thrust units of Peyrepertuse, which involve an overturned sequence of Albian marls and limestones

Fig. 10.11   North-Pyrenean Frontal Thrust around the Brézou palaeovalley. A View to the east from the GR367. At the main thrust front, Triassic beds override Upper Cretaceous units; but a subsidiary unit north of the NPFT, at Peyrepertuse, displays synclinal folds overturned to the north. Note how all of these tectonic structures have been cross-cut by palaeosurface S, a vestige of which is well preserved at Plateau de Saint-Paul. The abandoned Brézou valley (generation P1?) extends 250 m below the tread of range-top surface S and strikes in the direction of the Roussillon Graben. **B** The Brézou dry valley, looking west. **C** Alluvial quartz pebbles typical of the region's Miocene dry valleys (here from the Bedeau-Malabrac dry valley, west of Pic de Bugarach). The source rock for these quartz clasts is Campanian sandstone and conglomerate

followed by isoclinal folds in marls and limestones of Cenomanian to Santonian age, all overriding continental shelf sequences of the Mouthoumet massif. Note how the entire north-Pyrenean Mesozoic sequence, which forms a tight anticline at Pla de Saint-Paul, has been bevelled off by erosion: this is an example of the ancestral paleoplain, S. Another vestige of S also occurs at the top of Pech d'Auroux, where the limestone exposure has weathered into karren.

From the summit of Pech d'Auroux you will get a 360° panoramic view of the eastern Pyrenees: Axial Zone, Roussillon Basin, North-Pyrenean Zone, Corbières, and Mouthoumet. In the middle ground to the west (Fig. 10.9B), note the large vestige of summit surface, S, at Roc Paradet (900 m); it distinctly truncates the steeply-dipping (40–50°S) Mesozoic sequence, which in that area spans the Triassic to the Lower Cretaceous. Pic de Bugarach (1230 m) forms the overturned northern limb of a frontal anticline associated with the NPFT. The origin of this isolated peak, which rises 200–300 m above all other ridgetops in the Corbières, is unknown: (i) Bugarach could be a residual monadnock standing above erosion surface S, although this seems improbable given that it has been carved out of the lower limb of the anticline, where the rock is furthermore highly fractured, whereas the beds of the upper limb are cross-cut by topographic surface S. Alternatively, (ii) Bugarach could have been generated by an episode of late Neogene convergent tectonics, also recorded along the faulted boundaries of the Roussillon Basin (see Itinerary 2), and which would have attained unusually large magnitudes here because of lubrication by the anticline's gypsum core (which originally assisted in overturning the fold). The resulting structure is well preserved at this location because it is situated in a transfer zone between two segments of the NPFT.

The trail descends into an anticlinal valley eroded out of Lower Jurassic marly limestone. At the first junction you reach, follow the GR 367, which goes up to Col das Souls through Lower Jurassic marls before reaching a spectacular section above Galamus gorge (Figs. 10.9B and 10.12). Over on the right bank of the gorge, note how the topography cross-cuts the fold structures around 680–700 m: this is an occurrence of P1, here just a flat-floored corridor ~1 km wide smeared by a trail of quartz gravels reworked out of Upper Cretaceous beds (Fig. 10.11C). The flat floor is itself incised by a meandering palaeovalley 50 to 100 m deep and known as Coume Tiols (concealed by the foreground but clearly apparent on maps), with a steep slope (> 3%) and a narrow flat floor also strewn with quartz pebbles source from the Sub-Pyrenean Zone (sandstones from around La Bastide de Camps). This dry 'valley-in-valley' can be ascribed to the generation of landforms known in the Pyrenees as P2, widely represented farther west and visited later along Itinerary 5 (Col Saint Louis, Pays de Sault) (Fig. 10.12). The trail descends to Moulin de Cubières, and you can rejoin your vehicle by walking along the D7 through the gorge.

Return to Saint-Paul-de-Fenouillet and follow signposts to Maury, thereafter driving along the syncline's axis. At Maury, turn left onto the D19. The road takes you through the black shales, which in this area host vineyards that produce a pudding wine comparable to Banyuls. Stop at the parking area (spot elevation: 350 m).

Stop 8. La Mouillère

From this vantage point you gain a panoramic view of the Fenouillèdes syncline in an area where its southern limb has been overthrust by the overturned syncline of La Tourèze (Mesozoic cover of the Agly massif; additionally, in this area P1 cross-cuts massive Jurassic–Lower Cretaceous limestones). The Tourèze thrust continues eastward towards the Tautavel–Vingrau Thrust (see Itinerary 1), visible from here, and marks the beginning of the Corbières orocline. Further to the west, notice a large number of Pliocene to Pleistocene travertine beds at a range of altitudes, each situated where springs come out at constant temperatures of ~20 °C. The larger spring is situated 280 m south of the stopping point, made conspicuous by the bench of Holocene and recent tufa. Residual outcrops of much more intensely lithified travertine, often displaying leaf impressions, also occur at the tops of surrounding hills and buttes, locally grading to aprons of equally massive ancient slope breccia cemented in a red matrix. These travertine formations form

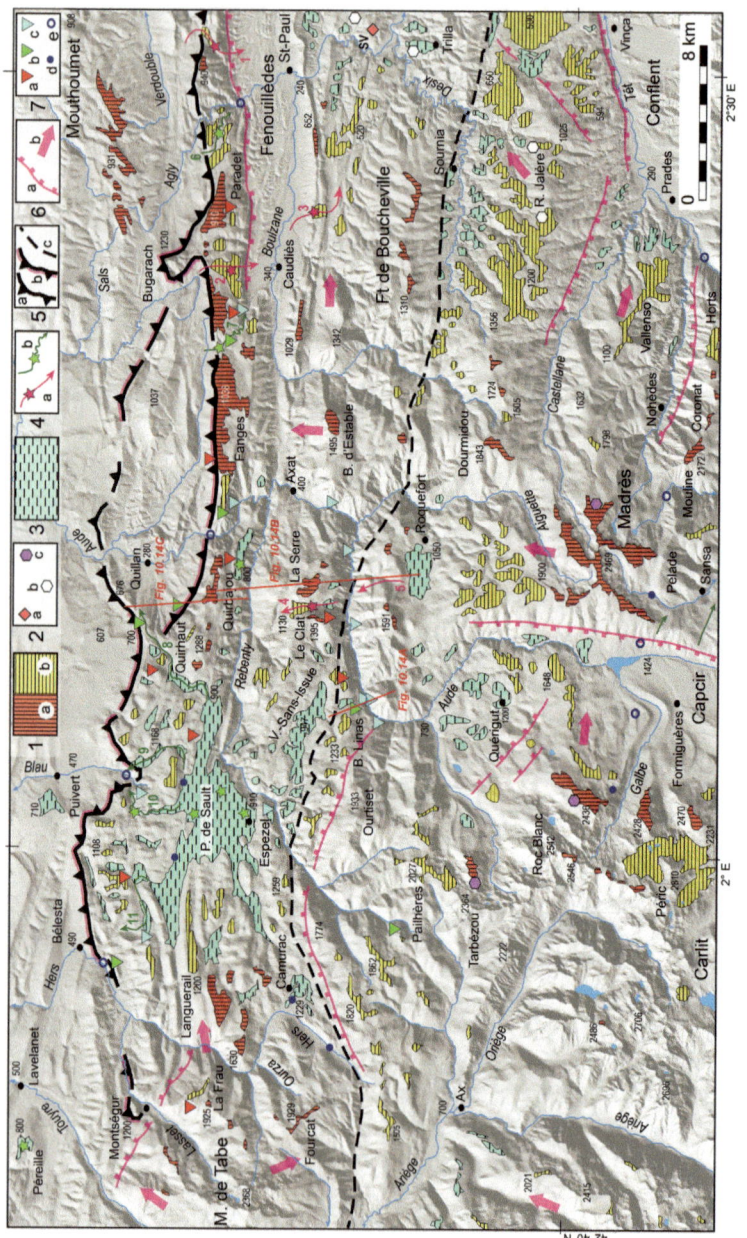

◀**Fig. 10.12 The north-east Pyrenees: a geomorphological map** (after Calvet 1996, modified). Key to symbols and ornaments—**1**: main post-orogenic erosion surfaces; **1a**: late Oligocene to Aquitanian paleoplain, S; **1b**: middle Miocene pediment population (generation P1). **2**: age constraints; **2a**: palaeontological fissure fillings (SV: Serre de Verges); **2b**: AFT cooling ages; **2c**: AFT and (U–Th)/He cooling ages. **3**: late Neogene partial erosion surfaces (generation P2: pediments, bedrock straths, poljes, wind gaps / hanging valleys, flat-floored etch basins in granite). **4**: abandoned (dry) valleys in limestone containing allochtonous silicate pebbles (flow directions indicated by arrows); **4a**: flat-floored, low-sinuosity erosional corridors of P1 affinity (1: Pla de Brézou; 2: Col del Bedau–Malabrac; 3: Fosse; 4: Le Clat–Pas del Corps; 5: Col de la Malagrède); **4b**: meandering bedrock palaeovalleys of P2 affinity (6: Coume Tiols; 7: Col Saint-Louis; 8: Lapeyre–Coudons; 9: La Malayrède; 10: Col du Chandelier; 11: Col de la Croix des Morts). **5**: major tectonic features and/or landforms; **5a**: major thrust-front escarpment relating to the NPFT; **5b**: site of suspected neotectonic reactivation of the NPFT; **c**: trace of the NPF. **6**: direct and indirect evidence of neotectonic deformation; **6a**: normal fault scarps, flexures; **6b**: direction of tilt of surfaces S and P1. **7**: limestone karst: abandoned horizontal passages, often containing relict allogenic alluvium; **7a**: associated with surface P1; **7b**: associated with surface P2; **7c**: associated with post-P2 valley incision. Limestone karst: active landforms; **7d**: losing streams; **7e**: main karstic springs. All altitudes in metres

overall 2 to 3 distinct levels between 300 and 380 m, the most elevated draping the topographic surface of a pediment ascribable to generation P2 (butte above the road, and also Roc de Natousque further west; Fig. 10.13).

The D19 takes you to Grau de Maury—a palaeovalley similar to Pla du Brézou, here forming a wind gap through the Galamus razorback ridge ca. 430 m a.s.l. and containing a few quartzite pebbles. From here you can also access the Medieval fort of Quéribus.

Stop 9a. Quéribus

For a panoramic view to the north, walk up to the edge of the escarpment at the north end of the car park. The Cucugnan depression is cut in Albian shales of the Peyrepertuse thrust unit, which here corresponds to an overturned Albian stratigraphic sequence dipping 70°S. The depression is thus nested on the backslope of the *Floridae* limestone thrust front at Roc Pounchut, which normally would form the top of this Albian sequence. The Medieval castle of Peyrepertuse stands out to the NW on folds at the leading edge of the NPFT thrust front involving Lower Santonian, rudist-bearing limestones. North of Roc Pounchut thrust front, the limb of the Mouthoumet anticline (alternating Cenomanian to Campanian marls, sandstones, and rudist-bearing limestones, intensely faulted by

Fig. 10.13   Staircase of older travertine accumulations at La Mouillère–Las Fountètes (road to Grau de Maury). Here, the younger (Holocene) travertine deposits are not visible. The limestone scarp forming the southern edge of Galamus ridge and Saint-Paul-Queribus plateau could be of neotectonic origin

synsedimentary normal faults) is carved up by consequent streams into a succession of gorge-and-flatiron features. To the NE, Mt. Tauch is a slab of Urgonian limestone resting unconformably on beds of Triassic gypsum, its structure still incompletely understood.

The D123 takes you down to Cucugnan through the Mesozoic sequence, i.e., from the Lower Jurassic shales and limestones to the Upper Triassic gypsum at the sole of the NPFT. Stop at elevation spot 374 m, where the road curves around at Pech Marty.

Stop 9b. Pech Marty
In this area the NPFT overrides a Garumnian sequence (Maastrichtian–Danian) of bright red marl containing beds of sandstone and conglomerate, covered further east by lacustrine white limestone (Rognacian facies) forming a rock face on the far side of the valley, below elevation spot 522 m (when referring to sedimentary

rock formations of late Cretaceous to early Cenozoic age with continental rather than marine characteristics, note that the geological sheets of Quillan, Limoux, Capendu and Tuchan use the lithostratigraphic terms 'Rognacien' and 'Vitrollien', which were defined in the context of Provence, rather than the term 'Garumnian', which is endemic to the Pyrenees. To avoid too much confusion, Garumnian will be used consistently in this GeoGuide). This entire sequence was warped into a syncline beneath the NPFT during Eocene time. The unconformable contact between the Triassic beds and the overlying Garumnian, added to the abundance, within the Garumnian conglomerates, of clasts of Albian green sandstone, together suggest NPFT activity and growing relief in the south during the late Cretaceous. Broader tectonic deformation of the area is nonetheless chiefly the result of Eocene tectonics.

Via Duilhac, then Rouffiac, follow the D14 around the thrust imbricates of Peyrepertuse (Fig. 10.11A). *The road here keeps to the autochtonous Sub-Pyrenean unit. You can reach Peyrepertuse castle from Duilhac up a winding road which cuts through slipped masses of rudist-bearing limestone. From Soulatgé (aim for Col de Redoulade following the D10, then the D212), it is also possible to reach the edge of the Mouthoumet massif, where outcrops of Paleozoic (Hercynian) basement consist of Lower Carboniferous limestones, pelites, and coarse-textured flysch (Milobre de Massac, 907 m). As a consequence of extensional tectonics occurring along the European continental margin during Cretaceous time, the Cenomanian beds rest unconformably on all of the older Mesozoic rocks. Summit areas in this landscape cut across basement and cover rocks indiscriminately, thus suggesting the land surface could be a degraded vestige of erosion surface, S. From Soulatgé, the road takes you to Cubières and the headwaters of the Agly catchment, after which (Col du Linas) you enter the Aude catchment. From Col du Linas you get good views of the NPFT at Pic de Bugarach, whose summit can be reached up a trail (3.5 hrs to get there and back). An apron of strongly cemented talus deposits containing large boulders, probably at least of early Pleistocene or Pliocene age given their position 'hanging' 270 m above the valley floor, suggests a neotectonic origin for this unusual mountain front (see also* **Stop 7**). *Before reaching the village of Bugarach take the D45, which loops around the mountain and provides good views of it from different angles. At St-Louis-et-Parahou (on the D46), join up with the D109 at Col Saint-Louis, where you will again cross the NPFT.*

Stop 10. Col Saint-Louis
This col is actually a dry valley, with a longitudinal slope of 5% (696 m upstream, 600 m downstream). Its floor hangs 260 m above the Boulzane valley to the south,

which presumably used to be its collector stream. The Col Saint-Louis palaeo-valley forms a series of entrenched meanders, and its upper catchment area was beheaded and diverted towards the Aude downstream of Quillan. The valley floor displays a distinctive spread of quartz pebbles (diameter: 2–10 cm), also present in caves occurring at a similar altitude (Pla Cribeillet Cave). Beneath the dry valley at the col lies the vast epiphreatic cave system known as Cthulhu Démoniaque. It hangs ~100 m above the floor of the Boulzane valley and is filled with sand and rolled quartz pebbles (Ournié 1987). The dry valley belongs to generation P2 (i.e., Pliocene). The Cthulhu subterranean network thus records an episode of landscape incision that occurred during the early Pleistocene. Rewatering of the karst occurs frequently, drowning the cave systems to depths of up to 150 m because the limestone outcrop is flanked by Albian shales on one side and by Lower Jurassic or Triassic marls on the other, in both cases acting as impermeable barriers to groundwater freeflow (confined karst).

Box 10.1 Dry valleys and pediments in the fold-and-thrust belts north of the Axial Zone

The population of dry valleys encountered thus far along Itinerary 5 all suggest palaeodrainage towards the S or SE, i.e., towards the Roussillon Basin, which became a depocentre from late Oligocene time as a result of the new extensional stress field in the Western Mediterranean. These streams probably started off on the ancestral paleoplain, S, whose deformations initially controlled the drainage network in the Agly watershed (Figs. 6.13 and 10.12). The first generation of incisional features, which involved a pause in base-level change, is recorded by the regional population of flat-floored corridors and palaeovalleys up to several kilometres wide already encountered at previous stops. Such bedrock corridors are fluvial straths typical of areas upstream of where more extensive pediments belonging to generation P1 often occur, with the palaeovalley widening to a fanhead embayment before broadening out to the full pediment (see also Box 8.1). A second generation of landforms, resulting from a resumption of fluvial incision, consists of narrower and usually meandering valleys. This population of landforms belongs to generation P2, and their altitudes are compatible with the more elevated travertine tabletops auround Maury. The Brézou and Grau de Maury occurrences may represent intermediate, more transient legacies of the landscape incision history. Early valleys of this kind, striking across ridges in resistant bedrock such as the Galamus razorback ridge, are numerous in the region; but the network became increasingly well integrated in the headwater areas as a result of piecemeal drainage captures facilitated (i) by the presence (and length) of the Sub-Pyrenean shale furrow, and (ii) by the widespread Mesozoic limestone outcrops which promote losing streams (sinking rivers), swallow holes, and bedload deposition. The Agly today still benefits from an exit through Galamus gorge, but this river has lost a small part of its catchment headwaters to the Aude in the west (dry valley at Col Saint-Louis) and to the Verdouble in the east (Brézou dry valley), thus potentially falling prey to processes that have been ongoing across this north-Pyrenean piedmont continually since the late Neogene.

*Drive on to Caudiès-de-Fenouillèdes and turn off onto the D117 towards Lapradelle. The road follows a narrower continuation of the Fenouillèdes syncline in Albian black shale, here constricted between the Fanges (1041 m) and Bac Estable (1495 m) anticlines to the north and south, respectively. The Medieval castle at Puilaurens stands on an outcrop of steeply-dipping Upper Aptian limestone, overturned towards the north. At the top of the Fanges massif, range-top surface S has been transformed by solution processes and is an eggbox mosaic of coalescent dolines and conical buttes. The hydrogeological drawdown responsible for these landforms probably occurred during the Miocene in relation to the base level under which generation P1 landforms developed throughout the region. After Col Campérié, you begin to descend into the Aude valley through outcrops of black shale. At the road junction for Axat (roundabout), turn towards Quillan and Pierre-Lys gorge. At the river junction between the Aude and Rebenty, take the D81 for a return trip to Quirbajou (Fig. 10.14C). You eventually reach erosion surface P2 around 780–800 m, which here is the lower end of the Sault plateau surface (Fig. 10.12; see **Stop 15**). As you exit the tunnel at its western side, note also the solutional fissures filled with alluvial sand and quartz and quartzite pebbles, also present across the land surface as you go westward. The small cave at the site is a dried-up Vauclusian spring displaying erosional features typical of groundwater karst on the roof of the cave. Just below the road, 5 m after the tunnel exit, also note the dome of recrystallised travertine (42°49′43.6″N, 02°11′03.3″E). Woodland vegetation blocks the view down onto the spectacular Pierre-Lys gorge, only properly appreciated from the top of a rocky ridge (Fig. 10.15) west of the transmission tower, 1.3 km before the tunnel (somewhat difficult access up a deerstalker path, which follows the escarpment from the tower,*

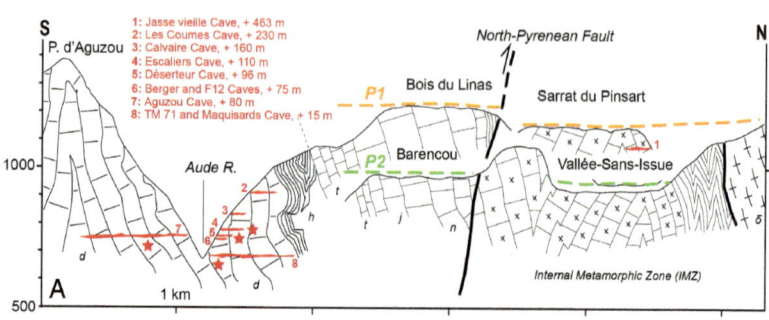

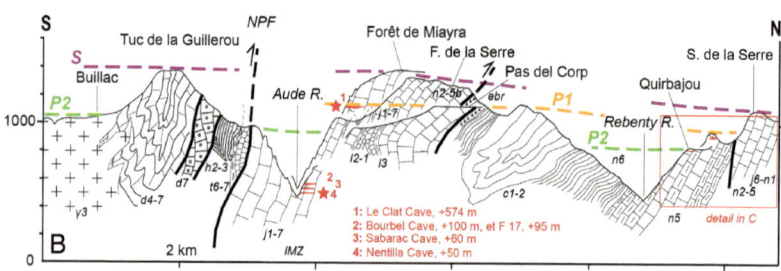

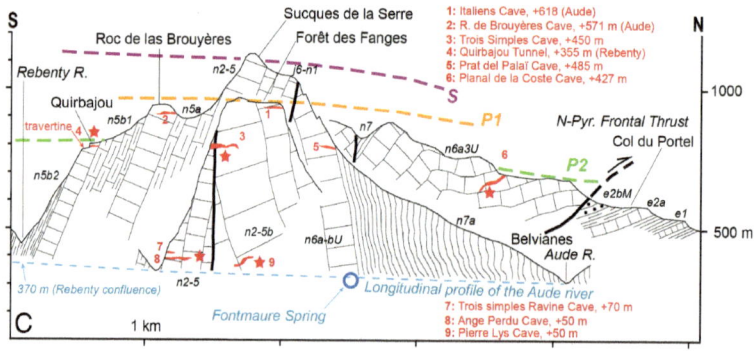

◀**Fig. 10.14** **Groundwater karst cave levels in the Aude river catchment and their links with surface landforms**. All three sections are located in Fig. 10.12. Chronostratigraphic and petrographic indices are given by the geological sheets of Quillan, Lavelanet, and (forthcoming editions of) Ax-les-Thermes and Saint-Paul-de-Fenouillet; γ: granite, δ: gneiss; d: Devonian, h: Carboniferous, t: Triassic, l: Lower Jurassic, j: Middle and Upper Jurassic, n: Lower Cretaceous, c: Upper Cretaceous, e: Eocene. Note magnitude of post-P2 valley incision. All the groundwater karst cavities mentioned belong to subhorizontal passage systems generated in environments coinciding with the top of the water table. Red stars: caves containing allochtonous pebbles and quartz sand deposits (quartz, granite, quartzite, Paleozoic schist). **A** Vertical cave sequence in the upper Aude valley between Usson and Gesse. **B** Cross-section through the Axial Zone, here passing through Le Clat and Quirbajou. **C** Details of vertical cave sequence in the lower Aude River (Pierre-Lys gorge)

Fig. 10.15 **Aude gorge at Pierre-Lys**, here seen from the east end of Roc de las Brouyères. Post-P2 incision amplitudes reach ~450 m. Trois Simples Cave contains pebbles deposited by a late Neogene (ancestral) Aude River

then cuts across beneath the overhead powerlines). The incision history of Pierre-Lys gorge was punctuated by a vertical succession of epiphreatic cave systems, each containing sand and gravel deposits from an ancestral Aude River (quartz, quartzite, granite, schist), particularly in the Trois Simples cave, which occurs ca. 780–790 m exactly across from the transmission tower (Figs. 10.12, 10.14C, 10.15). Return to the D117 and cross the gorge. It cuts through the entire Mesozoic sequence, here arched up into an asymmetrical anticline with steeply downturned limbs. Parking is equally possible at the entrance or at the exit of the gorge. The Vauclusian spring at Fontmaure collects water from about half of the Sault plateau. The road reaches Quillan syncline, in the midst of Albian black marl and sandstone outcrops.

From Quillan, two possibilities: an excursion down the Aude valley to the Paleogene retro-foreland conglomerate sequences (Stops 12–14): this corresponds to the small region known as Razès; and/or an exploration of the Pays de Sault (in which case go straight to Stop 15), which expands on themes previously touched upon while crossing the Fenouillèdes but consolidates the evidence with new landforms in a different landscape.

10.3   Razès

Cross-section through the Paleogene piedmont from Quillan back to Quillan (70 km)

Leave Quillan on the road to Carcassonne (D118). At L'Espinet (holiday village), 2.5 km after Quillan, you will be crossing the northern branch of the North-Pyrenean Frontal Thrust, also named 'accident de Saint-Ferriol'. The system of imbricates rising northward through the stratigraphy only displays outcrops of Albian marl and sandstone, and discontinuous thrust units of (Urgonian facies) Cretaceous limestone (Florideae limestone: Albian, Prealveoline limestone: Cenomanian). The topography of the thrust front here is relatively subdued (elevations <675 m).

Thereafter you enter the Sub-Pyrenean fold belt, where landscape features almost exclusively consist of structural landforms sculpted out of simple anticlines and synclines. The harder lithologies in the stratigraphy (limestone, sandstone, conglomerate) form either intact or faulted anticlines, or breached anticlines with inward-facing homoclines around an anticlinal valley; the synclines and outcrops in the anticlinal valleys coincide with the softer rocks (marl, mudstone). On the west side of the Aude valley, the white lacustrine and palustrine limestone (late Maastrichtian to Danian, mapped as 'Rognacian'), sandwiched between beds of continental red marls (overall sequence known as 'Garumnian' in Pyrenean stratigraphy), has been folded into an asymmetric syncline; its southern limb is upturned vertically (vegetation may conceal these features). The road subsequently cuts across-strike through the broad Campagne-sur-Aude brachyanticline in lacustrine limestone. The anticline is densely faulted and the stratigraphy contains the K–T stratigraphic boundary, including a clay-rich interlayer hosting dinosaur eggs. The lower red marls and their sandstone interlayers (late Campanian to Maastrichtian) host the rich dinosaur remains of Campagne-Bellevue (museum at Espéraza).

The road strikes through the north limb of the Campagne anticline at two locations with good exposures of its steeply upturned beds; first before reaching Espéraza: 'Rognacian' limestones, followed by 'Vitrollian' (Selandian to Thanetian) red and yellow marls containing very thin beds of sandstone and occasional conglomerate, followed by early Thanetian limestones and marls (partly marine); then just before reaching Couiza, at Pastabrac shopping centre: late Thanetian continental red marls, sandstones, conglomerates and evaporites, followed by Ilerdian marine limestones and marls, vertically upturned near the traffic roundabout. At Couiza, take the D52 to Rennes-le-Château. The road cuts once again through the entire upturned rock series; the limb of the anticline transitions eastward to a reverse fault along which appear outcrops of the Paleozoic basement; these form the rocky ridge in Devonian limestone (Cardou; spot elevation: 795 m). The road then reaches the flat-lying beds of the Campagne anticlinorium, wherein a gentle syncline has preserved a butte hosting the town of Rennes. At the base of the butte, note the exposure of early Paleocene ('Vitrollian') red marl containing sandstone beds and limestone stringers, resting on late Maastrichtian–early Danian ('Rognacian') white limestone exposed further to the east; the butte (518 m) is capped by pale yellow cross-bedded sandstone of Thanetian age.

Stop 11. Rennes-le-Château

This site is remarkable as a 360° viewing platform. The views to the north are excellent from the parking area (Fig. 10.16B), and even better from the local museum terrace (entrance fee, and otherwise profitable if you are interested in Cathare mythology). On the far side of Couiza syncline you can see the Paleozoic core of the Mouthoumet massif (Visean flysch and Devonian limestone) rising gently eastward out of its Mesozoic and Cenozoic cover sequence. Its structure as a large anticline with an axial plunge to the west is clearly apparent, particularly highlighted by the homoclinal ridge in Thanetian limestone below the line of wind turbines, and farther west by the scarp carved out of Ilerdian sandstone and limestone. The epigenetic gorge of the Aude is conspicuous, but its history difficult to reconstruct given the absence of chronological constraints and diagnostic landforms. It is likely that the period of base-level stability conducive to generating S or P1 set the conditions for eroding and truncating the fold structures of the basement and its cover sequence; however, palpable vestiges of these land surfaces, which can only be guessed at from this distance, are difficult to distinguish in that area from the sub-Campanian erosion surface, i.e., the stratigraphic unconformity which is currently still in the process of being

Fig. 10.16 Landscape views from Rennes-le-Château. A Looking south. Sub-Pyrenean fold belt in the foreground, and Campagne box anticline (Upper Cretaceous and Paleocene sequence); marine Ilerdian outcrops near Esperaza, covered in the west by synorogenic Palassou foreland units 1 and 2 (Pique l'Ordy). Middle ground: North-Pyrenean Zone, cross-cut by erosion surfaces S and P1, and its tectonic boundary: the NPFT. Far distance: summits in the Axial Zone. **B** Looking north, the Mouthoumet massif and its folded cover rocks (Sub-Pyrenean Zone). The broad, basement-cored anticline plunges west beneath the Palassou conglomerates. The Paleogene fold structure has itself been truncated by the Saint-Salvayre erosion surface, where S and P1 are indistinguishable (perhaps because they converge and merge at low angles). Chronostratigraphic lettering follows nomenclature of the Quillan geological sheet

exhumed from beneath the overlying Campanian sandstones (Alet Fm.). The only unambiguous occurrences of Cenozoic erosion surfaces S or P1 are at Saint-Salvayre (see **Stop 13**), where the land surface cross-cuts deformed Devonian as well as Campanian bedrock; and around the north edge of the Mouthoumet massif, where the low-gradient topographic surface forms a continuum cross-cutting the Palassou-filled syncline all the way up to the vast Lacamp plateau (700 m; see Itinerary 1, Fig. 6.12), where it impinges on Devonian limestone outcrops at Milobre de Bouisse (878 m). Until further evidence to the contrary, incision by the Aude River of Alet canyon can only reasonably be explained by geologically recent uplift in the Aquitaine piedmont zone.

Looking towards the south and west (Fig. 10.16A), the town-hall terrace and ramparts offer extensive views of the Pyrenean mountain front, embracing everything from the pyramidal landmark of Pic de Bugarach to the northern edges of Pays de Sault, via the overturned box anticline of the Fanges massif. In the far ground, the Axial Zone displays the dome-like profile of the Madrès, and the peaks of Roc Blanc and Saint-Barthélemy (Montagne de Tabe)? In the foreground, the outcrops of white Garumnian limestone and micaceous marls make the Campagne box anticline stand out quite conspicuously. The dark, wooded outline on the western horizon (Pique l'Ordy, 772 m) coincides with the first (Ypresian–late Lutetian) and second (late Lutetian–Bartonian) Palassou conglomerate units, which display progressively decreasing synsedimentary dips to the NNW, from 30° at the base of sequence 1, to 1–2° at its top. This 2-km-thick sequence (Palassou units 1 and 2) fills a large syncline (Eocene foredeep), which starts near Couiza in the east and broadens westward.

Return to Couiza and take the D52 along the Aude's left bank towards Antugnac. The road follows the northern limb of Couiza syncline and cuts obliquely through the Eocene sequence. Before Antugnac, a roadside quarry provides good exposures of Lower Ilerdian alveolines limestone. Marl outcrops of early to middle Ilerdian age (exposures in roadside ditches) have gained favour among wine growers. The scarp face overlooking Antugnac from the SW exposes late Ilerdian coastal sandstones (containing nummulites) and the base of the continental Palassou conglomerate sequence. Roughly 2 km after Antugnac, the nummulitic sandstone can be examined at close quarters at a section by the junction with the D152; 1 km further along, at a bend in the road (spot elevation: 352 m), a conglomerate-filled palaeochannel has been interpreted as a feeder to the basal Palassou sequence.

Stop 12. Lower Palassou sequence around La Serpent

Unlike the Iberian side of the Pyrenees, where Paleogene conglomerate beds have been eroded into spectacular rock pillars and escarpments, the Palassou conglomerates are difficult to observe because the cementing matrix is often silt and clay, and thus of moderate strength. Furthermore, given the oceanic climate, all Palassou outcrops are obscured by lush vegetation, particularly in the Pique l'Ordy massif, where access to the stratigraphy would otherwise be ideal. The base of the sequence ('Fa beds') lines the farmed depression of La Serpent, where marl containing interlayers of pebble beds (D_{99}: 10–25 cm) is dominant. The clasts consist exclusively of Mesozoic limestone and Albian black sandstone. A roadside outcrop (42°57'48"N, 02°11'30"E) illustrates this, with an additional content of small, poorly rounded pebbles of Paleozoic quartzite (beige with a yellow patina; Fig. 10.17B). The main mass of conglomerates, which contains a much larger proportion of pebble beds, underpins the steep, wooded scarp above the village (Fig. 10.17A; 'poudingues des Serres inférieures' in Crochet 1991). D_{99} increases to 40–60 cm towards the summit. Compositional assemblages comprise almost exclusively of Mesozoic source rocks: various limestones including marble, and Albian marl and sandstone—the latter abundantly represented in some beds, and all of this documenting early denudation of the uppermost outcrops of the North-Pyrenean Zone. Debris from the Paleozoic basement, including granites, were reported by Crochet (1991) to appear in the topmost beds. You can inspect sections along the forest track rising towards Le Léouc and Pech Sarda (poor midslope exposures NW of the village), or at the end of the road to Bauzeille Haute, just before reaching the farm (lenses with exclusively Albian content). This lowermost Palassou sequence has been dated on the basis of its vertical stratigraphic continuity with marine Ilerdian beds and, where those same beds extend farther west, by the inclusion of a fossil *Lophiodon* attributed to the Cuisian or Lower Lutetian (palaeontological sites of Sibra and Saint-Quentin, biozones MP 9–10, in the Hers River catchment; Astre 1958; Sudre et al. 1992) (see Part I, Fig. 2.3 therein).

From La Serpent, aim for Bouriège; at the bridge over the Corneilla River and elsewhere below the village, note the south-dipping marine nummulitic sandstones. The D121 cuts downward into the stratigraphy, with the village of Roquetaillade built on a hogback ridge in Ilerdian limestone. Below Roquetaillade, drive down the small country lane signposted to Borde Longue, which takes you to Pic de Brau

Fig. 10.17   Outcrops of Palassou unit 1 (Ilerdian–Lower Lutetian). A Homoclinal scarp in conglomerate beds at La Serpent. Below the village, floodplain siltstone (marl) is dominant and gravel beds are rare (e.g., 'Fa beds'); above the marl, the 'Serres inférieures' conglomerate (Crochet 1991) forms a discontinuous but massive escarpment. **B** Conglomerates of the 'Fa beds' along the D52. Note dominance of grey and beige limestone (L) and Albian sandstone (Sd). Chronostratigraphic lettering follows nomenclature of the Quillan geological sheet

(654 m, also conspicuous because of its windfarm). After reaching the summit, walk to the edge of the escarpment in Thanetian beds 150 m east of the watch tower.

Stop 13. Pic de Brau

The view from the top is 360°, but the escarpment edge is the best vantage point for appreciating the structure of this anticline. Its north limb is a steep-dipping reverse fault where the Sub-Pyrenean frontal thrust intersects the modern topography; the entire Paleogene sequence is upturned against a core of Paleozoic outcrops, particularly the belt of Ilerdian sandstone and marine marls (Fig. 10.18). To the north, a scarp-and-vale landscape locates the southern limb of the Limoux–Villefloure syncline, which is filled with conglomerates belonging to Palassou unit 2. The substantial relief of the northernmost escarpment hints at Neogene faulting in addition to scarpfoot stripping of the conglomerate sequence, outcrops of which reach maximum altitudes of 475 m among the Malepère hills, north of Limoux.

The horseshoe-shaped ridge below the wind turbines is a hogback facing east in Thanetian limestone. Bed outcrops across the summit plateau form conspicuous stripes on aerial photographs and show that the low-gradient land surface cross-cuts the upturned Thanetian beds almost entirely. This erosion surface continues on the other side of the Aude gorge across the undulating Saint-Salvayre plateau (~750 m), where outcrops of the Paleozoic Mouthoumet basement, sub-Campanian silcretes, and Campanian fluvial sandstone beds (Alet Fm.) are all erosionally truncated. The slope of the exhumed sub-Campanian unconformity is 11%, whereas the gradient of the post-Eocene erosion surface rising eastward from Pic de Brau to Saint-Salvayre is ~2.6%. Projecting this gradient to the summit envelope of the Mouthoumet massif, it becomes clear that the Mouthoumet range-top surface is post-Eocene rather than a vestige of the exhumed sub-Campanian stratigraphic unconformity. It remains difficult, however, to establish whether the Mouthoumet summit surface belongs to generation S or generation P1. The town of Alet-les-Bains sits at the centre of a NE–SW graben, which has preserved the Cretaceous and Paleogene sequence. Outcrops of Paleozoic basement occur on the left bank of the Aude, with estimated fault throws of ~150 m. This extensional structure affects the entire Eocene sequence, Palassou included, in the SW. Faulting was thus post-orogenic, but nonetheless relatively ancient given that its deformed rock sequence is cross-cut by the Saint-Salvayre erosion surface. From this it can be inferred that this graben formed coevally with the Oligocene grabens of the Western Mediterranean seaboard, explored along Itinerary 1 through the eastern Corbières.

Return to the D121 via the track going down Bac de Brau ravine to the Corneilla River, which (like the Aude) cuts an epigenetic gorge through the anticline. A quarry at the road junction offers good exposures of the Thanetian beds. Geological maps (sheet names: Limoux and Quillan; different authors, different dates) deliver conflicting interpretations of the detailed stratigraphic continuum in this area (discussion in Crochet 1991*). The valley widens out in outcrops of Ilerdian marine marls below a homocline in nummulitic sandstone (the Sub-Pyrenean thrust is not visible at this location). The most recent interpretation for this area (*Crochet 1991*)*

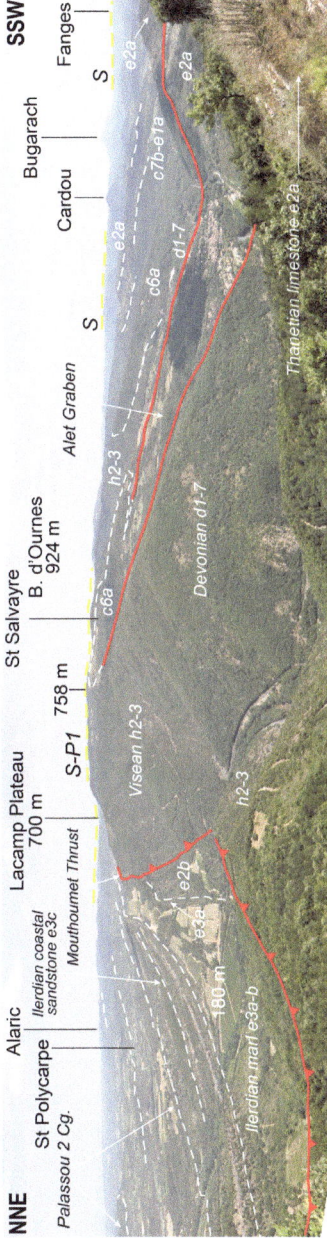

Fig. 10.18   Landscape views from Pic de Brau looking east. Basement-cored fold; north limb is subvertical and overthrust. Contact with cover sequence is highlighted as a continuous band of Ilerdian marine sandstones, with underlying blue marls. The Limoux syncline to the north is filled with synorogenic conglomerates of Palassou unit 2. Note that the summit surface (S and P1 indistinguishable) cross-cuts all tectonic structures, synorogenic Paleogene folds and post-orogenic extensional basin (Alet graben) included. Chronostratigraphic lettering follows nomenclature of the Quillan geological sheet

suggests synsedimentary progressive unconformities within the Palassou sequence, with the middle sequence onlapping the lower one and entirely covering it. This can be traced all the way to existing outcrops of the marine facies of confirmed Ilerdian age. Conglomerate and sandstone beds carved into a succession of homoclinal ridges northward from Magrie would thus correspond to Palassou unit 2. Good exposures are rare, but nonetheless more common that for the lower unit. Three of these are presented as part of **Stop 14** *(Fig. 10.19A).*

Stop 14. Middle Palassou sequence around Limoux
The single most notable feature of this middle unit is its abundance of pebbles from the Paleozoic basement, and thus from an Axial Zone source area: arkosic sand, pebbles of granite, schist, and vein quartz. The sequence is thick (at least 1 km) in the Limoux–Villefloure syncline, with individual sandstone and conglomerate beds sometimes exceeding 10 m. D_{99} attains 40 cm, but falls to 2–3 cm in distal areas north of Carcassonne. The age of this sequence (Lutetian to middle Bartonian) is documented by a number of palaeontological sites containing *Lophiodontidae* (Depéret 1910, Sudre et al. 1992; see also Robinet et al. 2015). Thus the Saint-Benoît-L'Ambrone (Lambrolle) site near the base of sequence 2, 15 km west of Limoux, belongs to biozones MP 11–MP 12, i.e., early to middle Lutetian (Sudre et al. 1992), although other authors link it to the Issel faunas (MP 14, in Crochet 1991; MP 14 is now placed in the Upper Lutetian: see Vandenberghe et al. 2012). Near the top of the sequence, in the hills south of Carcassonne and particularly at Métairie Grande, the age correlates with MP 15–16 (Bartonian; *Lophiodon Lautricense*). In more distal positions, except at Issel (MP 14) and Azillanet (MP 10) near the base, poor age constraints for the Palassou sequences come from the sites of Douzens Nord (MP 12–13), and from La Livinière 1, Pépieux, Cesseras, Siran, Olonzac, etc. (MP 10–14; see map in Danilo et al. 2013).

A first exposure occurs at the col just before reaching Magrie. The abandoned quarry at Palajo, situated on the dipslope of the Ilerdian sandstone homocline, was exploiting yellow marl and clay—possibly the 'Fa beds' of Palassou unit 1. However, similar-looking yellow marls with thin interbeds of grey sandstone, exposures of which (dipping 37°N) occur in other abandoned quarries north of the col, could be basal outcrops of Palassou unit 2. This inference is based on the observation that, at the top of the hill and thus higher up in the stratigraphic sequence, three sandstone palaeochannel fills (particularly the middle one: 42°01'38"N, 02°12'96"E; Fig. 10.19B), contain pebble lenses (pebble diameters: 1–4 cm) documenting abundant inputs from Paleozoic basement source rocks (poorly rolled quartz and quartzite, schist, aplites, etc.).

Fig. 10.19 Outcrops of Palassou unit 2 in the Limoux syncline. A Homoclinal scarpland topography in the conglomerate sequence at Magrie. Background: Mouthoumet massif and its summit surface. **B** Conglomerate beds south of Magrie; close-up shows an abundance of small pebbles (2 to 3 cm in diameter) of Paleozoic material including granite (Gr), schist, quartz, and black jasper. **C** North of Magrie, bed of coarse conglomerate containing abundant pebbles of granite (Gr), quartz (Q), Paleozoic schist, but also limestone (L). **D** Beds of sandstone and gravel NW of Limoux, Bois Grand cuesta scarp face, in a section along the D623. **E** Base of palaeochannel fill at the D623 exposure, showing an exclusive content of unrolled granite clasts

The scarp above Magrie to the north offers a large number of good exposures. They can be reached by following the trail up to La Serre, which starts after the bridge (Fig. 10.19A, C). Make sure you branch off to the left at the first junction on your way up the trail. Conglomerates start to appear from the slope base, presenting a succession of three upward-coarsening beds, where D_{99} rises from 10 to 20–30 cm. Granite clasts are frequent, petrographically diverse, and associated with Paleozoic schist and an abundance of grey, and often pitted, limestone pebbles. Midslope, three interbeds of white lacustrine or palustrine limestone occur amid thick sandstone palaeochannel fills and overbank floodplain siltstone (these siltstone beds are light ochre and grey, with dark ochre and red streaks and mottles). Dips to the north attain ~24°. The D121 cuts through this sequence 50 m before the junction with the D118. At this location, pebble sizes often exceeding lengths of 10 cm, including among clasts of coarse-textured granite exhibiting a facies typical of the Madrès–Quérigut batholith—and thus certainly sourced from that area.

A roadside section along the D623 occurs at Côte de la Carestie, NW of Limoux, on the way out towards Castelnaudary. The facies here is more distal and from further up in the stratigraphy than at Magrie. Several palaeochannel fills one to several metres thick contain greyish white, cross-bedded arkose interbedded with light grey to pink marls (Fig. 10.19C), with the entire sequence dipping 15° to the NNW. Pebbles on the channel floor are small (2–4 cm), poorly rolled, and consist almost exclusively of granite. The most remarkable roadside section occurs off the main itinerary, ~8 km north of Limoux on the D118, after the village of Cépie. Over a distance of ~1 km, the exposures display perfectly horizontal outcrops of Palassou conglomerate, here >50 m thick.

The abundance of debris from the Axial Zone at such distances from the mountain belt north of the Mouthoumet massif strongly suggest that the Mouthoumet anticline itself was inexistent or subdued at the time, and thus the massif remained buried beneath Palassou units 1 and 2 throughout their depositional histories. Within Palassou unit 2, decreasing dips from the base to the top of the sequence document synsedimentary tectonic deformation of the foreland. The most intense deformation of this basement fold, of its cover sequence, and of the overlying Palassou foreland sequence thus occurred most likely in post-Bartonian time. The uplift was subsequently bevelled by an erosion surface during early Neogene time (generations S and/or P1, both here probably indistinguishable because their planes merge at a very shallow angle along the hinge zone between the rising mountain range and the Aquitaine Basin depocentres), before regional uplift of both the mountain belt and the retro-foreland drove the incision of canyons by the Aude and other rivers.

A return journey along the D118 allows you to cut through the Alet anticline via *the Aude gorge. At La Tuilerie, some good exposures of the upturned Ilerdian sandstone can be seen along the right bank of the river. An outcrop of Paleozoic basement occurs in the large meander loop at Bois de Cazes, but lacks exposures of its contact with the cover rocks. The road through the gorge stays within outcrops of Visean schist, whereas on the right bank you can observe an anticline in Devonian limestone. Several bedrock-channel reaches suggest that the river is incising is bed, probably in response to crustal uplift, even today. Before you reach Alet, note that normal faults have made the Cretaceous cover reappear among roadside outcrops. From Alet-les-Bains, a 12-km return trip to Saint-Salvayre should allow you to appreciate features of the erosion surface on the flank of the Mouthoumet massif until now only appreciated from afar (e.g.,* **Stop 11**). *As you rise up, you begin to get good views over the Alet graben. Aim beyond Saint-Salvayre as far as the telecommunications towers to capture the vast panorama and get a sense of the erosion surface (*Figs. 10.18 and 10.20*), which cuts across structures in massive Devonian limestone as well as across the base of the Campanian fluvial sandstone and quartz conglomerate unit. From the plateau, views reach across to the Montagne Noire (north), to Naurouze (west), and over the full sweep of the Pyrenees from Mt. Canigou to Mt. Valier. Return to Alet, then onwards to Couiza and Quillan. After Alet, the road crosses a small canyon cut in white Rognacian/Danian limestones.*

Fig. 10.20 Views from de Saint-Salvayre erosion surface, looking west. Foreground: the erosion surface cross-cuts the basement rock fabrics and structures as well as its unconformable Campanian envelope of conglomerate beds. Middle ground: Thanetian limestone beds at Pic de Brau. The Mouthoumet fold disappears westward beneath the mass of Palassou conglomerate ('Cg', units 1 and 2), which form the topography at and around Pique l'Ordy. To the SW, the north-Pyrenean mountain front (Sault Plateau) is backed by the high ranges of the Axial Zone (Carlit and Aston massifs). The Montagne de Tabe (also named Saint Barthélemy massif), which forms a large enclave of Hercynian basement within the North-Pyrenean Zone, is obscured by the clouds

10.4   Sault Plateau

Leave Quillan on the D117 signposted to Foix. At Col de Portel, you should get a good view of the Quillan basin and the relief around the NPFT: Forêt des Fanges and homoclines in Albian sandstone (including Serre de Bec, 1037 m). The north branch of the NPFT strikes straight past the western side of the col, where Aptian limestones override the Upper Thanetian red marls and their conglomerate interlayers; 8 km farther west, onwards from Puivert, the same limestone unit overrides the Ilerdian nummulitic sandstones, red marls, and lower Palassou conglomerates ('Col de la Babourade beds', Fig. 10.21B).

From Col de Portel, take the D613 to Espezel. The road soon reaches Combe des Bans, an erosional bench in Urgonian limestone. This low-gradient surface is the distal extremity of the Coudons palaeovalley, which correlates altitudinally with generation P2. Planal de la Coste, a cave with its entrance nearby at 700 m (i.e., 427 m above the Aude river channel; Fig. 10.14C), is a former Vauclusian spring in which speleologists have discovered a plug of quartz sand at a depth of –50 m below the entrance. Such quartz-rich material can only originate from basement outcrops to the south of the limestone plateau. Between Coudons and La Peyre, the road follows a sinuous dry valley, its floor itself hosting a number of karstic depressions (large sinkholes, and a small polje at Centenières), and with its valley head situated at 895 m on the Sault Plateau.

Stop 15. Espezel and the Sault Plateau

Espezel offers good views of the landscape from the D1029 near the stadium, which is also a good starting point for a roam across the plateau (1 hr minimum) in order to look at the karstic landforms as well as the clastic deposits summarised and illustrated in Box 10.2.

Box 10.2 The Pays de Sault: flagship limestone karst and clastic piedmont
The Pays de Sault is the most extensive limestone karst plateau of the Pyrenees (Goron 1941a; Birot 1937; Lagasquie 1963; Calvet et al. in press). Its very flat surface, currently incised ~200 m by the V-shaped Rebenty canyon, is more complex than its prima facie appearance as just an ancient karst polje because the low-gradient land surface cross-cuts not just the Urgonian limestones but also the Albian marls and sandstones, as well as a few outcrops of Paleozoic basement such as the micaschist and migmatite outcrops around Rodome (Figs. 10.12, 10.14 and 10.21). South of the Aude River, analogues of this erosion surface occur in granite ca.

1050–1100 m at Escouloubre and Roquefort-de-Sault–Buillac, a very flat etch basin (Fig. 10.14B). They all belong to generation P2 (Pliocene) and indicate deep inroads into the Axial Zone of relief-reducing (rather than relief-enhancing) processes a that time. The Sault Plateau nonetheless also functioned as a karst polje during the Quaternary, and still today does so residually, with two small areas still drained out seasonally by active swallow holes: such is the case in the east with the small Rébounédou stream, which flows past the village of Belvis; and in the west around Coumeilles, where flat floors extending north and west of Espezel are countersunk 20 m below the tread of the main erosion surface.

The main plateau surface (Fig. 10.12) is riddled with dolines and covered by the remains of an ancient alluvial fan of the Rebenty containing two generations of deposit, both of which are in most ways comparable to the topmost deposits of the Lannemezan megafan. The main alluvial unit (Fig. 10.22A–C) is gravel (D_{99}: 10–20 cm) containing rolled Silici-clastic pebbles transported from Paleozoic outcrops in the upper Rebenty catchment, and/or from imbricate thrust units in the North-Pyrenean Zone (schist, sandstone, quartz, quartzite, rare gneiss), interbedded with thick lenses of beige-coloured clay. These deposits can be examined in the road embankment below Espezel stadium as well as over the surface of arable fields. They were analysed by Calvet et al. (in press) in a 4-m-deep trench ~885 m a.s.l. at the north end of the plateau, below the track to Belfort at Montplaisir (the top of the formation on the plateau occurs at 902 m; its minimum thickness is thus ~20 m). On the Rebenty valley side, thicknesses are at least 20 m below spot elevation 904 m. These gravels have produced a $^{26}Al/^{10}Be$ maximum burial age of 3 Ma, and magnetostratigraphic measurements in the clays record inverse polarity (Calvet et al., in press). The existence of these alluvial deposits documents a space on and around the Sault Plateau where river directionality was hesitant because valleys at that time were un- or weakly incised. Uplift-driven incision of the piedmont thus occurred after 3 Ma. To the north, bedload would enter the dry valley at Col du Chandelier, ~855 m a.s.l., above the town of Puivert; the abandoned valley strikes in the direction of the Blau River valley (which flows to the Ariège and thus the Atlantic). To the east, bedload was funnelled towards the Rebenty, a tributary of the Aude (thus flowing to the Mediterranean); deposits have left a trail also identified at Quirbajou, ~800 m a.s.l.

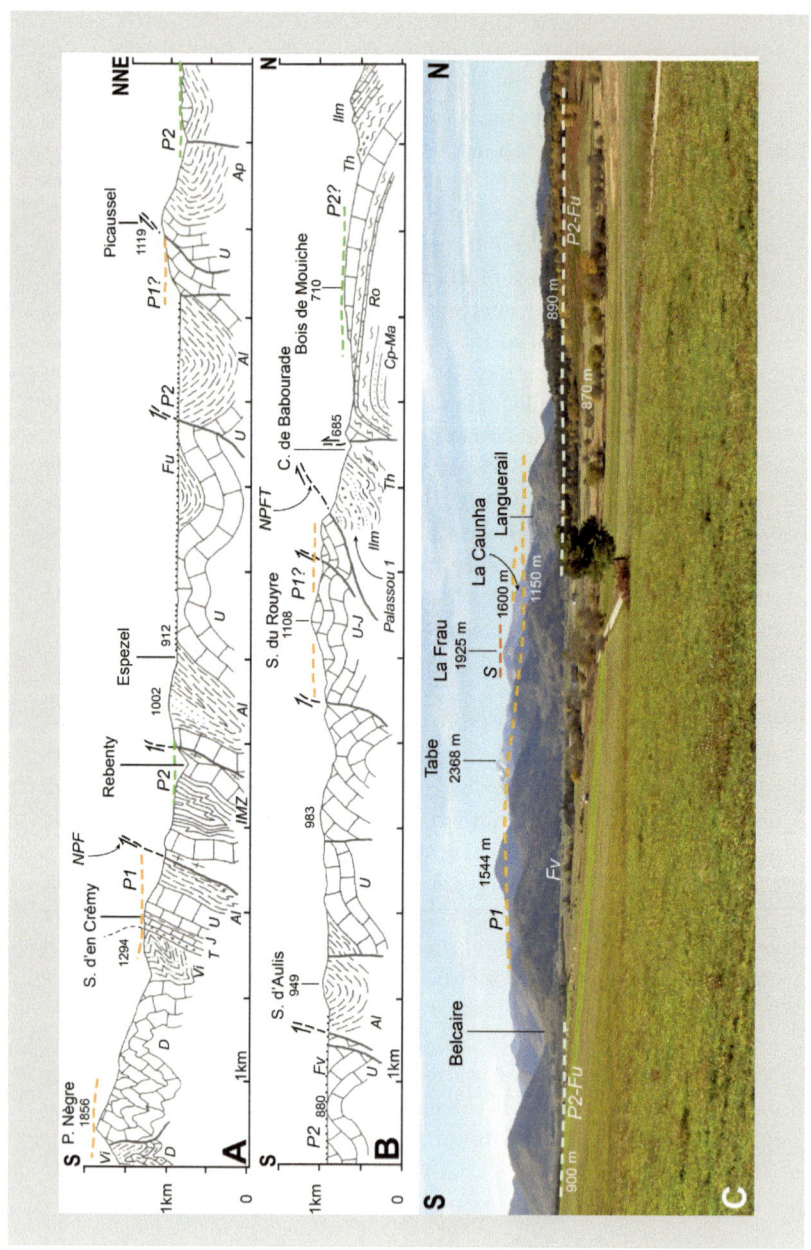

◄**Fig. 10.21   Sault Plateau: structure, landscape, and landforms. A, B** Geological cross-sections through the central area of the plateau. Note how surface P1 rises up towards the Axial Zone, and spatial extent of low-gradient erosional landforms belonging to generation P2. Stratigraphy and lithology: D-Devonian limestone; Vi-Visean flysch; IMZ- Internal Metamorphic Zone (marble, breccia, hornfels); T- Triassic; J- Jurassic; U- Cretaceous limestone (Urgonian facies); Al- Albian marls; Ap-Aptian marls; Cp–Ma- Campanian–Maastrichtian ('grès de Labarre'); Ro- Rognacian (i.e., Garumnian) facies; Th- Thanetian limestone and red marls; Ilm- Ilerdian (marine), capped at Col de Babourade by Palassou unit 1. **C** Looking west from the D613 north of Espezel. Foreground, and below the forest to the NW: surface P2 and its allochtonous, silicate-rich alluvial cover (noted 'Fu' on geological maps), here deposited by an ancestor of the Rebenty River. Background: floor of a young polje lined with locally-sourced alluvium ('Fv' on maps). All the hills in Urgonian limestone have been bevelled by P1 at approximately uniform elevations (1100–1200 m), but P1 rises to the west, as its tread approaches the Montagne de Tabe. Note possible vestiges of summit surface, S, on the limestone massif of La Frau. All altitudes in metres

A second alluvial unit, consisting of much coarser gravel, covers the previous unit near the fan apex (Fig. 10.22D). It has been substantially thinned, but quartz boulders 1.5 to 2 m in length, displaying thick, yellow weathering rinds, are still present. Large numbers of such boulders can be seen resting on limestone around Bouyches, 500 m NE of the stadium; access is possible via the trail through farmland which crosses the plateau.

The higher-relief topography bordering the plateau to the north and west is a Neogene cone karst landscape (Figs. 10.21C, 10.22A). As a result, remains of range-top erosion surface S are hard to detect, at least other than by imagining tangential planes to the highest limestone ranges, where nonetheless small bevels have been preserved. The tread of pediment P1, at places preserved at elevations of ~1050 m in the landscape, forms the floor of the oldest generation of karst cones; very ancient tunnel caves (Trou du Vent de Pédrou, Belvis, Lacaune de Coudons, etc.; Figs. 10.12, 10.14) appear to record the first stages of its incision before the subsequent lull in base-level change, which promoted the development of surface P2 prominent around Espezel.

From Espezel, the D29 takes you down into the Rebenty valley, then the D20 leads you back up to the eastern portion of the plateau. Here, generation P2 of low-gradient landforms is a wide, hummocky bedrock corridor devoid of alluvial deposits, cut in marble and hornfels outcrops of the Internal Metamorphic Zone (IMZ) as well as in basement lithologies of the Bessède massif (hosting the villages of Mazuby, Rodome

Fig. 10.22   Alluvial fan sequence capping the Sault Plateau surface around Espezel. **A** Tread of P2 and its alluvial cover ('Fu' on geological maps) NE of Espezel; note in the background the cone karst landscape typical of the Sault Plateau. **B** Section through alluvial formation 'Fu'; alternating clay beds with gravel-filled palaeochannels. **C** Deep weathering of pebble beds showing the clay-rich matrix and contours of internally rotten allochtonous schist pebbles. **D** Quartz boulder from the upper boulder beds at the fan apex, near Les Bouiches

and Galinagues). At Aunat, a small polje (named Vallée-sans-Issue, or blind valley) has been etched into the tread of P2, suggesting a further increment of uplift and lowering of groundwater flowlines (Fig. 10.14A).

When following the D29 towards Fontanès-de-Sault, the road cuts through the small massifs of Bois du Linas and Sarrat du Pinsart, whose summits have been bevelled by P1 (1230–1160 m; Figs. 10.12, 10.14A). The bevels cross-cut the IMZ marbles, the NPF, and Mesozoic strata (Triassic to Aptian) belonging to a non-metamorphic portion of the Axial Zone cover sequence, here crushed into a steeply upturned syncline. After Col des Aychides, the road crosses the NPF and descends into the small polje of Barencou (970 m). Its flat floor is consistent with generation P2 and hanging above the Aude gorge.

Stop 16. Successive generations of endokarstic cave levels in the Aude gorge
Up to 6 epiphreatic cave levels have been counted in the Devonian limestone along the gorge below the tread of surface P1. The caves are distributed over a vertical distance of 570 m (Figs. 10.12, 10.14A). The uppermost cavities only contain bedded clay and silt deposits, suggesting a paragenetic evolution of that network. Downward from Escaliers Cave (+110 m), in contrast, the sediment fill clearly consists of alluvium deposited by an ancestral Aude River, comprising a diversity of quartz, schist and granite pebbles with an abundance of sand. Cavity fills of this kind can be observed along the D29 in the last limestone rockface after the tunnels, e.g., (i) in Berger Cave; (ii) in the small F12 Cave (+75 m, in the road embankment itself), which contains large granite cobbles; and (iii) on the southern backslope of the limestone ridge at Déserteur Cave (+95 m; Fig. 10.23). On the right side of the gorge, Aguzou Cave (+80 m) is a kilometre-long cave system parallel to the modern river; it formed as a result of the Aude River recapturing its own channel. It contains thick sand- and gravel-rich fluvial deposits, and can be visited as part of an organized 'subterranean safari'. Cave TM 71 (+15 m) documents a similar scenario on the left side of the gorge (another scenario of self-piracy by the Aude River of its previous channel), but is a lot more recent (late Pleistocene) and still hydrologically functional in its downstream part. TM 71 is protected as a wilderness area and is thus off limits to all visitors.

Once you have reached the Aude valley, follow the D118 towards Axat as far as Gesse, then turn left onto the D20 to join up with the Sault Plateau. At Bessède, a small road will take you directly to Le Clat over Col de Triby and Col de Pradel.

Fig. 10.23   **Déserteur Cave (+95 m, Devonian limestone) between Usson and Fontanès-de-Sault. A** Epiphreatic cave and former (early Pleistocene) swallow hole of the Aude River. Water solution features and scallops on the cave ceiling; mean passage diameter: 2 m. **B** Large alluvial cobbles preserved on the cave floor (granite, quartz, schist)

After Col de Pradel, the route takes you across the tread of an undulating surface—arguably P1—ca. 1100 m a.s.l. Stop at the signposted viewing point, just below Col de Dent.

Stop 17. Panoramic view from Col de Dent

Towards the west you will recognise the Sault Plateau area described in the context of **Stop 15** and **Box 10.2**, with the treads of P2 extensively managed as cropland. On the horizon, altitudes increase around the Tabe massif (Saint-Barthélemy, 2368 m), where surfaces S and P1 are poorly preserved but appear to have been raised substantially by late Neogene uplift to altitudes of ~1900 m in the case of the Frau massif (Urgonian limestone; Fig. 10.12). Towards the WSW, the Bessède depression in the foreground is an anticlinal valley cut in basement gneiss, micaschist and albitites below the Devèze–Forêt de Miayra (1395 m) homocline in Jurassic limestones. Farther back in the middle ground, vestiges of P1 are evident at the top of Sarrat du Pinsart–Bois du Linas (1200 m), two massifs previously crossed en route before **Stop 16**. In the background, the skyline of the Axial Zone, rising abruptly to more than 2000 m, includes Pic d'Ourtiset (1933) and Pic de Bentaillole (1965) in Devonian limestone, and the plateau around Col de Pailhères, where occurrences of P1 occur quite extensively ca. 2000 m a.s.l. (see Itinerary 7). Such an abrupt vertical offset between vestiges of P1 at Bois du Linas (1200 m) and at the Pailhères plateau (2000 m) implies uplift of the Axial Zone during the Neogene along a fault or flexure at the boundary between the Paleozoic basement and its cover rocks. This tectonic structure has not yet been identified or named, but farther west it becomes indistinguishable from the NPF, where a similar Neogene vertical offset has been inferred on the basis of well-preserved triangular faceted spurs along the Camurac–Montaillou scarp face (Calvet 1996).

After Col de Dent, the road soon reaches Pas del Corps and Le Clat.

Stop 18. Le Clat: Capturing links between cave systems and dry valleys
Pas del Corps is a 300-m-wide, flat-floored palaeovalley occurring at 1133 m a.s.l. (Figs. 10.12, 10.14A, 10.24A), and situated at the contact between the massive breccia of the IMZ and the Jurassic and Cretaceous limestones forming the envelope of the basement-cored Bessède massif. Small, allochtonous vein-quartz pebbles can be found strewn across the Pas del Corps valley floor. The valley extends to the other side of the Aude valley towards the Madrès massif—from where it originated and to which it connects via the large wind gap at Col de la Malagrède. This fossil landform ostensibly belongs to generation P1 because, on either side of it, geological structures in the Devèze–Miayra and Forêt de la Serre (1300 m) massifs display erosional bevels which can only be correlated with range-top surface S given their summit positions (Fig. 10.12). The two very open, concave-sloped valleys under cultivation, which extend at elevations of ca. 1000 m on either side of Le Clat village, are sculpted out of massive IMZ breccia. Given their elevations relative to the vestiges of S (300 m) and P1 (150 m) mentioned above, these landforms around Le Clat could represent specimens of generation P2.

Fig. 10.24 A Miocene endokarstic cave system: Clat Cave and its allochtonous alluvial fill. A View to the southeast from eastern Sault Plateau. Ancestral surface, S, bevels the summits of the Serre and Miayra massifs, cross-cutting and thinning the Urgonian limestones and IMZ breccia. Pas del Corps wind gap correlates altitudinally with generation P1. **B** Entrance to Clat Cave in pink IMZ breccia, + 574 m above the Aude River thalweg. **C** Inside Clat Cave: epiphreatic solution features on the ceiling, and thick, allochtonous alluvial deposits on the floor. **D** Ceiling karren deeper in the cave. **E** Detail of the alluvial fill, showing deeply weathered granite and schist pebbles

Optional trailwalk to Clat Cave (1.5 hr return)

Clat Cave is located at the site marked 'Pas en Toulouse' on IGN maps (42°47′02.3″N, 02°09′13.3″E). Like others in the area, this cave is carved out of massive IMZ pink breccia, and is situated at exactly the same altitude as the Pas del Corps dry valley. Its characteristics thus indicate that it belongs to the same generation of landforms. You can reach the site from the road below the village of Le Clat (park in the first bend in the road, from where you also get a good view of the Aude gorge). From there, follow the dirt track descending into the hollow then taking you across it and, after the first bend, ascends obliquely across the meadows towards the forest edge; from here, the track is better marked and will take you to the foot of the rock face ca. 1100 m, then directly to the cave entrance situated 574 m above the modern Aude channel bed (Fig. 10.14B). Clat Cave is a large, horizontal cavity with an 8-m-high oval entrance and is abundantly filled with sediment. It displays ceiling channels and spectacular pendants, both of which indicate a protracted paragenetic / epiphreatic history of semi-immersion (Fig. 10.24B, D). Entering past the abandoned lime kiln you will soon see the abundant pebble deposits, mostly of silicate rocks (dark schist, granitoids, some unrolled and pitted quartz clasts; Fig. 10.24C, E). Mean pebble sizes range between 2 and 5 cm, but a minority reach 8–10 cm or more. The material is clearly of fluvial origin, with vertical exposures also showing sand lenses. Total alluvial deposit thicknesses are about 3 m. Post-depositional clast weathering was intense, with a number of kaolinized—and therefore totally friable—pebbles. Depositional conditions, at first indicative of vadose and fluvial environments, continue upward in the stratigraphy with thick, bedded clays displaying red and ochre patterns. This paragenetic stage of evolution filled the cave to its ceiling. The sediment was nonetheless partly flushed out at a later date, until the lag became covered and sealed by a speleothem which is steeply inclined towards the cave entrance. The clays were extracted during the Middle Ages for tile production or pottery.

Box 10.3 Regional landscape evolution in the Pyrenean retro-foreland
Pas del Corps palaeovalley and Clat Cave complement our understanding of regional landscape evolution already sketched out in Galamus gorge (**Stop 7**) and on the Sault Plateau (Box 10.2). The dry valley and cave are vestiges of a Miocene landscape, and understanding its evolution provides a good basis for recapitulating and summarising, as follows, the successive stages of regional landscape evolution in the Pyrenean retro-foreland (Fig. 10.25):

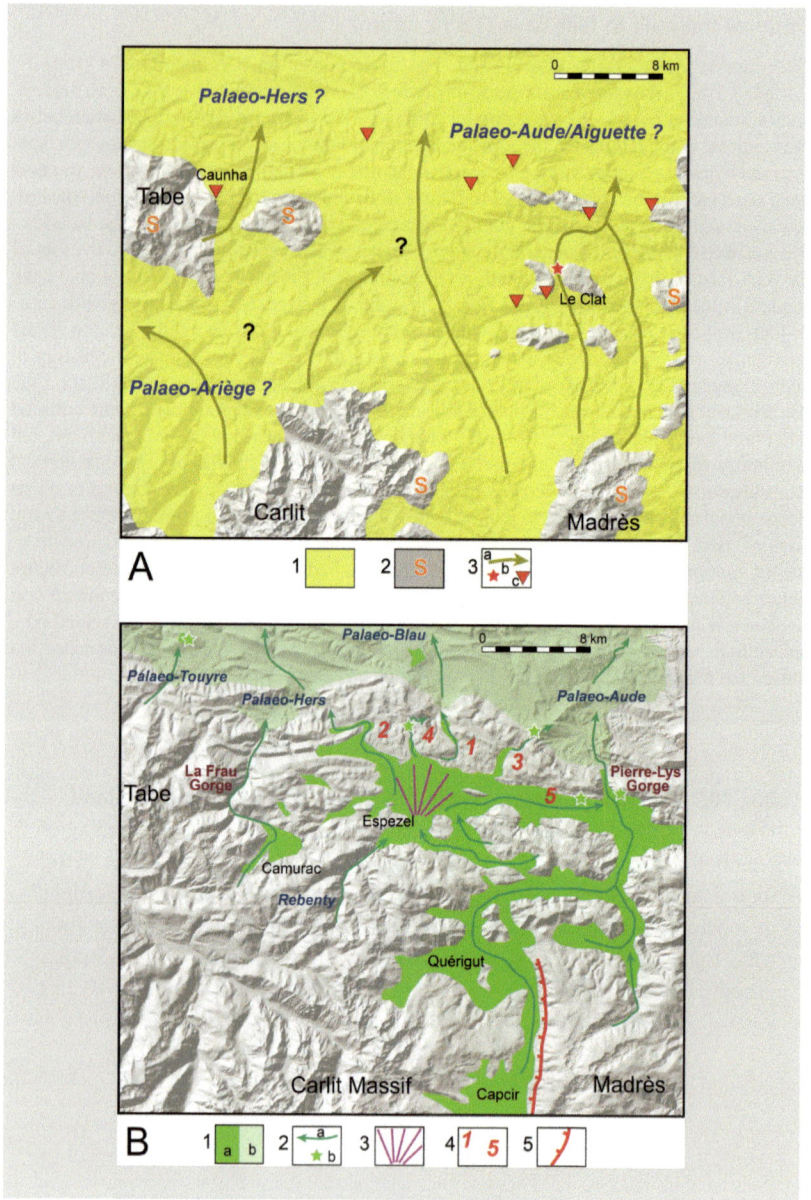

◄**Fig. 10.25   Palaeogeographic evolution of the Sault Plateau landscape.** A Population of middle Miocene landforms, generation P1. Excepting Le Clat, reconstructed palaeofloodways are hypothetical. Key to symbols and ornaments—**1**: pediment P1, which connected at the time to a base level set by the marine and continental molasse deposits of the Aquitaine Basin (see also Part I). **2**: upstanding residual topography above the treads of P1, some of it retaining vestiges of summit surface, S. **3**: some diagnostic features of P1; **3a**: hypothetical palaeodrainage trajectories through the epigenetic gorges of the Aude at Pierre-Lys and the Hers at La Frau (both rivers already in existence at that time); **3b**: residual presence of allochtonous alluvial bedload deposits, usually enriched in siliciclastic debris because of the long-term resistance to postdepositional weathering of such clasts (e.g., at Pas del Corps; likewise on the Frau massif around La Caunha, and on the northern ridgetops flanking the Sault Plateau); **3c**: near-horizontal epiphreatic cave systems connecting altitudinally to P1 or lying at elevations slightly lower than P1 (in the latter case thus recording a resumption of post-P1 uplift and incision). Some of these cave systems contain allochtonous alluvium transported from the North-Pyrenean Zone, and perhaps from the Axial Zone. **B** Population of Pliocene landforms, generation P2. Key to symbols and ornaments— **1**: pediments; **1a**: landforms preserved or restored to their positions in the present-day montane setting; **1b**: extension of P2 to the confines of the Aquitaine Basin; only the mass of Palassou conglomerates around Pique l'Ordy appears to have formed a group of residual hills. **2**: palaeodrainage trajectories; **2a**: main river floodways (the modern drainage network was already mostly established by that time); **2b**: whether in caves or on the land surface, allochtonous gravels rich in silicate rocks from the Axial Zone. **3**: Rebenty mountain-front alluvial fan (burial of the Espezel erosion surface). **4**: palaeovalleys striking across the Sault Plateau by decreasing order of antiquity (1: La Malayrède; 2: Col de la Croix des Morts; 3: Lapeyre–Coudons; 4: Col du Chandelier; 5: Quirbajou). **5**: Late Neogene fault scarps guiding the upper course of the Aude River. Floodways progressively converged on the two old-established water gaps of Pierre-Lys (Aude, joining the Mediterranean) and La Frau (Hers, joining the Atlantic), and the drainage conformation observed today progressively stabilised thereafter

1. where present, the treads of pediment P1 lie 400 to 500 m below the summit surfaces (generation S) on the Madrès, clearly visible to the south. P1 found opportunities to widen out on certain lithologies such as susceptible granitoids of the Axial Zone, but was restricted to flat-floored corridors and palaeovalleys in massive limestone outcrops, whether Devonian to the south of the Aude valley, or Mesozoic here and elsewhere. P1 is also usually surrounded by structural landforms (varyingly bold and steep homoclinal scarps), some of them retaining patchy vestiges of summit surface, S.

2. The rivers, which, at the time of relative base level stability when P1 was expanding, transported bedloads of relatively modest calibre from the more elevated Axial Zone, would lose discharge to the limestone karst, contributing to form epiphreatic tunnel-caves (such as at **Stop 18**, Clat Cave), just above the water table. The very low topographic gradients in these piedmont landscapes promoted sediment aggradation within the cavities of the endokarst but also most likely over the eroding surfaces themselves.

3. A similar conformation of circumstances and landscape geometries was repeated at the time of generation P2, generating planar landforms of lesser extent and typically occurring ~200 m below the treads of P1 at any given location.

4. The cone karst around the Sault Plateau developed under the warm and varyingly subtropical conditions of the Neogene, through cycles P1 and P2.

5. The continuity and energy of post-P2 canyon incision demonstrates a radical change of geomorphic and tectonic regime, which can be explained by regional crustal uplift and climatic oscillations during the late Neogene and Quaternary—particularly the glacial cycles prevalent in the Axial Zone.

Get back on the D83 to return to Axat over Col de Nadieu and through Artigues. The road descends into Rebenty syncline, where the thick sequence of Albian black shales displays additional outcrops (e.g., around Pech de Nadieu) of Cenomanian to Turonian flysch (limestone, sandstone, occasional breccia). From Axat, the journey back to Perpignan on the D117 follows the Rebenty–Fenouillèdes syncline, which extends uninterrupted for 80 km from the Pays de Sault all the way to Vingrau (an area explored in Itinerary 1).

References

Astre G (1958) Découverte de Lophiodon à Queilhe et remarque à propos des terrains lutétiens au sud de Mirepoix. Mémoires de l'Académie des Sciences, Inscriptions et Belles-Lettres de Toulouse, pp 23–32

Birot P (1937) Recherches sur la morphologie des Pyrénées orientales franco-espagnoles. Baillière Édit., 318 p

Calvet M (1996) Morphogenèse d'une montagne méditerranéenne: les Pyrénées orientales. Documents du BRGM, Orléans, 255:1177 p

Calvet M (1999) Régime des contraintes et volumes de relief dans l'Est des Pyrénées. Géomorph Rel Proc Environ 3:253–278

Calvet M, Gunnell Y, Farines B (2015) Flat-topped mountain ranges: their global distribution and value for understanding the evolution of mountain topography. Geomorphology 241:255–291

Calvet M, Monod B, Regard V, Bès C, Gayet JC (in press) L'évolution géomorphologique du Plateau de Sault au Neogene. Karstologia

Crochet B (1991) Molasses syntectoniques du versant nord des Pyrénées: la série de Palassou. Documents du BRGM, 199:387 p

Danilo L, Remy JA, Vianey-Liaud M, Marandat B, Sudre J, Lihoreau F (2013) A new Eocene locality in southern France sheds light on the basal radiation of Palaeotheriidae (Mammalia, Perissodactyla, Equoidea). J Vertebr Paleontol 33:195–205

Depéret C (1910) Note sur quelques gisements nouveaux de Lophiodontidés de la région de Carcassonne. Bulletin de la Société d'études Scientifiques de l'Aude, 104–127

Fallourd S, Poujol M, Boulvais P, Paquette J-L, De Saint Blanquat M, Rémy P (2014) In situ LA-ICP-MS U-Pb titanite dating of Na-Ca metasomatism in orogenic belts: the North Pyrenean example. Int J Earth Sci 103:667–682

Fonteilles M, Leblanc D, Clauzon G, Vaudin JL, Berger GM (1993) Carte géologique de la France au 1:50,000, sheet Rivesaltes (1090), Orléans, BRGM. Handbook by Berger GM, Fonteilles M, Leblanc D, Clauzon G, Marchal JP, Vautrelle C (1993), 119 p

Goron L (1941a) Les Pré-Pyrénées ariégeoises et garonnaises, essai d'étude morphogénique d'une lisière de montagne. Privat, Toulouse, 886 p

Gunnell Y, Calvet M, Brichau S, Carter A, Aguilar JP, Zeyen H (2009) Low long-term erosion rates in high-energy mountain belts: insights from thermo- and biochronology in the Eastern Pyrenees. Earth Planetary Sci Lett 278:208–218

Lagasquie JJ (1963) Le relief calcaire du Plateau de Sault. Rev Géogr Pyrén Sud-Ouest 34:11–32

Lagasquie JJ (1984) Géomorphologie des granites, les massifs granitiques de la moitié orientale des Pyrénées françaises. CNRS Édit., Toulouse, 374 p

Laumonier B (2008) Les Pyrénées pré-hercyniennes et hercyniennes. In: Canérot J, Colin JP, Platel JP, Bilotte M (eds) Pyrénées d'hier et d'aujourd'hui. Atlantica, Biarritz, pp 23–35

Messaoudi H, Debat P, Lelubre M (1993) Structure et mode de mise en place du complexe plutonique hercynien de Millas (Pyrénées-Orientales, France). C R Acad Sci Paris II 316:145–150

Meurisse M, Michaux J, Sigé B (1969) Un remplissage karstique à Micromammifères du Miocène inférieur à la Serre de Vergès, près St-Arnac (Pyrénées orientales). C R Somm Soc Géol Fr 5:166–168

Motus M, Denèle Y, Mouthereau F, Nardin E (2020) Mechanisms of continental break-up: tectonic, stratigraphic and structural constraints from a preserved distal rifted margin (Agly massif, eastern Pyrenees). EGU 2020–Sharing geoscienceonline

Ournié B (1987) Présentation spéléologique du massif des Fanges et du chaînon du Roc Paradet (Fenouillèdes, Aude et Pyrénées Orientales). Karstologia 10:1–6

Philip H, Bousquet JC, Escuer J, Fleta J, Goula X, Grellet B (1992) Présence de failles inverses d'âge quaternaire dans l'est des Pyrénées: implications sismotectoniques. C R Acad Sci Paris II, 314:1239–1245

Robinet C, Remy JA, Laurent Y, Danilo L, Lihoreau F (2015) A new genus of Lophiodon-tidae (Perissodactyla, Mammalia) from the early Eocene of La Borie (Southern France) and the origin of the genus Lophiodon Cuvier, 1822. Geobios 48:25–38

Sudre J, de Bonis L, Brunet M, Crochet JY, Duranthon F, Godinot M, Hartenberger JL, Jehenne Y, Legendre S, Marandat B, Remy JA, Ringeade M, Sigé B, Vianey-Liaud M (1992) La biochronologie mammalienne du Paléogène au Nord et au Sud des Pyrénées: état de la question. C R Acad Sci Paris II 314:631–636

Sylvander M, Monod B, Souriau A, Rigo A (2007) Analyse d'un essaim de sismicité (mai 2004) dans les Pyrénées orientales : vers une nouvelle interprétation tectonique du séisme de Saint-Paul-de-Fenouillet (1996). C R Geoscience 339:75–84

Ternois S, Odlum M, Ford M, Pik R, Stockli D, Tibari B, Vacherat A, Bernard V (2019) Thermochronological evidence of early orogenesis, eastern pyrenees, France. Tectonics 38:1308–1336

Vandenberghe N, Hilgen FJ, Speijer RP, with contributions by Ogg JG, Gradstein FM, Hammer O (2012) The Paleogene period. In: Gradstein et al. (eds), The Geologic Time Scale, Elsevier, pp 855–921

Yelland A (1991) Thermo-tectonics of the Pyrenees and Provence from fission-track studies, PhD thesis (unpubl.), University of London

Itinerary 6. Elevated Ranges
and Interior Basins of the Axial Zone:
The Conflent Basin and its
Surrounding Massifs

11

From Perpignan to Font-Romeu (164 km; additional 160 km; 2 days, and 4 or more when trailwalks included). See Fig. 11.1.

Itinerary 6 follows the Têt valley and, by doing so, opens a window onto the Paleozoic geological structure of the Axial Zone (lithostratigraphy, Hercynian tectonics). Paleogene tectonics, in contrast, has left limited direct evidence of the Alpine orogeny in this part of the Axial Zone, suggesting that the Hercynian basement in the eastern Pyrenees was not very intensely deformed by basement nappes and thrusts during the Alpine orogeny. The tectonic record can mostly only be inferred by proxy methods rather than direct observation, mainly from rock cooling histories based on low-temperature thermochronology. The main focus is thus on Neogene tectonics and geomorphological evolution. Both aspects are conspicuous in the landscape of the Conflent half-graben, its sediment fill, the remarkable faceted spurs along its boundary fault scarp, and the successive generations of erosional benches and pediments cut into the flanks of the surrounding footwall massifs. Another focus will be the incision chronology of the Têt valley, which can be measured by dating successive generations of alluvial deposits trapped in limestone caves on canyon flanks near Villefranche, as well as alluvial terrace deposits that extend deep into the Axial Zone along the Têt River. These have been used as proxies for estimating rates of neotectonic uplift of the mountain range.

© Springer Nature Switzerland AG 2022
M. Calvet et al., *Geology and Landscapes of the Eastern Pyrenees*, Geoguide,
https://doi.org/10.1007/978-3-030-84266-6_11

A
Pays de Sault
Axat
St-Paul
Fenouillèdes
Agly
Aude
2000
1000
500
Elevation in m
Perpignan
Têt
detail B
Ille
Roussillon
Réart
Madrès
Capcir
Conflent
Prades
Carlit
Aspres
Olette
Têt
Tech
Mt-Louis
Font-Romeu
Cerdagne
Canigou
Céret
Carança
Vallespir
Albères
Roc de France
Prats-de-Mollo
0 10 20 km
42° 50'
42°30'
2° 00'
2° 20'
2° 40'
3°
B
C. Jau
Tribes
Tarerach
Mosset
Marcevol
S.2
S.1
Rodès
Madrès
Vinça
S.3
Nohèdes
Prades
Sansa
Coronat
Villefranche
St-Michel
Serrabona
S.10
S.4
Taurinya
Aspres
Serdinya
S.11
S.7
S.5
Olette
Escaro
S.6
Vernet
Canaveilles
S.8
S.12
S.9
C. Fins
Cortalets
St-Martin
Thuès
Sauto
S.13
S.15
Fontpédrouse
Mt-Louis
S.14
Py
Mariailles
Canigou
Mantet
Pomerola
Carança
1 a b
2 a b c
0 5 10 15 km
42°40'
42°30'
2° 10'
2° 20'
2° 30'

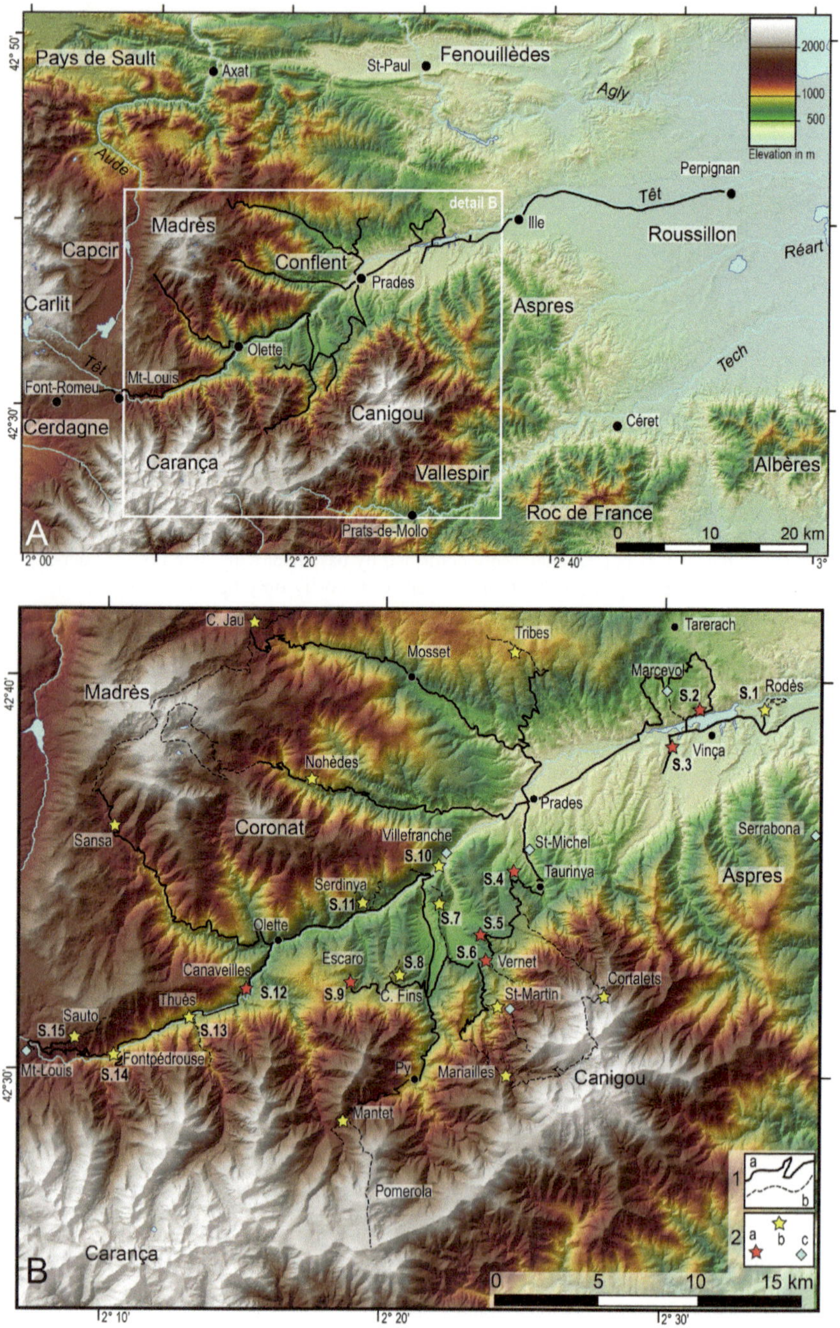

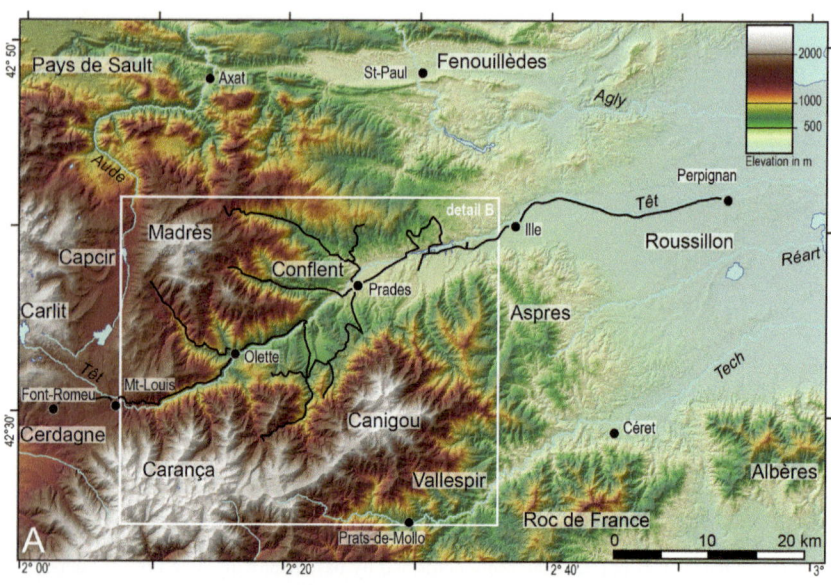

◀**Fig. 11.1** **Itinerary 6: route map and overview. A General location map. B Detail of routes**. Key to symbols and ornaments—**1**: roads and trails. **1a**: roadways, irrespective of road rank, size, or quality. **1b**: trailwalks. **2**: stopping points (outcrops, exposures, landforms, or landscapes). **2a**: Roadside stops. **2b**: roadside stops involving a short walk, or marking the start of a more major trailwalk (potentially including stopovers in a mountain refuge). **2c**: cultural stops. Labels S.1 to S.15 locate the sequence of roadside stops

---

Box 11.1 The Conflent half-graben: general overview

The Conflent is a half-graben bounded by a bold fault scarp on its southern margin (Têt Fault) where steep range-front river catchments have carved out triangular faceted spurs (Figs. 11.2 and 11.6). The relief of these tectonic landforms increases westward along the mountain front. Pediments belonging to generation P1 occur at elevations of 2300–2400 m in the Canigou–Carançà massif, but surface S is absent from the most elevated summit areas (Fig. 11.4). Minor faults separate the northern margin of the half-graben from the Madrès and Roque Jalère massifs (Millas–Quérigut granite). Among these massifs, vestiges of surfaces S and P1 are well preserved and form comparatively gentle topography sloping southward towards the Conflent depocentre. The Devonian limestones of the Hercynian Villefranche syncline, which divides the Conflent graben into two depocentres (upper and lower Conflent), are cross-cut around Mt. Coronat by a well preserved erosion surface known as Pla dels Horts. This pediment surface has been tilted to the SE and thus extends across elevations ranging between 1450 and 1250 m. It is a specimen of the population of Miocene pediments P1, which in this area form a bench extending 200–400 m below the summit surface, S (Fig. 11.2B, 11.4, see also Fig. 10.12, Itinerary 5).

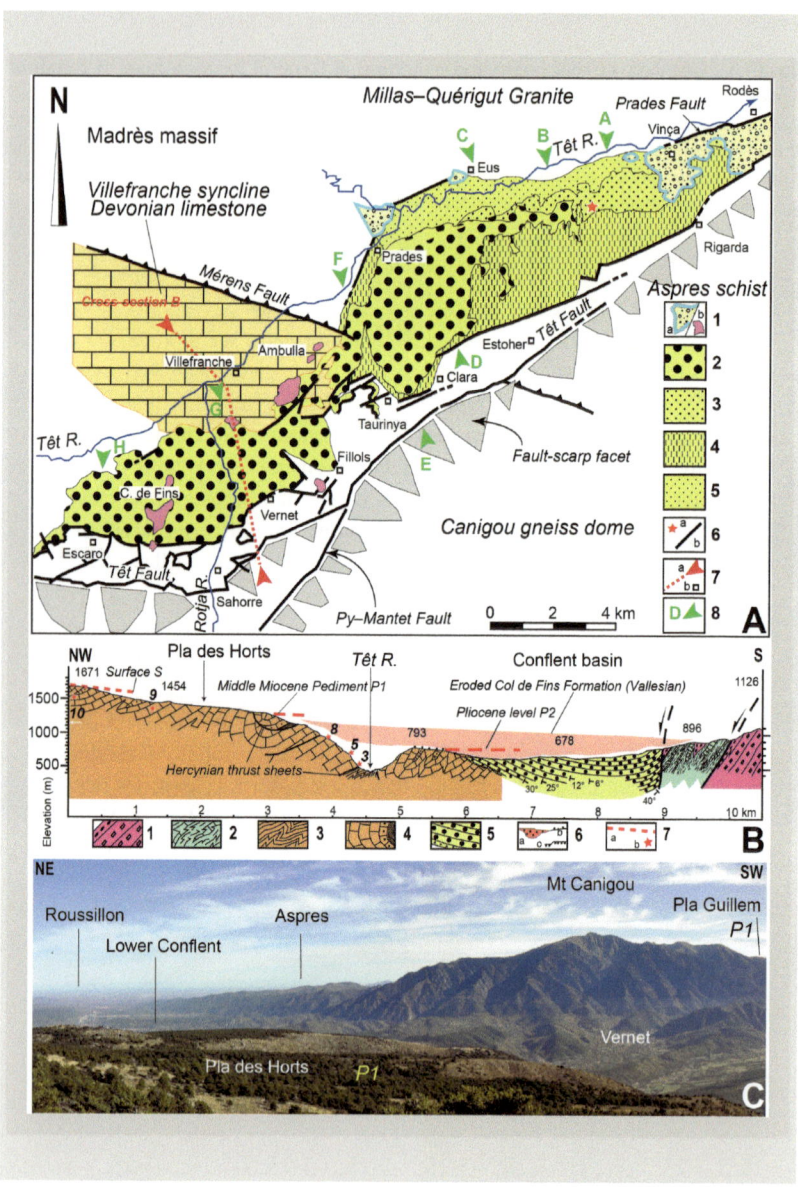

◄**Fig. 11.2 The Conflent, a Neogene half-graben. A** Geological sketch map. Key
to symbols and ornaments – **1a**: Pliocene sequences (continental and marine), with
outer limit of incision features (ravinement surface) into the Messinian sequence
boundary; **1b**: Coll de Fins Formation, probably Vallesian (i.e., ~ Tortonian); **2**: Escaro
Formation (contains large boulders of augengneiss), Lower Burdigalian; **3**: Lentilla
Formation (fluvial), Lower Burdigalian; **4**: Codalet Formation (contains schist), Aquitanian to Lower Burdigalian; **5**: Marquixanes Formation (arkose with granite cobbles),
Aquitanian; **6a**: Espira mammalian fossil site, Lower Burdigalian; **6b**: Main Neogene faults (named); Mérens Fault is indicated with triangles (Hercynian and Alpine
thrust fault, with evidence of tectonic inversion under an extensional regime during
the Neogene); **7a**: section in panel B; **7b**: main towns; **8**: geological cross-sections
shown in Fig. 11.3. **B** Geological cross-section of the Conflent Basin at Villefranche;
Key to symbols and ornaments—**1**: augengneiss; **2**: micaschist and marble; **3**: schist
(Cambrian to Ordovician); **4**: Limestone (Devonian) and schist (Visean); **5**: Escaro
Formation (Lower Burdigalian), with thin layer of limestone breccia and calcrete at
base; **6a**: Coll de Fins Formation (probably Vallesian); **6b**: boulders (either part of
Vallesian Coll de Fins Fm., or Pliocene age) resting on the limestone plateau surface;
6c: Quaternary terraces T1 and T2. **7**: key landforms; **7a**: Neogene erosion surface; **7b**:
main groundwater karstic cave system containing exogenous alluvial deposits (key to
caves: 10: Fornells, 9: Roquefumade, 8: Balcon, 5: Notre-Dame de Vie, 3: Faubourg).
Altitudes in metres. See also Figs. 11.23 and 11.24. **C** Panoramic view of the Conflent
Basin from the griotte limestone quarry at Roquefumade (situated below elevation
spot 1671 m in panel B). Note asymmetric morphology of the half-graben, considerable extent of erosion surface P1 at Pla dels Horts, faceted spurs along the southern
master fault, with the Aspres and Canigou footwall blocks rising above it. Source:
after Calvet (1996) and Calvet et al. (2015), modified

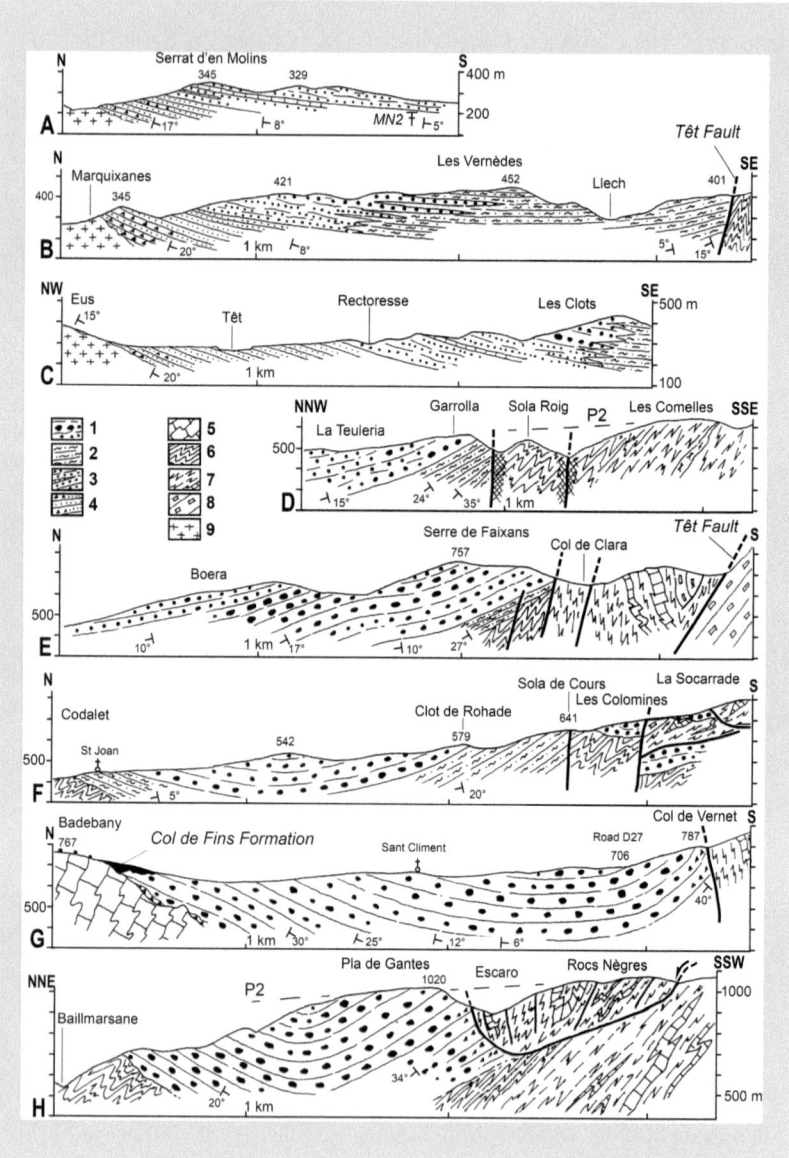

◀**Fig. 11.3 Detailed geological cross-sections through the Conflent Basin.** Neogene sequences – **1**: Escaro Formation. **2**: Codalet Formation. **3**: Lentilla Formation. **4**: Marquixanes Formation; Coll de Fins Formation in black. Axial Zone basement outcrops – **5**: limestone (Devonian). **6**: low-grade schist (Cambrian to Ordovician). **7**: micaschist and marble (Ediacaran to Cambrian). **8**: augengneiss. **9**: granite. Rajouter: Cross-sections a to h ar located in Figure 11.2A. Altitudes in metres. Source: after Calvet (1996), modified

Three different fill sequences occur in the Conflent Basin. The main depositional units were defined by Pannekoek (1935), whose terminology was maintained in subsequent literature (Oele et al. 1963; Bandet 1975; Calvet 1996).

1. The Lower Miocene sequence is ~1 km thick. At its base, the Marquixanes Series is a 200–500-m-thick sequence of Aquitanian age (23–20.4 Ma). It consists mainly of arkose, red clay, and granite boulders supplied by the Millas massif immediately to the north; the sequence is very similar to the lower 'Série Rouge' of the Céret Basin (see Itinerary 2). The overlying Lentilla Series (200–400 m minimum thickness) is an alluvial parasequence supplied by catchments situated in the Canigou massif and in the upper reaches of the Têt River. Observable facies include cobble-filled palaeochannels and flood-basin sand and silt sequences containing thin limestone beds. An early Burdigalian age (~20 Ma) based on the biostratigraphic characteristics of a mammalian fossil assemblage has been obtained for the lower third of the sequence (Baudelot and Crouzel 1974; see **Stop 2**). The Lentilla and Marquixanes units transition laterally to the Codalet Series. The distinctive feature of this third series is its reddish colour, its fabric typical of processes involving hyperconcentrated flow, and clast assemblages almost exclusively supplied by proximal schist outcrops located in the Aspres massif (with a few secondary outcrops to the NW of Prades, supplied by catchments in the schist and hornfels outcrops around Catllar, Nohèdes and Urbanya). Progressing westwards upstream, the Codalet and Lentilla units transition to the Escaro Series. This unit is a 900 m-thick debris-flood sequence containing huge boulders. It was supplied by the gneiss outcrops of the adjacent Canigou–Carança footwall upland. Large olistoliths, some of them 1 km across or more, descended the northern flank of Mt. Canigou and currently form interlayers within the Escaro Series. The Miocene deposits are faulted and tilted towards the

Conflent Basin, with dips generally steepest on the southern limb (~30–40°) (Figs. 11.2B and 11.3). Unlike the south side of the Conflent Basin, the north side was comparatively inactive during the early Miocene and underwent limited uplift. This is documented by clast provenance among the Miocene sedimentary sequences of the Conflent, which indicates limited contribution from the northern catchments.

►**Fig. 11.4 Geomorphological map of the Axial Zone, eastern part.** Key to symbols and ornaments—**1**: main post-orogenic erosion surfaces; **1a**: late Oligocene to Aquitanian range-top surface (or paleoplain), generation S; **1b**: middle Miocene pediment population, generation P1. **2**: Age constraints; **2a**: AFT cooling ages; **2b**: AFT and (U–Th)/He cooling ages (double dating). **3**: late Neogene partial erosion surfaces (generation P2: pediments, bedrock straths, flat-floored etch basins in granite). **4**: abandoned (dry) valleys (flow directions indicated by arrows); **4a**: flat-floored, low-sinuosity erosional corridors of P1 affinity; **4b**: palaeovalleys (generation P2). **5**: major structural landforms; **5a**: homoclinal scarps in the folded marine cover sequence (Paleogene); **5b**: Neogene normal fault scarps, flexures; **5c**: direction of tilt of surfaces S and P1. **6**: limestone endokarstic landforms (abandoned passages, often containing relict allogenic alluvium); **6a**: landforms accordant with the tread of subaerial surface P1; **6b**: landforms accordant with the tread of subaerial surface P2; **6c**: landforms associated with post-P2 valley incision. Limestone karst features: active landforms; **6d**: sinking river; **6e**: main karstic springs. Altitudes in metres

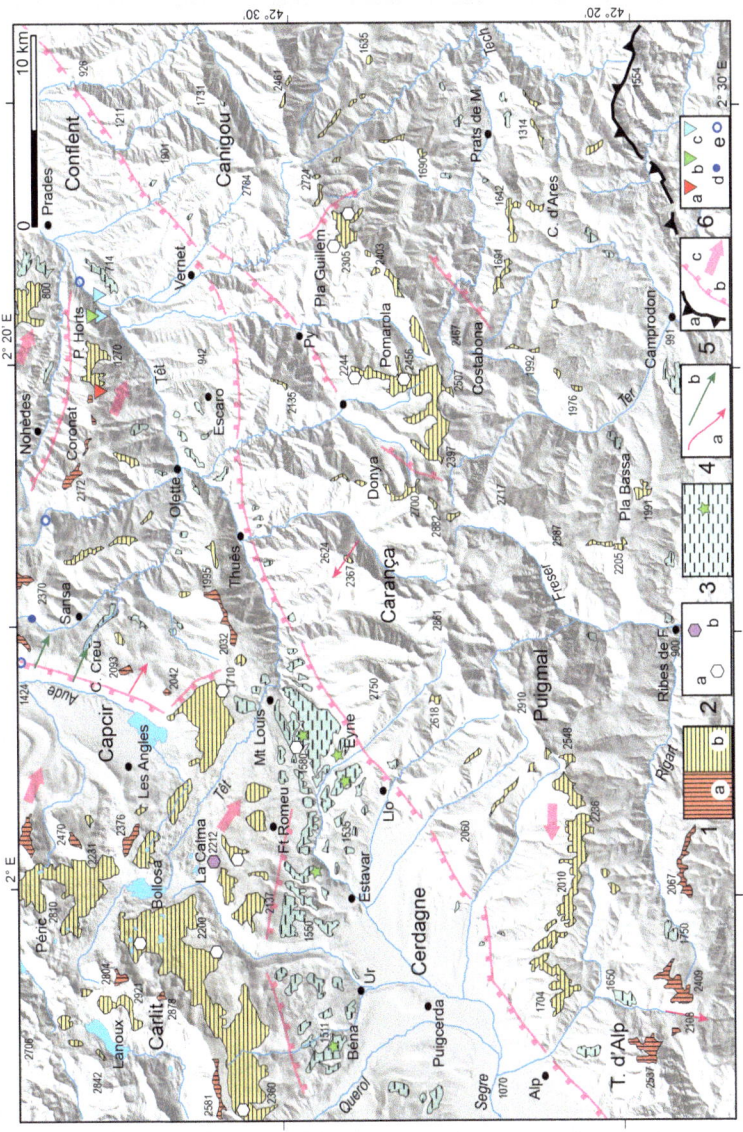

2. A sequence of probable late Miocene age has also been identified and was named Coll de Fins Formation. by Calvet (1996) and Calvet and Gunnell (2008). It was largely stripped away at the time of the Messinian Salinity Crisis sea-level lowstand, and its extant remains rest unconformably on the Lower Miocene units and the Paleozoic basement (Fig. 11.2A, B). Although never directly dated, these deposits are comparable to those encountered in the Cerdagne Basin, another half-graben farther to the west, also controlled by the Têt Fault but younger (late Miocene age, containing mainly Vallesian and Turolian fill deposits, i.e., ~11–6 Ma), and examined along Itineraries 7 and 8. At the Coll de Fins type locality (**Stop 8**), outcrops are rich in clay and sandy clay beds, with few bedload channel fills. Unlike the Escaro Series, which contains more feldspar, the sand fraction is anomalously rich in quartz (70–75% compared to 45–55% in the Lower Miocene sequences). Among the clay mineral assemblages, kaolinite is dominant over smectite (20–40% smectite, compared to 60–80% in the Lower Miocene units). Where present, quartz pebbles and cobbles are well rounded and abundant, whereas they are small, weakly rounded and infrequent in the Escaro Formation. Two-mica granite clasts eroded from the core of the Canigou metamorphic dome are present in this formation but absent from the Lower Miocene deposits. This provenance chronology age-brackets the onset of fault-scarp growth along the northern boundary of Mt. Canigou, where the two-mica granite is currently exposed from the base of the scarp to elevations of ~2 km.

3. The Pliocene sequence of the Conflent Basin is a westward continuation of the Roussillon fill sequence examined in Itinerary 2. The segment of the Têt palaeovalley that formed as a result of stream incision during late Messinian sea-level fall was subsequently buried by a Zanclean fill sequence, with extant foreset and topset beds. The topset beds, as at Ille-sur-Têt (see Itinerary 2), are high-energy fluvial deposits containing rounded pebbles and boulders, mainly gneiss and granite. In the Conflent Basin, no Pliocene marine or deltaic deposits are known to occur west of Vinça. The northern shores of this drowned Pliocene valley, or ria, have since been disturbed by tectonic activity along the north branch of the Têt Fault, known here as the Prades Fault (Calvet 1996). The marine-to-continental boundary within the Pliocene stratigraphy, detected 200 m below sea level in a borehole at Canet in the Roussillon Basin (see Itinerary 2), occurs above sea level at +240 m near Rodès, and +280 m near Vinça.

The regional slope of this major sequence boundary, which itself is probably diachronous from west to east, provides a minimum estimate of the magnitude of post-Pliocene uplift of the Axial Zone. Magnitudes of uplift are also documented by the current altitude of the top of the Neogene deposits in the Conflent Basin, which rise from 360 m near Vinça, to 750 m near Prades, to 1100 m at its westernmost termination (Fig. 11.2).

11.1 Madres Massif and Lower Conflent Basin

Leave Perpignan on the N116. This dual carriageway follows the treads of alluvial terraces T0 (Holocene and historical, along the base of terrace riser T2), then T1 upstream of Saint-Féliu-d'Avall. After Ille-sur-Têt, the road cuts through exposures of T1 (poorly weathered and very coarse-textured at L'Ermita), and a roadcut embankment through a vestige of T2 (weathered and rubified debris at Els Escatllars; see Fig. 7.26, Itinerary 2), which has yielded two imprecise ESR ages (174 ± 44 ka and 259 ± 90 ka) broadly compatible with MIS 6 (Delmas et al. 2018). Roughly 500 m before reaching Coll de Terranera (Col de Ternère), a low-elevation pass as the topography steps up from the Roussillon Basin to the Conflent Basin, the road progresses through low hills in Devonian limestone (local name: Les Pedreres) buried by residual patches of detrital Pliocene deposits in the south, and further north by marine and continental deposits (local name: Les Terralbes) corresponding to Zanclean fill of the drowned Messinian estuary (Clauzon et al. 1990; Calvet 1996). Disused quarries at the col show the deltaic foreset beds consisting of fine grey sand capped by sandy topset beds; these become more gravel-rich (fluvial facies) above 250 m a.s.l. Fossil oyster shells (often still attached to bedrock) and marine foraminifera have been reported from here (Bandet 1975; Magné 1978), but the key outcrop was probably destroyed when the road was widened. After Coll de Terranera, turn right into the village of Rodès, where you can use the parking area outside the school.

Stop 1. Rodès and the lower epigenetic gorge of the Têt (1.5 hrs round trip)

The Têt River initially flowed across the top of the Pliocene fill sequence, now substantially thinned by fluvial erosion. The active Conflent boundary fault to the south and the predominance of sediment influx from southern tributaries pushed the Pliocene ancestor Têt channel to the northern edge of the basin. Pediment P2, which at the time was forming on the flanks of the northern massifs (Itineraries 2, 4 and 5), graded to a local base level corresponding to the top of the Pliocene fill sequence. The asymmetry of the basin fill in the Conflent Basin explains the epigenetic character of the Quaternary Têt valley along the northern edge of the basin, where it cuts a sequence of four gorges through the underlying granite, leaving proud the buttes at Marquixanes and el Castelló upstream of Vinça, the bedrock hill at Sant Pere de Bell Lloc (where Vinça dam is positioned), and Rodès gorge. These last two are the narrowest, and still actively incising; they were entirely cut during the Quaternary. The other two gorges farther upstream are wider, cluttered by Pleistocene alluvial deposits, and display an alluvial rather than a bedrock channel. Here they are probably reoccupying the Messinian palaeovalley that was subsequently buried by the Pliocene deposits.

Walk into the village and, before you cross the bridge, follow the trail signposted to 'Gorges de la Guillera'. It follows the Corbère irrigation canal. The spectacular potholes in the granite (Fig. 11.5) reveal exposures of rounded enclaves of darker diorite in the lighter biotite granite groundmass (magmatic melange). Arches of the Perpignan Canal, rebuilt in the 14th century, are anchored to the rock face on the left bank. The canal used to cross the river over an aqueduct, one residual arch of which still stands on the right bank. At the gorge exit, take the trail to your right, which follows the Prades Fault (no fresh exposures here of a fault plane, however), and walk up to the castle. From the castle you will get good views of the Roussillon Basin and Ille-sur-Têt badlands to the east (Itinerary 2), and of the Conflent Basin to the west, with Vinça gorge in the foreground (closed off by Vinça dam) and the adjacent Romanesque chapel of Sant Pere de Bell Lloc.

Continue your journey along the N116, where the road skirts around a group of hills with exposures of deltaic Pliocene deposits forming cut-and-fill features into the underlying Miocene deposits. An optional 30 min return trip to the chapel of Sant Pere offers a good, wide-angle view over the Conflent Basin (Fig. 11.6A). The Vinça reservoir is used for autumn floodwater retention and summertime irrigation. By September it is practically empty. Escomes beach, on the shores of the reservoir, provides good views of the Pliocene deltaic deposits (Fig. 11.7): the fluvial topset beds are rubified, the foreset beds are pebble-rich, and the bottomsets consist of

Fig. 11.5   Epigenetic gorge of the Têt at Rodès (la Guillera), and its bedrock channel knickpoint. A Bedrock channel in granite during a small flood event (October 2019). Note ruins of the Perpignan Canal, rebuilt in the 14th century. **B** Same site during the moderate flood of January 2020 (700 m³/s)

Fig. 11.6 Panoramic views of the lower Conflent Basin. A View from Sant Pere chapel. Note structural asymmetry of the half-graben, highlighted by tilted erosion surfaces P and S descending from the Madrès–Coronat massif as opposed to the abrupt, fault-controlled range front in the south. The drainage pattern shows a reverse asymmetry, with the Têt cutting its succession of epigenetic gorges along the structurally less abrupt north edge of the basin, and the longest tributary rivers flowing in from the Canigou–Carança massif, across the steep mountain front in the south and northward across the basin. These counter-intuitively opposing patterns suggest a complex geomorphological history. **B** View from the path leading up to Marcevol, at elevation spot 430 m above the D13 bridge. Note the triangular faceted spurs and continuous range front, whether in schist (Aspres massif) or gneiss (Canigou massif). Red arrow points to El Castelló epigenetic gorge. Star locates the Lentilla/Espira Miocene mammalian fossil deposit

blue clay containing lignite and are best observed in the south near the lake by the campsite, at Escomes. This area is officially the westernmost outcrop of Pliocene deposits, but further upstream it is likely that boulder-bed outcrops below the village of Eus and between Prades hospital and Catllars, where they appear as channel fill (ravinement surface) in the Codalet Fm., are also of Pliocene age (Fig. 11.2A). After Vinça, the Serrat d'en Molins (a small hill) marks where Miocene outcrops begin. After crossing the Lentilla River, turn left towards Espira-de-Conflent on the

Fig. 11.7   Inner-deltaic marine Pliocene deposits at Vinça. Very coarse-textured sediment (cobbles and pebbles). Exposures of light grey foreset beds show up in the local gullies

D25 and stop 400 m after the turning. Here you will get good views to the east by walking up among the orchards.

Stop 2. The Miocene sequence of La Lentilla

The view towards Serrat d'en Molins (Fig. 11.8A, and cross-sections A and B in Fig. 11.3) shows the successive outcrops of the Marquixanes and Lentilla series, both dipping at an angle of ~8° to the south. The Marquixanes unit (Fig. 11.8B) can be accessed under the N116 bridge or at the eastern base of Castello Hill, where an outcrop occurs in a gully beneath terrace deposits T1, but is only visible at times when the Vinça reservoir is empty. Outcrops of the Lentilla unit can be inspected while ascending a dirt track up the NE flank of Serrat d'en Molins, and along the ridgetop.

The mammalian fossil deposit of Espira is not easily reached. It occurs 1 km upstream below elevation spot 277 m, in the left bank of the Lentilla River, ca. 260–264 m a.s.l. The fossils are dispersed among layers of fine silty sand interspersed with greyish-green clay lenses (Fig. 11.9), and cut-and-filled in the north by gravel. Its early Burdigalian age (biozone MN 2, see Fig. 11.11, Itinerary

Fig. 11.8 Continental Lower Miocene at La Lentilla. A General view of the exposed sections at Serrat d'en Molins. Note steady southerly dips, unlike the Pliocene sequence at Vinça, which has remained horizontal. **B** Marquixanes Fm., arkose sequence containing beds of unrolled granite clasts; this section is located under the N116 bridge over the Lentilla River. **C** Exposed section further upstream in the Lentilla riverbank, at Correc de Botás ('correc': ravine, in Catalan). Note lateral transition from the Lentilla to the Codalet (reddest tones) formations. This exposure lies above the Espira fossil mammalian deposits

1), has been established on the basis of a dozen indicator species (a carnivore: *Broiliana nobilis*; a bear-like species: *Amphycion major*; some artiodactyls: *Coenotherium miocoenicum, Dorcatherium crassum, Amphitragulus aurelianensis, Lagomeryx meyeri*; lagomorphs, *Prolagus vasconiensis, Lagopsis peñai, Amphilagus*; and rare rodents: *Sciurus fissurae, Steneofiber depereti*; see Baudelot and Crouzel 1974). Two other sites are known among the hills higher up, ca. 330 m a.s.l. (Els Baixos), allowing the age established at Espira to be extrapolated to the remainder of the overlying sequence up to those levels.

This depositional package transitions laterally, ca. 500 m further south (exposure in Botas gully) to more proximal facies, where the hillside at Les Vernèdes allows you to follow a succession of red beds belonging to the Codalet Fm. and other beds containing gneiss pebbles and boulders. The beds (dip angle: 3°S) are transitional between the Lentilla Fm. and the Escaro Fm. (Fig. 11.8C).

Fig. 11.9 **Lentilla/Espira Miocene mammalian fossil deposit**. The fossils were collected from fine-textured blue-grey facies of the Lentilla Fm. Key to abbreviations—Fm: clay-rich silt containing dispersed grains of coarse sand, occasional calcrete stringers, and reed stem impressions; FS: sandy silt; S: coarse sand, and sporadic gravel clasts; P: pebble-filled palaeochannel; PB: boulder- and cobble-filled palaeochannel

Return to the N116 and turn into the D13, which crosses the Têt over Vinça bridge. From the bridge head on the left bank of the river, a well-worn trail allows you to walk up to Marcevol monastery (2 hrs return) and observe the successive textures and facies of the Millas granite batholith during the ascent. Half-way up the scarp face (~350–370 m) you should also encounter a residual deposit of rubified and patinated quartz pebbles probably belonging to Quaternary alluvial generation T5. Views over the Conflent Basin are remarkably good. If you choose to drive up the road rather than walk up the trail, stop at the granite quarry at the entrance of

Tarerach valley (parking area) and enjoy the view over the lower Conflent Basin from there.

Stop 3. Panoramic view at the eastern edge of the Conflent Basin (Fig. 11.6B)

Here, the northern Conflent boundary fault was reactivated during the Pleistocene (see Itinerary 2), but most of the scarp relief results nonetheless from fluvial incision by the Têt along the fault. Note the numerous dry-stone walls all the way up the granite escarpment; most date back to the 19th century, a time when this landscape was extensively pastoral, agricultural and densely populated. Some vestiges of 17th–18th century and even Medieval terraces also occur (Passarius et al. 2009).

At the centre of the basin, the hilly terrain results from fluvial dissection of the Neogene sediment fill: Pliocene east of Vinça, Miocene west of the village. An apron of coalescing Quaternary alluvial fans forms a bajada at the basin edge. The Vinça fan is the most conspicuous and is coeval with alluvial sequence T2. Vestiges of T3 and T4 occur above it in the landscape, and a topographically constricted specimen of T1 occurs below.

Whereas the northern Conflent escarpment is predominantly a fault-line scarp, its relief having been generated during the Quaternary by partial stripping of the fill sequence in the hangingwall, the southern boundary of the Conflent Basin is mainly a young fault scarp. Whether in softer schist outcrops of the Aspres or in Canigou gneiss further west, its triangular faceted spurs appear uniformly fresh.

Continue northward along the D13. Opportunity is provided along the way to observe the steep bedrock knickpoint on the Tarerach channel. The stream flows out of Tarerach etch basin (generation P2) and joins the river Têt at the base of the north Conflent escarpment. As you enter Tarerach basin on the plateau, turn left towards Arboussols on the D35c. The road rises up to vestiges of pediment P1, here an extension of Montalba plateau featured in Itinerary 5.

Cultural stop: Marcevol priory. Panoramic view of the Conflent Basin and Mt. Canigou. Good opportunity to take a close look at the twelfth century Romanesque church, which was severely damaged by the 1428 earthquake. The window and porch are made of pink marble (Devonian limestone quarried at Villefranche-de-Conflent or Bouleternère), the walls and front of granite and green hornfels.

Return to the N116 via *Arboussols along the D35. During the descent towards Marquixanes, you once again get good views of Serrat d'en Molins and the Miocene exposures of the Lentilla drainage catchment; on the way down, also note vertical blades of granite cleaving apart along a late Hercynian band of mylonite striking E–W. Successively following the treads of T1 and T2, the N116 eventually reaches Prades; to the left, the small plateau around Llonat is a well-preserved vestige of terrace T4, here against a backdrop of hills dissected in the Escaro Fm.* (Fig. 11.3, cross-sections C, D, E). *Prades is a good base camp for exploring the Conflent and its surrounding massifs.*

Madrès massif and northern edge of the Conflent Basin

Entering the Madrès massif takes time because many of its forest tracks are closed to traffic, but the scenery and geological features are rewarding. Three optional excursions are described below. Each involves a hike requiring a full day. All three aim for the range-top erosion surfaces typical of the eastern Axial Zone, but only one of these walks, starting from the village of Nohèdes, will be described in detail (Figs. 10.12, 11.4).

Optional excursions and trailwalks through the Madrès massif from Prades

1. **Col de Jau and Castellane valley: erosion surfaces and glacial landforms** (64 km return) From Prades, take the D619 towards Catllar. At the junction with the D14, continue on the D619 towards Sournia. The road cuts through a hornfels outcrop (metamorphic aureole) before ascending into the Roque Jalère granite pluton, also offering wide views over the Conflent Basin and across to the Canigou massif. Stop at the junction with the unpaved road to Comes (spelling: Coma on recent maps). A 1 h walk will lead you up to the pediment surfaces of Roque Jalère (P1), studded with granite tors and small boulder inselbergs (see Itinerary 5, Fig. 10.4B). When returning towards Catllar, take the D14 to Molitg and Mosset. After passing a major channel knickpoint at Molitg spa, the road crosses the floor of Molitg etch basin (granite, vestige of P2 or P3), then follows the Molitg Fault as far as Col de Jau. At this point you enter an area more sharply influenced by Atlantic weather systems and, above the 1000 m elevation contour, a landscape that was glaciated during the Pleistocene. Weathered glacial till is exposed near the small bridge, and a highly degraded, pre-Late Pleistocene, lateral moraine occurs at Clot del Pasquer (1100 m). The road eventually ascends through the boulder fields of the Late Pleistocene frontal moraine before reaching Col de Jau (1506 m). From the col, you can drive up a badly maintained track to Refuge de Callau, after which the range-top erosion surface, S, of the Madrès will be reached on foot in ~3 h via the Castellane valley—a minor glacial trough (total 5–6 hrs return). The trail passes a disused talc quarry, which also exposes two stratified generations of till; the lowermost is weathered, the topmost unweathered.

From Col de Jau, another track will take you to the lowest Pleistocene glacial cirques of the eastern Pyrenees (2 h return). Scalloped out of the eastern flank of El Dormidor (1843 m; Dourmidou on some maps), and corresponding to an equilibrium line altitude ca. 1600–1550 m, their formation benefited from an exposure to moisture advection from the Atlantic and from wind-blown snow accumulation supplied by the plateau-like summit surface—a process still observable today from seasonal snowdrift patterns (Fig. 11.10; Calvet 1996, 2004).

Fig. 11.10 Small glacial cirques on east-facing side of Mt. Dourmidou (Dormi-dor). These are the lowest-elevation cirques of the eastern Pyrenees (summit: 1843 m; cirque floors: 1600–1500 m). Large snow patches illustrate process of snow accumulation downwind of the prevailing NW winds, thus also presumed to be a contributing factor of cirque-glacier growth during the Pleistocene

From Col de Jau, down the Aiguette River valley and through a granite etch basin (generation P2) at Roquefort de Sault, you can also join up with Itinerary 5 (Aude River gorge), or with Itinerary 7 through the Capcir Basin.

2. Sansa valley: highlights of Hercynian geology, Neogene palaeodrainage (40 km return)

Driving up the Têt valley from Prades, turn off at Olette onto the D4, which takes you through the very wild area of Garrotxes. Expect to be driving through a 3–4-km-thick schist sequence of Ediacaran–Cambrian age, displaying cordierite–andalusite- and biotite-zone metamorphic features at its base and chlorite-zone features above it, in both cases totally devoid of fossils (Guitard et al. 1992; Laumonier et al. 1996; Laumonier 2008; Laumonier et al. 2017). The road at first winds through the Cabrils Fm. (upper part of the Ediacaran Canaveilles Group: dark to black schists), then through the Évol Fm. (lower part of the Cambrian Jujols Group: psammites and pelites, here bordering on hornfels), before entering the Mont-Louis granite not long before reaching Sansa. Up to this point, the traverse amounts to a cross-section through the metamorphic country rock on the east side of the granite pluton. After the village, when approaching Coll de Sansa, the track reaches the Jujols Fm. (Middle to Upper Cambrian Jujols Group: greyish-green banded schist, generated from clay and silt protolith, see Fig. 11.27L, M), here in a position corresponding to the pluton roof. After Sansa, you drive up a well-maintained track that reaches the Late Pleistocene frontal moraine (Pla de l'Orri, 1650 m a.s.l., is a kame terrace). By continuing up the valley you will reach the top of the Madrès in about 2.5 hrs; along the way you will be cutting through the Devonian limestone core of the Villefranche syncline, which here is very narrow and forms the Esquena d'Ase–Pelada ridge. The Sansa River loses part of its discharge to the subterranean karst, later reappearing at the Réal karstic spring in the Capcir Basin and thereby supplying the Aude River catchment instead of the Têt (see Itinerary 7).

You can join up with Itinerary 7 via Col de Sansa (forest track), or via Railleu and Col de Creu (road D4), each time through segments of palaeovalleys which previously were tributaries of the Têt but were beheaded by active faulting along the Capcir boundary fault (see Itinerary 7).

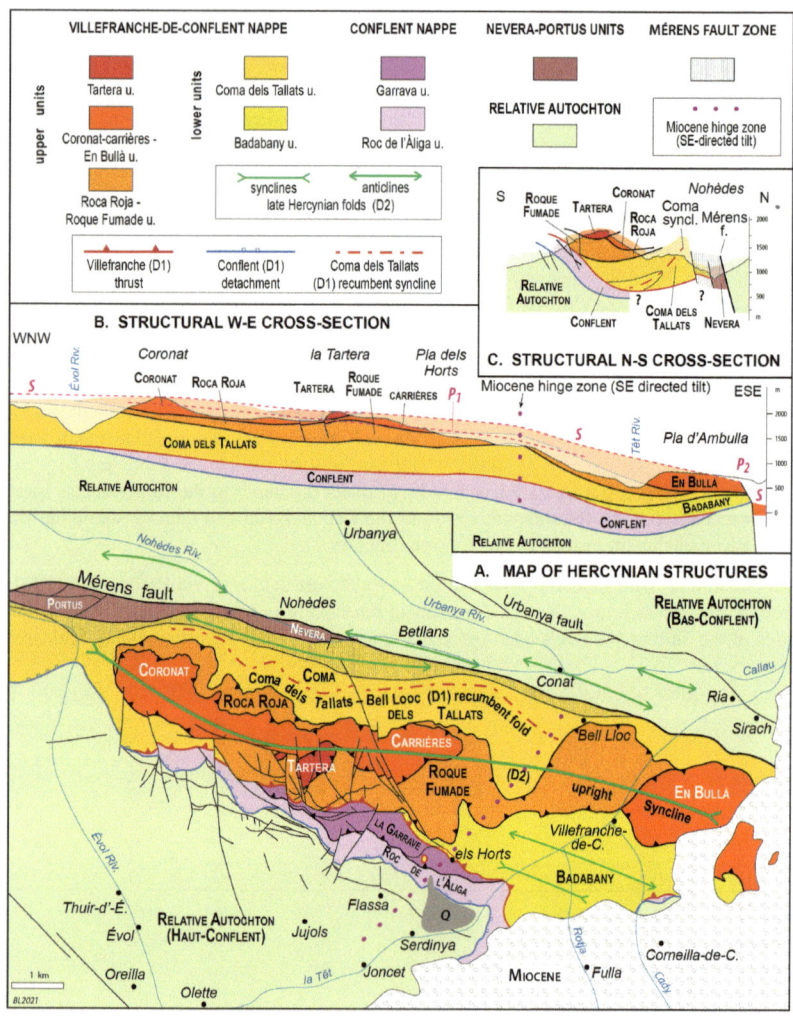

◀**Fig. 11.11 Structural map and cross-sections of the Villefranche syncline (Hercynian orogeny).** A Structural map. Hercynian tectonic deformation phases are noted D1 and D2 (D3 not shown). The syncline (D2) is a compressed older nappe stack in which the vertical succession of S- or SSW-vergent overturned to recumbent folds (D1) display identical strike directions (N110°E) to D2. The three vertically stacked structural units include: (i) a relative autochton, (ii) the Conflent nappe (two units: Garava, overridden by Roc de l'Àliga), and (iii) the Villefranche nappe. The Villefranche nappe contains five units: Badabany, Coma dels Tallats, Roque Fumade–Roca Roja, Coronat–En Bullà (quarry exposures), and La Tartera. The Coma dels Tallats unit is the largest; it presents itself as a recumbent syncline ~1 km in amplitude south of La Coma dels Tallats, but subsequently uptilted and thus spuriously making it appear as an anticline. A last, and almost vertical, structural unit (Nevera–Portus unit), in connection with the Mérens Fault, was generated by late Hercynian thrust deformation (D3). The Mérens Fault dips steeply to the north, and additional increments of thrusting were probably generated during Alpine collision. Some of that throw was diminished, however, by inverted (extensional) motion along the fault plane during the Neogene. The relative autochton consists of Upper Ordovician and Jujols Group outcrops (further to the SW of Olette, it consists of Canaveilles Group outcrops). The Conflent and Villefranche nappes consist of Silurian–Devonian and Devonian–Carboniferous units, respectively. All the unit boundaries, and most clearly so in the case of the Villefranche nappe, are S- or SSW-vergent thrusts. The only exception is the Conflent nappe, which displays the attributes of a detachment fault. Q: Marinyans rock slope failure. **B** E–W cross-section following the axis of syncline D2. Two generations of erosion surface, S and P1, are indicated. Note that they have been tilted to the SE, as if grading to the adjacent Miocene basins. **C** N–S cross-section passing through Nohèdes and La Tartera

3. **Nohèdes valley (30 km return) and the flat-topped summits of the Madrès massif** (7 hrs return)
Driving up-valley from Prades, turn off at Ria on the D26, which winds all the way through the Évol Fm. and follows the Mérens Fault, here a reverse fault thrust southward over the Villefranche syncline (Fig. 11.11). The Mérens Fault is a late Hercynian, and partly also Alpine, mylonitic fault zone. It dips steeply to the north with a throw of ~2–4 km, and can be followed for 80 km from Prades in the east, to Pique d'Estats in the west. Tectonic inversion almost certainly reactivated it as a top-to-the-north normal fault during the Neogene. Evidence for this is the ~400 m offset between vestiges of P1 at Pla dels Horts on Mt. Coronat in the south, and Pla de Vallensó (Balençou) in schist to the north (Figs. 10.12, 11.4, 11.23B). Details of the fault planes can be inspected locally in road embankments at two locations roughly 1 km and 300 m, respectively, before the entrance sign to the village of Nohèdes (Fig. 11.12A). The southern valleyside exposes picturesque limestone and dolomite spurs and crags along the steep, north-facing edge of the Coronat massif, which in structural terms consists of a vertical succession of south-dipping overturned folds and south-vergent thrusts in Silurian–Devonian–Visean stratigraphy (see also **Stop 11** for structural detail, and Figs. 11.11 and 11.12B). Beyond Nohèdes, a poorly maintained forest track should allow you to drive as far as Montellà (1239 m), where you will be entering thereafter a National Nature Reserve and must therefore find a space to park your vehicle—for example where the track widens alongside the light-coloured granite boulder field at Montellà (this is a Late Pleistocene till formation deposited by a valley glacier descended from the Madrès).

◀**Fig. 11.12 North face of Coronat massif near Nohèdes. A** Outcrop of Mérens Fault. Exposure in embankment of the D26, 1 km before reaching Nohèdes. s-d: Siluro-Devonian of Nevera unit (Conflent nappe); kÉ: Jujols Group (schist, Évol Fm.). **B** Villefranche syncline near Nohèdes. Extending from Ria to Pla dels Gorgs, bedrock along the north face of the Coronat massif is mostly concealed by forest or buried beneath talus cones, but two deep N–S ravines, clearly visible from Nohèdes (Coma dels Tallats, Coma de Pitxó), have cut through the Quaternary slope debris and expose the underlying bedrock. A footpath rises up into Coma dels Tallats and allows you to reach the limestone cliffs of Roca Roja. Alternatively, the view from Pic Lloset, above Nohèdes village, is excellent, particularly in the morning. Most of the massif consists of a stack of recumbent folds in limestone (see Fig. 11.11), and the Tallats syncline exposes those vertically stacked units separated by thrust planes, with a gentle dip to the south (see Fig. 11.11C); the upper half of the valley side, from Tapalrec to Mt. Coronat at the summit, consists of a succession of inverted limbs of the recumbent folds. The slope base coincides with the smaller structural units (Nevera, Portus) associated with the Mérens Fault. Key – kÉ: Évol Fm. (Lower Cambrian), relative autochton; s-d: alternating pelitic and carbonate beds (Silurian to Devonian); d1: grey dolomite (Lower Devonian); d2-3: light-coloured calcschist and white limestone (Lower to Middle Devonian); d4: cherty limestone and marl containing silicified corallites (Middle Devonian); d5: infra-griotte limestone (Middle to Upper Devonian); d6-7: griotte and supra-griotte limestones (Upper Devonian)

This walk is only recommended by clear weather. It focuses on landscape evolution over the last ~20 Ma, and on its legacies still encoded in the topography. Much of the itinerary lies within a National Nature Reserve (check the countryside code at the Reserve headquarters at Nohèdes). From Montellà, follow the forest footpath (yellow signposting), which shortcuts several of the loops in the forest road and will lead you up in a fairly straight line (initially through a beech forest) to Coll de Portus. From this mountain pass, head off northwest towards Pic de la Creu. Until you reach spot elevation 1903 m you will be following a segment of the disused, 12-km-long irrigation canal that descends to the village of Jujols from one of the glacial lakes farther up the mountain—a somewhat startling reminder that 150 years ago these mountains were ten times more densely populated than today, with self-contained and self-sufficient agricultural and pastoral economies largely cut off from the Conflent and Roussillon basins down below.

Stop A. Approximately on the 2000 m elevation contour, a panoramic view to the south discloses the interplay between structural geology and differential erosion in shaping the first-order topography. The valley of Nohèdes roughly follows the strike of the Mérens reverse fault (Figs. 10.12, 11.11A and 11.13A), with, to its south (footwall) the steep walls of Devonian limestone (Mt. Coronat); and to its north (hangingwall, valley included) an extensive outcrop of Cambrian schist. Towards the south, the upper portion of Évol valley hosted a short valley glacier during the Pleistocene. The glacier was fed by the cirques around Gorg Negre, and the glacier descended as far as Molina (Mouline) water gap 1500 m)

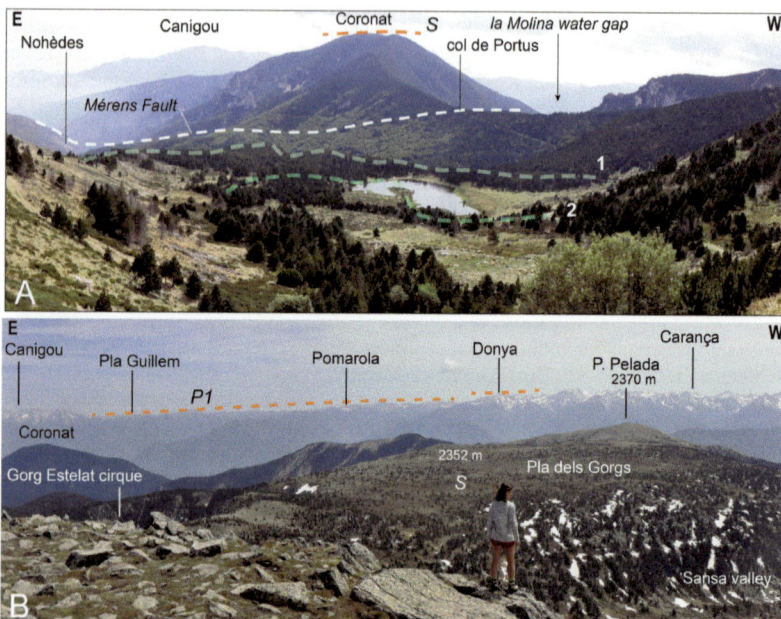

Fig. 11.13 Landscapes of the Madrès massif. A Valleys of Nohèdes and Évol, and Molina water gap, seen here from the path at Gorg Estelat (ca.1800 m a.s.l.). See text about the origins of the Molina water gap cutting through the Villefranche syncline. The Mérens Fault strikes through Col de Portus and continues westward up the Évol valley. Note the frontal moraines in the upper valley of Nohèdes: the terminal moraine (1) lies in the woodland around 1640 m; a recessional moraine (2) closes off Estany del Clot (the small lake) ca. 1660 m. **B** Pla dels Gorgs, seen from Madrès summit plateau edge. A vestige of range-top surface, S, also occurs around the summit of Mt. Coronat (the Molina water gap separates these two, previously probably continuous, vestiges of S). The limestones of the Villefranche syncline extend westward (Pelada ridge). On the horizon, note extensive vestige of pediment P1, here preserved between the Canigou and Carança massifs (see Figs. 11.4 and 11.22, and Itinerary 4, Fig. 9.17)

The large Molina water gap ('cluse de La Molina') cuts through a fin of resistant Devonian limestone (Figs. 10.12 and 11.11A) but its origin remains elusive. A pair of alternative explanations can nonetheless be put forward, each typical of many 'chicken or egg' interpretations of drainage evolution in plateau regions across the world:

(i) piracy of the Nohèdes catchment headwaters by the Évol River: this hypothesis is consistent with the steeper thalweg gradient of the Évol River (12%) compared to that of the Nohèdes River (9%), the sharp bend in the Évol stream at the water gap, and headward erosion facilitated by endokarstic processes within the limestone.

(ii) the Évol River established itself on the low-gradient range-top paleoplain, S (which you will be discovering at **Stop C**), and later cut into the limestone in similar fashion to the water gap cut by the Sansa River, during early Miocene time, a period when the nearby Conflent half-graben was forming a new local base level and the Madrès Massif was tilting to the SSE. The Évol River catchment during Miocene time may also have encompassed the entire Madrès summit area, with runoff from the Estelat subcatchment also at that time joining the Conflent Basin via Coll de Portus. This landscape configuration changed at some time after the middle Miocene when the Mérens Fault underwent reactivation, this time undergoing extensional displacement as a result of tectonic inversion (regionally consistent with the extensional regime of the Western Mediterranean), thus (i) reversing its initial synorogenic throw, (ii) lowering as a result the schist outcrops of the northern block relative to the limestone, and (iii) promoting the westward ingress of the Nohèdes River through the weaker schist and the capture by the Nohèdes stream of the Estelat headwaters.

Stop B. Continuing from Pic de la Creu, you will reach a saddle at Collada del Gorg Negre. This flat-topped spur in granite connects uphill to Pla dels Gorgs via a gentle topographic ramp, and is interpreted as a local vestige of the regional population of Middle Miocene pediments P1 (here: the upper, steeper slope of the pediment). The altitude of this pediment surface correlates laterally with the floors of the local glacial cirques scalloped into the edge of the high plateau between Roc Negre and Pla de la Roqueta. Such altitudinal congruence between cirque floors and the upper termination of pediment surfaces is far from unique in the eastern Pyrenees (see Itinerary 7, Carlit plateau), and emphasises a key fact: rather than landforms generated purely from the sheer power of ice flow and gravity, many of the glacial cirques in this relatively dry climatic region of the Pyrenees are pre-Quaternary fluvial landform assemblages marginally retouched by valleyhead (now: cirque-headwall) steepening and pediment (now: cirque-floor) scouring. Also note the numerous recessional moraines forming boulder ribbons around and across the lake at Gorg Negre; a pollen core has shown the cirque has been deglaciated since the Oldest Dryas (Reille and Lowe 1993).

Stop C. Ascend the residual spur between the two glacial cirques, and you will reach Font de la Perdiu refuge. Here begins Pla dels Gorgs, one among the most ancient landforms of the eastern Pyrenees, with coeval equivalents at the tops of the Carlit and Campcardós massifs. This plateau is a vestige of erosional plain, S, which has been mapped across extensive portions of the orogen, cutting across tectonic structures and lithological boundaries, attenuating structural landforms, and thus generating an advanced state of low relief in the eastern Pyrenees as early as 25 million years ago (Fig. 11.13B). The flat topography at the summit can be explored at will; the ascent of its residual summit hill (2352 m a.s.l.) offers a spectacular 360° view of the range-top paleoplain—both locally on granite (La Roqueta, to the NE), and further afield on Devonian limestone (summit zone of the Coronat massif: Mt. Coronat–La Tartera, to the ESE). The erosion surface was probably never perfectly level, but its low-energy morphology is nonetheless out of character with the expected topography of a narrow, glaciated mountain range produced during the Paleogene by plate-boundary collision. Depending on local lithology and the vicissitudes of fluvial erosion in the past, a few residual landforms displaying convexo-concave slope profiles have been preserved as monadnocks, such as at Roc Negre (granite) or La Pelada (Devonian limestone). Pla dels Gorgs is at the same time a blockfield consisting of granite debris shed by tors and distributed across low gradients by gelifluction during the Quaternary. These shallow periglacial processes have probably contributed to enhance the short-wavelength smoothness of the initial topography. In keeping with the sporadic presence of tors, however, the plateau also bears a mantle of weathered granite up to 50 m thick in places. Regolith thickness can be appreciated along the upper rim of cirque walls such as above Gorg Blau (Fig. 11.14B). The presence of rounded granite corestones, such as along the trail rising up from Gorg Blau to the plateau edge, testifies to the action of spheroidal weathering in a low-energy environment, seemingly at odds with glaciation in a high-energy mountain range. The in-situ saprolite is undated but most likely pre-Quaternary, and it was covered by the angular debris of the Pleistocene blockfield.

Stop D. Ascend Roc Negre and join Pla de la Roqueta to the east. This link of the trail affords good plunging views into the glacial cirques of Madrès, La Castellane and Gorg Estelat. Note that the deeper 'armchair' cirques with steep bedrock headwalls usually face east or southeast, suggesting that the erosive power of cirque glaciers was enhanced here by supplies of wind-blown snow from the summit plateau to the west and northwest. Note also the Lateglacial rock glacier on the slopes to the south of Lake Estelat, with its outermost lobe reaching the water's edge (Fig. 11.14D). Around the plateau rim (e.g., Roc Negre), the granite displays a number of arcuate rock-slope failure scars parallel to the cirque headwalls (Fig. 11.14C; see also Jarman et al. 2014). On Pla de la Roqueta (Fig. 11.14A), the landform also bearing that name (2345 m) is a good example of an isolated granitic tor. Whereas its SW face still preserves edges blunted by geochemical weathering, the more angular NE-facing side has been severely degraded by frost action.

▶**Fig. 11.14   Range-top erosion surfaces and glacial cirques in the Madrès. A** Pla de la Roqueta, and its granite tor rising out of deep grus. **B** Thick vestiges of in situ grus with corestones around the rim of Pla dels Gorgs. **C** Pla du Roc Nègre, dinted by rock slope failures around the cirque headwall above Gorg Estelat. **D** Gorg Estelat cirque viewed from Roc Negre. Note the small recessional moraine impounding Gorg Blau, and the protalus rampart transitioning to a rock glacier at the base of the rockface

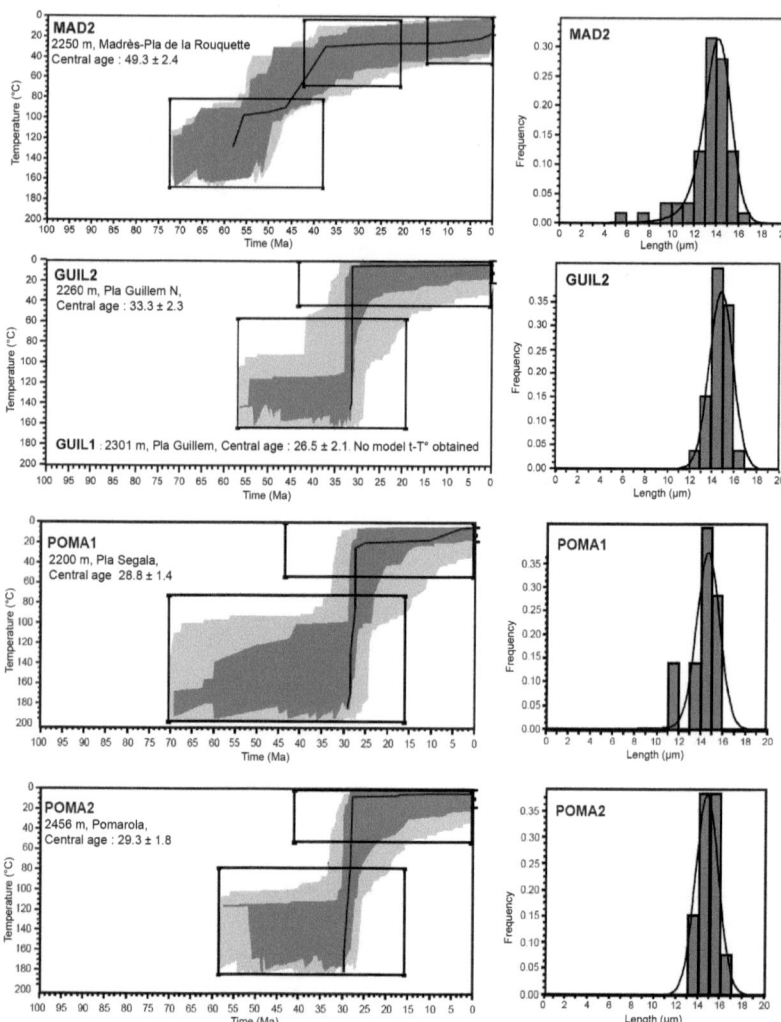

◄**Fig. 11.15** **Rock-cooling histories recorded by elevated vestiges of erosion surfaces located north and south of the Conflent Basin** (after Gunnell et al. 2009). Note the time lag in denudation between massifs to the north and south of the Conflent Basin, respectively, suggesting a landscape organised (as today) in discrete tectonic blocks. In the Madrès massif, denudation seems to decline significantly between 40 and 35 Ma (late Eocene), in keeping with the timing of Pyrenean tectonic convergence. However, denudation occurs much later and much faster in the Canigou–Carança block, apparently in response to the onset of Oligocene rifting ca. 30 Ma. Denudation during Aquitanian and Burdigalian time, also recorded by the thick clastic sequence of the Conflent Basin, obviously remains undetected in the thermal record of these long-surviving erosion surfaces (areas of the landscape least eroded since ~25 Ma)

Granite samples collected from Canrec cirque edge (2250 m), ca. 500 m east of La Roqueta, have yielded apatite fission-track and (U–Th)/He cooling ages of 49 and 39 Ma, respectively (Fig. 11.15; Gunnell et al. 2009). These samples record the period of rapid syn-orogenic crustal denudation (i.e., rock cooling) prior to their subsequent long-term residence at shallow depths beneath the low-gradient topography (i.e., at temperatures of ~40 °C or less, where the helium radiometric system ceases to record the thermal effects of rock cooling).

From La Roqueta, begin your descent to Gorg Estelat. This glacial lake is impounded by a small recessional moraine. Follow the well-worn trail down the glacial valley, which descends through a succession of boggy basins ('plas') and steeper bedrock steps. Below the bedrock step that separates Pla d'Amunt from Pla del Mig, you will pass the ruins of the water supply to the Jujols canal. After exiting the granite pluton and its hornfels rim, the trail winds down the mountainside, overlooking Estany del Clot. The lake today is an artificial reservoir engineered in the 1970s for a microhydro plant at Nohèdes, but it occupies the site of a natural peatbog impounded by the frontal moraine of the LGM glacier (elevation: 1650 m; Fig. 11.13A). Note the recessional moraines forming low boulder ridges extending across the lake and further back. The track eventually joins up with the forest track and its signposted pedestrian shortcuts back to Nohèdes.

11.2 Canigou Massif and Southern Edge of the Conflent Basin

Starting from Prades, drive up the Llitera River valley on the D27 towards Mt. Canigou. The Canigou horst (2785 m), as can be appreciated from Prades, is the dominant feature of the scenery. The hills on either side of the road consist entirely of Escaro Formation sediments, which contain boulders of augengneiss supplied by the Canigou–Carança metamorphic dome. At Codalet, and further south at Taurinya, this formation overlies a red-bed sequence exclusively containing schist and micaschist debris (Fig. 11.3E, F). This provenance stratigraphy has implications for understanding the geomorphological history of the mountain range.

Cultural stop at Saint-Michel-de-Cuxa (Sant Miquel de Cuixá)

The Saint-Michel benedictine monastery was founded in the 10th century by monks who had survived a catastrophic flood along the Têt, which in 878 CE had destroyed their original settlement at Saint-André d'Eixalada ('Bains de Thuès', see **Stop 13**). The abbey of Saint-Michel was a major religious, political and cultural centre during the early years of the Catalan principalities, and even hosted the Venetian doge Pietro Orseolo, who retired here at the end of the 10th century. The church, which was consecrated in 974 CE, is pre-Romanesque (Wisigothic style, with typical horseshoe arches). It was expanded (choir and belltowers) at the beginning of the 11th century by abbot Oliba, and a pink marble cloister and gallery were added during the twelfth century (part of the cloister has remained on site, another part has been incorporated to the Cloisters Museum in New York; the fountain is now at the Philadelphia Museum of Art).

Drive through the village of Taurinya and follow hairpin bends through a complex tectonic boundary zone in which large olistoliths of schist, marble and gneiss are found interlayered within the Escaro Formation (Fig. 11.3F). The entire package has been impacted by younger tectonic movements. The olistoliths contain iron ore (hematite and siderite) which was mined continuously from Roman times to the early twentieth century. Stop at a bend in the road, facing Mt. Canigou, elevation spot: 713 m at Colomines de Dalt.

Stop 4. The Têt Fault at Taurinya

The triangular faceted spurs of Mt. Canigou are the most spectacular of their kind to be found in France (Fig. 11.16A). The stop provides a good view of Les Costes, a facet 500 m high, which is particularly well preserved because its plane coincides with the foliation plane of a steeply dipping sheet of mylonitic gneiss belonging to the outer envelope of the Hercynian Canigou–Carança metamorphic dome.

The NW-facing Canigou range-front escarpment formed in several stages. The morphology exhibits a vertical succession of facets that become progressively more degraded towards the top of the spur (Fig. 11.16B). These facet stairways are typically interrupted by narrow topographic benches, e.g., at Llaceres (1365 m), and display low-gradient (erosionally bevelled) spurs such as at Roc Mosquit–La Soccarada (1900–2050 m), which is laterally accordant with the bench at Les Cortalets. These may be interpreted as vestiges of incipient pediments that were granted sufficient time to form notches into the rising mountain flank during intervals of relative tectonic quiescence. The base of the facets has

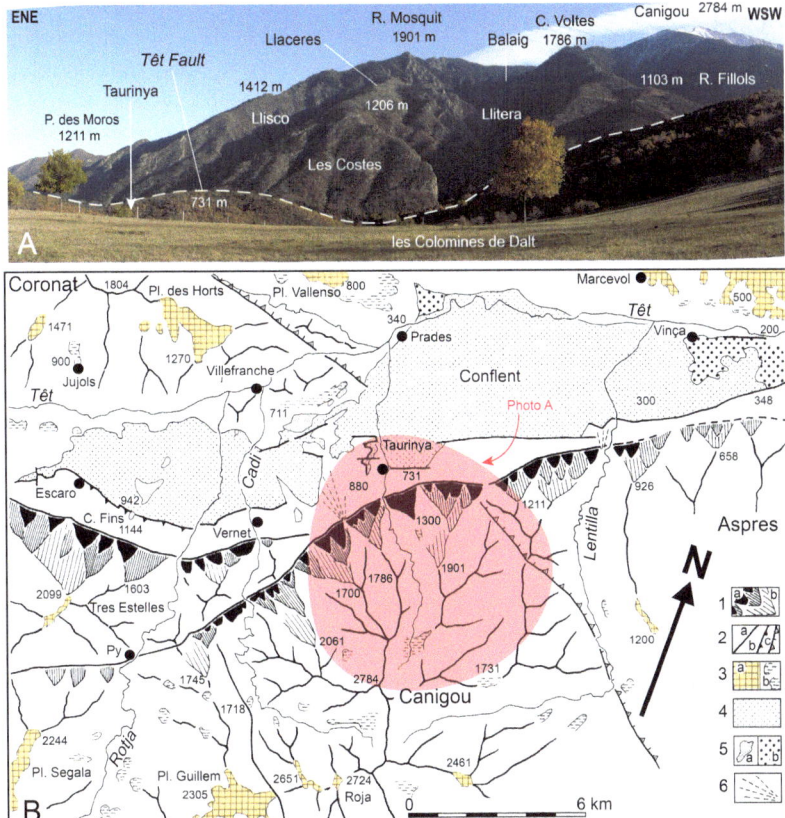

Fig. 11.16   Faceted spurs along the north-facing Canigou fault scarp. A Large scarp-face facets around Taurinya. The best-preserved faceted spurs (les Costes) are 500 m high; their upper vertices host erosional benches or display staircases of older bevelled spurs (Llaceres, Mosquit, Col des Voltes…). The master fault (Têt Fault) occurs exactly at the scarp base. This north-facing slope of Mt. Canigou was weakly glaciated, but is first and foremost deeply incised by fluvial gorges. Note the Fillols avalanche corridor (1.4 km of local relief), known to produce infrequent, high-magnitude avalanches. **B** Geomorphological map of the faceted fault scarp (after Calvet 1999, modified). Key to symbols and ornaments; **1**. Scarp-face facets; 1a: well preserved; 1b: degraded. **2**. Tectonic features; **2a**: normal fault; **2b**: possible Neogene reverse fault (as at the foot of the Albères massif); **2c**: Paleogene thrust fault (Mérens Fault), which underwent throw reversal under the prevailing Neogene extensional regime. **3**. Planar landforms; **3a**: vestige of pediment P1 (middle Miocene age); 3b: younger benches and pediments (generation P2; several levels). **4**. Early Neogene sedimentary formations (Aquitanian to Lower Burdigalian), high-energy fluvial facies. **5**. Late Neogene sedimentary formations; **5a**: Coll de Fins Fm. (probably Vallesian); **5b**: high-energy debris fans and delta (Pliocene). **6**. Early to middle Pleistocene debris cone (T4). Altitudes in metres

been exhumed over a maximum height of 100 m at places where differential erosion has peeled off the softer schist outcrops that form the outer metamorphic envelope. Close inspection of those exposures shows normal fault planes displaying several generations of oblique grooves and lineations, all indicating normal to left-lateral displacement. It is also possible that the freshness of the basal facets is a result of recent exposure after evacuation of the Coll de Fins Fm. (late Miocene, probably Vallesian), which previously filled the Conflent hangingwall to greater elevations than currently observed. However, this hypothesis is unlikely because most of the Coll de Fins Formation. was stripped out of the Conflent Basin during the Messinian eustatic fall (~5.3 Ma). It is also unlikely that the facets have preserved their current fresh appearance for as long as ~5 Ma. The preferred interpretation is thus that the better-preserved basal faceted spurs result from recent tectonic activity (Pliocene and Quaternary), with a total throw of 0.4 km or more.

This and other tectonic escarpments in the region (Albères: see Itinerary 2; Cerdagne: see Itineraries 7 and 8) are interpreted as having remained tectonically active until recent historical time (Briais et al. 1990). However, careful inspection of the alluvial deposits belonging to generations T1, T2 and T4, which all seal the fault plane at the base of the Canigou horst, has so far failed to reveal any clear indications of anomalous tectonic deformation—whether of the Quaternary deposits themselves (intraformational offsets) or of the river terrace treads (e.g., postdepositional tilt; Calvet 1996, 1999). This evidence tends to confirm that dip-slip motion on the Têt Fault declined significantly after the early Pleistocene. A Surface Process Model simulating erosion of the faceted spurs of Mt. Canigou through anisotropic diffusion processes has recently supported this conclusion (Petit and Mouthereau 2012). Provided the geological fabric remains parallel to the hillslope at an angle of 45° and that the bedrock is resistant to weathering and to slope failure (as appears to be the case with the mylonitic gneiss at Les Costes), model outputs showed that structural conditions of this nature were capable of ensuring the preservation of a fresh-looking faceted spur for at least 500 ka without requiring any fault reactivation. The model, however, is likely only valid for the larger Costes and Thuès facets (for Thuès, see **Stop 13**). It cannot be extrapolated with confidence to the entire escarpment because the fault cross-cuts a wide range of lithologies, including softer schist and micaschist to the east (Aspres massif) and to the west (Cerdagne), as well as mylonite, gneiss and granite. Furthermore, some facets cross-cut at an acute angle the entire metamorphic envelope and its foliated fabric, for example immediately to the east of Les Costes, suggesting that the computer model provides a restricted version of more diverse field conditions.

The road avoids Roc del Ram by following its western face, cutting through the complex structure of this olistolith. At the col, the track descending northwards cuts through a particular facies of the Escaro Fm. that contains a few boulders and pebbles of Rudist-bearing limestone (implying Upper Cretaceous source rock; Fig. 11.17). Added to the small pebbles of red Triassic or Garumnian sandstone encountered in the Lentilla Fm., these indicator clasts suggest the former presence here of a sheet of Alpine cover rocks similar to the one previously encountered at Amélie-les-Bains (Itinerary 4), now entirely eroded. This imbricate was most likely a southwesterly continuation of the Aspres Thrust, perhaps lodged at the boundary between the Costes gneiss unit and the rest of the Canigou dome.

On the left, the road passes through the formerly mined iron ore outcrops of Fillols. At Coll de Millères, the road to Cortalets mountain refuge (2100 m) is restricted to accredited tourist shuttle vehicles; from the refuge, the summit of Mt. Canigou can be reached in 2.5 hrs on foot (round trip: 5 hrs; if you prefer to walk all the way

Fig. 11.17 Hippurites in Upper Cretaceous limestone boulder. Outcrops of Escaro Fm. north of Roc del Ram, ~740 m a.s.l

◄**Fig. 11.18 Present and past flash-flood deposits at Vernet-les-Bains. A** Exposure of a historical debris cone, right side of Saint-Vincent channel, upstream of the D27 bridge. The 4–5-m-high cut was produced by bank erosion during the October 1940 flood. The tread of the 1940 cone thus extends from the base of the cut. **B** Tread of the historical debris cone, on left side of the Saint-Vincent channel, viewed here from the trail signposted to Cascade des Anglais. **C** Les Conques: catchment area of the Saint-Vincent valleyside gorge. The catchment consists of a cirque filled with 200 m of unconsolidated till (age: Late Pleistocene) contained behind a bedrock step ca. 1900 m a.s.l. Note tentative hillslope engineering works dating back to the 1960s. **D, E** The spa resort at Vernet before and after the October 1940 flood (credits: Genovèse, after Batlle and Gual 1990). **F** Lichenometry results (based on calibrated *Rhizocarpon geographicum* growth curve). Ages are based on the mean of the five largest thalli (age of the largest is indicated in brackets); black star symbol: age of a younger rockfall deposit covering the debris cone. Age estimates derived from a calibration curve established by Jacob et al. (2002); interpretation after Calvet (2006), and unpublished data. **G** General map of the 1940 Vernet flood setting. Key to ornaments—**1**: extent of 1940 flood deposits (after IGN aerial photograph from 1942). **2**: historical boulder fan. **3**: other Pleistocene or Holocene alluvial deposits. Altitudes in metres

from Coll de Millères, allow 12 hrs). The road then crosses the large Fillols alluvial fan (generation T4). At the western end of the fan, note the Romanesque chapel of Sant Pere, which collapsed in the 1428 earthquake. Past the chapel, a road cut reveals the thick debris accumulation of T4 (post-depositional weathering is characteristically intense). The village of Fillols is itself located on torrential bedload deposits belonging to generation T1. These were reworked during the Holocene as well as in the context of the 1940 flood. Following exceptionally high snowfall in January 1986, a 100-year snow avalanche swept down the Ravin de Fillols and halted 1200 m from the village (elevation: 970 m). After Fillols, stop at Coll de Sant Eusebi and walk up to the ridge to the north of the road (spot elevation 805 m).

Stop 5. Fillols–Coll de Sant Eusebi

This site (i) allows an appreciation of the Escaro facies, with its content dominated here by augengneiss clasts, some over 10 m in diameter and corresponding to olistoliths released during fault activity; and (ii) provides a panoramic view of the SW face of Mt. Canigou, with its stairway of faceted spurs. The range front is breached by the Saint-Vincent and Cadí hourglass catchments, both major contributors to torrential floods such as those of October 1940.

Stop 6. Vernet–Saint-Vincent debris cone: Legacies of recent and historical torrential processes

This steep, hourglass-shaped range-front catchment ('torrent', in French, thus named because of its potential to generate torrential floods), is the largest and most spectacular in the French eastern Pyrenees (Figs. 11.18 and 11.19A) because of its steepness, its local relief, and the intensity of autumn rainstorms on Mt. Canigou (see Itinerary 4, Fig. 9.10A, B). The active debris cone is currently situated upstream of the bridge on the D27 as you enter Vernet. The cone produced during the October 1940 flood has since been substantially transformed by the construction of channel embankments and by intensive mining for aggregate. Archives show, however, that the debris at the time buried the entire meadow situated on the right bank near the bridge. A vertical cut in an older (historical) debris cone can be inspected above the meadow (Fig. 11.18A). Cone morphology can also be appreciated by walking up the left bank from Vernet cemetary (trail towards Cascade des Anglais, 1 hr return). The cone exhibits an apical thickness of 30 m and a typically chaotic surface topography across which debris-flow channels and their boulder-rich levees are well preserved (Fig. 11.18B). The debris have been accumulating for at least 2000 years, with a paroxysm during the Little Ice Age consistent with the evidence of alluvial aggradation observed in the coastal plains (see Itinerary 3; Calvet et al. 2002). Ruins of the Sant Vicenç de Campllong Romanesque church, built in 898 CE, have been buried beneath 2 to 3 m of cobbles and boulders. Historical archives record torrential floods in 1421 and during a series of meteorological events between 1763 and 1777. Lichenometric dating based on *Rhizocarpon* sp. has been attempted (thalli 1.9 cm in diameter have grown on 1940 boulder deposits; a calibration curve was elaborated by Jacob et al. 2002) and has yielded a wide array of ages lacking a spatially consistent pattern (Fig. 11.18F). The median age is 16th century, with a range extending from the 14th to the 17th. The largest isolated specimens have yielded older ages (12th to 14th centuries), with one unexplained outlier from the first century BCE (Calvet 2006, and unpublished data).

Drive through the Medieval village of Vernet. The lower end of the village, partic-ularly the spa hotels, was entirely destroyed by the 1940 flood (Fig. 11.18D, E). A brief stop at the bridge (Pont des Thermes) provides a view towards the Cadí River valley to the south. From there you can catch a glimpse of the belltower of the abbey of Saint-Martin-du-Canigou, which stands at 1055 m amid imposing rock forma-tions along the fault scarp. This very early Romanesque edifice was consecrated in 1009. Its founder, Count Guifred II of Cerdagne, was buried in it. The church once harboured the relics of Saint Gaudéric, a Wisigoth patron saint renowned for averting natural hazards such as droughts and floods.

Optional trailwalks through the Canigou massif from Vernet

Two additional excursions will guide you through the structural architecture of the Hercy-nian Canigou–Carança metamorphic dome (cross-section in Fig. 11.19B), and particularly the outcrops of its inner core (Guitard 1970; Guitard et al. 1992). The augengneiss form-ing the outer envelope of the dome (Fig. 11.27K), preserved as 2–3-km-thick carapace, were for a long time interpreted as Precambrian granite basement, unconformably covered by the marine sequence of the Canaveilles Group and compressed into a large recumbent fold during the Hercynian orogeny. Recent dating of the orthogneiss has revealed that its protolith was actually a granite intrusion of Ordovician age (477–470 Ma; Deloule et al. 2002; Cocherie et al. 2005). Beneath the orthogneiss envelope, the core of the dome displays high-grade sillimanite micaschists and marble (Balaig micaschists), which form the base of the Ediacaran Canaveilles Series. The micaschists also include other intrusions, which were metamorphosed to orthogneiss both at the base (Cadí gneiss) and further up in the metamor-phic stratigraphy (Quazemi gneiss, leucocratic leptynites and finer grey banded leptynites, which form the highest summits of Mt. Canigou). A medium-textured, two-mica Hercynian leucogranite, effectively the deeper granite intrusion of the Canigou dome, also suffused the pre-existing structures; outcrops can be observed up to modern elevations of ~2000 m, and veins and dikes attain the summit area.

1. **Saint-Martin-du-Canigou and beyond**
Follow the D116 to Casteil. From there it is possible to walk up to the monastery in 2 hrs return. The track cuts across the major Py–Mantet Fault (which joins the Têt Fault), then through the Cadí gneiss, and reaches the two-mica leucogranite while also displaying enclaves of sillimanite micaschist and marble. Above and beyond the monastery (another 5 hrs round trip), the trail provides access to the same sequence, i.e., the deeper granite, then the sillimanite micaschists, and finally the Quazemi gneiss when you reach Coll de Segalers.

2. **Mt. Canigou via Mariailles**
The track from Casteil follows the Py–Mantet Fault up to Coll de Jou (1125 m), which most often is closed to traffic beyond the col. The Canigou summit can be reached in 10 hrs return, and exhibits the same rock sequence as the previous excursion.

At the Pont des Thermes in Vernet, take the D27 and ascend to a drainage divide marking the separation between the Vernet and Sahorre drainage catchments (spot elevation 706 m).

Stop 7. Panoramic views of the Canigou massif and northern edge of the Conflent Basin

A ridgetop walk northwards as far as the Devonian limestone massif of Badabany and Citerne de Vauban (2 hrs return, but access by car part of the way is also possible) affords even better views over the Conflent Basin than from the initial roadside stop (Fig. 11.19A), and the track embankments reveal key exposures of the Escaro Fm. along the way (Fig. 11.20B). At Citerne de Vauban, you will also get a view of the epigenetic gorges of the Têt and its Cadí and Rotja tribu-taries, all cut in Devonian limestone outcrops on the north side of the half-graben (Fig. 11.23B).

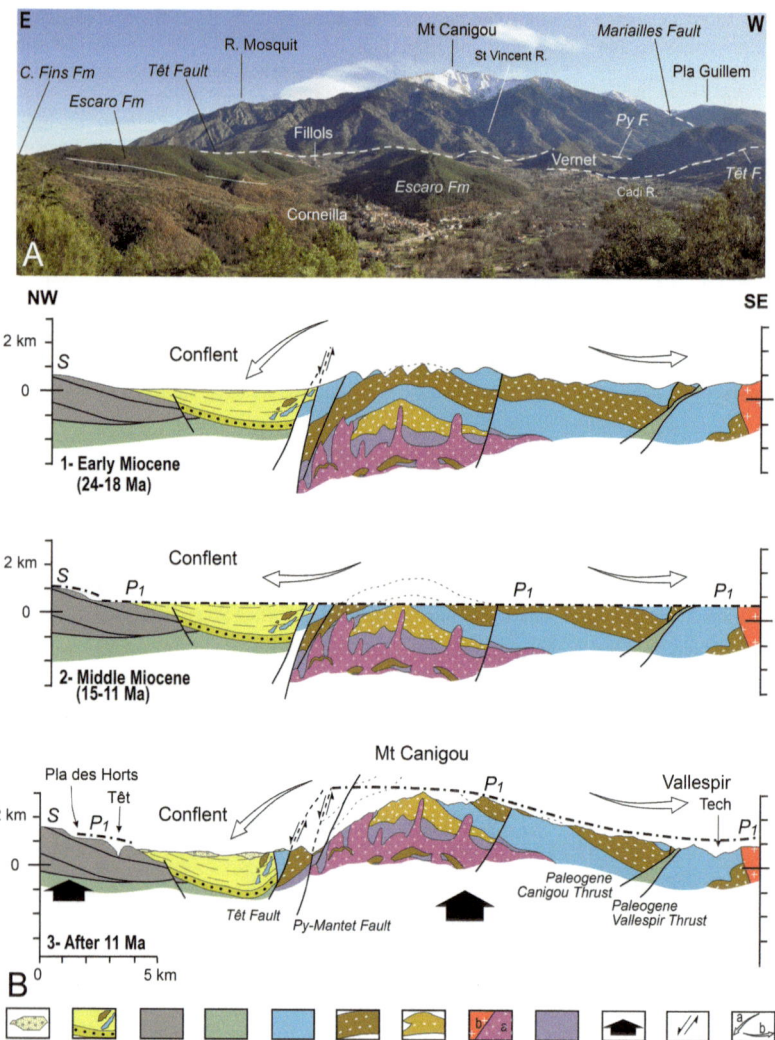

◀**Fig. 11.19**   The Canigou horst and its Neogene denudation history. **A** Mt. Canigou and the catchment above Vernet. View from Badebany. West of Fillols, a fork in the boundary fault gives rise to the Py–Mantet Fault, which cuts obliquely through the massif with diminishing throws to the SW (pivotal fault); and the Mariailles Fault, which has thrown up the Canigou block with respect to Pla Guillem (a vestige of pediment P1; see Fig. 11.4 and Itinerary 4, Fig. 9.17). Note southerly dip of Escaro Fm. east of Corneilla. **B** Post-orogenic evolution of Mt. Canigou during the Neogene (after Calvet et al. 2008, 2013, basement geology after Guitard 1970, modified). Key to symbols and ornaments—**1**: Coll de Fins Formation. **2**: Aquitanian and Lower Burdigalian. **3**: Silurian–Devonian–Lower Carboniferous (Conflent and Villefranche Hercynian nappes). **4**: chlorite and muscovite schist; Jujols Group and Upper Ordovician (Conflent), Jujols Group (Vallespir). **5**: biotite schist, cordierite–andalusite schist (middle Canaveilles Group). **6**: Canigou biotite augengneiss (Ordovician metagranite). **7**: Quazemi biotite–muscovite leucogneiss (Ordovician meta-leucogranite). **8a**: Hercynian biotite–muscovite leucogranite ('granite profond du Canigou' on geological maps). **8b**: Saint-Laurent biotite granodiorite. **9**: sillimanite micaschist (lowermost Canaveilles Group). **10**: vertical uplift during the Neogene. **11**: active fault. **12**: clastic flux directions. **12a**: clastic input to the Conflent Basin; **12b**: clastic outflux towards the Roussillon Basin and Mediterranean rifted margin. S: range-top surface; P: middle Miocene range-flank pediment (P1)

Box 11.2 Mt Canigou: indicators of its Neogene uplift

The 2-km-high Canigou range front was entirely generated by tectonic uplift in post-Burdigalian time, i.e., after 16 Ma. Evidence in support of this is provided by the Escaro Formation: whereas, from top to base, outcrops across Mt. Canigou represent the inner core of the Canigou–Carança metamorphic dome (two-mica granite, high-grade micaschist, grey leptynite, migmatite), none of these lithologies occur among the clasts inventoried in the Escaro Fm. The Escaro sequence consists instead entirely of debris supplied by the outer metamorphic envelopes of the dome, i.e., augengneiss and low-grade micaschist. Lithologies from the inner core of the dome only begin to appear as pebbles and boulders in the younger clastic beds of the regional stratigraphy, i.e., in the Coll de Fins Fm. and in the Pliocene sequences further down the Têt valley. Given the known age of these sedimentary sequences, it can be inferred that the fault scarp began to form during the late Miocene. Mt. Canigou as an elevated massif is therefore no more that 10–12 million years old, and is the result of fairly rapid crustal and topographic uplift. Figure 11.19B summarises the uplift and denudation scenario for the Canigou massif:

1. Kilometre-scale uplift between 24 and 18 Ma was accompanied by coarse terrigenous sediment aggradation in the Conflent Basin. Sediment input from the north (Madrès massif) was comparatively subdued after Aquitanian time (Marquixanes Fm.), consistent with the more limited uplift of this northern footwall block. Meanwhile, the basin recorded an influx from the south of schist debris and olistoliths, followed by a massive input of boulders and olistoliths of augengneiss. The sedimentary record totally lacks input from rocks forming the core of the metamorphic dome (granites, leptynites), thereby suggesting that the Canigou block was not yet deeply eroded at that time.

2. Relative tectonic stability and opportunities for pediment expansion (18–10 Ma). Erosion during that time interval supplied fine-grained sediment to the Roussillon and to other basins of the Mediterranean seaboard (sandy and argillaceous molasse).

3. Kilometre-scale uplift occurs (10–0 Ma). Denudation has now reached the deeper magmatic roots of the Canigou dome structure, breaching the outer metamorphic envelopes and eroding into newly exposed core outcrops. Debris of micaschist, granite, migmatite, and Quazemi leucogneiss, which together form the assortment of outcrops on the present-day mountainside, thus all occur in the younger Coll de Fins Formation (Upper Miocene) and in younger (Pliocene) sequences of the lower Conflent Basin.

Apatite thermochronology has consistently failed to produce analytically reliable results for Mt. Canigou, thus failing to document its most recent stages of uplift. The Aquitanian and Burdigalian paroxysm, which generated the thick Codalet and (primarily) Escaro formations, is documented by three AFT ages of 25.1 ± 1.0, 22.6 ± 2.2, and 21.6 ± 2.0 Ma, and likewise by helium ages ranging from 23.0 ± 1.2 to 17.8 ± 0.9 Ma. A single age of 17.7 ± 1.2 Ma was obtained from an intraformational sample of gneiss within the Escaro Fm.; track length modelling documents hydrothermal heating near the fault zone, thus testifying to its activity during the middle to late Miocene (Maurel et al. 2002, 2008). Milesi et al. (2020) documented two major periods of tectonic activity along the Têt Fault: before 20 Ma, and after 10 Ma, the two separated by an intervening period of relative quiescence which provided opportunities for generating pediment P1.

The elevated erosion surface known as Pla Guillem (2300 m), just visible to the west side of Mt. Canigou (see Itinerary 4, Fig. 9.17) and vertically disconnected from any recognisable base level, has survived rapid uplift of the upthrown block. A planar landform of this kind would have formed in closer connection and proximity to a local base level, most likely after the cycle of Aquitanian to Burdigalian aggradation in the Conflent Basin, and likewise of the 'Série rouge inférieure' in the Roussillon Basin (see Itinerary 2). This is consistent with the fact that Pla Guillem is an extensive bevel cut into the top of the metamorphic dome, which was progressively eroded to its core and produced the two early Neogene sedimentary sequences just mentioned. By analogy with the other age-bracketed erosion surfaces in the Corbières, Agly, Aspres and Montalba previously explored along itineraries 1, 2, and 5, Pla Guillem is thus a vestige of P1 rather than of the older summit surface, S. The ancestral surface, S, may have existed on Mt. Canigou, as it still does on the Madrès to the north and elsewhere further west, but it must have been destroyed by the more intense denudation rates, which appear to be regionally unique to the Canigou horst, in response to its late Neogene tectonic instability, well recorded by AFT and (U–Th)/He data.

Box 11.3 The northern boundary of the Conflent Basin
The Devonian limestones of the Hercynian Villefranche syncline appear clearly truncated by an erosion surface known as Pla dels Horts. It has been tilted to the SE and thus extends from elevations of 1450 down to ~1250 m. This is a specimen of Miocene pediment P1, notched barely 200 m below the summit interfluves among which vestiges of the summit surface, S, have also been mapped (Figs. 10.12, 11.4 and 11.11B). Unlike the south side of the Conflent Basin, this north side was comparatively inactive during early Miocene time and underwent limited uplift. This is evidenced by clast provenance among the Miocene sedimentary sequences of the Conflent, indicating very limited contribution from the northern catchments.

The river Têt and its tributaries cut a succession of epigenetic gorges into the limestone mass (see **Stop 7**). In the absence of stratigraphic constraints from unconformable Pliocene deposits (unlike in the Roussillon and lower Conflent), epigenetic superimposition (followed by gorge downcutting into the underlying limestone) most likely began with an ancestral Têt river flowing across the top of the Coll de Fins Fm., with incision thus beginning during the late Miocene, briefly accentuated in response to the Messinian Salinity Crisis (eustatic fall), and continuing thereafter. Epigenetic gorge incision thus probably began earlier at Villefranche than among the epigenetic gorges previously decribed in the lower Conflent Basin, i.e., from Marquixanes to Rodès (**Stop 1**, **Stop 2**). The successive gorges are diachronous rather than coeval. The position of the gorges on the northern edge of the upper Conflent Basin arises because this trunk river was forced northwards by the fan systems supplying the Coll de Fins Fm. from the actively rising Canigou–Carança massifs in the south. The river met with few options other than to superimpose itself onto this basin fill sequence and eventually incise it and the underlying bedrock in several stages. Successive generations of bedload deposited by the Têt during the incision process were trapped in groundwater karst cave systems, the most elevated of which occurs at the (modern) altitude of 1350 m, followed by a sequence of younger cave levels all the way down to the present thalweg ca. 1000 m below (see **Stop 10**). The current course of the Têt is thus chronologically antecedent to the vertical uplift and tilting motion undergone by P1 at Pla dels Horts. Furthermore, the tectonic movements were sufficiently gradual for the Têt to maintain is course along the northern edge of the Conflent graben and not shift southward to a more axial position within the clastic basin.

*Continue along the D27, which crosses the boundary fault (the gneiss directly abuts the Escaro Formation. at col de Vernet, ~400 m south of **Stop 7**, spot elevation 706 m). At Sahorre, stay on the D27 and continue towards Escaro. The road cuts through a vestige of alluvial terrace T4 in the first hairpin bend, and reaches Coll de Fins (897 m), which lies on the boundary fault.*

Stop 8. Late Neogene sequence at Coll de Fins

About 100 m south of the col, and resting on the gneiss, you should encounter some huge, rounded quartz boulders; these are interpreted as apical vestiges of a Coll de Fins alluvial fan (Fig. 11.20C). North of the col, walking up the slopes of Roc Colomer you will observe equally large boulders of augengneiss that clearly belong to the Escaro Formation (Fig. 11.20A). Walking down the NNE side of the hill and over towards the col at elevation spot 897 m, note that the interfluve on either side of the track is strewn with rounded pebbles and boulders of white quartz, and thus of a kind typically absent from the underlying Escaro Formation (Fig. 11.20D). The type exposure for the Coll de Fins Fm. was originally described along a fire-break trench north of elevation spot 904 m, roughly on the 880 m contour (Calvet 1996), but the site is now degraded and overgrown. The section displayed a few pebbles of two-mica leucogranite, quartz-rich sand, and clays dominated by kaolinite.

Looking south, note the large triangular facets at the main fault scarp. Here they expose fully 500 m of orthogneiss at the base of Pic de Tres Estelles (2099 m), and display embryonic pediment benches around 1600 m. Leucogranite veins and dykes occur exclusively below the 1800 m contour.

After the col, the D27 crosses the former mining area of Aytua, still active during the early 1950s, and continues to Escaro. The iron ore occurs in micaschist containing limestone beds beneath large gneiss olistoliths; the micaschist units are themselves displaced and were transported down the scarp face into the basin, where they form allochtonous masses amidst the Miocene sediment fill to the west of Escaro (revealed by detailed mining reports: Huard 1972*). Escaro hosts a small museum about the mining history. The last iron mine to close in Conflent was Escaro Nord, in 1963. Head north of the village to inspect the abandoned opencast excavations.*

Stop 9. **Neogene exposures at Escaro**

Fluorite mining took over iron extraction from the 1970s until 1993, the deposit turning out to be among the largest in Europe at the time. The ore could only be reached by removing the overburden, i.e., the Escaro Formation, destroying the hamlet of Escaro d'Amont, its church, its cemetary and its farmland in the process. The section (Fig. 11.20E) shows a 200 m vertical exposure of the pebble and boulder beds of the Escaro Formation, dipping 30–35° ENE and displaying rare interlayers of ochre or grey claystone containing iron pisoliths. An attempt at $^{26}Al/^{10}Be$ burial dating of some pebble levels has yielded highly dispersed results, including two highly improbably burial ages of 1.6 and 3.3 Ma (Sartégou et al. 2018). If confirmed by further studies in the future, such exceedingly young ages for the Escaro Formation would accentuate beyond reason the notion of very active neotectonics and intense erosion occurring during the Pliocene and Quaternary, thereby propounding an implausibly major geodynamic event in the eastern Pyrenees at that time and overturning currently established evidence of more progressive Neogene uplift and erosion. Claims that the Escaro Formation was a Pliocene sequence had also previously been proposed on the basis that the sediment facies were similar to the coarse-textured Pliocene deposits of the Rousillon; by extrapolation, it was postulated that the Escaro Fm. was a Zanclean sequence filling the Messinian estuary which, as Clauzon et al. (2015) advocated at the time, would have reached far into the Axial Zone ca. 5.3 Ma, and in the present case as far west as the upper Conflent. For reasons of regional consistency and current knowledge based on published geological maps, the biostratigraphic

◄**Fig. 11.20   Coarse-textured terrigenous deposits of the upper Conflent Basin. A** Boulder of augengneiss, Escaro Fm. Outcrop in a proximal position, on SW face of Roc Colomer, near Coll de Fins. The large boulder rests on a bed of small rolled pebbled. **B** Escaro Formation. Typical facies consisting exclusively of gneiss, here on the track to Badebany near Sant Climent de la Serra. **C** Coll de Fins Formation; large boulder deposit (quartz) to the SE of the col. **D** Coll de Fins Formation, exclusively quartz on the surface, along a track north of the col (topographic benches between elevation spots 897 and 904 m). **E** Deep exposures of the Escaro Fm. in opencast fluorite mine of Pla de Ganta. This photograph from 1984 shows a ~200 m vertical exposure of fresh sections (mine still in operation at the time). The excavation is floored by micaschist and marble olistoliths. The NE to ENE dips, uniform from base to top, indicate that tectonic deformation was postdepostional rather than synsedimentary. The tread of pediment P2 cross-cuts the upturned Miocene beds as well as the adjacent basement outcrops to the south and west

ages obtained at Espira de Conflent (previous **Stop 2**) are until further notice the most robust available, thereby ascribing an early Miocene rather than a Pliocene age to the Escaro Formation.

Optional trailwalk: Coll de Mentet and the summit erosion surfaces

Since almost all mountain tracks are now closed to traffic in the area, driving up to Coll de Mentet is the only easy access to the erosion surfaces of the footwall highlands surrounding the Conflent Basin. From Sahorre to Py, the D6 follows the Rotja gorge, which hosts the large knickpoint situated on the boundary fault. From Py to the col, the road follows the Neogene Py–Mantet fault line, its plane and crush zones in full view at the parking area in Mantet (village).

◀**Fig. 11.21** **Recent and historical flash flooding in the Mantet mountainside catchment.** **A** Ressec channel after 1992 flood. On both sides of the channel, note remains of debris-flow deposits, here utilised as a meadow, with debris recycled into the construction of several walls and a paved footpath (see text about a radiocarbon age obtained near the top of the deposit). **B** Junction of the Ressec and Alemany streams. Unlike the Ressec, the Alemany was unresponsive in 1992; the preserved riparian strip is rooted in deposits from the 1940 flood

Walking down from the village to the Ressec stream (1 hr return) reveals the damage caused by the September 26, 1992 flood, which here was nearly as intense as in 1940 (Fig. 11.21). Meanwhile, the neighbouring Alemany catchment remained unscathed (Fig. 11.21B). This illustrates the highly nonuniform spatial distribution of rainstorm hazards in Mediterranean environments. Older debris-flow deposits are distinguishable as boulder beds among the floodplain meadows on both sides of the Ressec. One of the deposits was chosen for establishing a charcoal smouldering pit, radiocarbon-dated to 1450–1635 cal ^{14}C (Ly-15912; in Laumonier et al. 2015) and thereby suggesting that the deposit hosting it was emplaced by the historically documented flood of 1421.

From Coll de Mentet (1750 m), a trail through the forest allows you to reach the summit erosion surfaces (5 hr return), first passing over Pla Segalà (2200 m) on orthogneiss, then Cim de Pomerola (2456 m) and the vast Pla de Campmagre (porphyritic granite of the Costabonne batholith), conspicuously injected with quartz dikes such as at Les Esquerdes de Rotja in the south (Fig. 11.22A, B). Two AFT ages were obtained from Pla Segalà (28.8 ± 1.4 Ma) and Cim de Pomerola (29.3 ± 1.8 Ma), completing those of Pla Guillem and the top of the Madrès (Fig. 11.15) and suggesting a post-Oligocene age for this low-relief topography (Gunnell et al. 2009). The sweep of the other erosion surfaces can be appreciated from Pomerola over a distance of ~15 km, from Pla Guillem to Coma Ermada. The plateau is covered by a deep mantle of grus containing corestones, at places up to 50 m thick and clearly visible around the rims of Rotja cirque in granite and Rocs Blancs cirque in gneiss (Fig. 11.22C). The massifs were lightly glaciated and the summit surfaces unscathed by ice dynamics. Cirque glaciation was likewise not intense, except—as in the Madrès massif and around Pla Guillem (see Itinerary 4)—in east-and south-facing environments.

*Return to Sahorre on the same road and continue to Villefranche-de-Conflent along the D6. The road follows the treads of alluvial terraces T1 and T2, then enters the epigenetic gorge of La Rotja in Devonian limestone. Along the way, note the vast entrance to Fuilla Cave, which is the upstream access to the vast underground cave network of Les Canaletes (Les Canalettes) (26 km); this groundwater karst system cuts through the entire massif, although only the eastern part is open to visitors for guided tours. At Villefranche, the parking area outside the city walls at the top of town will serve as an observation platform for **Stop 10**.*

◀**Fig. 11.22** Elevated vestiges of pediment P1 around Cim de Pomerola. **A** View from Pla Segala. The Pomerola monadnock rises against the skyline, with quartz dykes standing proud of its mostly convex slope profile. **B** Panorama of Pla Segala towards the south. Erosion surfaces rise to ~2450 m on the Franco-Spanish border (Pla de Campmagre, Pla de Coma Ermada, Rocs Blancs). The Py–Mantet Fault appears to have vertically offset P1 in the far distance, hence its occurrence at 2650–2700 m in the Donya massif. **C** Vestige of P1 at Rocs Blancs. Note thick, continuous, in situ grus and saprolite formed at the expense of the coarse-grained gneiss. The headwalls of the small glacial cirque (Jaça del Callau) cut indiscriminately into the weathered mantle and the underlying fresh bedrock

Limestone karst of the upper Conflent Basin

Box 11.4 The Têt canyon at Villefranche

Stages in canyon incision are inferred from the sedimentary record of sub-aerial fluvial terraces and alluvial deposits trapped in the groundwater karst system (Figs. 11.23 and 11.24). Fluvial incision began after deposition of the Coll de Fins Fm. (see **Stops 7 and 8**), i.e., after 10–8 Ma. An indication of the position of the Pliocene valley-floor base level is provided by vestiges of P2, which occur among the limestone massifs along the south side of the Têt at En Bullà (Embulla) (karst polje floor at 711 m, clearly detectable on maps, and which can be explored on foot). An indication of the early Quaternary base level has been identified just below the foundations of the fort situated above Villefranche (Fort Liberia), where a vestige of terrace T5 occurs (Fig. 11.24A, B; the base of the alluvial formation lies at 540 m a.s.l.). This alluvial level has provided ESR ages of 1133 ± 159 ka and 1099 ± 179 ka (Delmas et al. 2018). The alluvium is itself covered by 70 m of brecciated limestone talus deposits. The slope debris are cemented together and exhibit the fabric of periglacial scree in the upper part—an indication of cold-climate processes prevalent at these low elevations during the Pleistocene. At this location, the entire canyon wall is mantled by debris aprons forming slopes close to the angle of repose. The valley floor is also lined with specimens of T2 (tread occurs + 25–30 m above the Têt channel bed) and T1 (+15–20 m).

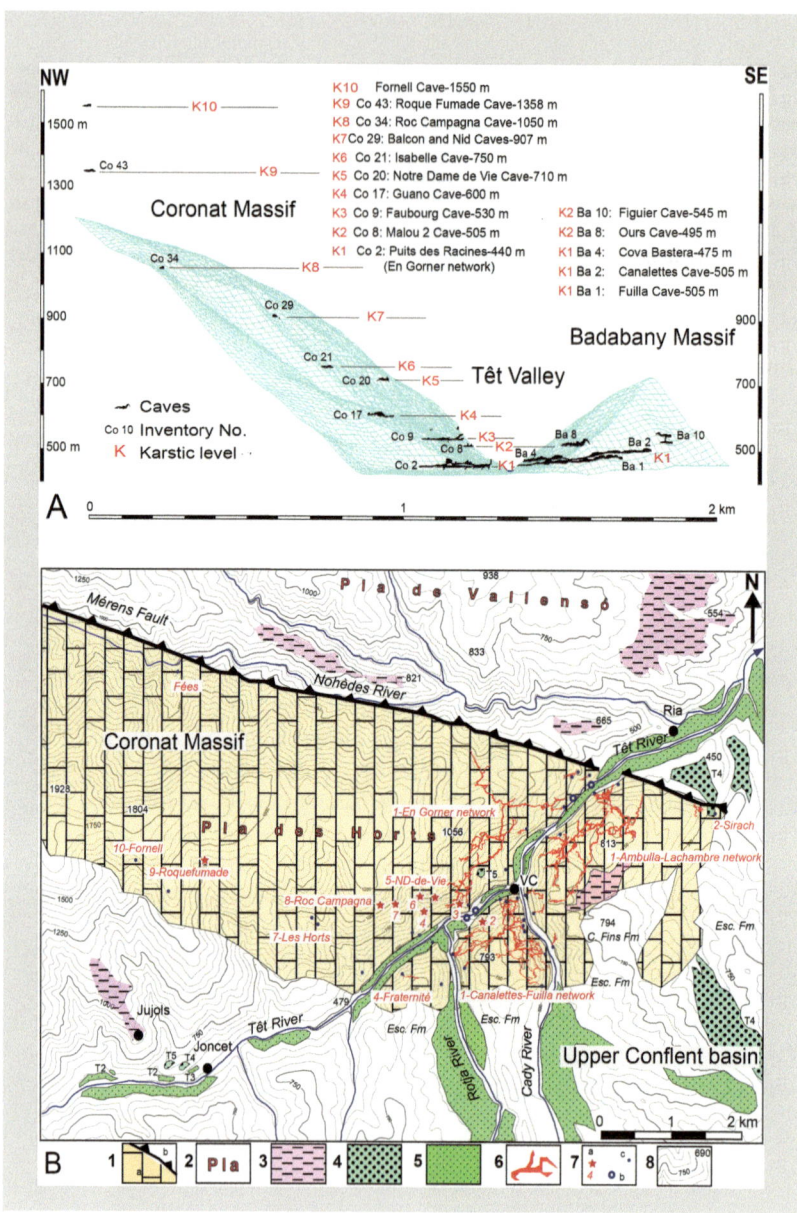

A

NW

SE

K10   Fornell Cave-1550 m
K9 Co 43: Roque Fumade Cave-1358 m
K8 Co 34: Roc Campagna Cave-1050 m
K7 Co 29: Balcon and Nid Caves-907 m
K6 Co 21: Isabelle Cave-750 m
K5 Co 20: Notre Dame de Vie Cave-710 m
K4 Co 17: Guano Cave-600 m
K3 Co 9: Faubourg Cave-530 m
K2 Co 8: Malou 2 Cave-505 m
K1 Co 2: Puits des Racines-440 m
     (En Gorner network)

K2 Ba 10: Figuier Cave-545 m
K2 Ba 8:   Ours Cave-495 m
K1 Ba 4:   Cova Bastera-475 m
K1 Ba 2:   Canalettes Cave-505 m
K1 Ba 1:   Fuilla Cave-505 m

Coronat Massif

Badabany Massif

Têt Valley

Caves
Co 10 Inventory No.
K   Karstic level

0   1   2 km

B

N

Pla de Vallenso

Mérens Fault

Nohèdes River

Fées

Ria

Têt River

Coronat Massif

Pla des Horts

1-En Gorner network

2-Sirach

10-Fornell

9-Roquefumade

1-Ambulla-Lachambre network

8-Roc Campagna

5-ND-de-Vie

VC

7-Les Horts

C. Fins Fm

Esc. Fm

4-Fraternité

1-Canalettes-Fuilla network

Jujols

Joncet

Têt River

Esc. Fm

Esc. Fm

Upper Conflent basin

Rotja River

Cady River

0   1   2 km

1   2 Pla   3   4   5   6   7   8

◄**Fig. 11.23   Upper Conflent groundwater karst in Villefranche syncline limestone.**
A Multilevel cave networks of the Villefranche karst. Ten epiphreatic cave levels have
been inventorised, but some subdivide locally into sublevels. Level 1 corresponds to
the Canalettes, En Gorner and En Bullà (Ambulla) systems, and can be subdivided
(see Fuilla–Canalettes) into currently active and abandoned passages, vertically distributed between the present water table and the treads of subaerial terraces T1 and T2,
respectively. Documented caves are indicated by their inventory number. Digital elevation data from Institut Géographique National (25 m ground resolution), processed
using Visual Topo (after Hez et al. 2015, and Calvet et al. 2015b, 2019, modified). **B**
Map of the Villefranche syncline karst. Key to symbols and ornaments—**1a**: Devonian
limestone outcrops; **1b**: Mérens Fault; **2**: main Miocene pediment, P1 (note widely
spaced contour lines); **3**: Pliocene En Bullà polje and other low-gradient erosional
landforms of P2 affinity; **4**: upper fluvial terraces T5 and T4; **5**: middle and lower
fluvial terraces T3, T2 and T1; **6**: large underground karstic networks (mainly lower
levels); **7**: karstic landforms. **7a**: older karstic levels, with their level (K) number (2:
Ours; 3: Faubourg; 4: Guano; 5: Notre-Dame de Vie; 6: Isabelle; 7: Nid, Balcon and
Horts; 8: Roc Campagna; 9: Roquefumade); **7b**: main karstic spring (active); **7c**: other
cave entrances. Altitudes in metres. After Calvet et al. (2015b, modified)

◀**Fig. 11.24** Features of the epigenetic Têt canyon through the Villefranche syn-cline, linking the upper and lower Conflent half-grabens. **A** View from En Bullà plateau. The canyon has cut a 840-m-deep gorge below the local tread of pediment P1 (Pla dels Horts–Pla d'Auça); vertical sequence of alluvium-filled caves is plotted on the canyon wall, tracking the incision history by the Têt; ESR age of terrace T5: ~1.1 Ma (after Delmas et al. 2018). **B** Links between cave level K3 (Faubourg Cave) and subaerial terrace T5. The talus deposits (at their angle of repose) overlying T5 are stratigraphically younger than 1.1 Ma, but their strongly cemented character (com-pared to Late Pleistocene talus deposits elsewhere in the region) suggests they are at least Middle Pleistocene. Note numerous cave entrances opening out on the canyon wall

The groundwater karst system remains the most promising environment for understanding the history of fluvial incision by the Têt into the uplifting eastern Pyrenees. Cave scientists have mapped at least 10 main levels of subhorizontal groundwater karst cave networks (Fig. 11.23A). Their occur-rences range between 1550 and 400 m above sea-level, and were each formed during a flooding stage at an altitude that coincided with the top of the phreatic zone (Fig. 11.25). Some of these levels present closely-spaced sublevels (vertical offsets of 2–10 m; Hez et al. 2015; Calvet et al. 2015; 2019). Each cave generation formed in connection with a local base level dictated by the position of the Têt River channel and by the corresponding position of the water table (Fig. 11.25). Given the great cave-generating potential of processes operating within the epiphreatic zone, each of the 10 passage systems represents a pause in the canyon's incision history. These pauses provided opportunities for at least parts of the Têt channel sys-tem to enter the canyon walls through apertures produced by groundwater processes during the preceding phreatic phase, and to deposit exogenous bedload (Figs. 6.13C–F, 11.26D). Vestiges of alluvial sand, gravel, as well as thick sequences of finely laminated silts, which at places have occluded entire passageways, occur in most of the conduits explored so far. The most extensive underground networks occur among the lower levels, i.e., Fuilla–Canaletes, En Bullà–Lachambre, and En Gorner, which together add up to a 70-km-long underground cave network. These systems provide a robust reference model for understanding the cave sequence chronology because their underground treads are very clearly correlated with subaerial alluvial terrace deposits T2 and T1 (Hez et al. 2015). Access to the caves is often closed off by padlocked gates, particularly at En Bullà–Lachambre, which is a designated heritage site. Contacting the local caving clubs for further information is the best option should you wish to investigate further.

◀**Fig. 11.25** **Examples of cave interiors used for characterising the multilevel karstic sequence.** All the cave passages are horizontal; they display wall and ceiling morphologies typical of phreatic or epiphreatic environments (scallops, ceiling channels, pendants, cupolas…) but periodically also evolving under vadose conditions ('water table passage' environment, generating remarkably horizontal cave-wall benches and notches). **A** Roquefumade Cave. Explored and mapped as level K9, it is the most elevated horizontal passage of the sequence (level K10 is poorly documented). **B** Balcon Cave (level K7). Passage height: 1.4 m; note pendants. **C** Exogenous alluvial fill in Balcon Cave. Note imbricate structure in the pebble fabric, clearly indicating an in- rather than outflowing channel. **D** Main passage in the En Bullà–Lachambre network, nicknamed 'Boulevard du Canigou'. Note symmetric pairs of cave-wall notches and a lag deposit of allochtonous fluvial gravels (credits: S. Jaillet). The tread of this passage connects to the tread of subaerial terrace T2, but its upstream segment was rewatered during the last glacial cycle (sinkholes of the Cadí, a losing stream). **E** 'Passage Pagès', main entrance to the En Bullà–Lachambre network. This passage functioned alternately in drowned and vadose environments, probably as the main outlet to the entire system; the cave-wall bench here is covered by a coarse gravel deposit, conveyed by the nearby Têt at the time of aggradation of T2. **F** 'Galerie des Racines', upstream portion of En Gorner Cave system (Puits des Racines). This passage is situated just beneath the tread of terrace T1. The sand on the surface covers a sequence of pebble beds deposited by the Têt; as elsewhere, the cave environment was alternately epiphreatic (ceiling channel; wall scallops, including on upper walls) and vadose (sidewall notches). This level of the network gets locally rewatered during large flood events

Box 11.5 Dating a population of alluvium-filled caves

Calvet et al. (2015) studied two endokarstic levels (noted K in illustrations) of the vertical cave sequence above Villefranche (Fig. 11.23A): Notre-Dame de Vie Cave (Nostra Senyora de Vida; K5, + 270 m), and Faubourg Cave (K3, + 110 m). The latter is situated just below subaerial terrace level T5 (ESR ages given in Box 11.4), and both cave systems yielded consistent $^{26}Al/^{10}Be$ alluvium burial ages (methodology explained in Granger et al. 1997, 2001; Stock et al. 2005; Calvet et al. 2015). The weighted mean burial age for sediment contained in K5 was 5.14 ± 0.41 Ma, thus indicating an early Pliocene event. The cavity network itself is, of course, older; perhaps late Miocene. This inference is supported by the fact that the lower beds of very thick silt sequences deposited in some galleries contain broken stalactites. This stratigraphy documents a vadose environment prior to cave invasion by the fine sediment. The varved silts capping the gravel bedload deposits have been tentatively correlated with the Zanclean marine transgression which, after sharp valley incision driven by the Messinian Salinity Crisis, generated the drowned estuary environments of the lower Têt valley and other circum-Mediterranean rivers. The Zanclean sea-level rise almost certainly induced fluvial aggradation in the upper reaches of valleys such as the Têt, with rising water tables in the adjacent limestone masses at the same time generating epiphreatic environments conducive to the corrosion and upward expansion of the groundwater karst. Cave system K3 yielded weighted mean burial ages of 2.23 ± 0.23 Ma in its upper sub-level, and 1.20 ± 0.28 Ma in its lower. These mean ages were obtained from separately dated aliquots of sand and gravel, respectively. The youngest age from K3 (lower sub-level) matches the ESR ages obtained for alluvial terrace T5. On that basis, it becomes possible to infer a mean valley incision rate of 52.5 m/Ma since early Pliocene time, but involving an acceleration between the Pliocene (55 m/Ma) and the Quaternary (92 m/Ma) as well as an interval of base-level stability (perhaps lasting up to 1 Ma), without which K3 would not exist.

A proliferation of sediment burial-age determinations in the local cave systems, particularly among the lower cave levels (Sartégou et al. 2018), has considerably muddled the initial clarity of the chronology summarised above. Among several issues, one of relevance here is that widely diverging ages have been obtained from sediment samples buried in the same cave. K1 is a good example, where an alluvial sequence encountered at the entrance of Lachambre Cave can be traced continuously over a distance of almost 1 km. The deposit has buried an equally continuous erosional bench notched in the cave wall (Fig. 11.25E). Despite the stratigraphic and geomorphological simplicity of this setting, six pebbles collected from the deposit at intervals of less than 1 m from one another yielded a wide array of individual burial ages ranging between 0.26 and 3.59 Ma. The Law of Included Fragments, which governs the logic of relative dating in stratigraphy and states that rock fragments must be older than the rock formation containing them, dictates that the most likely burial age of an alluvial deposit must be the age of its youngest dated clast. An age of 0.26 Ma or less for the Lachambre Cave deposit is thus the mostly likely default solution. This middle Pleistocene age is entirely consistent with the age compatible with MIS 6 obtained for alluvial terrace deposit T2 in the Têt valley slightly further downstream (Delmas et al. 2018)—a terrace level which, furthermore, grades topographically to the tread of this cave network (Hez et al. 2015). This preferred scenario has a corollary: it leads to the conclusion that the other dated clasts, which spuriously suggest they were buried in the cave for much longer than may actually be the case, must have entered the cave with lower $^{26}Al/^{10}Be$ ratios, R, than the theoretical value of ~6.75. This situation can arise when sediments conveyed by the stream to the cave were previously buried at a variety of depths below the land surface in various parts of the catchment, a possibility confirmed by numerous departures from the standard 'R = 6.7' among the data reported by Sartégou et al. (2018) for modern alluvium from the Têt, Cadí (Cady), and Rotja stream channels. This indicates that the Têt River has been transporting and mixing sediment from highly diverse subcatchments where debris residence times and pre-burial exposure histories were not uniform—thereby also providing a challenging counterfactual to the assumption, clearly verified in only some alluvial settings, that in the context of cosmogenic dating of alluvial deposits "nature does the averaging" (Gärtner et al. 2020). Contrary to interpretations produced by Sartégou et al. (2018) for this and other cave levels facing age-scatter issues, any alternative scenarios to the simple option presented here would entail highly improbable complications to the tectonic and/or eustatic history of the area, and thus should be ruled out until further notice.

Stop 10. Villefranche-de-Conflent, with a short walk up to Notre-Dame de Vie
(Nostra Senyora de Vida; 2 hrs return, powerful torch lamps recommended). Dating of alluvial deposits in terraces and canyon-wall caves documents stages of canyon incision by the Têt River. Looking up the valley side towards Fort Liberia from the car park at Villefranche affords good views of terrace T5, and of the cemented talus deposits mentioned in Box 11.4 (Fig. 11.24B). From this car park, a fairly short walk also allows you to reach a number of low- to mid-elevation cave entrances on the south-facing canyon wall. Notre-Dame de Vie, in particular, offers perhaps the easiest access to a cave entrance for visitors unequipped for exploring the local underworld without a guide. Leaving Villefranche, walk up the N116, cross the bridge over the Têt and, soon after it, the railway level-crossing. Ignore a track going sharp right towards the northeast, and which would lead you to the Puits des Racines (only 160 m away, but fenced off; this precinct

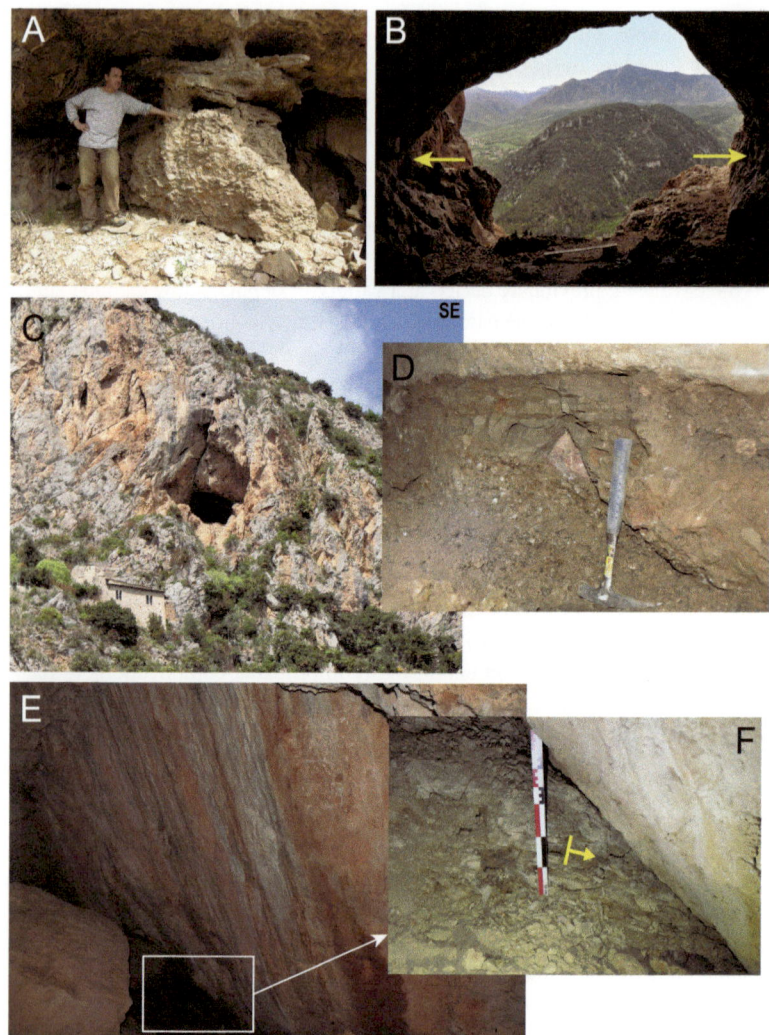

◄**Fig. 11.26 Trailwalk up to Notre-Dame de Vie Cave. A** Upper entrance to Faubourg Cave. This former entrance is filled up to the ceiling by ancient Têt alluvium and capped by speleothems. **B** Entrance to Notre-Dame de Vie Cave; note the sidewall notches. In the far distance: Serre massif (also karstic), Rotja valley, and Tres Estelles massif. **C** Notre-Dame de Vie. A number of cavities above the main entrance indicate the occurrence of a more elevated passageway; some passages, recognisable by their circular cross-section, were shaped in a phreatic environment. **D** Base of the cave's alluvial fill. It consists of a bed of small rolled pebbles (schist, quartz, granite) capped by laminated slackwater silts. The silt component is quite thick, and at one time probably plugged the entire passage. **E** Normal fault plane shuttering the far end of the passage. **F** Base of the fault plane, here offsetting the laminated silt stratigraphy

hosts an engineered access to En Gorner Cave network down a vertical shaft, and services Villefranche for its drinking groundwater supply. En Gorner corresponds to two cave levels coinciding with subaerial terraces T1 and T2; the two cave levels are plugged respectively by unweathered and weathered alluvial deposits; Fig. 11.25F). Aim instead for the well-marked trail which starts ~80 m to the west.

Before the trail begins its ascending loops, it is possible (but difficult) to branch off and reach either of the two entrances to Faubourg Cave. For this you need to clamber over a succession of abandoned dry-stone terraces. The upper cave entrance is particularly striking because it is entirely plugged by a cemented mass of gneiss, schist and granite pebbles; it corresponds to an ancient entry point for the Têt—at the time a losing stream—into the wall of the canyon (Fig. 11.26A). Back on the main trail and after a few hairpin bends, roughly at the same altitude as Faubourg Cave, the small entrance to Palangane Cave comes into view: allochtonous alluvial gravel occurs underfoot along the trail as you approach the cave and continues at least 2 m into the cave itself, where the bedload deposit is additionally capped by a sequence of grey silt. The main trail continues onward to Notre-Dame de Vie chapel. Immediately west of the chapel is a small cave still currently used as a warehouse; it is entirely plugged by grey-coloured fine silt. You gain from ascending to the porch of the main entrance, further up the slope (Fig. 11.26C). Inside the cave you will see the symmetric lateral benchlines characteristic of channel flow and corrosion in a vadose environment (Fig. 11.26B); and an equally characteristic ceiling channel and ceiling potholes, typical of an epiphreatic (i.e., almost entirely drowned) hydrological environment. The regular, oval shape of the passageway, here ~60 m long, is also an indication of phreatic dynamics. The end of the cave (torch lamps

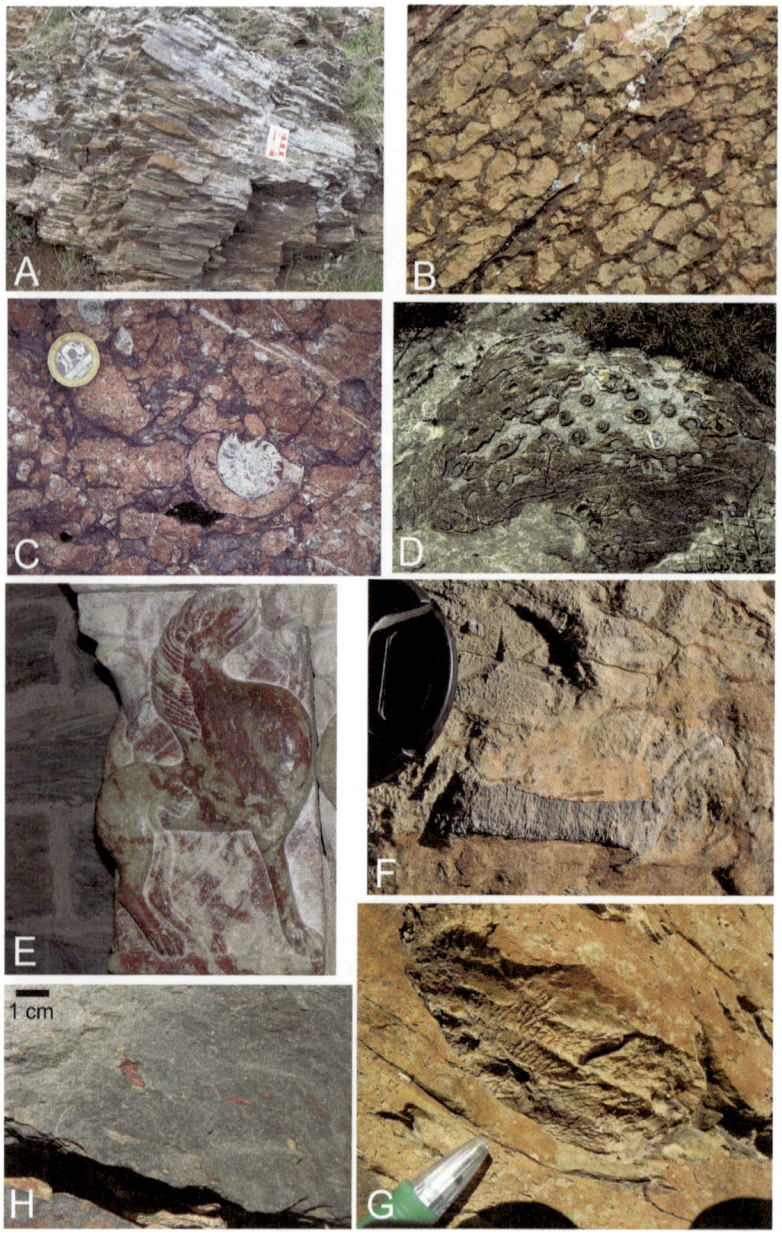

◀**Fig. 11.27** **Characteristic examples of Hercynian rocks making up the Axial Zone. A**
Flysch (Visean). Roc Sant Julia, Roquefumade unit. **B, C**. Griotte limestone (Upper Devonian: Famennian). White shape in C: Goniatite fossil. **D** Cherty reef limestone containing silicified corallites (Middle Devonian: Upper Emsian to Middle Eifelian). **E** Pink and white *Stromatactis* limestone (Devonian: Givetian–Frasnian). Known locally as 'marbre flambé de Villefranche', here part of a twelfth century capital at Serrabona priory (Aspres massif); greenish church walls in the background are in Jujols Group schist. The suite of rocks A to E is representative of the Villefranche nappe (i.e., relative allochton). **F** Silurian–Lower Devonian limestone containing fossil Crinoid stalks (Conflent nappe), seen here along the trail up to Els Horts near Costa de Campinya. **G** Greywackes containing *Orthis* fossils (Upper Ordovician: Ashgillian). Also known as 'schistes troués', these represent the oldest fossil-bearing Paleozoic rocks documented in the Pyrenees (they also contain other Brachiopods, alongside Bryozoans, Echinoderms and Trilobites). **H** Fucoid schists (Ashgill: Upper Ordovician). Fucoids remains are the red inclusions. **I, J** Puddingstone (Caradocian: Upper Ordovician), rich in quartz and quartzite pebbles, as encountered along the trail to Els Horts. **K** Canigou augengneiss (meta-porphyritic granite; Lower Ordovician). **L** Lower Jujols schists (banded quartz phyllites; Middle–Upper Cambrian). **M** Close-up of Jujols schists. Note sedimentary protolith made up of whiter (quartz-rich) silt and darker pelitic beds, cross-cut by Hercynian metamorphic cleavage (S1) through the microfold axial planes. **N** Canaveilles Group marble, intensely deformed (recumbent D1 microfold refolded by open D2 fold)

required) is shuttered by a N15°–60°E fault plane displaying vertical striations (Fig. 11.26E). These indicate a top-to-the-west dip-slip motion, and the fact that the varved silt deposit is also offset by faulting indicates that the last neotectonic event was geologically younger than the sediment fill (Fig. 11.26F). Two small galleries branching off on the left display vestiges of the silt deposits which, at one time, probably plugged then entire cave system; the base of the fill sequence also shows sand and small pebble layers (Fig. 11.26D) lying exactly at the level of the conspicuous corrosion benches observed back in the main cave. The pebble composition of this deposit is quite different from the lithological assemblages encountered among the lower cave levels, Grotte du Faubourg included, as here they consist overwhelmingly of schist and quartz, with only a minor presence of gneiss and granite.

Cultural stop: Villefranche, a UNESCO World Heritage site
Villefranche was built between 1088 and 1092 under the reign of Guillem Raimond I, count of Cerdagne–Conflent. Its purpose was to provide the Conflent with a capital and to drive economic development. It was thus originally a fortified market town. The two porches of its Romanesque church are sculpted out

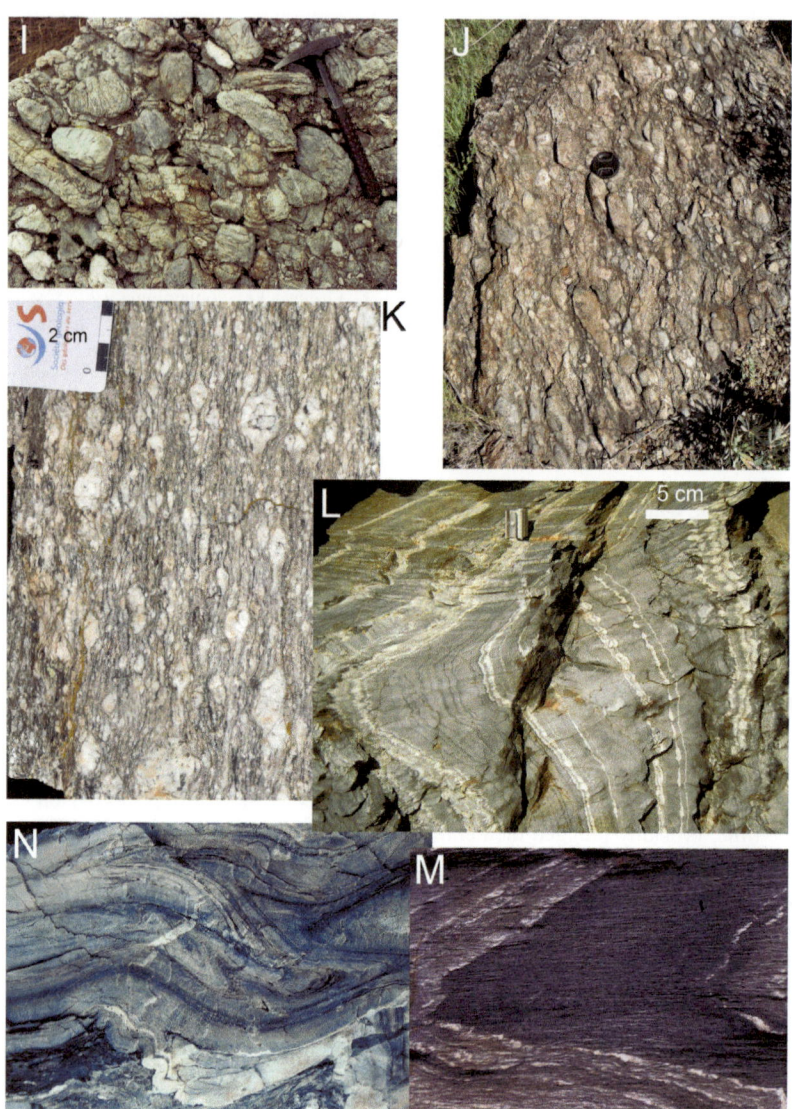

Fig. 11.27 (continued)

of Middle Devonian pink *Stromatactis* marble (from Villefranche) and Upper Devonian red griotte limestone (Fig. 11.27B, C); the quarries are situated at the mouth of the canyon. In the 17th century, the Marquis de Vauban took the initiative of reinforcing the Medieval fortifications by adding bastions to the curtain wall, and by constructing a new strategic fort above the town (Fort Liberia, just above the tread of terrace T5). Villefranche is one of 12 gazetted sites constructed by Vauban to have acquired World Heritage status (in the present case, in 2008). Another of these is the citadel of Mont-Louis, situated in the upper Têt catchment, farther up the N116 [Sébastien Le Prestre, Seigneur de Vauban (1633–1707), was a civil engineer, architect and military general who zealously served Louis XIV but was also a hydraulic engineer, urban planner and economist obsessed with eradicating poverty].

Get back on the N116 heading west. Before arriving at Serdinya, the road skirts around the slipped mass of a large landslide involving Devonian limestone and calcschist over more beds of Silurian red-and-black ampelitic schists (only few outcrops visible) and other varieties of Upper Ordovician schist. The Coronat massif, already encountered earlier from its north side in the valley of Nohèdes (see earlier trailwalks, and maps and cross-sections in Fig. 11.11 *and panoramic view of* Fig. 11.12B), *is explored here from its south side. It is arguably the best area in the Pyrenees for examining the intricacies of Hercynian tectonics in the Upper Paleozoic basement. As you enter Serdinya, the first bend in the road offers parking space.*

Stop 11. Serdinya and Els Horts (trailwalk: 3 hrs return).

The ascent provides opportunities to inspect outcrops of Upper Paleozoic rocks (Upper Ordovician and Siluro-Devonian) that stratigraphically overlie the Cambrian Jujols schists, which otherwise are not well exposed at this particular spot but have been presented in this GeoGuide at many other locations (note a few outcrops in the village itself; Fig. 11.27L, M).

Optional trailwalk: Paleozoic geology of the Villefranche syncline

The trail (signposted) starts from Serdinya before the curve in the road (Fig. 11.28A). As you ascend the valley side, the trail reaches conglomerate beds around 620 m a.s.l. containing light-coloured quartzite and white quartz pebbles, cemented by greenish or purple quartzite, and cross-cut by a dense network of veins of white quartz. These are the famous (Upper Ordovician) 'Caradoc Series' puddingstones (Fig. 11.28I, J). At regional scale, the Upper Ordovician units rest unconformably on Cambrian rocks (so-called 'Sardic unconformity', which cross-cuts a tectonically deformed and eroded Cambrian to earliest Ordovician Series, and to which are associated granitic intrusions that later became the protolith for large masses of Hercynian orthogneiss, such as those previously encountered in the Canigou dome,

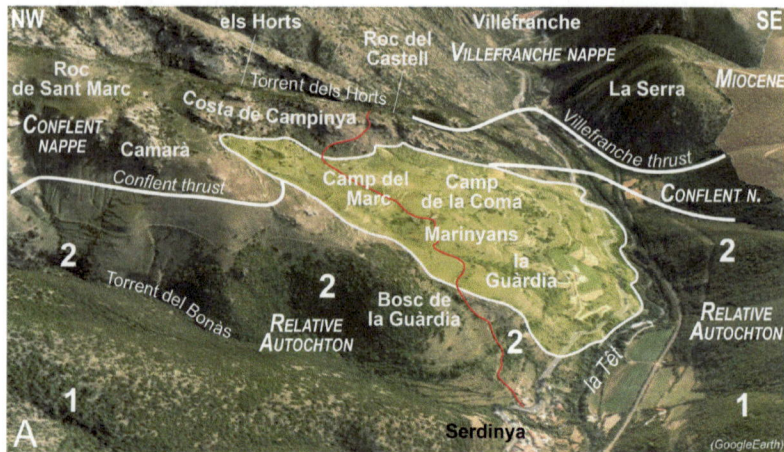

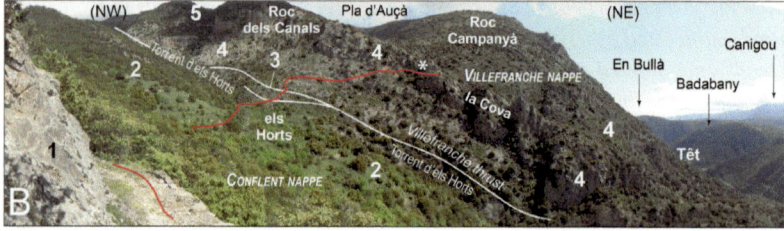

Fig. 11.28 Trailwalk to Els Horts. A Overview: the path trails across the large slipped mass of the Marinyans landslide (yellow). **1**: Cambrian (Jujols schists); **2**: Upper Ordovician (Caradoc puddingstone and Ashgill schist) forming the top of the relative autochton; the overlying Silurian to Middle Devonian Conflent nappe is separated from the autochton by the Conflent thrust (a detachment rather than an additive thrust); in the background: massive limestones (Villefranche nappe; imagery from Google Earth). **B** View of Els Horts from Roc del Castell. Trail in the foreground. **1**: partly dolomitic limestone (Lower to Middle Devonian) of the Conflent nappe. **2**: Silurian–Devonian pelite and carbonate beds of the Conflent nappe. **3**: Silurian beds at Els Horts, highlighting by their variegated colours the thrust sole of the Villefranche nappe. **4**: cherty limestone and marl containing silicified corallites (Middle Devonian), Villefranche nappe. This unit hosts most of the endokarst. White star locates tunnel-cave K7 above Els Horts. **5**: Roque Fumade unit (Devonian to Carboniferous)

Fig. 11.27K). The path eventually reaches La Guardia-Marinyans, a flat area corresponding to the upper portion of the slipped mass. This rock slope failure is quite ancient given that it has been reincised by the river below and rests on vestiges of terrace T2 beneath the N116. The landslide crown corresponds to the Costa de Campinya escarpment. It exposes in situ Devonian limestone and calcschist of the Conflent nappe. Continuing your ascent along the

Costa de Campinya, you will reach a first allochtonous unit (Conflent nappe) consisting of limestone containing large crinoids (Fig. 11.27F) and ochre patinations (Costa de Campinya), followed in the stratigraphy by dolomitic limestone (Roc del Castell). Where the trail reaches the abandoned farm of Els Horts (850 m), you enter another major allochtonous unit (Villefranche nappe; Fig. 11.28B). Its basal thrust occurs on the left bank of the gully, just above the larger ruined building, highlighted by black, purple and salmon-pink Silurian exposures. The Devonian Villefranche nappe here consists of Eifelian (Middle Devonian) cherty limestone with silicified corallites (Fig. 11.27D).

The Villefranche syncline is a highly complex Hercynian (Westphalian–Stephanian) structure. It consists of a vertical succession of recumbent folds and early thrusts cutting through the local autochtonous units (i.e., the Jujols schists), through the Silurian–Devonian Conflent nappe (itself consisting of two stacked units), and through the Devonian–Carboniferous Villefranche nappe (five stacked units). This entire structural system was subsequently warped into the shape of a syncline—the Villefranche syncline. The attendant Canigou–Carança dome to the south, a basement anticline, thus forms with the Villefranche syncline a tectonic continuum of Hercynian affinity, only more recently disrupted by Alpine tectonics and by the Conflent graben (Fig. 11.19B).

The limestone scarp above Els Horts displays a population of caves, many of which are plugged by allochtonous sand and gravel deposits. Their entrances lie just a few minutes' walk away. They occur at elevations between 900 and 950 m and correlate with groundwater karst level K7. The most remarkable cavity occurs in the east (~900 m a.s.l.), accessible from a path that follows the tops of a succession of dilapidated dry-stone buttressing walls. The path skirts around the limestone ridge, but another trail strikes directly through a cave in the limestone mass. This cave forms a tunnel 10 m long, and its walls are coated with vestiges of quartz-rich sand and gravel deposits (Fig. 11.29). From Els Horts, a 4 hr return trip along a trail rising up the right side of Querllong creek, then onward to the forest cabin of Roquefumade, provides the most direct access to Pla dels Horts (1450 m, pediment P1), which serves as an excellent natural viewing platform over the surrounding mountain landscape (Fig. 11.2C).

Continue on the N116 towards Olette. The road roughly follows alluvial level T2, here situated +30 m above the modern river channel and frequently buried by talus deposits (schist), particularly around Joncet. In this area, vestiges of T3, T4 and T5, the latter at +125 m, have been identified on the valley side as you ascend the track up towards Flaçà (Flassa). Turning up the steep road to Jujols, type locality for the Jujols Group, is an option. This remote village provides great views of Mt. Canigou, and is among the best ports of entry to exploring the upper regions of Mt.

Fig. 11.29 Els Horts tunnel-cave. A View from its eastern entrance. **B** Vestige of allochtonous alluvial fill pasted on passage wall. Note stratigraphy of alternating quartz-rich sand and gravel beds containing quartz, quartzite, and harder schist

Coronat and its Hercynian nappe stack. The N116 passes through the small town of Olette (with roads branching off to Évol and Sansa: see excursion options described earlier), before entering an increasingly narrow portion of the Têt valley. The tread of the road is still constructed just under terrace level T2, which rises progressively to +45–50 m.

11.3   Upper Têt valley

Stop 12. Canaveilles gorge ('Défilé des Graus') Parking possible before or after the tunnel.

Here you can descend to the 'Bains de Canaveilles' (ruins of a spa and hotel) located at the bottom of the gorge. The area provides good exposures of the contact between the orthogneiss and the base of the Canaveilles Group— easily identified by thick beds of marble (Fig. 11.27N). The gorge in this area forms a spectacular entrenched meander, a unique occurrence between otherwise straighter reaches up- and downstream. The extant alluvial terrace levels have also undergone substantial tilting, thereby suggesting a recently active tectonic hinge zone in this area (Calvet 1996; Delmas et al. 2018). Upstream of the tunnel, alluvial deposits belonging to generation T1 (unweathered pebble assemblages) occur at the anomalously high elevation of +48 m, and thus lie, for the first time since leaving Villefranche on the N116, above rather than below the road.

In the upper part of the gorge, you may want to stop off at the thermal baths at Thuès. A large number of thermal springs occur in this area along the Têt Fault. Whereas the opposite valley side to the north is highly unstable and subject to frequent slope failure (the road gets cut off for weeks at a time), the large faceted spurs in Hercynian mylonite along the south side of the valley are much more compact and stable despite the hot springs and the fault line. The faceted spurs appear particularly spectacular when observed from the village of Llar (Fig. 11.30); note the succession of cut benches above the spurs at 1400 m a.s.l., here corresponding to generation P2.

Fig. 11.30 **Faceted spurs along the Têt Fault between Thuès and Fontpédrouse. A**. Spurs on either side of the Carança River gorge, here seen from the trail linking Canaveilles to Llar. Note the topographic benches and older truncated spurs rising above the main facet. **B**. General view of fault scarp from the road to Llar. The low-gradient spur ridgetops (P2) can be followed in succession along the valley side all the way upstream to the Plateau de la Perche, also a very well-preserved vestige of P2 (see Itinerary 7)

Stop 13. Thuès and the Carança gorge

The top of the village and the church stand on the vestige of an alluvial terrace at +90 m. The sediment contains very large gneiss cobbles, displays limited weathering (T2 facies), but its exceptionally large thickness (45 m) suggests anomalous local conditions in this part of the valley. Similar thicknesses continue for a few kilometres farther upstream, where the deposit is continuous as far as Fontpédrouse. These features constitute strong indices of recent reactivation of the Têt

Fault. Faulting would have impounded incoming alluvial sediment (Delmas et al. 2018) and would also explain the fresh appearance of the faceted fault plane along the scarp base. Reverse faulting in Thuès gorge, or perhaps extensional slip along the Têt Fault, could explain trapping of the alluvium in a small N70E graben-like structure in the area. A rock exposure at Thuès railway level-crossing provides a clue in support of fault reactivation, revealing two vertical N15°E and N8°E shear zones at the boundary between the granite and the alluvium, with upturned pebbles along the fault plane suggesting fault drag rather than gravitational shear.

The Carança River gorge nearby has also recorded a similar history, detectable in the cross-section of its lower reach where the valley opens out at the level of the +90 m Thuès terrace but becomes a steep-walled vertical canyon below. The well-marked path at places provides dizzying views of the gorge, and allows the upstream portion to be reached within a 2.5-hr round trip. You will pass a series of knickpoints while walking through a geological succession from north to south displaying (i) Hercynian granodiorites (intensely mylonitised at the end of the Hercynian orogeny) resting on (ii) micaschist and marble containing injections of leucogranite, followed by (iii) Canigou–Carança gneiss. The entire sequence dips north. In the railway trench to the west, you should find exposures of the Têt Fault plane.

The N116 continues up to Fontpédrouse. Facing the Thuès hydropower plant, note the vestige of T1 at +35–40 m; at the Séjourné viaduct, the road reaches and follows a topographic bench still covered by the thick alluvium (T2) described at Thuès (Fig. 11.31A). The formation here consists of two facies: at the viaduct, you will encounter 25 m of weakly weathered alluvial boulder beds (gneiss and granite, Fig. 11.31B) filling a volume up to the level of the railroad track, i.e., +70 m; upstream of the Torrent de Canals and as far as Fontpédrouse, the material consists of debris-flow deposits (unsorted, angular clasts in a sand- and silt-rich matrix) suggesting high-energy breakout events descending from further up-valley (Fig. 11.31C).

Stop 14. Fontpédrouse and Prats Balaguer

The village of Fontpédrouse was a long-standing stagecoach stop on the road up to Cerdagne from Perpignan, its name alluding to the springs ('font': a fountain) rising from the thick and unstable talus deposits on the slopes all around. Fontpédrouse (1000 m) was also repeatedly damaged during the Little Ice Age by snow avalanches (44 fatalities between 1728 and 1822). These became more frequent because of hillside deforestation for pastoral land use.

Turn off onto the D28 and stop at the first junction near the cemetery. From there you get good views of the Têt Fault, which, here, cuts through Hercynian granitic mylonite. Unconsolidated gouge from Neogene fault slip crops out in

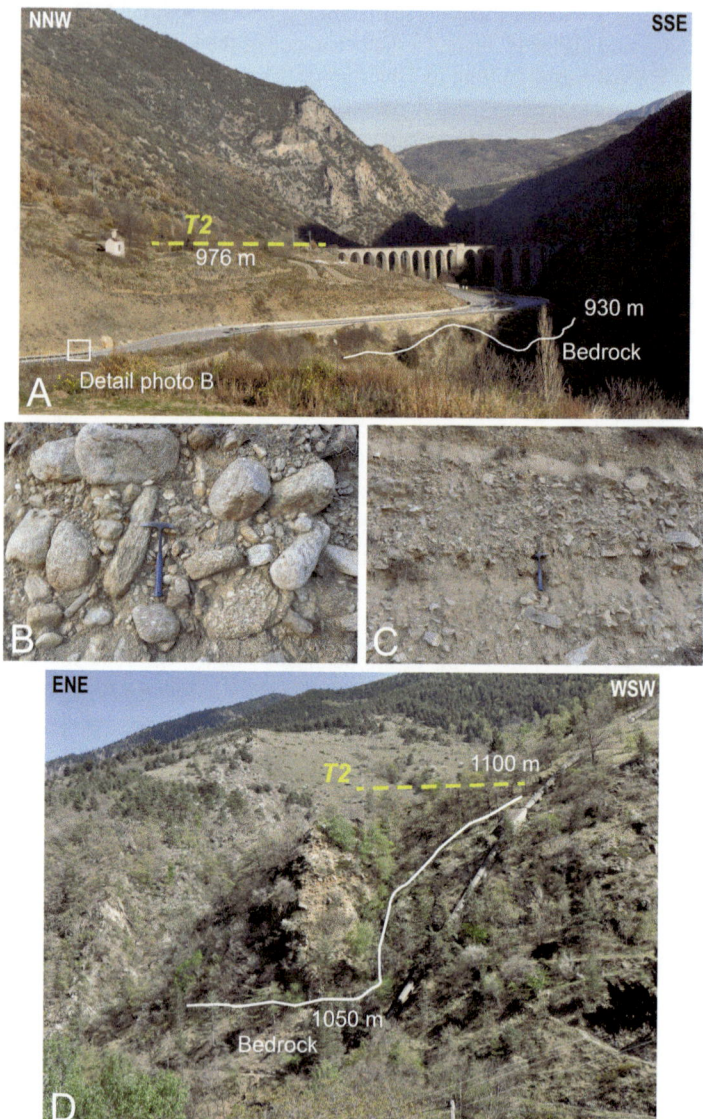

◀**Fig. 11.31 Thick aggradational deposit of terrace T2 in the upper Têt valley. A** Alluvial terrace T2 at Pont Séjourné. Here, the deposit is 45 m thick. **B** Basal boulder-rich bedload facies; section situated upstream of Pont Séjourné. Note limited post-depositional weathering, regionally typical of generation T2, throughout the entire sequence. **C** Debris-flow facies within alluvial deposit T2, here seen just downstream of Fontpédrouse (exposures in the N116 road embankments). **D** Section through T2 at Fontpédrouse, on south side of the Têt valley. Here, deposit thickness exceeds 50 m. Rolled boulders document high-energy flow conditions

the roadside ditch. By walking rather than driving along the D28 towards Saint-Thomas, over a distance of 100 m the area provides opportunities for observing at close quarters the strain gradient across the footwall boundary. The most spectacular feature is a 50-m-thick cataclasite (inner damage zone of the Têt Fault), with mylonite clasts and abundant white quartz cement and fractures. After the hairpin bend in the road, you should encounter outcrops of the much less fractured Hercynian mylonite (outer damage zone), followed by sillimanite micaschists intruded by non-mylonitic granitic sills. On the other side of the fault, symmetric observations of the hangingwall along the D28 will reveal poorly mylonitised orthogneiss and granite sills affected by cataclasis of the damage zone. The deformation at this point is composite: late-Hercynian mylonitisation and, probably, Neogene cataclasis. Alpine (Cretaceous–Paleogene) components of deformation may also exist in the area, but no known exposures exist to support the hypothesis.

Above the hydropower plant, the winding path rises up along an aggraded mass of boulder-rich alluvium, >40 m thick and still corresponding here to generation T2. It reaches relative elevations of at least +90 m (Fig. 11.31D). Continuing further up the road to Prats Balaguer, aim for the spur indicated on maps by spot height 1243 m. On the spur ridgetop you will find a number of large rolled boulders of gneiss and leucocratic granite, including a few pebbles (e.g., quartz) resting on the micaschist (Fig. 11.32B, C). These deposits belong to generation T5, here at + 200 m. Between the lower Conflent and this portion of the upper Têt valley, the offset between successive generations of terraces thus increases substantially and documents a persistence of regional uplift in the Axial Zone late into the Quaternary. Slope instability is accordingly greater in this area than farther downstream, as testified, for example, by the large slipped mass of Aussera, on the opposite valley side (Fig. 11.32A; the N116 bends around it). The landslide mass rests over components of T2 (outcrops of its pebbles and boulders can be seen just below the road; Delmas et al. 2018). It is likely that the landslide impounded the valley for some time, and that the surge subsequent to

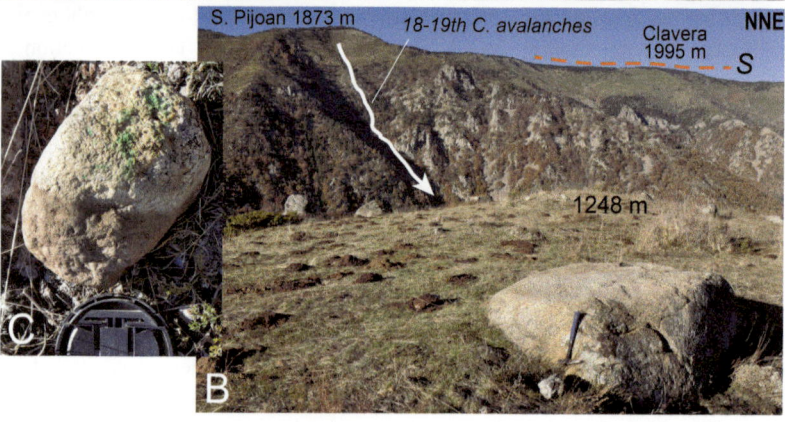

Fig. 11.32 Alluvial terrace staircase at Fontpédrouse. A Aussera landslide upstream of Fontpédrouse, here seen from the tread of terrace T5 on the road to Prats Balaguer. The displaced mass rests on the tread of T2. **B** Level T5 at Prats Balaguer, spot elevation 1243 m. Scatter of large rolled boulders and pebbles on the surface (gneiss, leucogranite). **C** Rolled quartz pebbles of alluvial terrace T5

the dam breach generated the debris-flow sequence observed further downstream between Fontpédrouse and Thuès.

The thermal springs of this area have been harnessed and commercialised as a small spa open to the public (Bains de Saint-Thomas), but an unguarded spot is also accessible below the '1243' spur. Water from the Aigues spring comes out at 60 °C, but recent man-made pool structures at the base of the talus apron allow it to reach more enjoyable temperatures.

After Fontpédrouse, the N116 rises rapidly in a series of loops up to Coll de la Perxa (Col de la Perche) and its plateau—gateway to the Cerdagne upland basin (Itinerary 7). On the way up, note the 'monument Gisclard' and the railway suspension bridge—historically the first of its kind in France. At Fetges, turn off towards Sauto (5 km return trip).

Stop 15. Panoramic view from Sauto

The views are good from the road before the village, and even better if you walk up above the village along the trail to Figama until you reach elevation spot 1762 (Fig. 11.33). The major channel knickzone of the Têt begins at Mont-Louis, below which the river has cut a valley 500 m deep over a distance of a few kilometres into the low-gradient erosion surface of La Perche (a cone-shaped pediment). The Perche pediment correlates with generation P2 (Itineraries 7 and 8 deliver supporting evidence). This generation of pediments and strath levels appears clearly in the upper part of the gorge in the form of small hanging benches on the right-hand valley side, above the base of the large, fresh-looking faceted spurs previously observed at Thuès–Fontpédrouse (Fig. 11.30). By continuing the walk up to Clavera (1995 m), or even just as far as Serrat de Pijoan (1873 m), the view of the Carança mountain front and its faceted fault scarp is even more stunning.

Back on the N116, after driving up to Mont-Louis (potentially another Vauban-related cultural stop), a recommended option is to aim for Font-Romeu, an advantageously situated gateway to Itinerary 7.

Fig. 11.33 **View towards Plateau de la Perche and the Têt channel knickzone downstream of Mont-Louis.** Here seen from the trail to Figama, above Sauto. Note faceted spurs along the Têt Fault. La Perche is an elevated rock pediment (generation P2; see Itinerary 7). In the far distance, note the Cerdagne Basin and its surrounding range-top and range-flank erosion surfaces

References

Bandet Y (1975) Les terrains néogènes du Conflent et du Roussillon nord-occidental. PhD thesis, Université, Paul Sabatier, Toulouse

Baudelot S, Crouzel F (1974) La faune burdigalienne des gisements d'Espira de Conflent. Bull Soc Hist Nat Toulouse 110:311–326

Batlle M, Gual R (1990) 1940, l'Aiguat. Rev Terra Nostra (Prades) 42, 200 p

Briais A, Armijo R, Winter T, Tapponnier P, Herbecq A (1990) Morphological evidence for Quaternary normal faulting and seismic hazard in the eastern Pyrenees. Ann Tecton IV:19–42

Calvet M (1996) Morphogenèse d'une montagne méditerranéenne: les Pyrénées orientales. Documents du BRGM, Orléans 255:1177 p

Calvet M (1999) Régime des contraintes et volumes de relief dans l'Est des Pyrénées. Géomorph Rel Proc Environ 3:253–278

Calvet M (2004) The Quaternary glaciation of the Pyrenees. In: Ehlers J, Gibbard P (eds) Quaternary Glaciations—Extent and Chronology, part I: Europe. Elsevier, Amsterdam, pp 119–128

Calvet M (2006) Accidents aléatoires ou crises morphogéniques : l'interprétation risquée des dynamiques fluvio-torrentielles holocènes et historiques en Méditerranée. Actes de la Table ronde "L'érosion entre société, climat et paléoenvironnement", 25–27 mars 2004, Presses Universitaires Blaise Pascal, Clermont-Ferrand, pp 401–406

Calvet M, Gunnell Y (2008) Planar landforms as markers of denudation chronology: an inversion of East Pyrenean tectonics based on landscape and sedimentary basin analysis. In: Gallagher K, Jones SJ, Wainwright J (eds) Landscape evolution: denudation, climate and tectonics over different time and space scales. Geol Soc Lond Spec Publ 296:147–166

Calvet M, Serrat P, Lemartinel B, Marichal R (2002) Les cours d'eau des Pyrénées orientales depuis 15 000 ans: état des connaissances et perspectives de recherches. In: Bravard J-P, Magny M (eds), Les fleuves ont une histoire, paléoenvironnements des rivières et des lacs français depuis 15 000 ans. Errance, Paris, pp. 279–294

Calvet M, Gunnell Y, Delmas M (2013) The Têt river valley: a condensed record of long-term landscape evolution in the Pyrenees. In: Fort M, André MF (eds) Landscapes and landforms of France, Springer, pp 127–138

Calvet M, Gunnell Y, Braucher R, Hez G, Bourles D, Guillou V, Delmas M, Aster Team (2015b) Cave levels as proxies for measuring post-orogenic uplift: evidence from cosmogenic dating of alluvium-filled-cave in the French Pyrenees. Geomorphology, 246:617–633

Calvet M, Hez G, Gunnell Y, Jaillet S (2019) Le karst du synclinal de Villefranche, enregistreur de l'incision de la vallée de la Têt. Bol Soc Esp Speleo Sci Karst 14:15–32

Clauzon G, Suc JP, Aguilar JP, Ambert P, Capetta H, Cravatte J, Drivaliari A, Domenech R, Dubar M, Leroy S, Martinell J, Michaux J, Roiron P, Rubino JL, Savoye B and Vernet JL (1990) Pliocene geodynamic and climatic evolutions in the French Mediterranean region. In: Iberian Neogene Basins, Paleontologia i Evolució, Mem. Especial n° 2, Sabadell, pp 131–186

Clauzon G, Le Strat P, Duvail C, Do Couto D, Suc J-P, Molliex S, Bache F, Besson D, Lindsay EH, Opdyke ND, Rubino J-P, Popescu SP, Haq BU, Gorini C (2015) The Roussillon Basin (S. France): a case-study to distinguish local and regional events between 6 and 3 Ma. Mar Pet Geol 66:18–40

Cocherie A, Baudin T, Guerrot C, Autran A, Fanning M-C, Laumonier B (2005) Evidence of the Lower Ordovician intrusion age for metagranites in the Late Proterozoic Canaveilles Group of Pyrénées and Montagne noire (France): new UPb datings. Bull Soc Géol Fr 176:269–282

Delmas M, Calvet M, Gunnell Y, Voinchet P, Manel C, Braucher R, Tissoux H, Bahain JJ, Perrenoud C, Saos T, Aster Team (2018) Terrestrial [10]Be and Electron spin resonance dating of fluvial terraces quantifies quaternary tectonic uplift gradients in the eastern Pyrenees. Quat Sci Rev 193:188–211

Deloule É, Alexandrov P, Cheilletz A, Laumonier B, Barbey P (2002) In-situ U–Pb zircon ages for Early Ordovician magmatism in the eastern Pyrenees, France: the Canigou orthogneisses. Int J Earth Sci 91:398–405

Gärtner A, Merchel S, Niederman S, Braucher R, ASTER-Team, Steier P, Rugel G, Scharf A, Le Bras L, Linnemann U (2020) Nature does the averaging–In-situ produced [10]Be, [21]Ne, and [26]Al in a very young river terrace. Geosciences 2020 10:237. https://doi.org/10.3390/geosciences10060237

Granger DE, Muzikar P (2001) Dating sediment burial with cosmogenic nuclides: theory, techniques, and limitations. Earth Planet Sci Lett 188:269–281

Granger DE, Kirchner JW, Finkel RC (1997) Quaternary downcutting rate of the New River, Virginia, measured from differential decay of cosmogenic [26]Al and [10]Be in cave-deposited alluvium. Geology 25:107–110

Guitard G (1970) Le métamorphisme hercynien mésozonal et les gneiss œillés du massif du Canigou (Pyrénées orientales). Mém BRGM, 63:1–353

Guitard G, Geyssant J, Laumonier B, Autran A, Fonteilles M, Dalmayrac M, Vidal JC, Bandet Y (1992) Carte géol. France (1:50,000 scale), sheet Prades (1095). BRGM, Orléans. Handbook by Guitard G, Laumonier B, Autran A, Bandet Y, Berger GM (1998), 198 p

Gunnell Y, Calvet M, Brichau S, Carter A, Aguilar JP, Zeyen H (2009) Low long-term erosion rates in high-energy mountain belts: insights from thermo- and biochronology in the Eastern Pyrenees. Earth Planet Sci Lett 278:208–218

Hez G, Jaillet S, Calvet M, Delannoy JJ (2015) Un enregistreur exceptionnel de l'incision de la vallée de la Têt : le karst de Villefranche, Pyrénées-orientales, France. Karstologia 65:9–32

Huard M (1972) Etude géologique du district à fluorine et sidérite de la bordure septentrionale du massif Canigou-Carança (Pyrénées orientales). Bulletin du BRGM 2:1–43

Jacob N, Gob F, Petit F, Bravard J-P (2002) Croissance du lichen Rhizocarpon geographicum l.s. sur le pourtour nord-occidental de la Méditerranée: observations en vue d'une application à l'étude des lits fluviaux rocheux et caillouteux. Géomorph Rel Proc Environ 4:283–296

Jarman D, Calvet M, Coromina J, Delmas M, Gunnell Y (2014) Large-scale rock slope failures in the Eastern Pyrenees: identifying a sparse but significant population in paraglacial and parafluvial contexts. Geogr Ann Ser B 96:357–391

Laumonier B (ed) (1996) Cambro-Ordovicien. In: Barnolas A, Chiron JC (eds), Synthèse géologique et géophysique des Pyrénées, vol 1: introduction, géophysique, cycle hercynien. BRGM-ITGE, pp 157–209

Laumonier B (2008) Les Pyrénées pré-hercyniennes et hercyniennes. In: Canérot J, Colin JP, Platel JP, Bilotte M (eds) Pyrénées d'hier et d'aujourd'hui. Atlantica, Biarritz, pp 25–35

Laumonier B, Le Bayon B, Calvet M (2015) Handbook to the Carte géologique de la France (1:50,000 scale), sheet Prats-de-Mollo-La-Preste (1099). Orléans : BRGM, 189 p. Geological map by Laumonier B et al. (2015)

Laumonier B, Calvet M, Delmas M, Barbey P, Lenoble JL, Autran A (2017) Handbook, carte géol. France (1:50 000 scale), sheet Mont-Louis (1094). BRGM, Orléans, 139 p. Carte géologique by Autran A., Calvet M., Delmas M. (2005)

Magné J (1978) Études microstratigraphiques sur le Néogène de la Méditerranée nord-occidentale. t. 1 : Les bassins néogènes catalans, 259 p; t. 2. : Le Néogène marin du Languedoc méditerranéen, 435 p, Éd. CNRS, Centre Régional de Publication, Toulouse

Maurel O, Brunel M, Monie P (2002) Exhumation cénozoïque des massifs du Canigou et de Mont-Louis (Pyrénées orientales, France). C R Geosci 334:941–948

Maurel O, Moniè P, Pik R, Arnaud N, Brunel M, Jolivet M (2008) The Meso-Cenozoic thermo-tectonic evolution of the Eastern Pyrenees: an ^{40}Ar/^{39}Ar fission track and (U–Th)/He thermochronological study of the Canigou and Mont-Louis massifs. Int J Earth Sci 97:565–584

Milesi G, Monié P, Münch P, Soliva R, Taillefer A, Bruguier O, Bellanger M, Bonno M, Martin C (2020) Tracking geothermal anomalies along a crustal fault using (U-Th)/He apatite thermochronology and rare-earth element (REE) analyses: the example of the Têt Fault (Pyrenees, France). Solid Earth 11:1747–1771

Oele E, Sluiter J, Pannekoek AJ (1963) Tertiary and Quaternary sedimentation in the Conflent, an intramontane rift valley in the Eastern Pyrenees. Leidse Geol Meded 28:297–319

Pannekoek AJ (1935) Évolution du bassin de la Têt dans les Pyrénées-Orientales pendant le Néogène. Étude de morphotectonique. Univ. Utrecht, 72 p

Passarius O, Catafau A, Martzluff M (eds) (2009) Archéologie d'une montagne brûlée, massif de Rodès, Pyrénées-Orientales. Éditions Trabucaire and Conseil Général des Pyrénées-Orientales, 504 p

Petit C, Mouthereau F (2012) Steep topographic slope preservation by anisotropic diffusion: an example from the Neogene Têt fault scarp, Eastern Pyrenees. Geomorphology 171–172:173–179

Reille M, Lowe JJ (1993) A re-evalutation of the vegetation history of the eastern Pyrenees (France) from the end of the last glacial to the present. Quat Sci Rev 12:47–77

Sartégou A, Bourlès DL, Blard PH, Braucher, Tibari B, Zimmermann L, Leanni L, Aumaître G, Keddadouche K (2018) Deciphering landscape evolution with karstic networks: a Pyrenean case study. Quat Geochron 43:12–29

Stock GM, Granger DE, Anderson RS, Sasowsky ID, Finkel RC (2005) Dating cave deposits for use in landscape evolution studies: insights from caves in the Sierra Nevada, California. Earth Planet Sci Lett 236:388–403

Itinerary 7. Elevated Ranges and Interior Basins of the Axial Zone: The Upper Cerdagne Basin, Capcir Basin, and their Surrounding Massifs

12

From Font-Romeu back to Font-Romeu (160 km; additional excursions: 52 km; allow for 2 days, 4 days or more when including trailwalks). Cerdagne is attractive for tourism and offers many options in terms of accommodation. Font-Romeu is a conveniently situated temporary base camp. See Fig. 12.1.

Itineraries 7 and 8 aim to extend the understanding of Neogene landscape evolution already partly gained from the Roussillon and Conflent basins. We focus on the sediment record contained in the more westerly basins of Cerdagne (Fig. 12.2) and Capcir, which exclusively display late Miocene stratigraphic sequences until now not encountered in the previously visited extensional basins of itineraries 1–6, but which also occur in the Empordà Basin of Catalonia (not included in this GeoGuide). Another highlight is the widespread population of erosion surfaces that either grade to the margins of these interior clastic basins (generations P1 and P2) or have been preserved in summit positions among the surrounding massifs of the Axial Zone. In either case, their occurrence and excellent state of preservation in this elevated core of the eastern Pyrenees can be explained by its relative remoteness from the Aquitaine, Ebro and Mediterranean base levels. A third focus is on the impact of Pleistocene glaciation (Fig. 12.3) among the northern massifs. Quaternary glacial chrononology is particularly well documented because sequences of moraines and glacifluvial terraces are well preserved across the Cerdagne and Capcir basin floors.

© Springer Nature Switzerland AG 2022
M. Calvet et al., *Geology and Landscapes of the Eastern Pyrenees*, GeoGuide,
https://doi.org/10.1007/978-3-030-84266-6_12

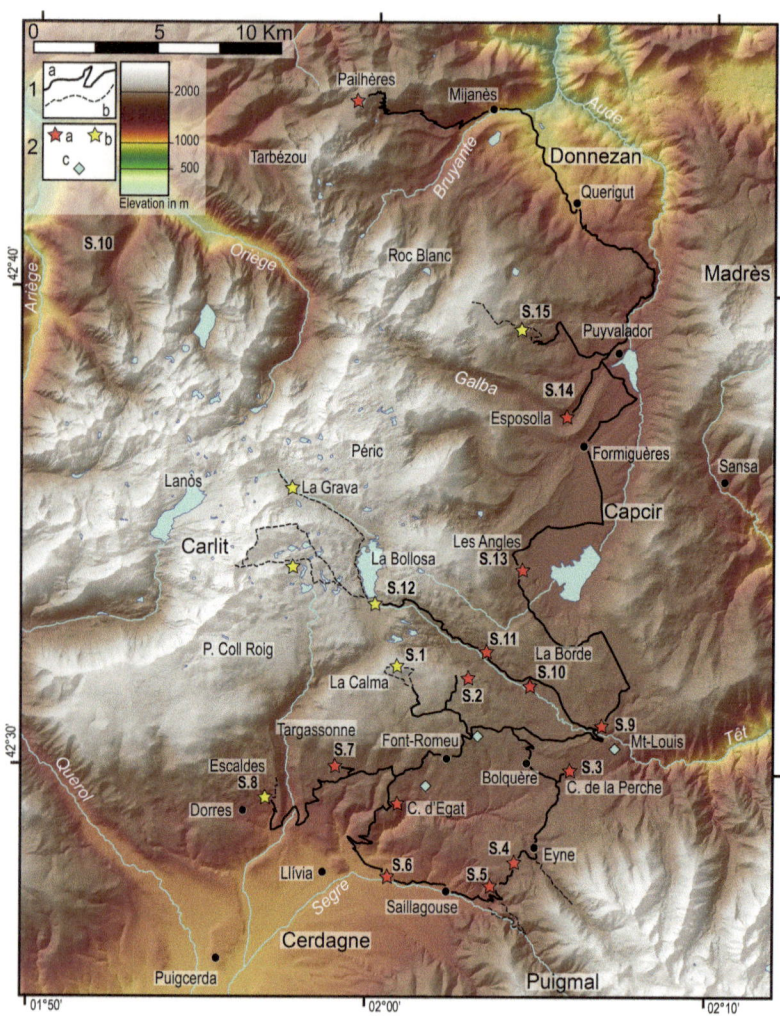

Fig. 12.1 Itinerary 7: route map and overview. Key to symbols—**1**: Roads and trails. **1a**: Roadways, irrespective of road rank, size, or quality. **1b**: Trailwalks. **2**: Stopping points (outcrops, exposures, landforms, or landscapes). **2a**: Roadside stops. **2b**: Roadside stops involving a short walk, or marking the start of a more major trailwalk (potentially including stopovers in a mountain refuge). **2c**: Cultural stops. Stops numbered S.1 to S.15 refer to the sequence of roadside stops

Box 12.1 The Cerdagne Neogene Half-Graben
The Cerdagne Basin has been attracting geologists for over a century (Depéret and Rérolle 1885; Chevalier 1925; Boissevain 1934). Like the Conflent, it is a half-graben with a master fault (the Cerdagne Fault, effectively a continuation of the Têt Fault) forming a succession of E–W and NE–SW segments along its southern boundary (Gourinard 1971; Fig. 12.2). The major erosion surfaces on the north-bounding massifs slope towards the basin and plunge beneath its clastic fill. The interpretation of fault throws has given rise to two plausible models of response to tectonic stresses: either they document normal to right-lateral slip on the NE–SW segments of the Têt Fault (Cabrera et al. 1988; Roca 1996a), or normal to left-lateral slip along its E–W segments (Pous et al. 1986). The basin was progressively filled by two late Miocene sedimentary sequences. The stratigraphy is distinctly non-marine, and its age has been documented by mammalian index fossil sites. These have yielded Vallesian and Turolian ages (11–5.5 Ma; Golpe Posse 1981; Agustí and Roca 1987; Agustí et al. 2006). The lower sequence contains an inventory of large mammalian fossils consistent with ELMA biozones MN 9 and MN 10, including *Hipparion primigenium catalaunicum*, *Dicerorhinus scheleiermarcheri*, *Tetralophon longirostris*, *Chalicomys jaegeri*, *Amphicyon major pyrenaicus*, and *Chalicotherium grande*. The upper sequence is only present in the western part of the basin. Almost at its base, the micromammalian deposit of Can Vilella correlates with MN 13, i.e., the Turolian (6.5–6 Ma). It contains *Apodemus gudrunae*, *Kowalskia* sp., *Epimeriones aff. austriacus*, *Sminthozapus janossy*, *Muscardinus aff. vireti*, *Prolagus michauxi*). The lacuna corresponding to MN 11 and MN 12 is probably ascribable to an erosional unconformity between the two sequences. Pollen assemblages have been studied from both sequences, and plant macrofossils are abundant among the lower units; in all cases, the taxa indicate heat-loving and wetlands plants (Menendez Amor 1955; Suc and Fauquette 2012; see Itinerary 8, **Stops 7, 8** and **9** therein).

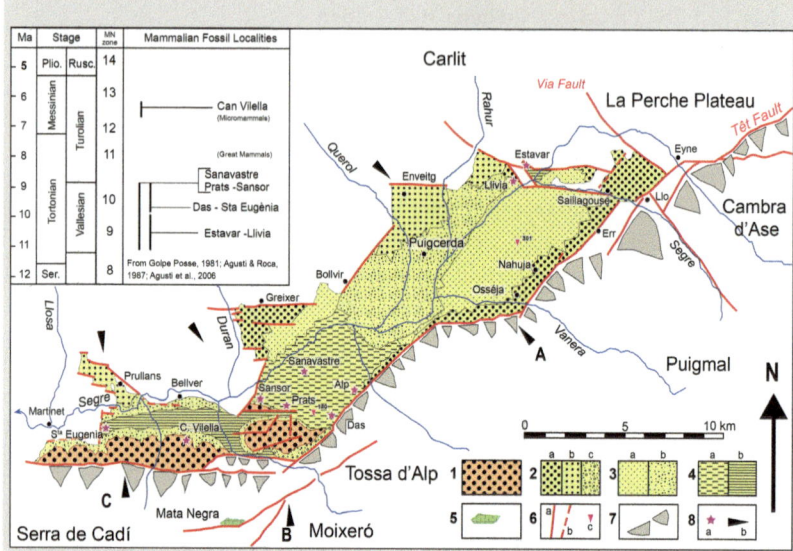

Fig. 12.2 The Cerdagne half-graben: geological structure, sedimentary facies and chronology of Mammalian fossil deposits (after Calvet 1996, redrawn and modified). Key to symbols and ornaments—**1**: Red siltstone and limestone conglomerate, upper Turolian series. **2**: High-energy fluvial deposits, lower Vallesian series; **2a**: Schist-dominated debris; **2b**: Granite-dominated debris; **2c**: Limestone-dominated debris. **3**: Fine-grained fluvial facies (sand, gravel, silt, clay, lignite); **3a**: Schist-dominated debris; **3b**: Granite-dominated debris. **4**: Distal facies; **4a**: Deltaic, fluvial and lacustrine facies (sand, clay, lignite); **4b**: Lacustrine facies (grey mudstone, diatomite). **5**: Upper Cretaceous. **6**: Other geological features of importance; **6a**: Neogene fault; **6b**: Probable Neogene fault; **6c**: Borehole. **7**: Faceted spurs along southern fault scarp. **8**: Other features; **8a**: **Mammalian** fossil site; **8b**: Location of geological cross-sections in Fig. 13.4 (Itinerary 8)

The proximal stratigraphy consists of coarse alluvial-fan debris, whereas the more distal facies correspond to silt- and clay-dominated flood-basin deposits with interlayers of sand and gravel (palaeochannels). Lignite beds occur sporadically, but lacustrine facies are only present in the western part of the Cerdagne Basin. The upper sequence mostly contains proximal range-front and distal alluvial fan deposits, exclusively fed by southern drainage catchments, with a few palustrine and lignite-bearing intercalations in the distal stratigraphy. Bedding dips steadily to the SE, and fill thicknesses reach their preserved maximum (~0.7–1 km) at the base of the southern boundary fault.

Box 12.2 Quaternary Glaciation in the Eastern Pyrenees

Glaciation in the Pyrenees was introduced in Part I, and Fig. 12.3 captures the maximum extent of the Late Pleistocene icefield in the eastern part of the range. In order to better understand age data provided in the context of various stops and trailwalks, here we provide some keys to Quaternary chronostratigraphy and terminology as used through these pages. The Cerdagne and Capcir basins are unique in having operated as raised base levels to glaciers that, elsewhere in the Pyrenees, would have descended to the foreland zones. They thus constitute a natural laboratory where Quaternary Pyrenean glacial chronology can be documented within well-contained boundaries, where sequences of well-preserved glacial landforms were produced by a population of fairly short glaciers. The classic Alpine glacial chronology (Würm, Riss, Mindel, Günz, Donau, Biber) was in usage in Pyrenean literature until recently, but it is increasingly being abandoned in favour of other reference frames that allow intercomparisons with independently established palaeoclimatic chronologies. These are explained below.

1. **The isotope-based marine stratigraphy.** The well-known scale of Marine Isotopic Stages (MIS) is calibrated on variations in $^{18}O/^{16}O$ ($\delta^{18}O$) ratios measured in ocean-floor sediments. Values of $\delta^{18}O$ are indirectly controlled by relative variations in global temperature, which themselves control relative amounts of ice stored in high-latitude ice sheets, and thus by global sea levels. During cooler periods, the volume of oceanic water is reduced and relative ^{18}O concentrations increase. Because $\delta^{18}O$ also depends on marine water temperatures, variations in ice sequestration on continents are best recorded by deepwater organisms, typically benthic foraminfera, which live in an environment where seasonal to secular variations in water temperature are negligibly small.

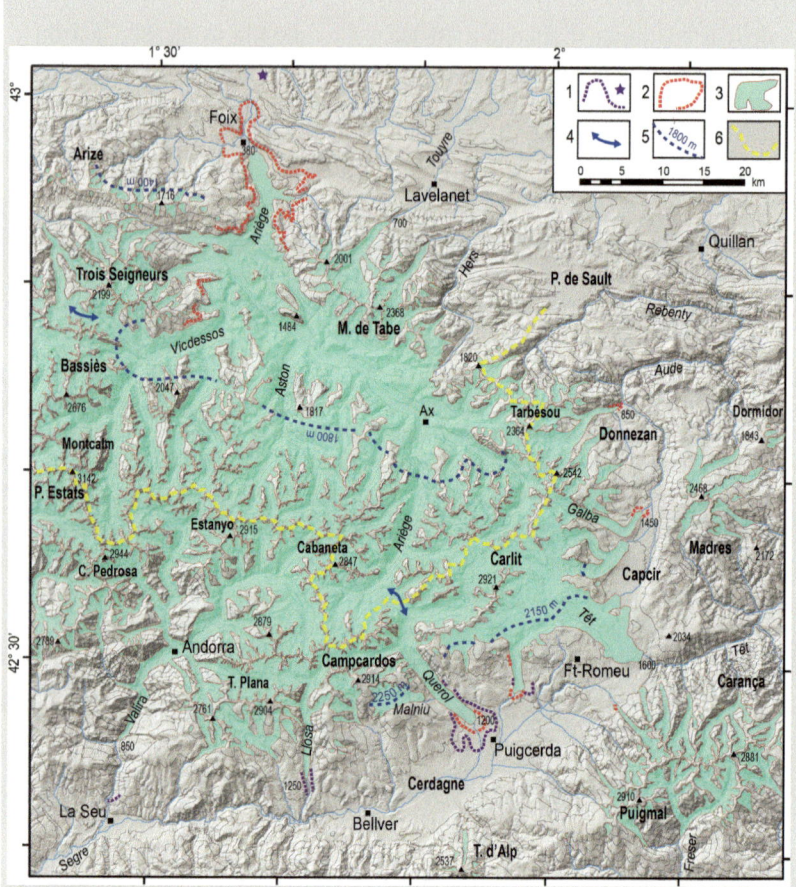

Fig. 12.3 **Exent of glaciation in the eastern Pyrenees** (after Calvet 2004; Reixach et al. 2021). Key to symbols and ornaments—**1**: early Middle Pleistocene deposits; star indicates isolated erratic boulders in lower Ariège valley. **2**: mid to late Middle Pleistocene deposits (probably MIS 6). **3**: Late Pleistocene deposits, here shown during the most extensive glaciation of that period. **4**: Main transfluence cols. **5**: Equilibrium line altitude. **6**: Drainage divide between Atlantic and Mediterranean catchments

The longest continuous MIS sequence available (LR04, Lisiescki and Raymo 2005) is based on 57 oceanic sediment cores, mostly from the

Atlantic, and spans the last 5.3 Ma. The δ^{18}O record on that basis has defined 104 MIS intervals for the Quaternary. Cooler periods are labelled by an even MIS number, and warmer periods by an odd number. From 2.56 Ma to 781 ka, the δ^{18}O periodicity of glacial–interglacial cycles was 40 ka. Subsequently, cycles lengthened to 100 ka. The last of these is known as the Late Pleistocene: it began ca. 126 ka, ended ca. 11.7 ka (this is the start of the Holocene, i.e., the current interglacial: MIS 1), and includes MIS 2, MIS 3, MIS 4, and MIS 5a, 5b, 5c, 5d and 5e (MIS 5e is effectively the last interglacial, also known as Eemian interglacial by reference to chronostratigraphies established in northern Europe). The colder glacial interval spanning MIS 5d to MIS 2 corresponds to the Würm.

2. **The Greenland icesheet stratigraphy**. Palaeoclimatic inferences made from marine sediment cores have also been independently corroborated from high-latitude ice cores (120 ka record from North GRIP in Greenland, 800 ka record from Vostok in Antarctica), where δ^{18}O this time records relative atmospheric temperature at the time of snow fall. The very high resolution obtained from Greenland was able to resolve millennial-scale subcycles nested within the longer-wavelength variations (Dansgaard et al. 1993; Grootes et al. 1993). These shorter Dansgaard–Oeschger cycles, or 'events', involve an abrupt warming phase (Greenland interstadial, GI: typically +5 to +16.5 °C in less than a century), followed by a slower cooling phase (or Greenland stadial, GS) lasting 1–3 ka (Severinghaus and Brook 1999; Landais et al. 2004; Kindler et al. 2014). Identified Dansgaard–Oeschger cycles, numbered 1 (most recent) to 25 (oldest) (North GRIP members 2004; Wolf et al. 2010), have become the basis for an 'INTI-MATE Event stratigraphy', or Greenland stratigraphy (Rasmussen et al. 2014), capturing millennial-scale oscillations.

3. **The Last Glacial Maximum (LGM) concept**. The LGM is a period during the Late Pleistocene defined by maximum storage of ice on continents, and thus by the lowest recorded global sea levels (Mix et al. 2001). This time interval is thus held as a global event and appears on the MIS scale as the lowest-recorded δ^{18}O values (falling roughly at the centre-point of MIS 2, and initially positioned around 18 ka BP by CLIMAP Project Members 1976, 1981). The timing of this low sea-level reference frame is now more finely calibrated by large numbers of U–Th and ^{14}C ages obtained around the world from coral reefs, which are precise indicators of sea level. The most recent and comprehensive revision of this

global calibration covers the last 35 ka and shows that the global ocean was at its lowest, i.e., lower than −125 m, during the interval lasting from 29 to 19 ka cal BP. The lowest trough in the curve, reaching −134 m, occurred between 20.9 and 20.4 ka cal BP (Lambeck et al. 2014). For calibrating the timing of glacier fluctuations in terrestrial environments such as the Pyrenees against a globally valid reference frame, this definition of the LGM (29–19 ka) currently prevails. Any geochronology-driven study of local mountain glacier behaviour in response to Quaternary climate change, however, is likely to encounter evidence of time lags and discrepancies with respect to the LGM-defined time frame. Rather than artefacts or anomalies, these discrepancies are routinely treated as real features, with environmental implications requiring interpretation based on local or regional context. It turns out, for example, that the timing of the local last glacial maximum (noted LLGM), which corresponded to the most extensive glaciation in the Pyrenees during the Late Pleistocene, was not synchronous with the global LGM. These offset are important for understanding palaeoclimate dynamics.

4. **The Last Glacial to Interglacial Transition (LGIT).** This time interval spans the progressive, but highly irregular, warming period that marked the end of MIS 2 between the LGM and the Holocene (Wohlfarth 1996; Hoek 2009; Denton et al. 2010). This eventful period recorded very abrupt climatic oscillations, with amplitudes of up to 10 °C in just a few centuries at mid latitudes (Anderson and Mackintosh 2006; Larocque-Tobler et al. 2010; Millet et al. 2012; van Asch 2012), and was initially documented in northern Europe by the four biozones known as Bølling, Older Dryas, Allerød, and Younger Dryas (Mangerud et al. 1974). These palynostratigraphic intervals are well corroborated by the Greenland stratigraphy (Rasmussen et al. 2014), which shows that the Bølling–Allerød interstade corresponds to GI-1 (14.7 b2k; b2k means 'before the Year 2000') and the Younger Dryas to GS-1 (beginning 12.9 b2k). The Oldest Dryas is a cooling interval older than the Bølling–Allerød. It was detected in pollen cores from mid-latitude mountains such as the Jura, Alps, Massif Central and Pyrenees, where the Older Dryas cooling event is not as conspicuous as in northern Europe. Although not very sharply defined by the North GRIP stratigraphy, where its ill-defined lower boundary (interval name: GS-2.1a) is provisionally set at 17.5 b2k (e.g., Rasmussen et al. 2014), the Oldest Dryas nonetheless appears very clearly in global palaeotemperature curves

based on sea-level data. Lambeck et al. (2014) thus show that post-LGM sea levels rose from −134 to −122.4 m around 17.9–18.1 ka cal BP, then dipped again to −123 m ca. 16.4–16.7 ka cal BP. The cooler, Oldest Dryas episode thus started ca. 18 ka and ended ca. 14.7 b2k, i.e., at the start of the Bølling–Allerød (GI-1). It was apparently driven in the North Atlantic by Heinrich event HE 1 (18–15.6 ka; see Sánchez-Goñi and Harrison 2010).

12.1 Cerdagne Basin and Carlit Massif

Heading out of Font-Romeu towards Mont-Louis on the D618, turn off towards the ski resorts signposted at the roundabout after L'Ermitage. Drive up to the parking area at Mollera dels Clots (2046 m), which lies at the base of Roc de la Calma (2212 m).

Stop 1. Roc de la Calma
Views of the summit surface, S, on the horizon, and Neogene pediment P1 in the middle ground; appreciation of glacial erosion rates during the Late Pleistocene glaciation by [10]Be dating and other methods (1 hr return on foot).

The path rises obliquely to the NW and reaches the western end of the ridgetop. On the way up you will get good views of the rolling plateau sloping gently to the ESE from 2100 to 1800 m—its lower portion above Font-Romeu covered by forest (Fig. 12.4D, E). This plateau is a tilted vestige of pediment P1. Short-wavelength undulations consisting of low bedrock hills and depressions lined with saprolite have accentuated the relief of this granitic land surface, thus probably roughening the originally smoother morphology of the pediment's tread. La Calma is a residual granite landform rising above it. At times of excess ice thickness, the Têt glacier sporadically spilled over onto the pediment, constructing a lateral moraine. The upper branch of the moraine tapers out at an altitude of 2100 m, thereby providing a fairly accurate estimate of the equilibrium line altitude (ELA) during the Late Pleistocene. From the ridgetop, you get a 360° view of the Cerdagne Basin and the Carlit massif, although forest regrowth today partly obscures the view to the north, where the Têt glacial trough and Bouillouses (Bollosa) reservoir (shown in Fig. 12.4A) will be encountered and explored later.

Box 12.3 Range-Top Surface, S, and Range-Flank Pediment, P1: the Carlit Massif as a Prime Showcase

The Carlit massif displays a strikingly well-preserved double tier of Cenozoic erosion surfaces (Figs. 11.4, 12.4A–D; Calvet 1996; Calvet and Gunnell 2008; Gunnell et al. 2009). The vernacular name given regionally to flat topography in the steep mountain landscape is 'pla', a Catalan word meaning 'flat'.

The summit surface, S, forms spectacularly flat vestiges (Puig de Solana Carnicere, 2878 m, and Ras del Carlit, 2804) on either side of Puig Carlit, a residual monadnock (choice of trailwalks, see later options). It also extends across the summit of the Campcardós massif, visible from here to the west (Fig. 12.4A), and across the top of the Tossa d'Alp, on the south side of the Cerdagne Basin (see Itinerary 8). On the SW horizon it also bevels the Eocene limestone ridge of Serra del Cadí. Other vestiges to the east and NE occur at Roc d'Aude and Puig del Pam, and in the Madrès massif situated east of the Capcir Basin (see Itinerary 6).

The lower surface is a vestige of pediment P1, which forms a wide topographic pedestal between 2200 and 2400 m at the base of the Puig Carlit–Puig Péric residual massif, which form a large residual mass sculpted by Pleistocene glaciers (Fig. 12.4A, D). Immediately south of La Calma, P1 extends as an undulating plateau overlooked by Pic dels Moros. It slopes gradually SE towards Col de La Perche (Coll de la Perxa), where P2 forms a small but perceptible step notched into P1 (see **Stop 3**). On the south of the Cerdagne Basin, along the mountain front between Puigmal and Osseja forest, P1 bevels the band of long sloping spurs that descend from 2700 to 1700 m (Fig. 12.4E). From E to W, examples include Pla de Gorra Blanc (2543 m), Pla de les Salines (2234 m) and Ras de la Basseta (2016 m).

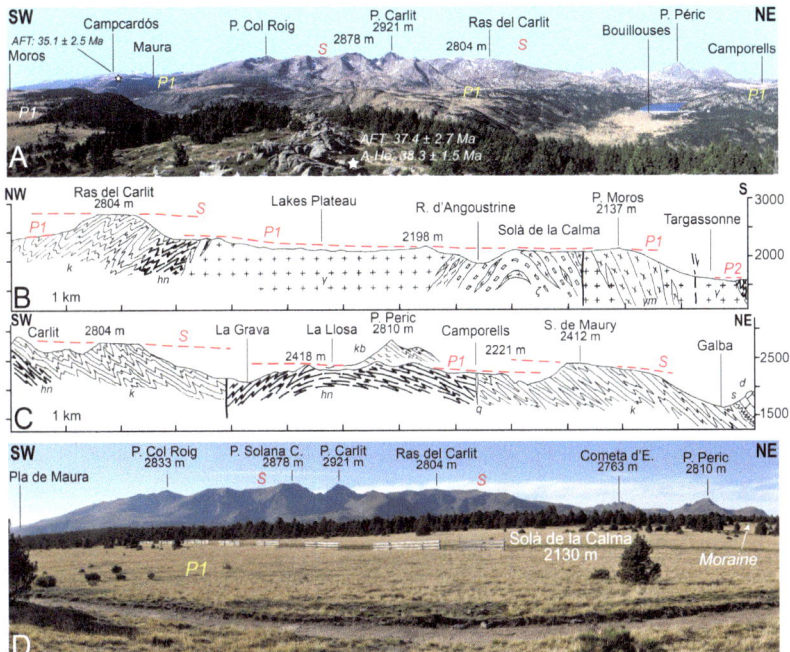

Fig. 12.4   Elevated erosion surfaces around the Cerdagne Basin. A Elevated erosion surfaces of the Carlit massif, viewed from La Calma (looking NW); note the two tiers, S and P1, and the residual nature of Carlit and Péric. **B** NNW–SSE geological sections. **C** SW–NE geological sections. Lithostratigraphy—γ: Mont-Louis granite; γm: biotite and muscovite leucogranite; *hn*: Hornfels; ζ: orthogneiss (Ordovician granite); *k*: schist (Jujols Group); *s*: schist (Silurian); *d*: limestone (Villefranche syncline, Devonian). **D** General view of the Carlit massif from Sola de la Calma plateau. **E** Cerdagne Basin and Puigmal massif, viewed from La Calma; note faceted spurs along the Têt–Cerdagne fault scarp **F** and P1 pediment on west side of Puigmal

Fig. 12.4 (continued)

As previously presented at sites in the Agly massif (Itinerary 5) and Conflent Basin (Itinerary 6), apatite thermochronology confirms that the erosion surfaces in this landscape are vestiges of S and P1. Regardless of their current elevation, a rapid decline in denudation rates occurred after 35–30 Ma (Fig. 12.5). Given the relatively small topographic offset between S and P1 (~400 m in the Carlit massif, often much less at other sites), apatite thermochronology is not, however, a sufficiently sensitive tool for detecting such correspondingly small differences in crustal denudation depth and topographic relief. As a result, the thermochronological ages of S and P1 are indistinguishable. The landscape, however, nonetheless clearly exhibits two generations of planar landforms, and we know from the thermochronology, biochronology and stratigraphy (see Itineraries 1, 5, 6) that they formed sequentially between ~ 30 and 15 Ma (late Oligocene to middle Miocene).

▶**Fig. 12.5 Rock cooling models for bedrock samples collected from S and P1 in the Carlit massif**. Note the uniform cooling histories among the samples, with rapid denudation around 35 Ma followed by residence at shallow crustal depths since late Eocene time. In the case of DONZ2, situated north of the Mérens Fault, cooling began earlier than among samples south of that fault. After Gunnell et al. (2009), modified

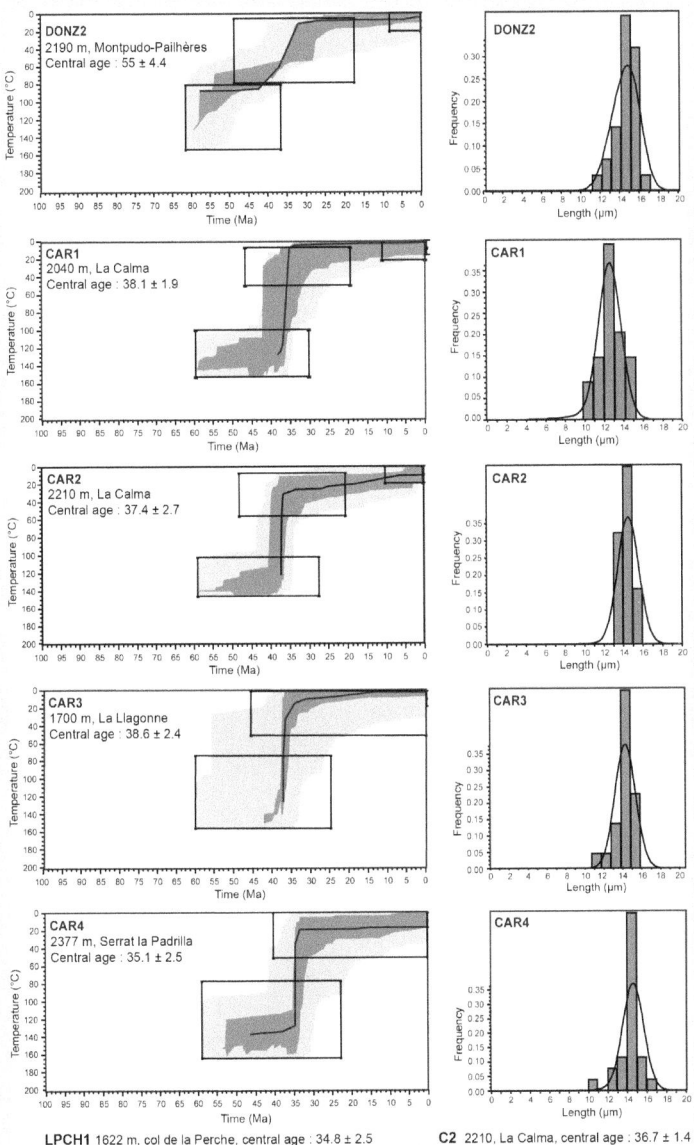

LPCH1 1622 m, col de la Perche, central age : 34.8 ± 2.5   C2 2210, La Calma, central age : 36.7 ± 1.4

Box 12.4 Estimates of Late Pleistocene Glacial Erosion Rates in the Carlit Massif

The SE-facing flank of the Carlit massif has provided constraints on spatial and chronological variation in depths of glacial erosion during the Late Pleistocene (Delmas et al. 2009). The unique advantage of this catchment is that the frontal and lateral moraines are remarkably well preserved in the landscape because the glaciers were contained within low-energy environments such as the Carlit plateau (P1), the Cerdagne Basin, and the Capcir Basin. These elevated intermontane base levels, all situated above the major fluvial knickzones of the area such as the Têt gorge below Mont-Louis (see itinerary 6), the Segre canyon (see Itinerary 8) and the Aude gorges (see Itinerary 5), have accordingly undergone comparatively limited postglacial denudation. The detailed chronology of glaciation in this area has been obtained from [10]Be exposure dating, locally aided by ^{14}C dating of peat deposits adjacent to glacial landforms. Three chronostratigraphic units have been reconstructed on that basis: (i) the Terminal Unit, which corresponds to moraines produced during the last most extensive glaciation, or local last glacial maximum (hereafter noted LLGM; see Box 7.2); the (ii) Recessional Unit, which collectively refers to the succession of moraines generated by receding glaciers between the LLGM and ca. 20 ka cal BP; the (iii) Cirques Unit, which represents the final glacial standstill (or local stade), by the end of which (i.e., by the end of the LGIT; see Box 12.2) the icefield had shrunk to a collection of cirque glaciers situated above 2400 m (see also Fig. 12.21).

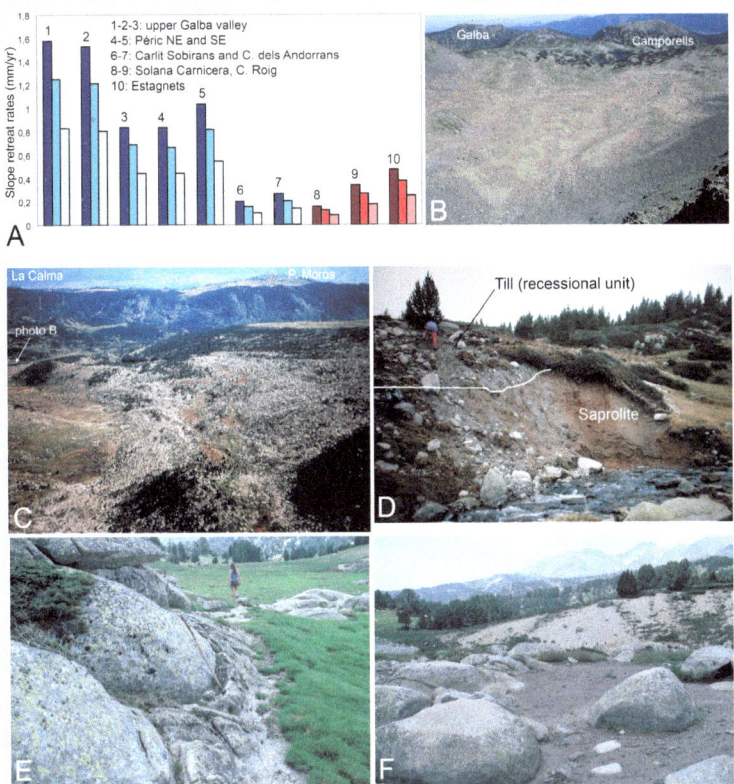

Fig. 12.6 Aspects of the glaciated landscape in the Carlit massif. **A** Post-LGM cirque headwall recession rates. Blue bars: cirques in schist; red bars: cirques in granite. The colour gradient represents headwall recession rates calculated for each cirque on the basis of three different density ratios between the debris and the bedrock (V63, V50 and V33, respectively). Cirques are numbered from NE to SW; units 6 and 8 coincide with a mixed lithology. **B, C** Cirque moraines (bouldery facies: ablation till). **B** La Coquilla cirque, in schist, NE of Pic Péric. **C** Soccarade moraines (granite boulders), at the base of Pic de Col Rouge. **D–F** In-situ saprolite and weathering front on pediment P1, covered by subglacial till, evidencing the weakness of Quaternary glacial erosion. **D** Section through rubified saprolite at Els Forats. **E** La Balmette bedrock step, barely stripped of its spheroidally weathered granite envelope. **F** La Balmette weathering front with corestones, covered by recent till further in the background (after Delmas 2009; Delmas et al. 2009, merged and modified)

By dividing the total volume of moraine preserved in the Terminal and Recessional units by the surface area of the glaciated catchment at each standstill (supraglacial nunatak areas included), it was possible to estimate mean Late Pleistocene landscape denudation during each of these intervals. Given its coarse-grained texture (boulders, cobbles), debris angularity, and openwork fabric, the Cirques Unit was mainly supplied by cirque-wall rather than cirque-floor material (Fig. 12.6A–C). This youngest unit consequently provides a quantitative estimate of headwall recession rather than of basal erosion. Results (Delmas et al. 2009) show that denudation was ~10 times greater during the shorter Recessional period (0.6 mm/yr) than during the protracted glacial advance and LLGM (0.05 mm/yr). This is ascribable to contributions from paraglacial processes, which in the Carlit massif were quite active during the post-LLGM deglaciation. The glaciers at the time were mostly evacuating debris produced in the thinning accumulation zone as the ELA gradually rose to elevations in excess of 2200 m (this contour coincides with the low-gradient floor of pediment P1). Denudation rates remained high during the LGIT given that the Cirques Unit has yielded headwall recession rates of 0.1–0.3 mm/yr in granite and 0.6–1.2 mm/yr in schist (Fig. 12.6A). These figures confirm the theory according to which cirques grow faster during deglacial and interglacial periods rather than during glacial maxima (Cook and Swift 2012; Crest et al. 2017).

In summary, Late Pleistocene denudation achieved a catchment-averaged minimum denudation depth of 4 m (basic calculation using the total volume of moraine currently preserved in the SE Carlit landscape); this value should be doubled in order to allow for the evacuation of suspended load in streams. Despite the inevitable uncertainty around such estimates, cold-climate denudation calculated on the basis of these benchmark values and cumulated over the last ten or so 100,000-year Milankovitch cycles has been modest, and mainly exerted on cirque and glacial-trough walls. The impact of Quaternary glaciation on the landscape, at least compared to wetter areas of the central Pyrenees more directly exposed to Atlantic weather systems, has thus remained limited. Accordingly, this explains why deep weathering profiles have been widely preserved even under till at elevations impacted by Pyrenean glaciation (Fig. 12.6D–F), and partly also why relict Neogene landforms such as high-elevation erosion surfaces are still so prominent in the landscape (see choice of trailwalks, below).

You can choose to return to the parking area by walking down the ski piste under the chairlift. Drive back to the main parking area (Les Airelles) and over to the eastern car park at Coll del Pam. Ca. 150 m NE of it, you get a fine view of the Têt valley.

Stop 2. View from Coll del Pam over the glacial valley of the Têt

The small plateau at the col is a kame terrace impounded by the right-margin moraine of the Têt glacier during the LLGM. The moraine has been degraded by the heavy machinery involved in piste moulding for the ski resort, but its upper extremity is still distinguishable upvalley ~ 2030 m a.s.l. Its original upper tip is no longer preserved because the upper catchment area is steep and the debris have been dispersed by postglacial slope processes, but its altitude is inferred to have reached 2100 m on the west face of Roc de la Calma. The Pam belvedere also provides a good angle on the U-shaped profile of the Têt valley and its succession of rock basins and transverse rock bars (or bedrock steps). Looking up (Fig. 12.7A) this glacial stairway, note the large Malpàs bedrock step (which from here conceals the Bouillouses reservoir previously mentioned at **Stop 1**); and the Avellans rock basin, glacially excavated 300 m below your feet. Looking down valley (Fig. 12.7B), note the meadows extending across the sediment-filled Borda glacial rock basin, here contained behind the massive Late Pleistocene terminal moraine complex of the Têt, now overgrown by forest. At the time of the LLGM, the Têt glacier produced an offshoot (diffluence lobe) towards the Capcir Basin (Figs. 12.3, 12.21).

Turn back down towards Font-Romeu and take the D618 signposted to Super Bolquère; then turn off onto the D10c, which cuts through the outermost frontal and lateral moraines and joins up with Bolquère and Coll de la Perxa on the D10. At the col, drive down the N116 for about 800 m towards Mont-Louis until you reach a large parking area.

Stop 3. Plateau de La Perche: an introduction to pediment P2

A glimpse of La Perche (La Perxa) was briefly mentioned at the end of Itinerary 6. Here its tread is quite pristine, dinted by the concave-sided Jardó valley, and contrasting with the deep gash of the Têt canyon in the far distance. Plateau de La Perche (Fig. 12.8) is a population of remarkably well preserved, fan-shaped rock pediments. These connect via a concave slope to the Cambra d'Ase fault scarp, which here displays a series of rather subdued faceted spurs compared to other occurrences encountered along the Têt Fault (see Itinerary 6). The pediment apex is situated at 1800 m, below the 'Eyne 2600' ski resort. Its tread cuts across contrasting outcrops of local bedrock such as the base of the Mont-Louis sheeted

Fig. 12.7 View of the Têt glacial valley from Coll del Pam. A Looking upstream: Avellans rock basin and large rock bars at Malpas. **B** Looking downstream towards the terminal moraine units

Fig. 12.8 Plateau de la Perche and its fan-shaped pediment ('rock fan'), backing onto the Cambra d'Ase–Puigmal horst. The tread of P2 is poorly distinguishable from P1 in this area, probably explaining the landform's large extent. The recent moraine (M), produced by a glacier descended from Cambra d'Ase, overrides the pediment. The faceted fault scarp is less conspicuous here than farther east (Conflent) or west (Cerdagne), suggesting smaller fault throws. La Perche is understood to be the vestige of P2 that has undergone the largest magnitude of geologically recent regional uplift (see Delmas et al. 2018, for palaeoaltimetric reconstructions)

granite intrusion (laccolith), and its underlying micaschist and orthogneiss floor. A residual scatter of pebbles has been reported 500–600 m SE of the col, but the topographic surface is largely clear of alluvium and regolith. Note how the Cambra d'Ase glacial cirque has cut back into the Carança gneiss dome and its envelope of micaschist and marble; its frontal moraines have spilled out onto the eastern tread of the pediment, now covered by forest (Saint-Pierre-dels-Forcats).

Return to Coll de La Perxa and take the D33 towards Eyne. The 70-m-deep Eyne valley has been cut into the tread of P2; a residual deposit of early Pleistocene alluvium can be traced along midslope benches ~ +55 m above the modern channel, for example where the road leads up to the ski resort. From the parking area serving as gateway to the Eyne valley, you gain access to a U-shaped glacial valley; its well-preserved terminal moraine is situated just ~ 250 m down the path from the car park (Fig. 12.9A). Eyne valley is a nature reserve, famous since the eighteenth century among botanists for its diverse flora and high proportion of endemic species (Fig. 12.9B, C). The road rises to the tread of the Eyne pediment, which is a replica of La Perche. Roadside parking options in the vicinity of Coma d'en Llanes should allow you to follow any number of tracks crossing the plateau on foot.

Stop 4. Eyne plateau: constraints on the age of pediment P2
Unlike Plateau de La Perche, here the pediment (Figs. 12.10A, 12.11A) is covered by a thin, patchy veneer of deeply weathered alluvial deposits, all supplied by the Eyne catchment, with obvious contributions from its widespread white vein quartz cutting through the Bosc del Quer orthogneisses. The small pebbles (quartz, various schist facies) show up in the ploughed fields, particularly north of the road around elevation spot 1614 m. Around Coma d'en Llanes, you will also find quartz boulders 1–2 m in diameter, usually rounded and polished by fluvial transport and displaying ochre-tinted postdepositional patinations. These boulders have often been cleared by farmers and piled in heaps on field boundaries (Fig. 12.11B, C), but a few still remain in the fields. The NW part of the pediment tread cross-cuts the NW–SE Via Fault (below the solar power station) and likewise some NE–SW boundary fault lines (Coll Rigat–Saillagouse Fault) of the Cerdagne Basin, as well as upturned beds of Vallesian sediment in the basin itself (Figs. 12.10B, 12.11A). Pediment P2 is thus clearly younger than late Miocene, and Itinerary 8 will show that it also cuts across structures in Turolian beds. Overall, P2 thus formed at some time during the Pliocene (Calvet 1996; Calvet and Gunnell 2008; Delmas et al. 2018). The large triangular faceted spurs at Bois d'Eyne and Pica del Quer highlight the fault scarp and its likely post-P2 throw. Evidence for post-P2 neotectonics is provided by the area situated above

Fig. 12.9 Eyne valley: its glacial and botanical heritage. **A** Frontal moraine (foreground), with P1 on the skyline. **B** *Xatardia Scabra* (Lapeyr.) Meisner, 1838, a unique endemic plant species named after botanist Barthélemy Xatart (1774–1846); it grows on active talus cones and is restricted to just a few valley sides of the eastern Pyrenees. **C** *Xatardia* as a geomorphological marker; *Xatardia* roots grow to lengths of up to 2 m during their 3- to 4-year (?) life span, thereby allowing intensities of talus cone activity, here in schist, to be quantified (Huc 2008, 2010)

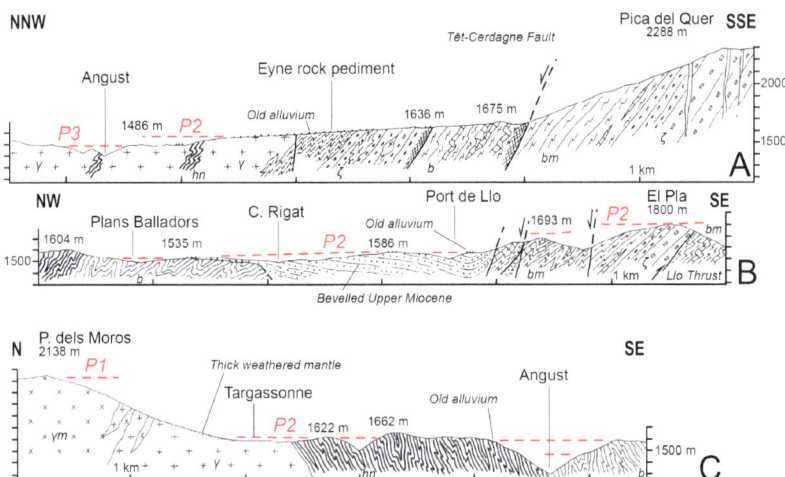

Fig. 12.10 **Cross-sections through pediment P2 at La Perche** (after Calvet 1996). **A** Section west of Eyne; pediment with its mantle of ancient alluvium and the benchland staircase involving P2 and the vestige of a pediment or broad palaeovalley ascribed to generation P3. **B** Section through Port de Llo and Coll Rigat; illustrates the topographic bevel across uptilted Upper Miocene beds and a possible tectonic offset by the boundary fault (presence of an anomalously elevated bench at El Pla). **C** Targassonne bench and its regolith (mature saprolite and ancient alluvium). Lithostratigraphy: γ, Mont-Louis granite; *hn*, hornfels; *b*, chlorite schist and marble (Canaveilles Group, middle series); *bm*, andalusite–cordierite micaschist (Canaveilles Group, base); ζ, metagranite (Ordovician)

the apex of P2, which forms a wide concave zone (fanhead embayment) above the Segre River valley upstream of Llo (El Pla, 1800 m) and is vertically offset by ~ 100 m by the Têt Fault and its offshoots (Figs. 12.10B, 12.11A).

Resume your journey and head towards Port de Llo. As you approach the thermodynamic solar power plant (9 MW, officially supplies the needs of 6000 households), the road crosses the local NW–SE fault line between the basement and the Miocene basin deposits; both compartments have been levelled off by erosion. Stop in the hairpin bend above Llo.

Stop 5. Panoramic view of Llo: lower Cerdagne and its Vallesian deposits
The roadside exposure shows an outcrop of proximal boulder beds (Saillagouse Fm.) attributed to the lower fill sequence of Cerdagne (Fig. 12.12A, B). At a more distal location, the base of this formation has yielded two fossil sites (Estavar and

Fig. 12.11   Fan-shaped rock pediment P2 west of Eyne. **A** General view from the D33 at La Coma d'en Llanes. Note faceted spurs along the fault scarp and probable vertical off-set of the pediment at El Pla; the pediment surface cross-cuts Hercynian basement in the foreground, and uptilted Vallesian beds in the background (to the SW). **B** Boulders of rolled quartz along the GR 36 trail; triangular faceted spur of Bosc d'Eina in background. **C** Rolled and patinated quartz boulders, here removed from neighbouring fields and dumped along the GR 36 trail

Llivia) bearing mammalian assemblages of lower Vallesian age. Here, the facies displays boulder- and cobble-filled palaeochannels cutting into beds of red and ochre claystone and siltstone. The entire package has been tilted towards the boundary fault, with apparent dips of 13° to the south. The clastic material consists exclusively of schist supplied by outcrops of the Canaveilles Group, with apparently very little andalusite–cordierite micaschist and no gneiss despite the fact that these last two lithologies coincide with the large faceted-spur expo-sures of Cambra d'Ase (2750 m) and Bosc del Quer (2288 m at Pica del Quer) (Fig. 12.13). It must follow from these clues that the fault scarp was generated in post-Vallesian (i.e., post-Tortonian) time, and probably during Turolian (i.e., Messinian) time—as will be revealed further west in the context of Itinerary 8.

The topographic floor of the lower Cerdagne Basin lies ca. 300 m below ped-iment surface P2, and is clearly visible from this point. The topographic basin

Fig. 12.12 **Vallesian flood debris at Col de Llo. A** Section along the D33 below Camp de la Paret; exclusively contains schist; note dip towards the boundary fault. **B** Close-up of depositional unit containing a large, poorly rolled schist boulder. **C** Badlands topography in the Vallesian outcrops; view from the D33 towards the west and Rec dels Torrents

was overdeepened during the Quaternary by rivers of the Segre catchment, which removed a substantial thickness of the stratigraphy. Erosional stripping occurred in 4 successive stages, with each stage punctuated by a population of alluvial fans and terraces. This staircase of landforms is still preserved in the landscape along the base of the mountain front (see Itinerary 8). The gully systems (Fig. 12.12C) cutting into the footwall hillslopes as well as into the hangingwall stratigraphy between Llo and Saillagouse—and elsewhere in Cerdagne—are believed archaeologists to have hosted ancient gold mine workings, where the sediment (rich in schist debris) could be washed and sorted (Cauuet et al. 2014).

The successive bends in the road cut across the structural boundary with the basement (schist), and also display components of large olistostromes (schist and marble from the Canaveilles Group). These appear to have been released into the late

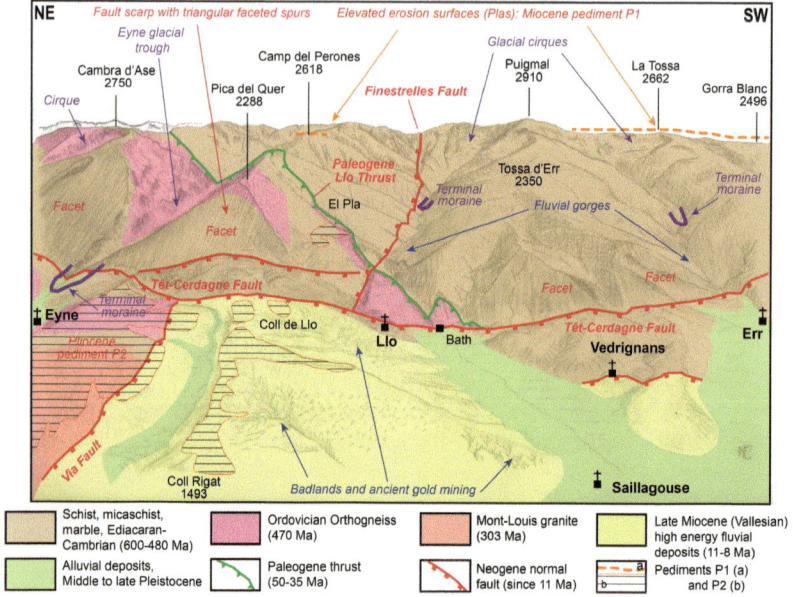

Fig. 12.13 Triangular faceted spurs along the Cerdagne master-fault scarp between Eyne and Err. Note position of pediment P2, which cross-cuts the Hercynian basement and the Vallesian sequence. Also note the palimpsest of directly observable geological structures in this area: Hercynian anticlinorium in gneiss, south-vergent Alpine thrust, and Neogene extensional fault

Miocene basin from outcrops of the Cortal Blanc massif (directly above) as a consequence of tectonic activity. The 'Parking des Bains' at Llo is a convenient stopping point. From there you can walk to the spa building and observe a segment of the Têt–Cerdagne fault plane cutting through orthogneiss (Fig. 12.14). Exploration of the area around Llo on foot should allow you to observe different rock facies of the Axial Zone basement and the effects of Hercynian and Alpine tectonics on the rock fabrics. Using Fig. 12.13 as a rough guide, special highlights are the Paleogene Llo reverse fault, thrusting the Sant Feliu orthogneiss over the micaschist, and the Neogene tectonic deformations. Continue towards Saillagouse and Estavar on the D33. Before reaching Estavar, stop and turn right into Chemin de l'Arbolosa, where you can leave your vehicle.

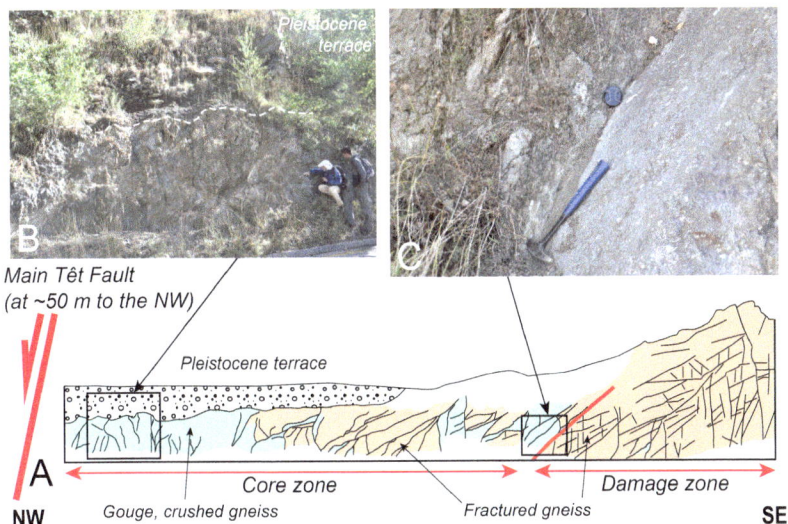

Fig. 12.14   The Neogene Têt–Cerdagne Fault at Bains de Llo. **A** Sketch from a photograph of the outcrop (after Taillefer 2017, redrawn and modified). **B** Detail of the crush zone, here affecting only the gneiss; alluvial deposit T2 rests over the fault line and remains unaffected by tectonic offsets. **C** Close-up of a fault plane parallel to the main boundary fault. The main fault, along which the schist and crushed gneiss are found in juxtaposition, is situated 50 m further down the valley to the NW and offers no exposures

Stop 6. L'Arbolosa: evidence of a Quaternary reverse fault

The exposure, on the left side of the track, is about 60 m long starting from the D33 but is threatened by overgrowth and may not be as fresh-looking as in the illustrations shown in Fig. 12.15. You are here in the immediate vicinity of the E–W Cerdagne boundary fault (Calvet 1996; Baize et al. 2002), which appears to guide the Segre River and has raised the base of the Neogene depositional sequence (resting over the schist bedrock) by about 60 m by comparison with its relative position at Estavar railway station. Trenching has revealed that the N130°E–40°NE listric reverse fault has brought grey and beige silty clays of Miocene age into abrupt contact with a pebble-rich bedload deposit of Middle to Late Pleistocene age (generations T2 or T1, linked to the Segre River). The alluvial deposit also fills palaeochannels cut in the underlying Miocene stratigraphy of the upthrust block (fault throw: ~5 m).

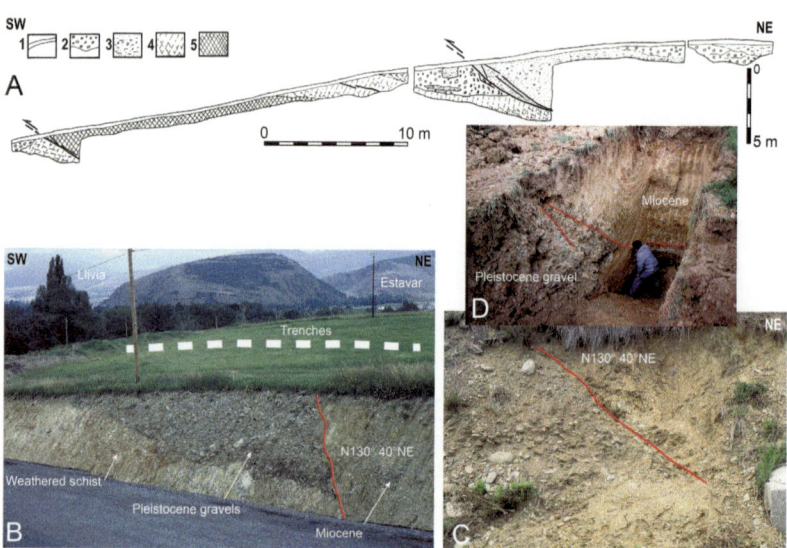

Fig. 12.15   Evidence of Quaternary reverse motion on the Segre Fault near Estavar. A.
Exposure of reconnaissance trenches cut in 1992. Key to symbols and ornaments—**1**: brown
soil; **2**: alluvial deposits T2, mostly containing weakly weathered schist pebbles; **3**: Vallesian beige and ochre sand-rich clay containing flakes of schist (either colluvial or in situ);
4: deeply weathered basement (red to ochre schist); **5**: unweathered basement (schist). **B.**
Embankment exposing the reverse fault in 1987. **C.** Close-up of the reverse fault plane in
1997. **D.** Trench exposure of the reverse fault in 1992

*The D33 crosses the Angust River at Estavar. A disused lignite mine, which originally
yielded an assemblage of fossil faunas, used to lie 400 m WNW of the church
(42°28′09″N, 01°59′28″ E), along the 15-m-high embankment rising above the
trail. Signs of the mining activity are no longer directly evident; nonetheless, at
the base of the embankment, residual mining spoils of greyish-white and blackish-
brown platy clay, still exposed in the late 1980s, should reveal a rich content of
fossil gastropods and lignite debris. The lignite beds were mined at depths of 12
and 18 m.*

*At Estavar, take the D33f towards Égat. The ditch in the last bend of the road on
the way up (~ 1570 m; 42°29′13.6″N, 02°00′49.9″E) show a vestige of deeply
weathered alluvial gravel and boulders, rich in smectite, filling gullies cut into the
underlying schist (Fig. 12.16). This alluvium is a continuation of the deposits previously encountered across the tread of P2 at Eyne (**Stop 4**). The road now reaches*

Fig. 12.16 **Ancient alluvial deposit at Col d'Egat associated with pediment P2. A, B** Overview and details in 1987. **C, D** Overview and details in 2021. Note intensely weathered state of pebbles and boulders

an extensive erosional bench, known as the Targassonne surface, that extends along the northern edge of the Cerdagne Basin and is a continuation of P2 (also previously visited at La Perche, **Stop 3***). At the first roundabout, take the D618 towards Targassonne. After the village, follow the road signposted to Thémis, a disused solar power station. Stop above Mas de Vilalta, from where you will get a good view of P2 and the wider basin.*

Stop 7. Targassonne pediment bench and its weathering mantle

The concave connecting slopes of P2 to the Calma massif are mantled by a $\sim$50-m-thick layer of mature in situ saprolite (9–14% clay-size fraction in the matrix material, abundant smectite). Exposures of the saprolite can be inspected in the ditches along the road to Thémis, for example in the first bend (42°30′00.75″N,

01°59′34.3″E) and locally elsewhere along the D618, between the numerous granite tors. This tor field (Fig. 12.17) in the southwest part of the Targassonne pediment occurs because meltwaters from the Angoustrine glacier and/or the small Clot cirque selectively stripped and gullied much of the weathered mantle, thus generating a spectacular rocky maze of tors and corestones. It is now mostly overgrown by woody vegetation, however, partly because of the sharp decline in sheep stocking throughout this area.

*The D618 passes the Targassonne tors, then over a bedrock step across the valley (Les Fontetes). This hornfels rock bar stood across the path of a Middle Pleistocene glacier, further documented at **Stop 8**. The road then curves around an*

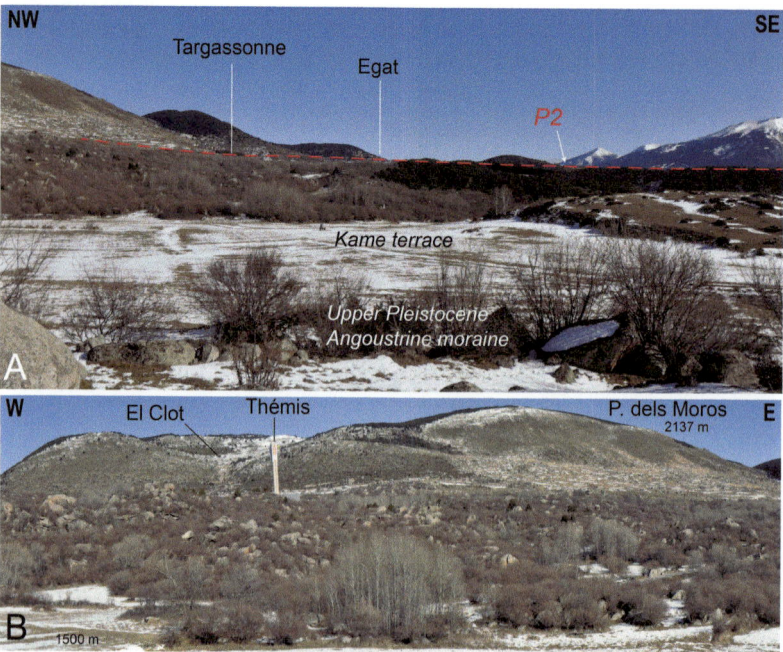

Fig. 12.17 Pediment P2 and granite weathering features at Targassonne; viewed from Angoustrine left-margin lateral moraine. **A** Landscape view showing concave slope linking the pediment to the steeper slope of Pic dels Moros (note its uniform gradient). The concave slope segment in the east is lined with substantial thicknesses of grus (i.e., 'rotting' granite). **B** Detail of the corestone boulder field exposed in the west, where the grus mantle has been stripped deeper to the weathering front

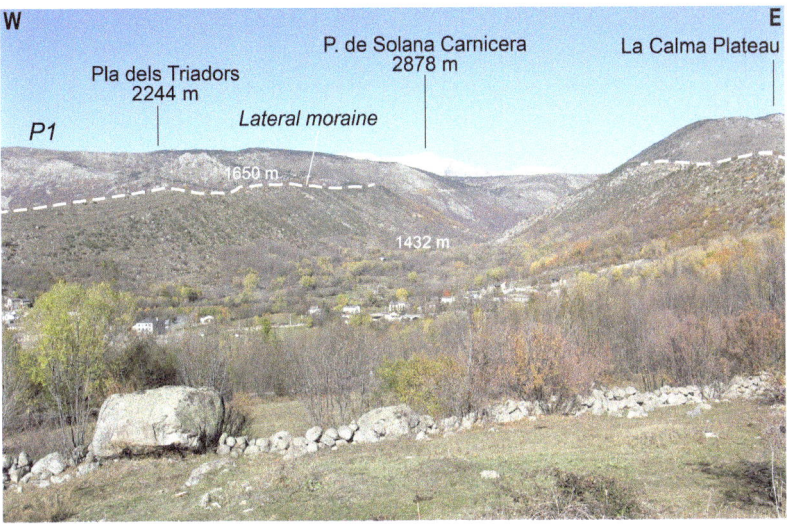

Fig. 12.18 Angoustrine glacial trough and lateral moraines. View from the frontal moraine on the D618. Carlit plateau in background

ice-marginal kame deposit obstructed by the LLGM left-margin moraine of the Angoustrine glacier; it reaches the terminal moraine at elevation spot 1384 m, from where you get good views of the Angoustrine trough and its pair of lateral moraines (Fig. 12.18). Towards the east and south, other residual ribbons of moraine and patchy till have been correlated with at least two generations of Middle Pleistocene glaciation, both of which extended by at most 1 km *beyond the LLGM frontal moraine (see also* Fig. 12.21*). At Villeneuve-les-Escaldes, take the D10 towards Dorres and park above the functional rehabilitation centre (also a spa, with water naturally heated at* 41 °C*). The restored and functioning 'Roman' baths nearby at Dorres are open to the public.*

Stop 8. Les Escaldes: stratigraphy of ancient saprolite and moraines (40 min walk return)

Follow the well-marked 'Tour du Carlit' trail, which passes behind some disused apartment blocks. A good section through thick saprolite containing corestones appears at the foot of the slope (Fig. 12.19C). The tor and boulder field on the catchment slopes of the Coma Ermada where exposed after stripping of this regolith mantle. This site became the focus of intense stone quarrying from

Fig. 12.19 Exposures at Les Escaldes hospital. **A** Deeply weathered ancient till. Exposure near the water tank. **B** Granite tor, initially partly exhumed and subsequently buried by the till. **C** Deep grus at slope base; section ~ 10 m high

the late nineteenth century to the 1950s, and its products contributed to many buildings, walls, bridges and railway viaducts throughout this region. A few tors emerge here and there, but the saprolite overall drapes the entire hills-lope as far as a track leading to a water reservoir. At this point (42°29′34.4″N, 01°57′18.2″E) you will encounter the first occurrence of ancient till capping the saprolite (Fig. 12.19A, B). The till contains granite boulders often 1 m in diame-ter, characteristically rotten to the core, in association with pebbles of dark schist and hornfels often displaying striation marks. This patch is the oldest among three local generations of glacial deposit which were inventoried (Fig. 12.20), classified, and regionally correlated on the basis of their weathering attributes

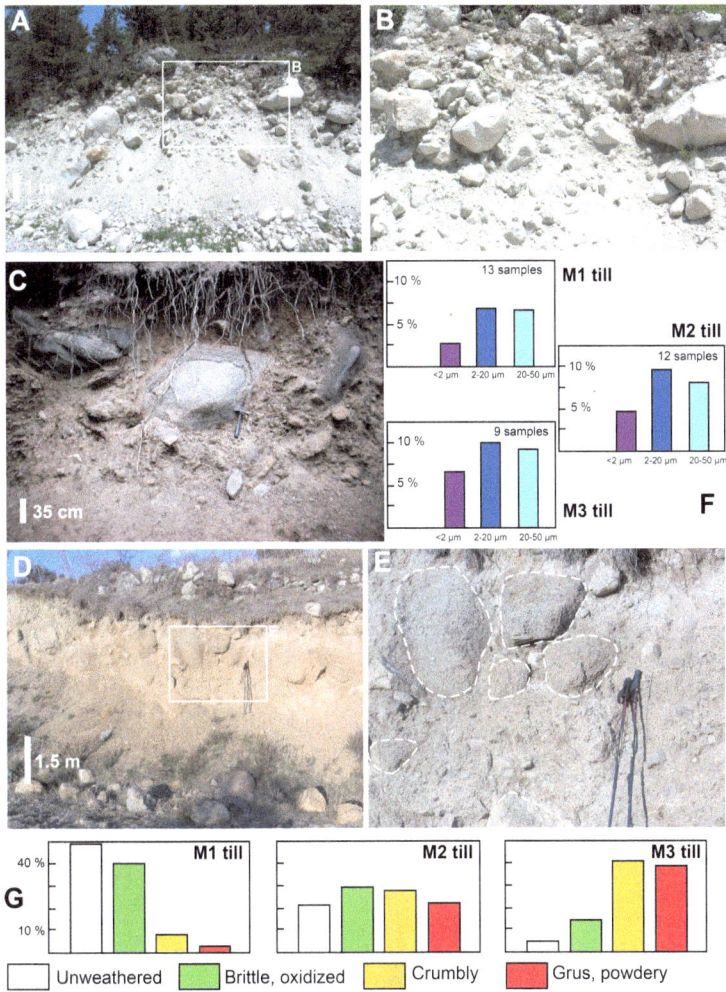

Fig. 12.20   Aspects of the three generations of glacial depositional material in the eastern Pyrenees. **A**, **B** Younger generation, M1 (Late Pleistocene); section through frontal moraine near Mont-Louis, D60 roadside. **C** Intermediate generation, M2 (mid to late Middle Pleistocene); boulders are spheroidally weathered on the outside, but with a fresh core; Esposolla section. **D**, **E** Older generation, M3 (early to mid-Middle Pleistocene); boulders are weathered to the core; Escaldes reservoir section. **F** Proportion of matrix fines among the three till generations. **G** Proportion of weathered clasts, and weathering intensity among granite pebbles 5–20 cm in length. The data are valid for till deposits from the Cerdagne and Capcir basins

(Calvet 1996). Based on systematic counts of granitoid clasts and a number of laboratory analyses, the attributes of this older generation of deposits are as follows: 7% of clay-sized fraction in the matrix, 80% of granite pebbles intensely weathered, 5% still unweathered. The till deposit is itself overridden by the right-margin moraine of the Late Pleistocene Angoustrine glacier, from which you can gain a more commanding view of the area. The stratigraphy overall clearly indicates that (i) the saprolite is pre-Middle Pleistocene, and probably mostly pre-Quaternary; (ii) the successive glaciers merely covered the saprolite without energetically ploughing into it or stripping it. This evidence corroborates conclusions previously drawn at **Stop 1** (La Calma) concerning the relative infirmity of glacial erosion in these transitional Mediterranean mountain environments where, additionally, the mountain hypsometry is attenuated by the elevated base levels of Cerdagne and Capcir and by the extensive occurrence of low-gradient topography (particularly P1) in the glaciated catchments.

Return to Font-Romeu on the D618 and the village of Égat, then continue towards Super Bolquère and Mont-Louis. After Super Bolquère, the road enters the frontal moraine complex of the Têt glacier and follows the topographic furrow lodged between the outermost (and most voluminous) terminal ridge and a series of smaller recessional ribbons known as the Barrès stadial moraines (Delmas 2009). *The Têt River exits its glacial valley through a narrow cut in the last lobe of the terminal moraine, and the road likewise cuts through it before reaching the fort of Mont-Louis, a citadel designed by Vauban in the seventeenth century (about Vauban, see also Itinerary 6). At the roundabout, follow the D118 toward La Llagonne and the Capcir Basin and park at the junction with the D60 signposted to Les Bouillouses (La Bollosa). Because this road leads to a designated site, access is regulated by summer opening hours and closed to private vehicles from 7 a.m. to 7 p.m. in July and August—a time when shuttle buses are the only option. Winter snow usually lasts from December to April, but the snow plough regularly clears the road as far as Els Avellans.*

Stop 9. The Late Pleistocene terminal moraine complex at Mont-Louis
During the Late Pleistocene glaciation, the LLGM Têt glacier was only 18 km long. As such, it was nonetheless the largest among the valley glaciers supplied by the Carlit icefield, which covered the massif between elevations of 2200 and 2400 m (Fig. 12.21) while leaving a few conspicuous nunataks rising above it (Viers 1961, 1968; Delmas 2005, 2009; Delmas et al. 2008).

No record of Middle Pleistocene glaciations, such as encountered at **Stop 8** on the basis of till weathering grade (Calvet 1996), have been identified in the vicinity of the Mont-Louis terminal moraine system, either because their vestiges are buried beneath the mass of more recent debris, or because their (speculative) position farther out near the Têt knickzone has made them vulnerable to stripping and destruction. The chronology of the Late Pleistocene glacial cycle in the Têt valley has benefited from a large number of ^{10}Be exposure ages from moraine boulders and glacially-polished bedrock steps (Delmas et al. 2008; Delmas 2009). In this guidebook, previously published exposure ages are updated on the basis of a uniform protocol incorporating revised decay constants, nuclide production rates, and corrections for snow cover (Reixach et al. 2021).

The LLGM frontal moraine at Mont-Louis (Fig. 12.22) is perfectly preserved, 60–90 m high. Its ridgetop actually consists of two to three elementary but closely bunched arcs. A ^{10}Be exposure age of 24.3 ± 3.8 ka on a boulder embedded in this frontal moraine indicates that the LLGM Têt glacier was synchronous with the global LGM. Another sub-population of less imposing, but equally fresh, moraine occurs ~ 900 m further down valley below a scatter of detached houses west of Mont-Louis (Les Artigues), as well as on the plateau to the east of the D618 opposite the parking area (spot elevation: 1633 m). These less conspicuous frontal ribbons could be vestiges from the earlier part of the Late Pleistocene glacial cycle, as suggested by a calibrated Schmidt hammer proxy age of 40.9 ± 1.1 ka from Les Artigues (Tomkins et al. 2018). A cut in the deposits along the D60, ca. 600 m to the west of the parking area ($42°30'55.1''$N, $02°06'24''$E), reveals the internal facies of this mostly unweathered Late Pleistocene till (Fig. 12.20A, B): ranker-type soil; grey to light beige silt- and sand-rich matrix containing less than 2% clay-sized particles; 50% of entirely fresh granite pebbles, with another 40% displaying oxidation features and vulnerable to hammering, and the remaining 10% characteristically friable—the latter exclusively consisting of granodiorite and biotite-rich monzogranite.

Driving up the Têt glacial stairway offers a chance to observe successive frontal moraines representing post-LLGM recessional standstills. These have been chronologically named Barrès (near the campsite), and La Borda (1 km upstream). The Barrès stadial deposits have not yielded a robust age record, but the Borda moraine has provided three ^{10}Be exposure ages of 20.4 ± 0.4, 21.6 ± 2.7 and 23.3 ± 3.3 ka. Stop at La Borda parking area and climb to the top of the rock bar (1690 m) situated just above the farm buildings.

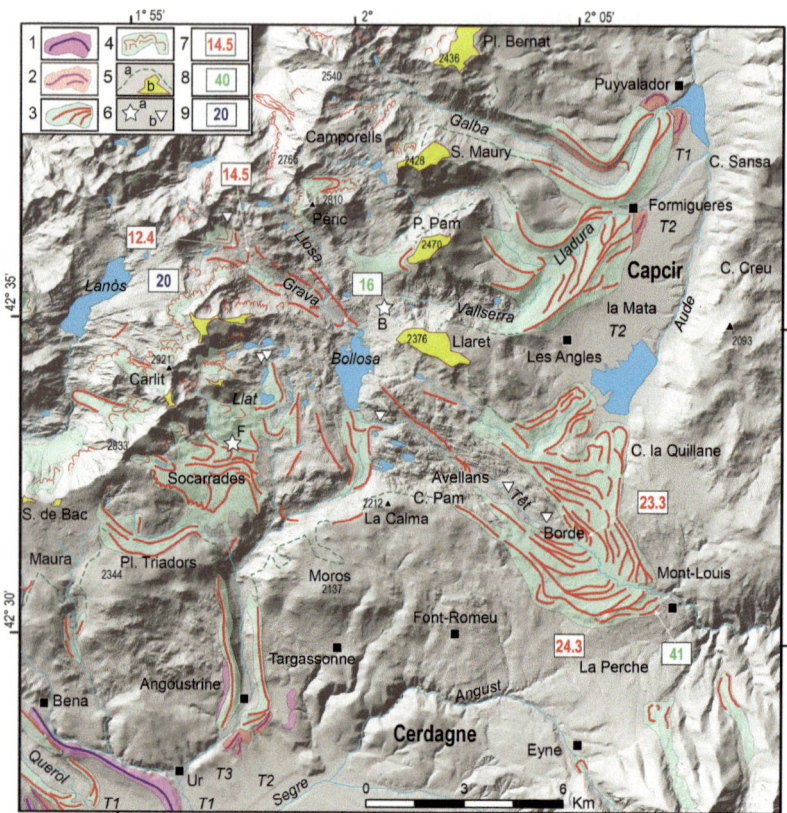

Fig. 12.21 Pleistocene glacier systems of the eastern Carlit massif. At the time of the most extensive glaciation, an icefield covering the Carlit massif was feeding several radially distributed valley glaciers: Galba, Lladura, Vallserra, Aude, Têt, Angoustrine, Brangoli, and Bena. Key to symbols and ornaments—**1**: Older till (M3) and moraines; **2**: Intermediate till (M2) and moraines; **3**: Younger till (M1) and moraines; **4**: Cirque moraines and rock glaciers; **5**: Glacial and supraglacial landforms; **5a**: Late Pleistocene ice limits; **5b**: Supraglacial, pre-Quaternary erosion surfaces hosting large stone polygons; **6**: Evidence of weak glacial erosion; **6a**: Granitic saprolite lining the Carlit plateau surface (pediment P1; B: la Balmeta; F: Els Forats); **6b**: Bedrock steps with a record of ^{10}Be nuclide inheritance, and thus indicating shallow depths of glacial erosion; **7**: ^{10}Be exposures ages (moraine boulders); **8**: calibrated Schmidt hammer ages (moraine boulders); **9**: ^{14}C ages (fossil peat)

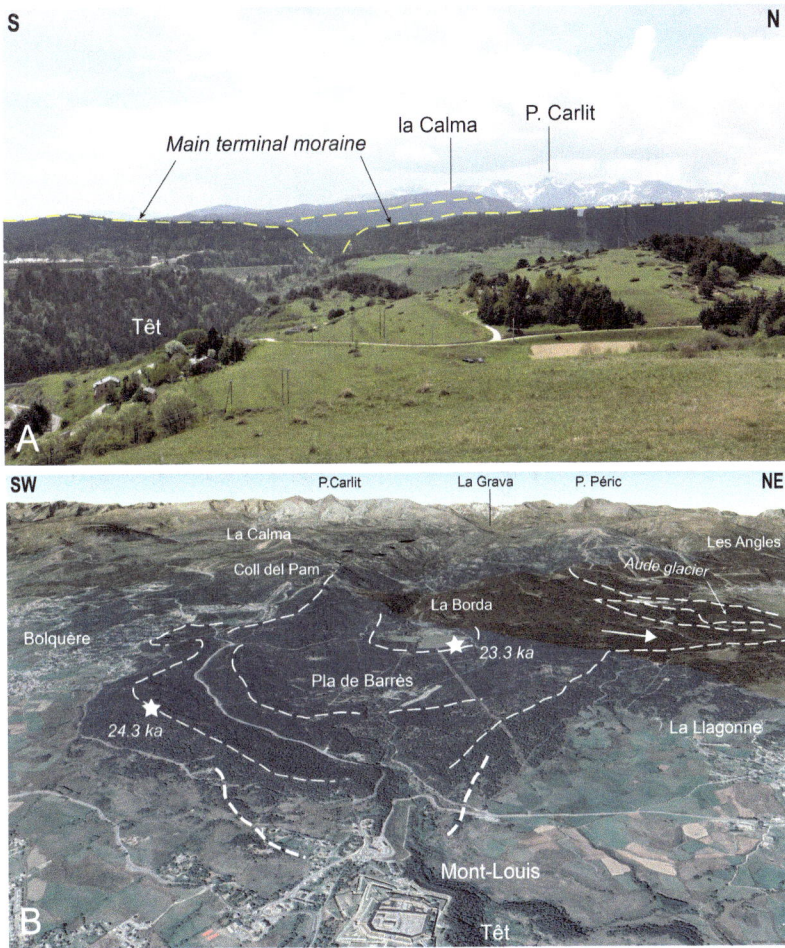

Fig. 12.22   Late Pleistocene frontal moraines of the Têt glacier at Mont-Louis. A Viewed from the east (Sauto plateau). **B** Oblique aerial view of the Têt frontal moraine sequence, denoting the outlines of a piedmont lobe (CNES/Airbus imagery, here provided by Google Earth); note glacier diffluence towards the Capcir Basin

Stop 10. Panoramic view of La Borda moraine and the Têt glacier terminal complex

The top of this granitic bedrock step provides a good vantage point from which to view the sequence of moraines, starting from the much taller LLGM terminal moraine at the far end of the valley to the smaller recessional occurrences in the middle ground (Fig. 12.23). The Borda moraine is covered by forest and closes off the large expanse of meadows and peatbog that currently fills the glacial rock basin. A 4.3 m sediment core from the peatbog has also sampled underlying lacustrine sediments. These have been chronologically tied to the LGIT on the basis of their pollen assemblages, and the '15,000 event' (Reille and Lowe 1993), now calibrated to ca.18 ka, ties in with the transition between GS2-1b and GS2-1a in the reference Greenland stratigraphy (Rasmussen et al. 2014).

The road next rises up a long break in slope where the glacial valley constricts into a gorge less than 300 m wide. Note the dominant NW–SE fracture pattern (Bollosa Fault) across this glacially-polished and plucked bedrock step. Stop as you enter the

Fig. 12.23 La Borde (La Borda) rock basin, here looking SE from La Borde bedrock step. Glacial and pre-glacial incision of the Têt valley is very limited in this area because of its location upstream of the large channel knickpoint below Mont-Louis. Incision depth into La Quillane–La Perche pediment levels (generation P2) does not exceed ~ 100 m

Fig. 12.24 Nuclide inheritance on ice-polished rock-bar surfaces in the Têt glacial trough. **A** Avellans bedrock step, also a rock-climbing spot. **B** Detail of ice-polished rock exposure sampled for [10]Be dating, Avellans rock outcrop. **C** La Grava bedrock step (upper site: 2350 m). **D** Detail of exposure-dating sample locations, La Grava (upper site)

Avellans rock basin (spot elevation: 1706 m; avellan: 'hazel' in Catalan). Here you can ascend the 'Rocher des Bouillouses' (summit: 1755 m), also a rock-climbing spot, by skirting round to its more gently inclined north-end rock face through the woods. Once at the top, aim for the south end (42°32′25.4″N, 02°03′25″E).

Stop 11. Summit of Avellans bedrock step

The imprint of glaciation on the pre-Quaternary valley morphology appears modest compared to the impact of the tectonics-driven fluvial knickzone below Mont-Louis (see Itinerary 6, Stop 15). The light touch of glacial denudation (see Stop 1) in this catchment (see Box 12.4) is confirmed by [10]Be exposure ages on several bedrock steps that are much older than the LLGM. An example is the summit of the Avellans rock mass (Fig. 12.24A, B), which has yielded an exposure age of 21.3 ± 3.6 ka, consistent with the deglaciation chronology, whereas perfectly preserved glacial polish on the flank of the rockface documents an exposure age of 74.0 ± 7.2 ka. This implies extremely limited bedrock erosion at this location during most of the Late Pleistocene glacial cycle (the anomalously old age is a composite of [10]Be atoms acquired since the last deglaciation of the glacially-polished bedrock step, i.e., after the LGM, but also of a preserved contingent of atoms produced prior to the Late Pleistocene, i.e., during MIS 5e). Other sites also record varying levels of nuclide inheritance, with anomalously old ages likewise indicating only partial resetting of the radiometric clock. They also, therefore, testify to shallow bedrock denudation depths during the Late Pleistocene—too shallow to attain depths of bedrock where the [10]Be radiometric clock would be entirely reset (i.e., much less than 3 m). Similar age dispersal has been observed at other bedrock steps, where a sensitivity test was applied by sampling various points at the top and on the front or sloping lateral face of the glacially-polished bedrock steps (Fig. 12.24C, D). The bedrock step at Les Bouillouses (1995 m), for example, yielded an age of 40.5 ± 4.6 ka, clearly suggesting nuclide inheritance at that location. Higher up in the catchment (see Trailwalks 2 and 3), the Sobirans step (2270 m) on the Carlit plateau likewise provided three ages of 22.3 ± 3.0 ka, 27.2 ± 4.0, and 27.3 ± 4.5; and the uppermost step of the glacial stairway, in the Grave cirque (2380 m; see Trailwalk 1), records one age at 27.7 ± 1.4 ka among other much younger results (Fig. 12.24D), suggesting a strong component of short-range spatial heterogeneity in subglacial erosion.

The road continues across the sediment-filled Avellans glacial rock basin. At Avellans, the glacial trough is deeper (300 m) than further down the valley. The frontal lobe of the corresponding recessional moraine is positioned at the top end of the Avellans basin floor. The road next winds along the edge of the lateral moraine *and finally rises up the large bedrock step at les Bouillouses. The latter is probably a glacially-retouched fluvial knickzone, similar to (but smaller than) the one previously encountered below Mont-Louis (Itinerary 6, **Stop 15**). The dead-end road terminates at Bouillouses dam.*

Stop 12. Panoramic view from Lac des Bouillouses (La Bollosa)

On the west side of the valley, the dam rests against the lateral branch of a recessional moraine; it joins its frontal lobe ~500 m further down the valley. The impounded reservoir fills a glacial rock basin mostly containing postglacial alluvium, and it was still blanketed by peat at the time of dam construction. To the west, the extensive Carlit plateau (presented at **Stop 1**) is a knock-and-lochan landscape, i.e., it was partly covered by Pleistocene plateau ice, unevenly scoured, and is dotted with variously sized lakes. To the north (Fig. 12.25), the Camporells plateau is also a vestige of middle Miocene pediment P1. Rising above it, the Puig Carlit and Puig Peric rock masses are Cenozoic inselbergs or monadnocks, lightly retouched by cirque glaciation during the Pleistocene (see **Stop 1**). Off to the south, the Angoustrine glacial trough previously encountered at **Stop 7** and **Stop 8** is a dry valley, its drainage having been beheaded by the Têt River in the vicinity of Avellans knickpoint—perhaps as recently as the LGIT and in relation to the massive recessional moraine accumulation situated just below Bouillouses dam (Fig. 12.21).

Optional Trailwalks Through the Carlit Massif

The Bouillouses car park is a good base for gaining access to the elevated erosion surfaces of the Carlit massif, and to a diverse mosaic of glacial, periglacial, and paraglacial landforms and deposits. Three excursions of varying duration and difficulty are proposed below.

1. **Upper Têt (Grava) glacial valley** (5 hrs return, easy hike)
The trail is the GR10, segment of a very well-marked, popular and spectacular trail spanning the entire Pyrenees. Through Bollosa orthogneiss, then through micaschist overlying the orthogneiss, in both cases pervaded by granite sills, the trail follows the west side of the reservoir, then begins to rise through granodiorites of the Mont-Louis pluton up the Grava valley, which is just another name given to the upstream reaches of the Têt River.
This catchment-head valley provides good opportunities for observing a closely-spaced succession of lateral and frontal moraines generated by fluctuating glaciers. At the bottom end of the Grava valley, a double ribbon of left-margin lateral moraines located between 2200 and 2100 m marks out the past existence of a 6.5-km-long glacier, its tip located around 2020 m. After walking another 1 km up the valley, you will encounter a frontal moraine ca. 2050 m produced by a shorter glacier. Still further by another 2 km, you will pass three other frontal and lateral moraines, in this case all generated at a time when glaciation was confined to the cirques, with successive glacier fronts at 2150 and 2160 m. [10]Be and Schmidt-hammer exposure ages on boulders embedded in these frontal and lateral moraines indicate 16.1 ± 0.5 ka for the 6.5-km-long trunk glacier, and 14.5 ± 1.9 ka and 12.4 ± 1.9 ka for the two cirque

Fig. 12.25 **Bouillouses Reservoir (La Bollosa), looking north**. **A** In winter, frozen lake. **B** In early summer. Note pediment P1 and its inselberg-like residual hills, Pic Péric and Pic de la Cometa d'Espagne

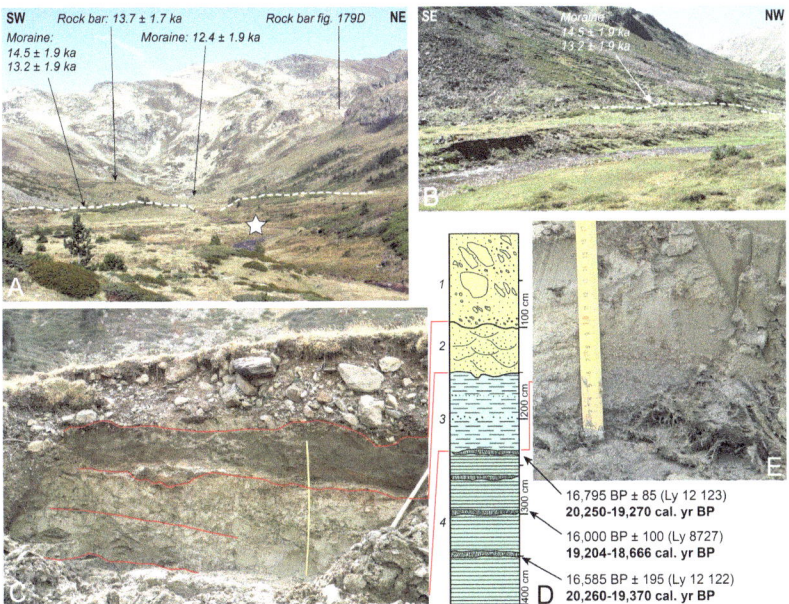

Fig. 12.26   Uppermost Têt valley (La Grava), site of a fossil peatbog (elevation: 2150 m).
A General view of the site; star indicates position of the peatbog, just below a small frontal
moraine (exposure-age best estimate: ~ 14.5 ka). **B** Fossil peatbog site viewed from the
stream's left bank; the vertical section cuts into a small outwash fan connecting directly to
the frontal moraine. **C** The fossil peat in 1997; section height: 2.1 m; note lower stratigraphic
units dipping slightly upstream. **D** Stratigraphy of La Grava deposit; **1**: high-energy outwash
alluvium; **2**: cross-bedded lenses of medium- to coarse-textured sand; **3**: beige lacustrine silty
clay containing occasional matrix-supported gravel beds; **4**: dark greyish-blue clay, rich in
organic matter and containing well-preserved layers of sphagnum. Stratigraphic boundary
between 3 and 4 appears conformable; ravinement surfaces are apparent below 1 and, likewise, between 1 and 2, although less conspicuously in the latter case. The lower part of the
stratigraphy was documented by coring (hand auger). **E** Contact between the lacustrine muds
and uppermost peat deposit. At one time, this boundary could be observed at the base of the
exposed section further to the east (usually below water, or level with the top of the water
table)

glaciers terminating respectively at 2150 and 2160 m. On that basis, the older stadial position correlates with the Oldest Dryas, and the younger positions with the Younger Dryas or earliest Holocene (9.4 ± 0.6 ka Schmidt-hammer age for the upper ridge ~ 2200 m, Fig. 12.27; Tomkins et al. 2018; Reixach et al. 2021).

A radiocarbon-dated fossil peatbog (~20 ka cal BP) situated around 2150 m a.s.l. on the upper Grava valley floor suggests that the three LGIT stadial deposits listed above were generated after an interval of pronounced deglaciation in this valley segment, thus allowing the peat to develop in a proglacial position before subsequent glacier readvance (Fig. 12.26). This peat-forming event highlights how fast post-LGM deglaciation of the SE Carlit massif occurred. This situation is at least partly ascribable to the hypsometry of the glaciated catchment, which contained extensive plateau surfaces (see Trailwalk 2 across the Plateau des Étangs). When contained within the accumulation zone, such as during the LGM, these plateau surfaces could host volumes of ice large enough to feed several valley glaciers 15–20 km long, such as the Angoustrine, Têt, and Vallsera (see also Fig. 12.21). Any small upturn in palaeotemperatures, such as occurred after the LGM, would raise the ELA, and even a small increment would drastically reduce the size of the accumulation zone and hasten the retreat of those valley glaciers far back into the cirque zone.

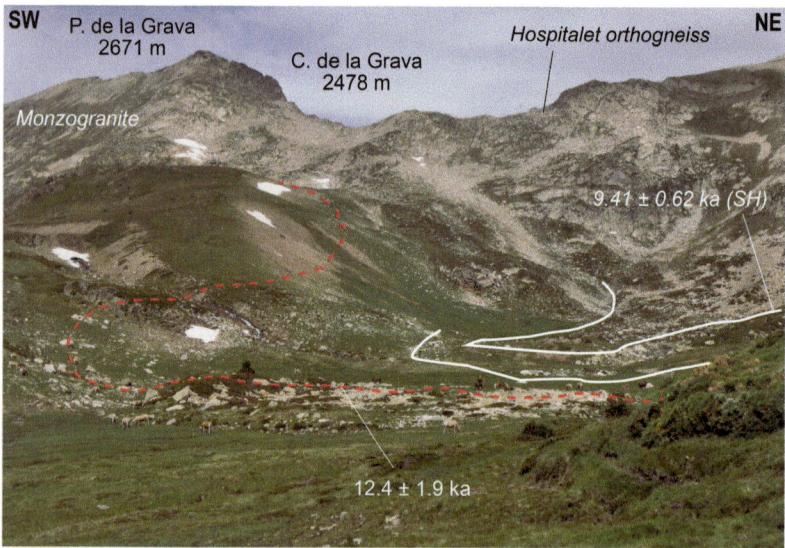

Fig. 12.27 La Grava glacial cirque hosting the youngest frontal moraines of the Têt glacial stairway (cirque floor elevation: 2160 m). The main moraine could be Younger Dryas. This upper valley receives abundant snowfall in winter because Atlantic weather systems from the NW tend to spill through Col de la Grava; this situation probably epitomizes the situation prevalent towards the end of the last glaciation, thereby sustaining this south-facing glacier late into the LGIT interval

2. **Plateau des Étangs** (2–3 hrs return; easy terrain, but to be emphatically avoided in foggy weather)
This marked trail starts at hotel Bones Hores, by the dam. It rises fairly steadily until you reach the tread of pediment P1, previously viewed from a distance from Stop 1. At the time of the LGM, the plateau surface was covered by plateau ice and, today, is analogous to the ice-scoured knock-and-lochan topography of NW Scotland, i.e., a mosaic of rocky knolls and lake-filled erosional hollows—at places also impounded by small post-LGM frontal or lateral moraines, overall covered by heathland and boggy vegetation (Fig. 12.28). Follow signs to Estany Llat, a lake impounded by a large frontal moraine ('estany' means 'lake' in Catalan).

Fig. 12.28   Glacial features of the Carlit plateau. View from Sobirans bedrock step. Neogene pediment P1, i.e., the plateau, was only lightly refashioned by cold-climate processes: components of knock-and-lochan topography, recessional moraines. Note another vestige of pediment P1 on the Puigmal skyline

From here, the infirmity of glacial erosion across the plateau can be appreciated by visiting one among several exposures of glacial till overlying—and thus preserving—a substantial thickness of rubified saprolite (Fig. 12.6D). To see this section (42°32′56.3″N, 01°58′0.6″E, elevation: 2094 m), aim to walk 1.5 km (return) towards the SSE from Estany Llat along the channel of the local stream (marked as Els Forats on maps). Otherwise, from Estany Llat continue the main itinerary by following the Llat lateral moraine to its upper tip, from where you will get a sweeping view over the knock-and-lochan topography of the plateau. Then veer to your right towards Estany de les Dugues. You can close the loop and return to Bouillouses dam by following the trail down from the outlet of Estany de les Dugues.

3. **Puig Carlit and the range-top surface**, S (6–7 hrs return; expect a few easy clambers over rock outcrops; check beforehand for at least 24 hrs of clear weather)

The first leg of the trail up to Llat is identical to Trailwalk 2. The path then rises over the large Estany de Sobirans bedrock step, where polished granite surfaces have documented substantial doses of nuclide inheritance (see **Stop 11**) and thus confirm the relative feebleness of Late Pleistocene glacial erosion in these catchments. Further along, the trail makes way through the hornfels outcrops of the metamorphic aureole, thus leaving the granite and entering the country rock (schist, Jujols Group), which also underpins the highest summits of the Carlit massif. Estany Sobirans is closed off by a recessional frontal moraine of undetermined age, but probably Oldest Dryas based on correlations with other dated occurrences in the eastern Pyrenees. Well-supplied but undated rock glaciers later spilled over the southern tip of the moraine deposits (Fig. 12.29A), perhaps during the Younger Dryas. The path continues up (Fig. 12.29B) into the only cirque of the Carlit massif devoid of rock glaciers—this clue suggesting the late existence of a cirque glacier with a more extended life span in this particular cirque than in any other in the vicinity. The trail rises up to Puig Carlit along the bedrock ridge, then up the main avalanche corridor (steep but easy). From the top, the views are stunning and secure a reality check on the otherwise poorly accessible vestiges of summit erosion surface, S, which here forms a perfect (unnamed) table-top surface (elevation: 2804 m, near Ras del Carlit) cutting across schist fabrics to the NE of Puig Carlit (2921 m). Around the residual monadnock (or inselberg), all the cirques host fresh-looking rock glaciers. These LGIT landforms were possibly reactivated during Holocene time, particularly given that persistent snow patches in current climatic conditions also suggest the (so far never investigated) occurrence of residual permafrost at these altitudes (Fig. 12.30). To the east and SE, the view overlooks the knock-and-lochan topography of pediment P1 encountered in Trailwalk 2. To the NW, the broad, ice-scoured topographic bench ~2350 m (Solà de Lanós) above the glacial trough hosting Lake Lanoux (Estany de Lanós) could also be a vestige of P1 (Fig. 12.31). The return trip to Les Bouillouses follows the same route, either directly back to Estany Llat, or via Coma dels Forats (but the trail is not as well worn, and thus less easy to navigate).

An additional possibility while near the summit is to walk to the table-top vestige 2804 m of range-top surface, S; the best route skirts around the east-facing cirque floor, then up a faintly worn trail to Puig de Sobirans (aim for the south face of the rocky peak, spot elevation: 2806 m) and finally across to the plateau. Its surface displays frost polygon patterns, with vertically churned up slabs of schist daylighting through the patchy soil (Fig. 12.32). From there, descend along the eastern spur down to Castellà (2550 m). The north side of this peak exhibits a succession of large open gashes in the schist bedrock (Fig. 12.33), typical of paraglacial rock slope failure—so-called deep-seated gravitational slope deformation, or DSGSD (Jarman et al. 2014). Continuing southward towards Estany del Castellar, you will pass a neat exposure of the boundary between granite and hornfels at elevation spot 2418 m (walk round the west side of the corresponding spur). From here, the trail back to Estany de les Dugues and Les Bouillouses is well marked.

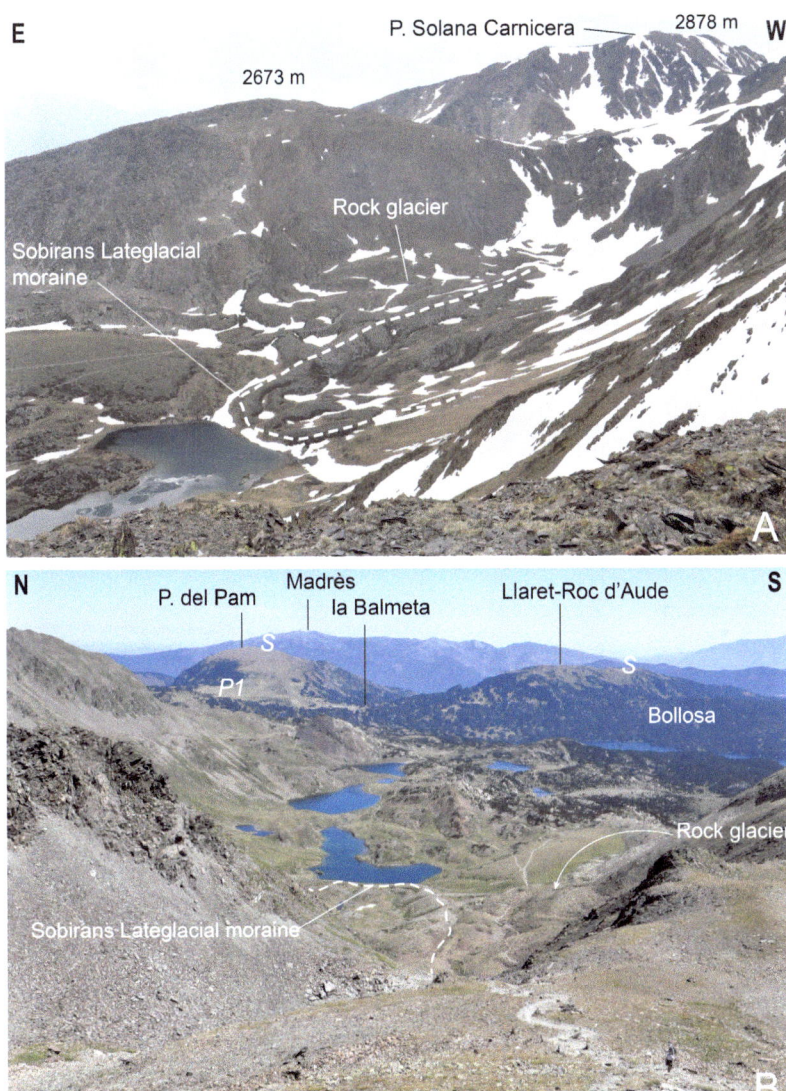

Fig. 12.29 Sobirans glacial cirque, on the edge of Pic Carlit. **A** View from the NE (spot elevation: 2690 m); the trail up to the Carlit summit rises between the small Lateglacial moraine and the rock glacier. **B** Plateau des Étangs and Sobirans cirque, here viewed from ~ 2700 m on the lower slopes of Pic Carlit

Fig. 12.30   Col des Andorrans cirque and its Lateglacial landforms, viewed from Pic Carlit. **A** Andorrans glacial cirque; note the two Lateglacial moraines forming a diffluence pattern around a bedrock obstacle, and surrounding the rock glaciers. **B** Close-up view of Andorrans rock glacier. The steep, totally unvegetated and unstable rock glacier front (elevation: 2600 m) suggests this landform is still undergoing periglacial dynamics; firn patches usually persist through the summer and could also indicate an occurrence of discontinuous permafrost under present climatic conditions. It is therefore also possible that such periglacial landforms, particularly when situated on north-facing slopes, were reactivated during the Little Ice Age

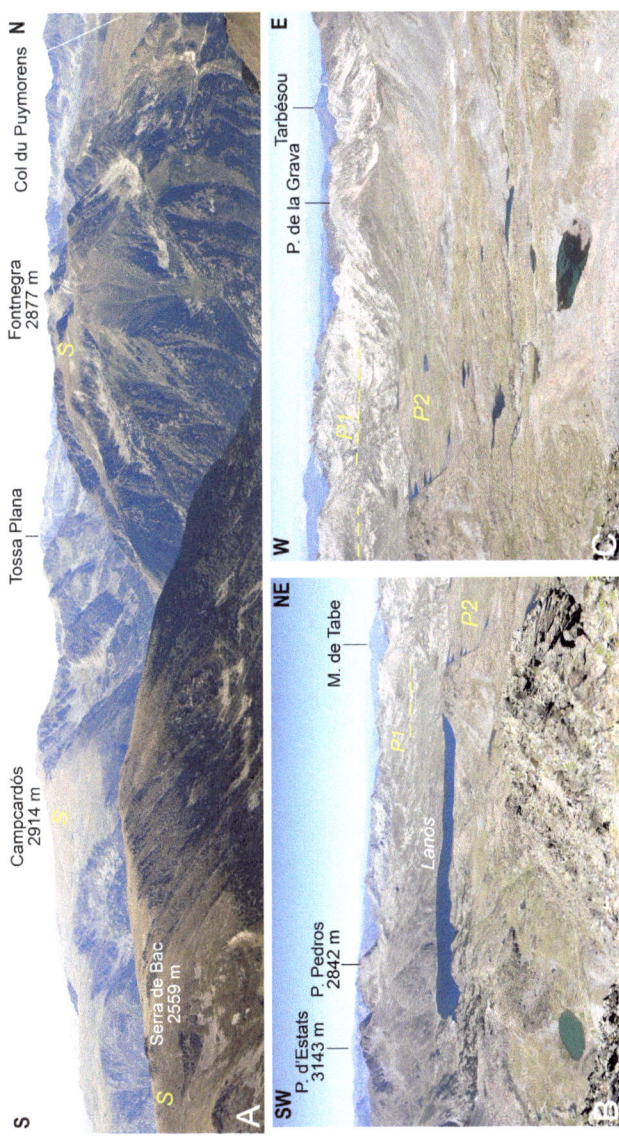

Fig. 12.31 Views from the summit of Pic Carlit. A Looking WSW. Note the population of range-top erosion surface vestiges, which are well preserved in massifs surrounding the Cerdagne Basin (Serra de Bac, Campcardós), more fragmented around Fontnegra, and their disappearance further west and north (Andorra, upper Ariège). **B, C** Looking NW and N. Note widespread glacial scouring in Lanoux basin (Lanós) and, despite this, the relatively unscathed treads of pre-Quaternary pediments P1 and P2. The Lanoux depression is probably an ancient etch basin, similar to many others encountered in previous itineraries (Vallespir, Pays de l'Agly, lower Conflent). Here, the basin floor lies 150–200 m below P1 but was overdeepened further down the valley by the successive Querol glaciers. Lake Lanoux fills this U-shaped valley today

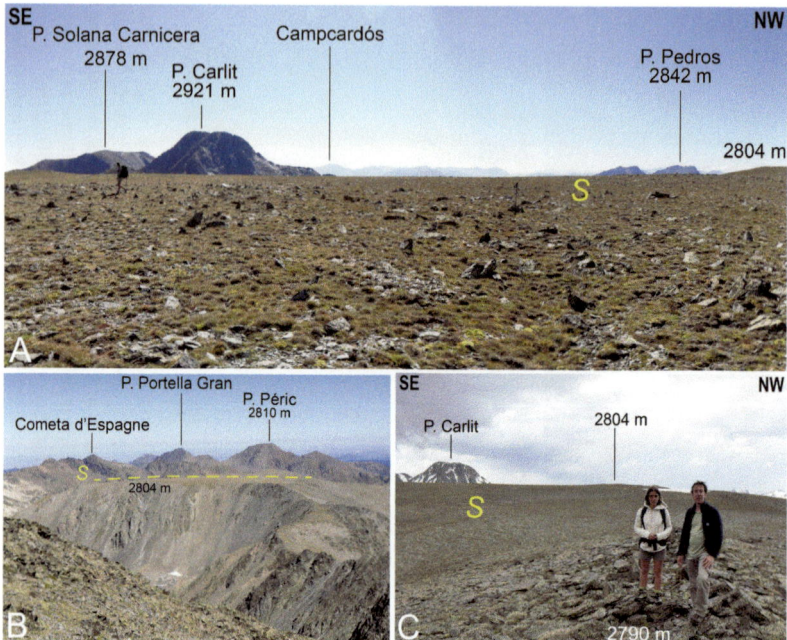

Fig. 12.32 Range-top erosion surface, S, at Ras del Carlit (elevation: 2804 m). Arguably the most spectacular vestige in the entire Pyrenees for its flatness and horizontal-ness. **A** Close-up view, looking SW; Pic Carlit could be interpreted as an inselberg; also note periglacial evolution indicated by nearly vertically upturned slabs of schist. **B** General view from Pic Carlit summit. **C** View from eastern end of ridge-top spur (elevation: 2790 m)

Fig. 12.33   Castella spur, split by a ridge-top graben caused by gravitational rock slope failure. Direction of displacement is to the north. Note stairway of erosion surfaces S and P1 in background

12.2   Capcir Basin

The road back from Les Bouillouses takes you to Mont-Louis. At the junction between the D60 and D118, turn left towards La Llagonne. The road soon reaches the undulating plateau around Coll de la Quillana (Col de La Quillane), which—like the Plateau des Etangs and Solà de Lanós—is a vestige of P1 covered by an eggbox mosaic of small basins etched in the deeply weathered granodiorite. The flat floors of these semi-enclosed basins could have formed and widened during the transient base-level stillstand regionally inferred from landforms labelled as generation P2 (see itineraries 4, 5, 6). An important focus at the next few stops will be a very straight NW–SE-trending scarp, which thus strikes obliquely to the Capcir half-graben's N–S master fault and overlooks the eastern part of the Quillane plateau. The scarp appears to have been generated by a previously unreported Neogene fault, but the published geological map has failed to define a fault despite depicting the scarp and showing that it cuts across the entire local succession of Axial-Zone Hercynian structures. At Col de La Quillane, take the D32f towards Les Angles through the forest of En Calvet (a placename with ancestral ties to one of the authors). The road cuts through two LLGM moraines, from a time when the Têt and Aude glaciers were coalescent (Figs. 12.21, 12.22B). *Aim to park at the eastern entrance to Les Angles, and walk up to the area above the chairlift departure building until you gain a suitable view of the basin.*

Stop 13. The Capcir half-graben from Les Angles

The Capcir Basin (for map views, see Figs. 10.12 and 11.4) is bounded in the east by a N–S fault, with the adjacent Madrès massif horst rising above it to the east (Fig. 12.34). Unlike the Cerdagne Basin, its fill sequences are poorly documented because they lie beneath Pleistocene deposits and are overall hidden from view by continuous vegetation cover. Close to Matemale hydroelectric dam in the south, some beds of deeply weathered gravel have been reported, grading farther downstream at first to sand, then to lignite-bearing lacustrine or palustrine clays (outcrops around Lake Puyvalador; Calvet 1996). None of these deposits have been dated, but the different facies are similar to those encountered in Cerdagne, including the clay mineralogy, which is dominated by kaolinite. A younger age than late Miocene for this half-graben cannot be ruled out, however, particularly given the totally atypical N–S strike of this tectonic structure compared to the other basins examined along previous itineraries. This E–W extensional rupture is compatible with transpression arising from roughly N–S applied maximum compressive stresses, a tectonic regime believed to have prevailed regionally during the late Pliocene, Quaternary, and to be still prevalent in present time (Goula et al. 1999; Rigo et al. 2015).

From the vantage point at les Angles, the boundary scarp in the east is clearly outlined below the Madrès despite the extensive forest cover. It cuts at right angles across the entire suite of Hercynian outcrops. The youthfulness of the escarpment is suggested by the scarcity of dissecting drainage catchments, even among more vulnerable outcrops such as schist around Coll de Sansa. The flat basin floor coincides with the depositional surface of a glacifluvial outwash sequence (around La Mata forest, in particular). Above the village of Les Angles, the extended ridge under forest is the LLGM lateral moraine of the Vallsera (Balcère) glacier (Fig. 12.35A). The till is characteristically unweathered and the moraine tapers up-valley to an elevation of ~1950 m, thus documenting a substantially lower ELA in this more northern position than previously encountered among the south-facing Carlit glaciers.

Follow the D32 and D52 towards Matemale, then the D118 towards Formiguères. The road takes you across the vast outwash terraces that appear to extend out

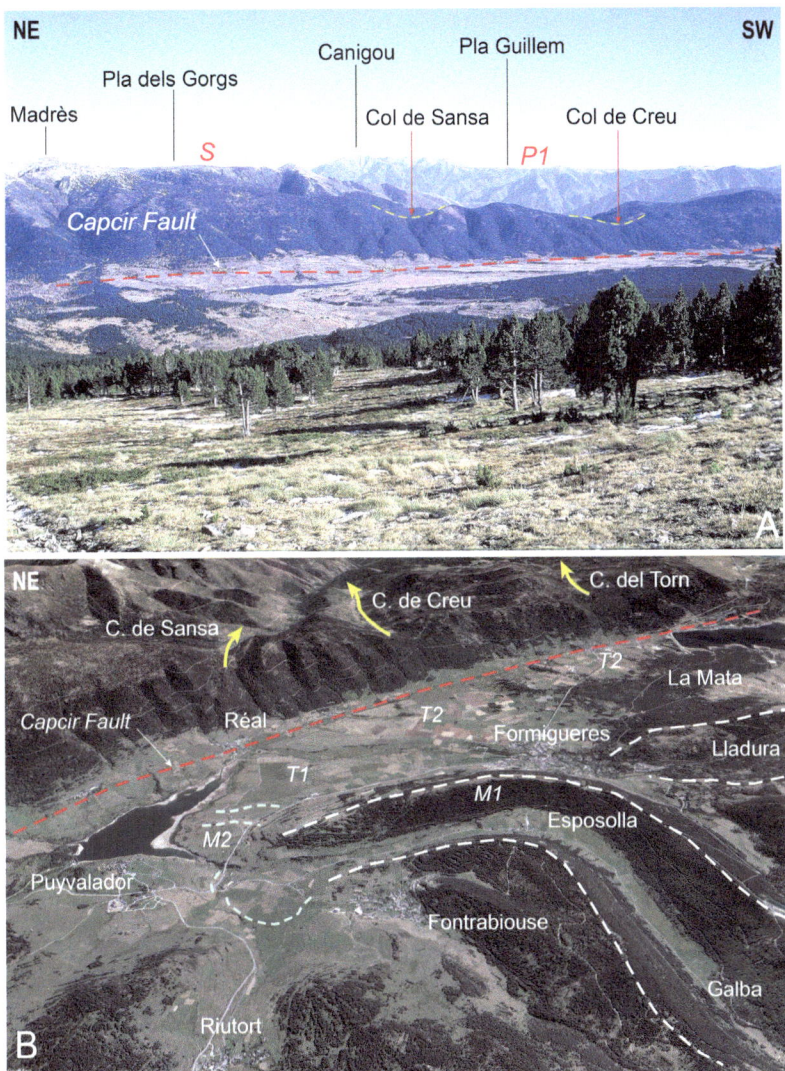

Fig. 12.34 The Capcir half-graben. A General view from eastern slopes of Pic de Ginevre. Note wind gaps (Col de Sansa, Col de Creu) in the footwall, forming broad cuts in the scarp skyline. **B** Oblique aerial view (Landsat/Copernicus imagery, here provided by Google Earth); note dry valleys and their wind gaps in the footwall block, and incurvation of the Galba lateral moraines as they enter the basin

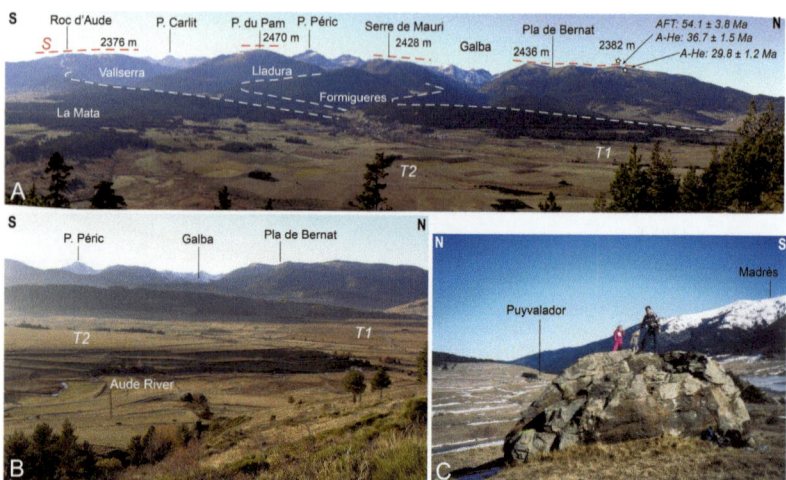

Fig. 12.35   The Capcir Basin and its Quaternary landforms. A View from the forest track
to Col de Sansa; note range-top surface, S, across the Carlit massif, and conspicuous bends in
the lateral moraines of both the Vallserra (Balcère) and Galba glaciers after they have exited
their respective valleys (i.e., no apparent valley-side bedrock constraints). In the foreground:
outwash terraces T1 and T2. **B** Respective treads of T2 and T1 at the junction between the
Aude and Lladura rivers. **C** Boulder of gabbro-diorite embedded in the Galba frontal moraine

*from the LLGM moraines. This first impression is deceptive, however, because the
deposits are much older than the moraines. The alluvial material is quite weathered
(usual list of features elsewhere compatible with generation T2) and capped by
thick and mature soils. The LLGM meltwaters from the Vallsera glacier cut box-
shaped valleys into the floor of T2 (Coma Llarga, Coma de Fontfreda, etc.), but
these valleys are currently dry. At Formiguères, turn into to the D32 towards Réal;
this road crossed the tread of a vast alluvial deposit inset ca.* 20 m *below the main
tread of T2. The alluvial deposit is the Late Pleistocene outwash fan of the Vallsera–
Lladura glacier* (Fig. 12.35A, B); *it connects to the LLGM terminal moraines. The
lakeshore of Puyvalador (right side mostly) exposes a few outcrops of Neogene
sandy clay. The unsurfaced forest road rising up to Coll de Sansa affords excellent
views of the LLGM moraine systems preserved on the floor of the Capcir Basin*
(Fig. 12.35A). *Drive back along the D32 towards Puyvalador and turn left onto
the D32b towards Fontrabiouse. This is a vast, almost level cave system open to the
public, with nonetheless several other levels of dry and drowned galleries, here once
again hosted by the Devonian limestone of the Villefranche syncline (see Itinerary*

6). Take the D32d (the road junction is situated among the Galba frontal moraines, Fig. 12.35C) and park when you reach Esposolla.

Stop 14. Esposolla and the anomalous Galba glacial trough

The Galba is a perfect U-shaped glacial valley, flanked by two perfectly parallel lateral moraines. In its lower reaches, however, the moraines undergo an unusually sharp bend at right angles without nonetheless losing their parallel geometry. The change in direction of strike is identical to that of the Lladura valley further south (Figs. 12.21, 12.34B). This evidence suggests the existence of a west-facing fault cutting through the Capcir half-graben, thus subdividing the Capcir basin floor into two compartments and raising the upthrown schist, hornfels and granite basement to shallower depths beneath the right-margin lateral moraines. This obstacle would explain why the valleys curve around in the vicinity of the fault line, as well as explain the positions and anomalously large sizes of their right-margin moraines. Sediment ponding caused by the bedrock obstacle over several Late Pleistocene and earlier glacial cycles would explain why disproportionately large accumulations of moraine appear to have been produced by such small catchments. Outcrops of weathered till, thus accrediting the notion that these moraines have aggregated several generations of glacial deposits, can be observed at the frontal lobe of the Galba moraine as well as at the base of its left-margin moraine. A good access route is the forest track rising NE behind Esposolla village (Fig. 12.20C), where exposures occur in the embankment, ~ 300 m from the village. On the regional scale of till weathering grade (Calvet 1996), weathering intensity ranks as intermediate: partly weathered large boulders, i.e., with a fresh core but spheroidally weathered outer layers; bright ochre matrix containing, on average, 5% of clay-sized fraction; and friable granite pebbles count for 50% of the stock, with 20% of unweathered granite pebbles.

Return to the D118 and follow the D32g towards Rieutort and the tourist resort of Puyvalador. The best views of the Capcir Basin (Fig. 12.34A) are gained from above the settlement and above the 1900 m contour (e.g., Serrat de la Cornera, 1 hr 15 min return on foot up a forest track), and farther up if time allows it (you can reach surfaces P1, and then S, at Pla de Bernat–Pic de Ginebre: 2382 m; Fig. 12.35A).

Stop 15. The unmapped neotectonic Capcir Fault from Puyvalador

The remarkably straight Capcir boundary fault scarp shows up very clearly from this vantage point. A series of wide, cradle-shaped cuts along the top of the scarp correspond to wind gaps opening out eastward into shallow palaeovalleys, e.g., Railleu–Coll de Creu in the middle (1715 m), Coll de Sansa (1775 m) in the north, and Coll del Torn in the south (1897 m) (Figs. 10.12, 11.4, 12.34). These are the vestiges of beheaded palaeodrainage feeding into the Sansa and Têt catchments (encountered in Itinerary 6) and previously descending from the Carlit massif before the Capcir graben existed. Note that these wind gaps align perfectly with the Galba valley in the case of Coll de Creu, with the Lladura in the case of Coll del Torn, and with the Rieutort in the case of Coll de Sansa. The valleys were thus cut off by recent tectonic activity along the Capcir Fault, thereby promoting drainage diversion to the benefit of the Aude watershed.

Optional excursion to the Quérigut Basin

An extended option involves driving north to visit the Quérigut Basin, a somewhat enigmatic enclosed basin due north of the Capcir and displaying a number of glacial deposits and landforms (52 km return from Puyvalador).

The floor of the Quérigut basin in granite defines a small geographical region also known as Donezan, and lies roughly 1200 m a.s.l. Its topography is an irregular mosaic of hills and hollows (Lagasquie 1984). This basin could correlate with other basin floors of generation P2 (many occurrences encountered throughout the region in the form of etch basins in granite, karst poljes in limestone, rock pediments at range fronts, and flat-floored palaeovalleys in limestone). Many occur as consistently mappable entities between here and the Plateau de Sault to the north (link with Itinerary 5; Fig. 10.12). The Quérigut Basin may also have hosted a Neogene depocentre similar to the Capcir but today entirely stripped clean of its fill sequence, perhaps because this north-facing area is under direct influence from Atlantic weather systems, and Quaternary glaciation as a result was comparatively more intense than on the south side of the Carlit massif. The Bruyante valley glacier thus descended to altitudes of 800 m, and the ELA stood at around 1700 m (based on the altitudes of lateral-moraine upslope tapers).

After Puyvalador, leave the D118 by turning off onto the D32 (becomes D16 after crossing the Ariège county line). The Aude River knickzone begins in this area, where the river rapidly drops into a deep gorge. The road follows a succession of benches above the gorge (perhaps generation P1, here degraded by more recent processes and lowered by tectonics in the vicinity of the Capcir Fault), then descends into the Quérigut Basin. At Bac de Rouze, note the Late Pleistocene lateral moraine of the Bruyante glacier (estimated terminal position around 850 m). At Rouze, the D16 on the way down to Usson cuts through exposures of weathered till of 'intermediate' grade, in an area where the corresponding glacier terminated just above Usson castle, i.e., ~800 m. On the way up out of Rouze, follow the D116, then the D25 at Mijanès: the road rises up the Bruyante glacial trough, then up the Pailhères valley, flanked on either side by well-preserved lateral moraines. In the first few bends up to Col de Pailhères, the road exits the Quérigut granite and enters outcrops of Cambrian schist and marble. Col de Pailhères (~2000 m) is an extensive vestige of pediment P1 (Fig. 12.36), which here cross-cuts the intensely refolded structures of the Cambrian rock fabrics and extends (i) below the inselberg-like residual of Pic de Tarbésou (2364 m); and (ii) below Pic and Pla de Montpudou (2190 m), which itself displays a small summit flat area—possibly a vestige of range-top surface, S. Extant glacial till deposits on the Pailhères plateau surface indicate that

Fig. 12.36 **Pediment P1 at Col de Pailhères–Pla de Mounégou,** looking south towards the residual peak of Tarbésou. Relative age of Pla de Montpudou, S or P1, is uncertain; it lies ca. 150 m above the low-gradient tread prevalent at Col de Pailhères (Google Street View)

the area was covered by a small ice dome that was not only feeding valley glaciers such as the Bruyante and Ascou (which joined up with the Ariège trunk glacier), but also generating a small local diffluence towards the Rebenty basin to the north.

Follow your outbound route back to Puyvalador, or alternatively drive back through the Aude gorge and Escouloubre (small spa). From Col de Pailhères it is also possible to link up with the Ariège valley via Ascou.

References

Agusti J, Roca E (1987) Síntesis bioestratigráfica de la fosa de la Cerdanya (Pirineos orientales). Estud Geol (Madrid) 43:521–529

Agustí J, Oms O, Furió M, Pérez-Vila M-J, Roca E (2006) The Messinian terrestrial record in the Pyrenees: the case of Can Vilella (Cerdanya Basin). Palaeogeogr Palaeoclimatol Palaeoecol 238:179–189

Anderson B, Mackintosh A (2006) Temperature change is the major driver of late-glacial and Holocene glacier fluctuations in New Zealand. Geology 34:121–124

Baize S, Cushing M, Lemeille T, Granier B, Grellet B, Carbon D, Combes C, Hibsch C (2002) Inventaire des indices de rupture affectant le Quaternaire, en relation avec les grandes structures connues en Francemétropolitaine et dans les régions limitrophes. Mém H S Soc Géol Fr, vol 175, 142 p

Boissevain H (1934) Étude géologique et géomorphologique d'une partie de la vallée de la haute Sègre (Pyrénées Catalanes). Bull Soc Nat Toulouse 66:33–170

Cabrera L, Roca E, Santanach P (1988) Basin formation at the end of a strike-slip fault: the Cerdanya basin (Eastern Pyrenees). J Geol Soc Lond 145:261–268

Calvet M (1996) Morphogenèse d'une montagne méditerranéenne: les Pyrénées orientales. Documents du BRGM n° 255, Orléans 1177 p

Calvet M (2004) The quaternary glaciation of the Pyrenees. In: Ehlers J, Gibbard P (eds) Quaternary glaciations extent and chronology, part I: Europe. Elsevier, Amsterdam, pp 119–128

Calvet M, Gunnell Y (2008) Planar landforms as markers of denudation chronology: an inversion of East Pyrenean tectonics based on landscape and sedimentary basin analysis. In: Gallagher K, Jones SJ, Wainwright J (eds) Landscape evolution: denudation, climate and tectonics over different time and space scales, Geol Soc Lond Spec Publ 296:147–166

Cauuet B, Olesti Vila O, Viladecavall M, Tamas C, Boussicault M, Christophoul F, García-Pulido LJ (2014) New research approach for identifying the mining landscapes. The exploitation of Au alluvial deposits in Cerdanya and tin in Morvan. In: García-Pulido et al. (eds) Preactas del VIII Congreso sobre minería y metalurgia historicas en el sudoeste europeo, SEDPGYM y Dpto. de Prehistoria y Archeologia de la UGR, Granada, pp 152–157

Chevalier M (1925) Note sur les terrains néogènes de la Cerdagne. Butlleti Inst Catal Hist Natural 126–138

CLIMAP (1981) Seasonal reconstruction of the Earth's surface at the last glacial maximum. Geol Soc Am Map Chart Ser c36

CLIMAP Project Members (1976) The surface of the ice-age earth. Science 191:1131–1137

Cook and Swift (2012) Subglacial basins: their origin and importance in glacial systems and landscapes. Earth-Sci Rev 115:332–372

Crest Y, Delmas M, Braucher R, Gunnell Y, Calvet M, ASTER Team (2017) Cirques have growth spurts during deglacial and interglacial periods: evidence from [10]Be and [26]Al nuclide inventories in the central and eastern Pyrenees. Geomorphology 278:60–77

Dansgaard W, Johnsen SJ, Clausen HB, Dahljensen D, Gundestrup NS, Hammer CU, Hvidberg CS, Steffensen JP, Sveinbjörnsdottir AE, Jouzel J, Bond G (1993) Evidence for general instability of past climate from a 250-kyr ice-core record. Nature 364:218–220

Delmas M (2005) La déglaciation dans le massif du Carlit (Pyrénées orientales): approches géomorphologique et géochronologique nouvelles. Quaternaire 16:45–55

Delmas M (2009) Chronologie et impact géomorphologique des glaciations quaternaires dans l'est des Pyrénées, Ph.D. thesis Université de Paris 1 (unpubl.), 523 p

Delmas M, Gunnell Y, Braucher R, Calvet M, Bourles D (2008) Exposure age chronology of the last glaciation in the eastern Pyrenees. Quatern Res 69:231–241

Delmas M, Calvet M, Gunnell Y (2009) Variability of Quaternary glacial erosion rates—a global perspective with special reference to the Eastern Pyrenees. Quat Sci Rev 28:484–498

Delmas M, Calvet M, Gunnell Y, Voinchet P, Manel C, Braucher R, Tissoux H, Bahain JJ, Perrenoud C, Saos T, Aster Team (2018) Terrestrial [10]Be and electron spin resonance dating of fluvial terraces quantifies quaternary tectonic uplift gradients in the eastern Pyrenees. Quat Sci Rev 193:188–211

Denton H, Anderson H, Toggweiler R, Edwards J, Schaefer R, Putnam J, Aaron P (2010) The last glacial termination. Science 328:1652–1656

Depéret C, Rérolle L (1885) Note sur la géologie et sur les Mammiféres fossiles du bassin lacustre miocène superieur de la Cerdagne. Bull Soc Géol Fr 13:488–506

Golpe Posse JM (1981) Los Mamíferos de las cuencas de Cerdanya y Seu de Urgell y sus yacimientos; Vallesiense Medio-Superior. Bol Geol Min España, Madrid XCII-II:91–100

Goula X, Olivera C, Fleta J, Grellet B, Lindo R, Rivera LA, Cisternas A, Carbon D (1999) Present and recent stress regime in the eastern part of the Pyrenees. Tectonophysics 308:487–502

Gourinard Y (1971) Détermination cartographique et géophysique des failles bordières du fossé néogène de Cerdagne (Pyrénées orientales franco-espagnoles). 96ème Congrès National des Sociétés Savantes, Toulouse 1971, sciences, vol. II:245–263

Grootes PM, Stuiver M, White JWC, Johnsen S, Jouzel J (1993) Comparison of oxygen isotope records from the GISP2 and GRIP Greenland ice cores. Nature 366:552–554

Gunnell Y, Calvet M, Brichau S, Carter A, Aguilar JP, Zeyen H (2009) Low long-term erosion rates in high-energy mountain belts: insights from thermo- and biochronology in the Eastern Pyrenees. Earth Planet Sci Lett 278:208–218

Hoek WZ (2009) The Last Glacial-Interglacial Transition. Episodes 31:226–229

Huc S (2008) Mobilité des éboulis supraforestiers des Pyrénées orientales (France): morphodynamique et marqueurs biologiques. Géomorph Rel Proc Environ 14:99–112

Huc S (2010) Éboulis mobiles et marqueurs biogéographiques: le cas de la haute montagne des Pyrénées orientales, PhD thesis, Université de Perpignan, 2 vols, 481 p and 155 p

Jarman D, Calvet M, Coromina J, Delmas M, Gunnell Y (2014) Large-scale rock slope failures in the Eastern Pyrenees: identifying a sparse but significant population in paraglacial and parafluvial contexts. Geogr Ann Ser B 96:357–391

Kindler P, Guillevic M, Baumgartner M, Schwander J, Landais A, Leuenberger M (2014) Temperature reconstruction from 10 to 120 kyr b2k from the NGRIP ice core. Clim Past 10:887–902

Lagasquie JJ (1984) Géomorphologie des granites, les massifs granitiques de la moitié orientale des Pyrénées françaises. CNRS, Toulouse, 374 p

Lambeck K, Rouby H, Purcell A, Sun Y, Sambridge M (2014) Sea level and ice volume since the glacial maximum. Proc Nat Acad Sci 111:15296–15303

Landais A, Caillon N, Jouzel J, Chappellaz J, Grachev A, Goujon C, Barnola JM, Leuenberger M (2004) Quantification of rapid temperature change during DO event 12 and phasing with methane inferred from air isotopic measurements. Earth Planet Sci Lett 225:221–232

Larocque-Tobler I, Heiri O, Wehrli M (2010) Late Glacial and Holocene temperature changes at Egelsee, Switzerland, reconstructed using subfossil chironomids. J Paleolimnol 43:649–666

Lisiecki LE, Raymo ME (2005) A Pliocene-Pleistocene stack of 57 globally distributed benthic $\delta^{18}O$ records. Paleoceanography 20, PA1003

Mangerud J, Andersen ST, Berglund BE, Donner JJ (1974) Quaternary stratigraphy of Norden, a proposal for terminology and classification. Boreas 3:109–128

Menendez Amor J (1955) La depression ceretana Española y sus vegetales fósiles. Características fitopaleontológicas del Neogeno de la Cerdaña Española. Mem Real Acad Sci Ex Fis Nat Madrid XVIII:345

Millet L, Rius D, Galop D, Heiri O, Brooks SJ (2012) Chironomid-based reconstruction of Late-Glacial summer temperatures from the Ech palaeolake record (French western Pyrenees). Palaeogeogr Palaeoclim Palaeoecol 315:86–99

Mix AC, Bard E, Schneider R (2001) Environmental processes of the ice age: land, oceans, glaciers (EPILOG). Quat Sci Rev 20:627–657

North GRIP Members (2004) High-resolution record of Northern Hemisphere climate extending into the last interglacial period. Nature 431:147–151

Pous J, Julià R, Sole Sugrañes L (1986) Cerdanya basin geometry and its implication on the neogene evolution of the eastern Pyrenees. Tectonophysics 129:355–365

Rasmussen SO, Bigler M, Blockley SP, Blunier T, Buchardt SL, Clausen HB, Cvijanovic I, Dahl-Jensen D, Johnsen SJ, Fischer H, Gkinis V, Guillevic M, Hoek WZ, Lowe JJ, Pedro JB, Popp T, Seierstad IK, Steffensen JP, Svensson AM, Vallelonga P, Vinther BM, Walker MJC, Wheatley JJ, Winstrup M (2014) A stratigraphic framework for abrupt climatic changes during the Last Glacial period based on three synchronized Greenland ice-core records: refining and extending the INTIMATE event stratigraphy. Quat Sci Rev 106:14–28

Reille M, Lowe JJ (1993) A re-evaluation of the vegetation history of the eastern Pyrenees (France) from the end of the last glacial to the present. Quat Sci Rev 12:47–77

Reixach T, Delmas M, Braucher R, Gunnell Y, Mahé C, Calvet M (2021) Climatic conditions between 19 and 12 ka in the eastern Pyrenees, and wider implications for atmospheric circulation patterns in Europe. Quat Sci Rev 260. https://doi.org/10.1016/j.quascirev.2021. 106923

Rigo A, Vernant P, Feigl KL, Goula X, Khazaradze G, Talaya J, Morel L, Nicolas J, Baize S, Chery J, Sylvander M (2015) Present-day deformation of the Pyrenees revealed by GPS surveying and earthquake focal mechanisms until 2011. Geophys J Int 201:947–964

Roca E (1996) The Neogene Cerdanya and Seu d´Urgell intramontane basins (eastern Pyrenees). In: Friend PF, Dabrio CJ (eds) Tertiary basins of Spain. The stratigraphic record of crustal kinematics. Cambridge University Press, Cambridge, pp 114–118

Sánchez-Goñi MF, Harrison SP (2010) Millennial-scale climate variability and vegetation changes during the Last Glacial: concepts and terminology Introduction. Quat Sci Rev 29:2823–2827

Severinghaus JP, Brook EJ (1999) Abrupt climate change at the end of the last glacial period inferred from trapped air in polar ice. Science 286:930–934

Suc JP, Fauquette S (2012) The use of pollen floras as a tool to estimate palaeoaltitude of mountains: the eastern Pyrenees in the Late Neogene, a case study. Palaeogeogr Palaeoclim Palaeoecol 321–322:41–54

Taillefer A (2017) Interactions entre tectonique et hydrothermalisme: rôle de la faille normale de la Têt sur la circulation hydrothermale et la distribution des sources thermales de Pyrénées Orientales, PhD thesis (unpubl.), Université de Montpellier

Tomkins MD, Dortch JM, Hughes PD, Huck JJ, Stimson A, Delmas M, Calvet M, Pallas R (2018) Rapid age assessment of glacial landforms in the Pyrenees using Schmidt hammer exposure dating (SHED). Quatern Res 1–12. https://doi.org/10.1017/qua.2018.12

Van Asch N (2012) Environmental response to Lateglacial climate change: reconstructions of temperature and vegetation changes in northwest Europe. Ph.D. thesis, University of Utrecht, 144 p

Viers G (1961) Le glaciaire du massif du Carlit (Pyrénées-Orientales) et ses enseignements. Rev Géogr Pyrén Sud-Ouest 32:5–33

Viers G (1968) La carte du relief glaciaire des Pyrénées. Sheet Mont-Louis, 1:50,000 scale. Rev Géogr Pyrén Sud-Ouest 39:429–434

Wohlfarth B (1996) The chronology of the Last Termination: a review of radiocarbon-dated, high-resolution terrestrial stratigraphies. Quat Sci Rev 15:267–284

Wolf EW, Chappellaz J, Blunier T, Rasmussen SO, Svensson A (2010) Millennial-scale variability during the last glacial: the ice core record. Quat Sci Rev 29:2828–2838

Itinerary 8. Elevated Ranges and Interior Basins of the Axial Zone: The Lower Cerdagne Basin and its Surrounding Massifs

13

From Font-Romeu, France, to Martinet, Spain (174 km; additional excursions: 70 km; allow for 2 days, or more when including trailwalks). Accommodation is widely available in Puigcerdà, Bellver, Martinet, Llès, among other options. See Fig. 13.1.

Leave Font-Romeu on the D618 until you reach Odeillo. Drive through the village on the D29; note to your right the large parabolic reflector of the solar oven (3500 °C) and its scientific installations dedicated to studying the behaviour of materials under very high temperatures (guided tours can be arranged). The road continues across the treads of extensive low-gradient erosion surfaces (generations P2 or P3) in weathered granite. After the village of Via, these planar landforms are abruptly incised by the (Quaternary) Angoust River gorge. After crossing the gorge at the junction with the N116, turn off towards Saillagouse. Roughly 1.2 km after the junction, the road crosses the NW–SE Via Fault, where the granites and the Vallesian basin fill are in direct contact. Aim to stop at Coll Rigat (1490 m), where space for vehicle parking is available.

Stop 1. Coll Rigat

Follow the trail to the north until you find yourself above the railway tunnel. Here the Vallesian gravels form a thin wedge resting unconformably to the north on basement schists (Canaveilles Group). Some exposures can be reached in sections above the irrigation canal and along the trail embankment. In this relatively distal position, and contrary to the features described at Port de Llo (Itinerary 7, **Stop 5**), the deposits are relatively fine-textured. The Vallesian series is exposed in the badlands around Torrent dels Andius, and thickens rapidly southward (200–300 m thickness at least), with shallow-angle dips to the SE. Pebble imbrications in the gravel beds indicate palaeochannel transport to the west. The fill sequence appears to have buried a steep scarp in the basement. The panorama from here

© Springer Nature Switzerland AG 2022

M. Calvet et al., *Geology and Landscapes of the Eastern Pyrenees*, GeoGuide,

https://doi.org/10.1007/978-3-030-84266-6_13

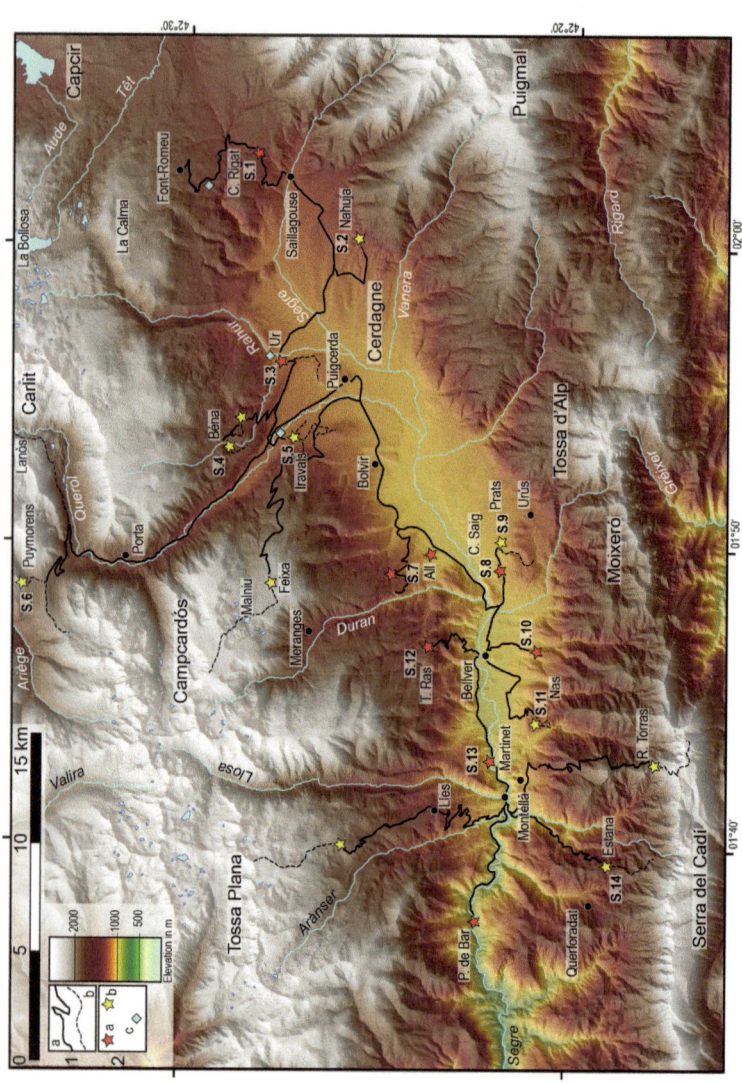

Fig. 13.1 Itinerary 8: route map and overview. Key to symbols—**1:** Roadways, irrespective of road rank, size, or quality. **1a:** Roads and trails. **1b:** Trailwalks. **2:** Stopping points (outcrops, exposures, landforms, or landscapes). **2a:** Roadside stops. **2b:** Roadside stops involving a short walk, or marking the start of a more major trailwalk. **2c:** Cultural stops. Stops numbered S.1 to S.14 refer to the sequence of roadside stops

Fig. 13.2 Lower Cerdagne Basin viewed from Coll Rigat. The Cerdagne half-graben cuts obliquely through the fabric of Hercynian Axial Zone. Its southern boundary, however, meets a Mesozoic to Cenozoic sedimentary marine series (Cadí unit), itself covered by the Pedraforca nappe, which also consists of Mesozoic material and was transported southward during the Alpine orogeny from a root area situated further north in the Axial Zone. Middle ground: stairway of Quaternary alluvial terraces. Foreground: Coll Rigat, with the Calvet farm and farmland dating back to the early nineteenth century

of the lower Cerdagne Basin (Cerdanya, in Catalan) is broad and far-reaching (Fig. 13.2), with good views of the Serra del Cadí (Eocene limestone) to the south. In the gap between the Cadí and Moixeró ranges, note the unmistakable outline of Pedraforca (2506 m, 'forked rock' in Catalan), a klippe of Jurassic to Lower Cretaceous limestone and effectively the leading edge of a south-vergent nappe originating from the northern part of the Axial Zone (see Part I). This viewpoint also provides a sense of the depths of Miocene sediment removed by post-depositional erosion and the extensive Quaternary range-front alluvial fans capping the older fill sequence (Itinerary 7, **Stop 5**).

The road loops down from the col to Saillagouse and cuts twice across the NE–SW-striking contact—presumed to be fault-related—between the schists and the Miocene sediments. After Saillagouse, the road continues over the Err alluvial fan, its deposits 8–10 m thick and corresponding to generation T2. After crossing the Err River, you then rise up to the tread of the older alluvial fan T3 (Sainte-Léocadie), used here as an aerodrome (Fig. 13.3A). Note to the south the ochre-tinted badlands dissecting the Vallesian outcrops, here previously exploited as a tile quarry (Serrat d'en Calvo), with once again gentle dips to the SE (Figs. 13.4A, 13.5C). This midfan to distal facies contains lignite-bearing palustrine grey clays. After driving through Rec de la Verneda and La Tuilerie, the road rises back up onto the tread of T3 (Pla de Medes). Turn off towards Nahuja on the D30. At the junction, note the good exposure of T3 fan deposits, here 4–5 m thick; a lignite bed (Vallesian), which used to be exposed in the ditch at the base of this section beneath the Quaternary gravels

Fig. 13.3 Quaternary alluvial terraces of lower Cerdagne. A General view to the south from Belloc chapel. **B** Section through level T3 at Pla de Medes, junction between the N116 and D30. The photograph, from 1984, shows the Quaternary alluvial deposit (matrix-supported schist pebbles) resting disconformably on the SE-dipping Vallesian beds

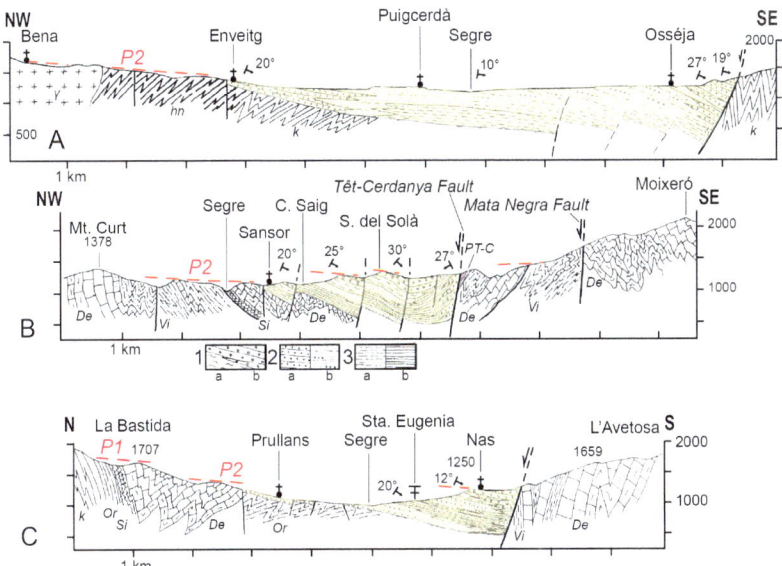

Fig. 13.4 Geological cross-sections through the Cerdagne Basin (from Calvet 1996, modified). **A** Eastern part; **A** Central part; **B** Western part; **C** Western part. The position of the sections is shown in Fig. 12.2 (Itinerary 7). Key to geological ornaments: γ, Mont-Louis granite; hn, hornfels; k, Or, Si: schist (Jujols Group, Ordovician and Silurian, respectively); De: limestone and calcschist (Devonian); Vi: flysch (Visean)

(Fig. 13.3B), *is now overgrown by vegetation. At the level crossing, take the D30a. At Nahuja, follow the asphalt track above the village, then continue on foot along the GR 36 trail to the base of the faceted spurs (spot height 1376 m).*

Stop 2. Nahuja

From here you gain excellent views of the Cerdagne Basin and its northern edge, of the Carlit and Campcardós massifs with their range-top and range-flank erosion surfaces S and P1 over to the north, and of generation P2 pediments in the middle ground (Fig. 13.6A), previously encountered in Itinerary 7 (Targassonne, Bena) (see also Box 13.1). In the foreground, the badlands topography exposes the ochre- and yellow-toned Vallesian stratigraphy, here displaying a proximal to midfan facies with thick beds of clay and siltstone interlayered with boulder and pebble channel beds. A narrow proximal belt of outcrops at the scarp foot exhibits unrolled boulders in a reddish matrix, including very large slipped boulders of schist; the fault, however, is nowhere apparent at this site. Stratigraphic dips (typically ≤10°) along this southern boundary of the basin are to the SE

Fig. 13.5 Granite-rich Vallesian deposits in eastern Cerdagne. **A** Enveitg, fanhead facies with boulders; note intraformational faults among the lower levels (here in 1993; today the exposure in partly shielded by some new buildings). **B** Bollvir–El Castellot, proximal to mid-fan facies with sand- and gravel-filled palaeochannels. **C** Ur–Els Empardells (abattoir), midfan fluvial facies containing lenses of sand, small pebbles, and dominated by thick silt-stone units. Dips are to the SE (here seen in 2010, before the new buildings were constructed; currently no free access)

(Fig. 13.6B). The hill at spot height 1348 m is capped by a vestige of alluvial fan T4, with its fan head probably grading to the Ravin del Racó, which cuts through the boundary scarp. The erosional benches extending 20–30 m above the tread of T4 to the south are cut in the schist of the footwall block, and there is no evidence in this area of Quaternary fault-related offset.

▶**Fig. 13.6 Nahuja hills and schist-rich Vallesian deposits**. **A** View from spot elevation 1376 m; gullies in the hillside reveal a proximal facies capped by an Early to Middle Pleis-tocene alluvial fan (generation T4); in the distance: Carlit massif and its range-flank surface P1; also note bench P2 at Targassonne (refer to Itinerary 7). **B** Section at Ravin del Botàs (900 m south of Nahuja), at the boundary with the Puigmal horst; proximal facies exhibits poorly rolled schist pebbles. The main boundary fault strikes across the gully head but is poorly exposed

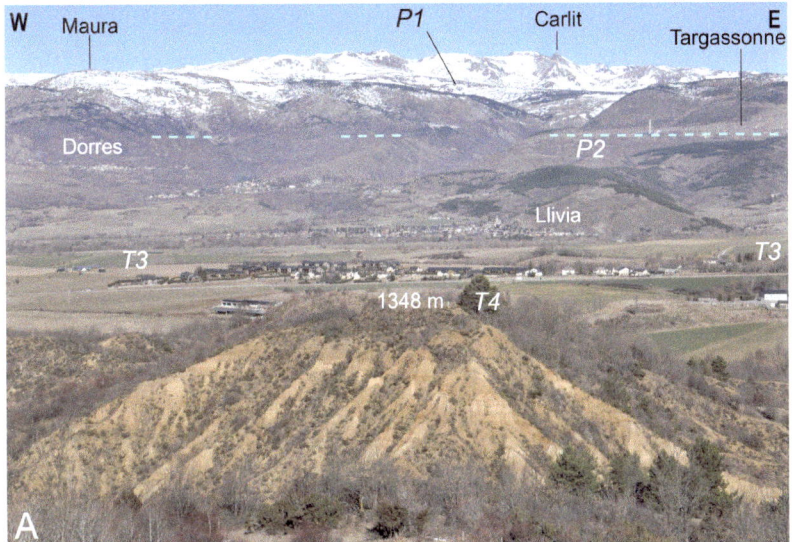

Box 13.1 Straths, Pediments, Erosion Surfaces: A Short Primer

Crustal movements and climate are continually driving gravity-dependent and radiation-dependent erosional processes. As erosion sculpts the Earth's surface, topographic evolution is commonly steered by contrasts in rock mass strength and narrowly controlled by patterns in the underlying geological fabric: harder rock outcrops thus generally become hills, scarps, and ridges, whereas softer outcrops and fault lines become vales, valleys and basins. On a dynamic Earth, differential erosion is thus the routine *modus operandi* and typically generates structural landforms. Structural geomorphology is intuitively easy to understand, and the diversity of structural landforms on continents has been methodically classified over the last 100 years or so.

Erosion surfaces are essentially the opposite of structural landforms. An erosion surface is an expanse of low-gradient, low-energy topography which, counter-intuitively, cross-cuts underlying geological structures and rock fabrics at various angles, thus also ignoring geological contrasts in rock hardness. As such, erosion surfaces contrast with other low-gradient topographic features such as depositional surfaces (e.g., alluvial fans, outwash terraces, etc.) and structural rock surfaces (e.g., a volcanic flow, a bedding plane). Mapping erosion surfaces in a mountain range is thus not a trivial exercise because a minimum acceptable topographic surface area for each identified planar land unit needs to be defined (rounded spurs or mountain ridgetops broad enough to host a wide footpath, for example, would be dubious candidates), and because the criterion of cross-cutting relations with geological structures and rock fabrics cannot commonly be verified by remote methods. The authentication of erosion surfaces ultimately occurs in the field based on multiple criteria.

Because we associate mountain environments with high-energy processes, the occurrence of low-gradient, low-energy land surfaces in mountains is profoundly counter-intuitive. Their presence is usually held as an indication that the mountain range, during its landscape evolution history, recorded relative stillstands in base-level change. These lulls were conducive to generating relief-attenuating planar landforms rather than relief-accentuating structural landforms or ridge-and-ravine landscapes. As such, vestiges of erosion surfaces can be used as tools (in ways that structural and depositional landforms cannot) for documenting successive regimes of landscape evolution and rates of base-level change. In methodological terms, the approach is analogous to using fluvial strath terraces to track the chronology of valley incision by rivers into a mountain or plateau. Bedrock straths are elementary erosion surface: they document a time interval during which a river was stationary relative to some reference base level, and was able to erode laterally across its floodplain instead of cutting vertically into the underlying bedrock as it responds to rock uplift. Itinerary 6 showed how this form of reasoning could be applied to stacks of alluvium-filled limestone cave levels, which were treated in that context as bedrock straths, in order to establish a chronology of canyon incision in response to Neogene crustal uplift.

This logic can be upscaled to planar landforms larger than bedrock straths, and applied across a regional population of massifs rather than just individual valleys. A relevant format of planar landform at that scale is the pediment (wash pediments in softer rocks, rock pediments in harder rocks). Pediments are typically backed by relief: a scarp or mountain front, which is why, in the Pyrenees, pediments have been described in this GeoGuide as 'range-flank' erosion surfaces. The literature on pediments is vast. In the Basin-and-Range (western USA), Bull (2007) has widely documented the fact that range-front pediment size is inversely proportional to the magnitude and frequency of range-front fault activity, with (all things being equal) wider and deeper embayments making inroads into the footwall block in the context of low tectonic activity. Bull (2007) also hints at the link between range-front pediments and alluvial fans: pediments form where bedload transport on steep channels exiting the range spreads laterally and abrades the bedrock. The eastern Pyrenees corroborate this scenario of elementary pediment evolution: Plateau de la Perche (La Perxa), investigated in Itinerary 7, is a fan-shaped bedrock pediment (occasionally called bedrock fan) with its apex at the mouth of a palaeovalley coming out of a steep catchment. The pediment surface is covered by tell-tale lag deposits (quartz gravels and boulders) that served as abrading tools across the pediment surface. When coeval pediments of this kind coalesce laterally along a mountain front hosting a succession of steep river exit gorges, eventually the range-flank pediment may take the form of a continuous bench-like feature. Subsequent regional uplift may once again activate fluvial incision, dissecting and fragmenting the pediment into a population of vestiges.

Apart from straths and pediments, erosion surfaces documenting a relative standstill in base-level fall take other forms and topographic expressions depending on lithology. The successive itineraries have shown that the eastern Pyrenees host a variety of low-gradient erosional landforms such as etch basins in granite (Itineraries 4, 5, 6, 8), karst poljes in limestone (Itineraries 5 and 6), flat-floored palaeovalleys and corridors through limestone ranges and plateaus (sometimes only surviving locally as wind gaps; Itinerary 5), and a variety of fanhead embayments encroaching on a footwall block across low-activity range-front boundary fault lines (Itineraries 7 and 8). Clearly, younger populations of planar landforms, such as those affiliated to generation P2 in the eastern Pyrenees, are often better preserved than older generations of erosion surfaces.

Scaling up from pediments, some erosion surface occurrences encountered in itineraries 1, 5, 6, 7, and 8, do not slope away from upstanding relief and belong to yet another, widespread but comparatively less extensive category. They present themselves as open, flat or rolling massif summits surrounded at 360° by steep escarpments. The formative environment of such 'range-top' erosion surfaces is inevitably speculative. Labelling them as vestiges of a regionally extensive peneplain, for example, is expedient but remains a semantic minefield. Some occurrences of range-top surface display residual summits that could, accordingly, qualify as inselbergs (e.g., in the Madrès and Carlit massifs, itineraries 6 and 7), but we also saw that neotectonics may have raised certain peaks above vestiges of S or P1 along active faults (e.g., Mt. Canigou: Itinerary 4; Pic de Bugarach: Itinerary 5; Pic Pedraforca: Itinerary 9); others, however (Tossa Plana, Campcardós), are low-gradient plateaus lacking residual relief. Rock-cooling age data from these various landforms (Figs. 11.15, 12.5, 13.19) are somewhat time-transgressive across the region, but nonetheless sufficiently consistent for the population of summit vestiges to be linked to one another, thereby defining across the eastern Pyrenees a late Paleogene ancestral paleoplain, S, with a relatively uniform history. This paleoplain was deformed and fragmented by extensional tectonics and valley incision during Neogene and Quaternary time, repositioning (and mostly raising) its vestiges to a variety of elevations in the Axial Zone, outer fold or thrust belts, and Corbières orocline.

The key value of the various erosion surfaces as tools for reconstructing landscape evolution is not so much the exact morphology or absolute elevation of the individual landforms as their relative elevations in any given massif. The itineraries of the GeoGuide each show how different generations of planar landforms have collectively produced distinctive topographic stairways in any given focus area of the landscape. The range-top surface, S, is the oldest tier because it is the most elevated, and thus also the longest-preserved. The fact that the three- to four-level stairway in east-Pyrenean massifs is a widespread leitmotiv across various parts of the mountain range, whether in the low-elevation Corbières (Itinerary 1), the innermost Axial Zone (itineraries 6, 7 and 8), or the outer tectonic belts (itineraries 5 and 9), lends consistency to the inference that the entire mountain range has undergone topographic uplift under a regime punctuated by episodic lulls during which relief-attenuating processes (generating several generations of range-flank pediments and straths) took pervasive and widespread precedence over relief-enhancing processes (more typically generating ridges, narrow spurs, serrated peaks, and deep V-shaped valleys). For a more detailed examination of erosion surfaces in the Pyrenees and in other mountain ranges around the world, see Calvet et al. (2021).

Return to the D30–N116 road junction, cross the Rec de la Verneda once again and turn towards Càldegues on the D30 towards Ur. After Càldegues, the road descends to the modern floodplain of the Segre, which, in this area, is indistinguishable from the tread of its alluvial terrace T1 because the river lacked the stream power required to entrain the coarse Pleistocene bedload and incise T1 during the Holocene. The road continues onward across the Late Pleistocene outwash fan (again T1) of the Rahur, fed by the Angoustrine valley glacier (see Itinerary 7, Stop 8). This glacial outwash fan extends ~60 m below the raised alluvial plain of Puigcerdà, in a topographic position compatible with generations T4 or T5, co-aggraded by the Querol (Carol on older maps) and Rahur rivers. The extensive high plain of Puigcerdà stands as the most elevated Quaternary fluvial landform of Cerdagne. At Ur (eleventh century Romanesque church of Saint-Martin, with a Roman milestone preserved under the main porch), take the N20 towards Toulouse and park 500 m after the bridge, at the first large parking area on the right.

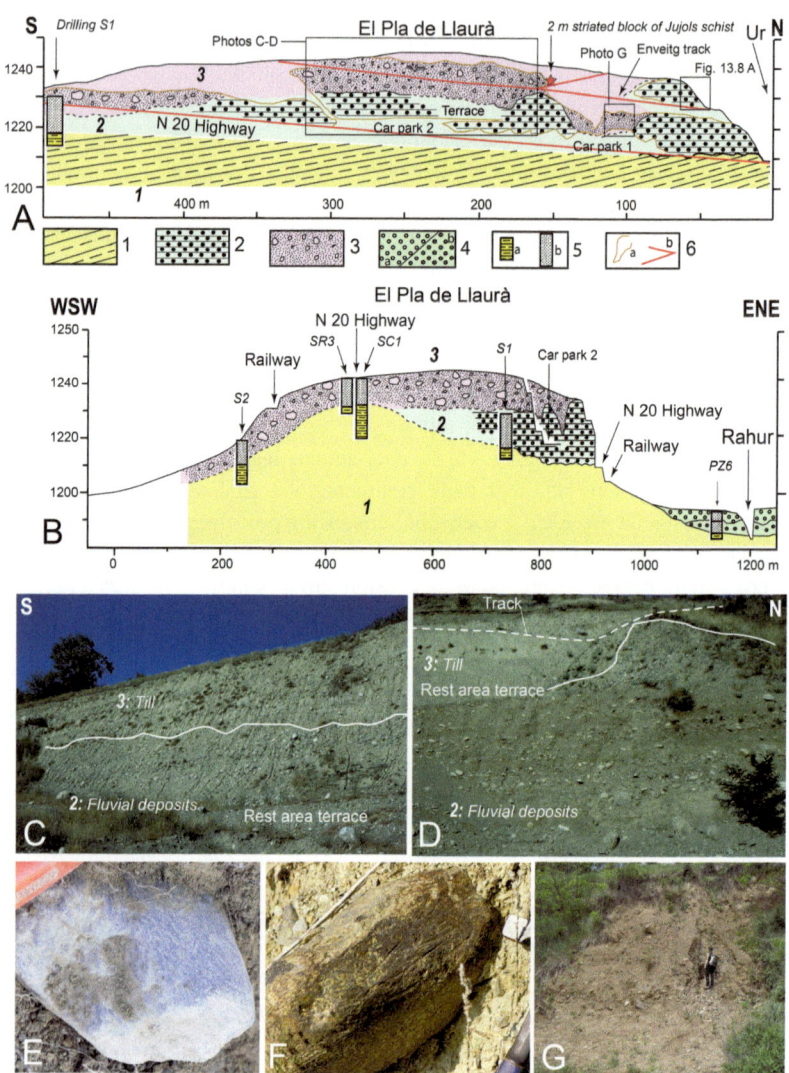

◀**Fig. 13.7** **Sections through Early Pleistocene deposits near Ur (highway N20).** A Road-parallel view of the section; positions of the footpath ('Enveitg track') and engineered terrace in the quarry face (a tourist rest area) are indicated. **B** Road perpendicular section. Key to ornaments—**1**: Vallesian sandy clay; **2**: High-energy fluvial deposit (lower unit); **3**: Deeply weathered ancient till; **4**: Low alluvial level, Rahur stream (**4a**: alluvium T1; **4b**: weathered older alluvium or Vallesian sand); **5**: Borehole data (**5a**: Vallesian clays; **5b**: alluvium or till); **6**: Other features (**6a**: outcrops; **6b**: roads and tracks). **C, D** View of the vertical section at Car park 2 in 1984. Note clear stratigraphy, with the unsorted till facies over the bedded fluvial deposits. 'Rest area terrace' indicates position of the engineered platform area recently cut into the quarry-face stratigraphy. **E, F** Glacially-striated schist pebbles observed in situ withing the fine-textured till matrix. **G** Recent section at car park 1 (in 2020) revealing the disconformity between the till and the underlying alluvium (figure, for scale, is standing on the alluvium). Positions of photographs (ph.) C, D, and G are indicated on cross-section A

Stop 3. Ur–Llaurà, a deep section into the elevated alluvial deposits of Puigcerdà

This abandoned gravel quarry face, now converted to a roadside rest area, has deteriorated substantially over the years (for comparison, Fig. 13.7C, D shows the state of the quarry face in 1984) but still displays some of its key features (Fig. 13.7A, B). The 30 m vertical exposure consists of two types of deposit resting unconformably on the tilted Vallesian beds (Unit 1, southerly dip, i.e., towards the Cerdagne Basin depocentre), which are faintly distinguishable in the north corner of the section. The base of the overlying sequence (Unit 2) is a high-energy deposit (rolled cobbles, imbricate structures, ochre sandy matrix) (Figs. 13.7D,13.8A) The upper part (Unit 3) is glacial till (abundant clasts of fine-grained, glacially striated grey schist (Fig. 13.7E, F), beige to grey silty sand matrix). These are the remains of a frontal or left-margin lateral moraine of the Querol glacier. The entire sequence is deeply weathered, with many granite boulders rotten to the core. The shallowest levels are enriched in quartz, quartzite, and quartz phyllite debris displaying thick weathering rinds. The colloidal fraction (fines) in these levels represents, on average, 7 to 8% (compared to 2% in the most recent generation of moraines), and 4% among specimens of the intermediate generation (Fig. 12.20F, G).

The Ur–Llaurà section was initially interpreted as a single outwash deposit (Unit 2) grading vertically and laterally to a coeval till deposit (Unit 3) (Calvet 1996, 2004; Calvet et al. 2011), and the Puigcerdà terrace (here part of the Segre catchment) was correlated with alluvial generation T4 in the Roussillon catchments. That interpretation has now been revised. What appeared to be a simple paraconformity between Units 2 and 3 is actually a sharp unconformity, with

◀**Fig. 13.8 Ancient moraines of the Querol (landscape views and deposit exposures). A** Enveitg section (located in panel E and Fig. 13.7A), showing high-energy fluvial deposits. **B** Brangoly stream, above Ur, showing disconformity between the till and the underlying alluvium (location in panel E). **C** Puigcerdà hospital terrace section, with a ravinement surface between the ancient till and Vallesian clays exposed in a cutting through a steep slope corresponding to the edge of the uppermost outwash terrace (here seen in 2013; see panel E for location). **D** Piedmont lobe at the mouth of the Querol valley, here seen from the terminal moraine at Puigcerdà hospital. **E** Outline of Querol piedmont lobe viewed from Belloc

a wide time gap between the two formations documented by electron spin resonance and cosmogenic burial age data (work in progress). A fresh examination of the section and borehole logs from the 'Banque du sous-sol' archive (Bureau de Recherches Géologiques et Minières) have likewise revealed that the glacial till fills two washout channels cut in the underlying glacifluvial debris (Fig. 13.7A, B, G). Similar gully fills have been observed further up-valley in a section exposed by the Riu de Brangoli channel (Fig. 13.8B; 42°27′42″N, 01°56′02.8″E) and in freshly cut exposures below the new hospital building at Puigcerdà, at the base of the terrace riser (Fig. 13.8C; 42°26′37.2″N, 01°55′28.5″E). At this site, the base of the deeply weathered till cross-cuts both the outwash formation and the Vallesian mudstone. Dating of the stratigraphy at Ur has so far revealed that the alluvium (Unit 2) is ancient (1–2 Ma; electron spin resonance and cosmogenic nuclide burial dating, work in progress): given that the alluvial deposit is vertically inset by 25 m here at Llaurà (see Fig. 13.7A) and by 35 m at Puigcerdà hospital, the till is likely Middle Pleistocene, probably coeval with outwash deposits labelled elsewhere as generation T3. When standing in the meadow at the top of the Ur–Laurà section, the convex shape of the ancient lateral moraine is distinguishable; the corresponding ribbon of glacial till rises gently to the NW, curving round to the WNW from Pla de Llaurà (SE of Enveitg) all the way to the hamlet of Bena, a further 5 km to the NW (Figs. 13.8D, E; 13.9).

Drive on to Enveitg. As you enter the village on the N20, behind the first few houses (42°27′34.4″N, 01°55′09.4″E) *note some sections in the base of the Vallesian sequence, here also exhibiting synsedimentary faults and dips (20°–25°) to the SE, and displaying proximal facies with palaeochannels containing large boulders of granite (~1 m) that were fed into the basin from the north* (Fig. 13.5A). *These coarse-textured facies become rapidly finer to the south, where palaeochannel deposits in the stratigraphy consist of sand and small pebbles* (Fig. 13.5B, C). *A little further along, on the left side of the road, the southern half of the Romanesque church of Saint-Saturnin was partly destroyed, probably by the 1428 earthquake* (Fig. 13.10). *Just 250 m after the church, turn off to the right and follow the small*

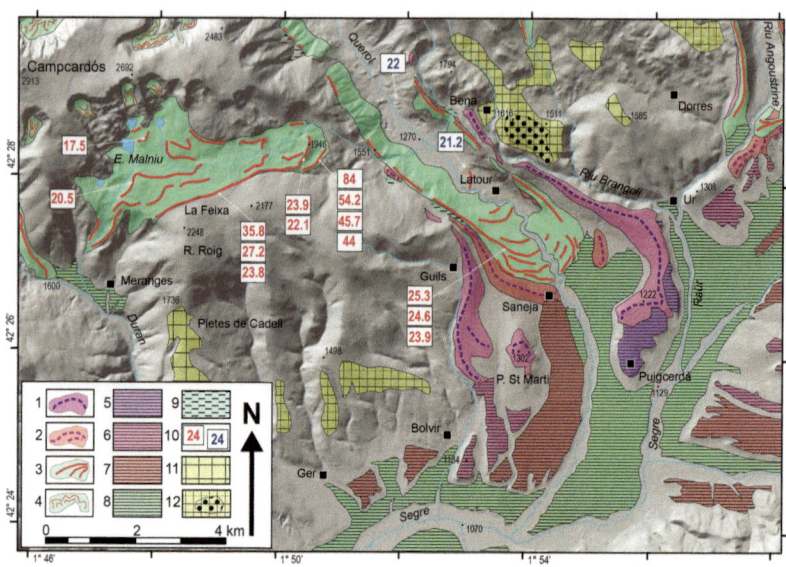

Fig. 13.9 **Geomorphological map of the Querol glacial system and surroundings**. Key to symbols and ornaments—**1**: early Middle Pleistocene till and moraine. **2**: late Middle Pleistocene till and peripheral moraine. **3**: Late Pleistocene till and moraines. **4**: Openwork Lateglacial bouldery ablation till of the cirque floors, cirque moraines, and rock glaciers. **5**: Uppermost outwash terrace (Early Pleistocene) of Puigcerdà. **6**: Outwash terraces and fans, generation T3 (Middle Pleistocene; several levels, one of which grades to the oldest moraine). **7**: Outwash terrace, generation T2 (late Middle Pleistocene, grades to the intermediate moraine). **8**: Outwash terrace, generation T1 (connects to the youngest moraine). **9**: Late Pleistocene kame terrace. **10**: Chronological data. **10a**: [10]Be exposure ages from granite boulders embedded in moraines. **10b**: [10]Be exposure ages from glacially-polished bedrock steps. **11**: Pediment P2 (its upper bench P2a: late Vallesian). **12**: Pediment P2 (its lower bench P2b: Pliocene), and associated alluvial deposits (Bena Fm.)

road signposted to Bena. The road passes over the ancient lateral moraine seen earlier from Ur and crosses the Riu de Bena. This stream follows the ribbon of moraine for 4.5 km all the way to Ur (Fig. 13.9). While probably the descendant of a juxtaglacial stream, the Riu de Bena has cut vertically into the underlying bedrock (schist, hornfels), and the depth of incision (~50 m) indirectly attests the antiquity of the moraine. The road then rises to the Empardines pediment surface. The area around Bena offers a choice of stopping points.

Fig. 13.10 **Enveitg Romanesque church.** Evidence of possible damage caused by the 1428 earthquake

Stop 4. Bena: pediment P2 and ancient moraines (see also Box 13.1)

A first recommended stop is at the col, at the junction with the road to Brangoli. Pediment P2 cross-cuts the hornfels outcrop to the south (Empardines) and the Mont-Louis granite to the north. Looking north, note the pediment embayments in the granite batholith, one coaxial with the Brangoli valley, the other with the Bena (Fig. 13.11C, H). The pediment slopes by 10–11% at Empardines, a slope in excess of most natural pediments and thus suggesting subsequent tectonic tilting. Vestiges of an older pediment (perhaps late Miocene) occur as a bench cut ~100 m above P2 (Els Plans, and interfluve summit at Belloc chapel in the east) (Fig. 13.11D). Pediment P2 is still covered by a thick alluvial mantle consisting of deeply weathered boulders of granite, hornfels, and quartz. This formation is preserved in a ~20-m-deep palaeochannel feature cut into the underlying hornfels. The deposit is fairly conspicuous in the road embankment after the bridge across the Riu de Bena (Fig. 13.11A, B; 42°28′13.5″N, 01°54′13.5″E); its clay-textured matrix contains mostly smectite (Calvet 1996), exactly like the in-situ

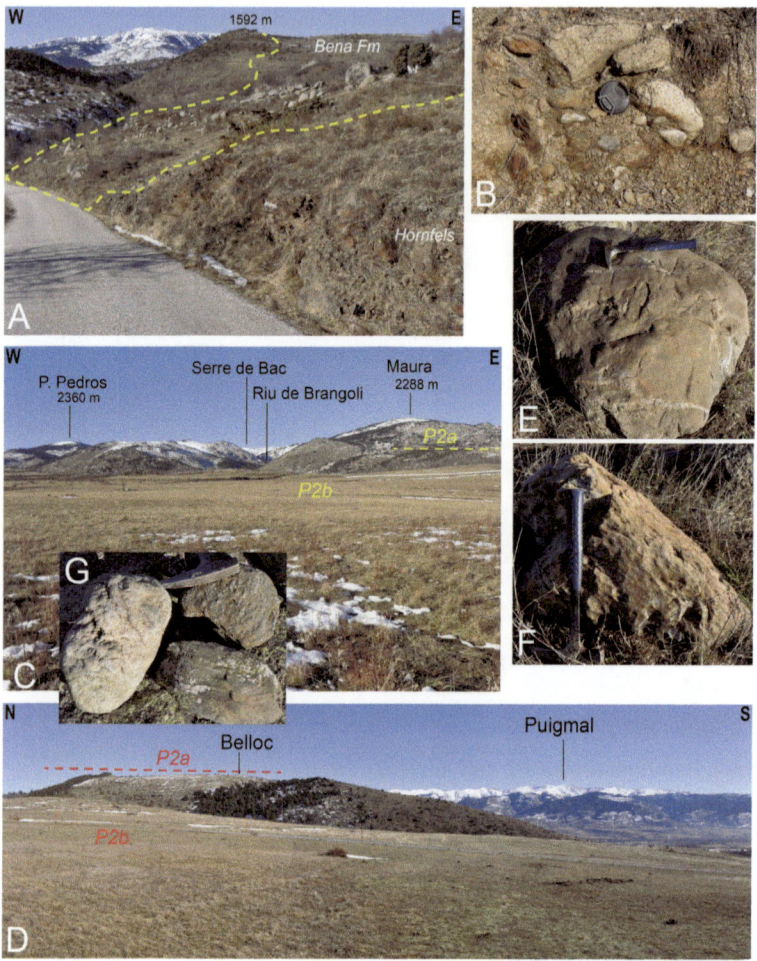

Fig. 13.11 Bena plateau and its ancient deposits. A Section in Bena–Les Empardines valley, granite boulder bed filling a bedrock channel cut in the underlying hornfels. **B** Bena Formation, detail of panel A (channel base). **C** Bena pediment (P2b), as seen from the first bend in the road (elevation: 1450 m). **D** Bena pediment near La Solana; looking east, showing two-tiered bench system (P2b and P2a). **E** Rounded boulder of Paleozoic quartzite displaying a brown weathering rind. **F** Sand-blasted and patinated quartz boulder. E and F are also at the first bend in the road, ca. 1450 m. **G** Rolled pebbles in topmost beds of the Bena Fm., here at the summit of Serrat dels Pins: quartz, hornfels, zoned schist (likely Jujols Group). **H** Fanhead embayment of the Bena pediment, its distal part in the middle distance (La Solana–Les Empardines) grading to the fan head from Serrat dels Pins (alluvial fan deposits 50 m thick). In the background: Puigmal massif and its crowning vestiges of surface P1

Fig. 13.11 (continued)

grus encountered at Targassonne and in the older alluvium of Egat (Itinerary 7). On the plateau, look out for quartz cobbles (Fig. 13.11E, F; 42°28′18″N, 01°54′20″E) quite similar to those previously encountered across the surface of rock pediment P2 at Eyne (Itinerary 7, **Stop 4**); they display red and brown patinations and remarkable wind-abrased facets.

From the Brangoli crossroads at the col, walk up to Serrat dels Pins (1616 m) from the south and west. The outcrop along the way consists of older alluvium, here ~50 m thick and resting unconformably on the hornfels/granite contact at the col itself, near spot elevation 1592 m. A large number of quartz and hornfels pebbles, encountered all the way to the top, testify to the alluvial character of the formation under foot (Fig. 13.11G). Spot elevation 1592 m offers good views of Bena and the ancient moraine (Fig. 13.12A).

A second stop at Bena parking area (elevation: 1596 m) should allow you to reach the top of the moraine in just a few minutes' walk from Bena to Coll de Bena. Because of its antiquity, this lateral moraine is quite 'threadbare', i.e., it has lost its matrix and only exhibits a trail of large granite boulders resting on the bedrock (initially granite, then hornfels further down) (Fig. 13.12B). Unsurprinsingly, the destruction of this moraine by long-term erosion has yielded very young [10]Be exposure ages (Calvet et al. 2011), more compatible with boulder exposure by postdepositional erosion rather than with the true depositional age

Fig. 13.12 Older moraine at Bena. **A** General view from Serrat dels Pins. The contact between the metamorphic aureole (hornfels) and the granite is a few steps away from the viewing point; note field of granite corestones stripped bare of overlying grus in the small valley below, and on the spur overlooking the hamlet of Bena. **B** Top of the older moraine. The landform is largely reduced to a scatter of residual and conspicuously weathered boulders (intense desquamation)

of the moraine. This vantage point also offers good views over the Querol terminal moraines, examined in greater detail at the next few stops (Fig. 13.9). Note in particular the right-margin lateral moraine of Iravals, cloaked in forest, of Late

Pleistocene age. It ends just below Latour-de-Carol–Enveitg railway station. Further in the distance, the raised alluvial plain of Puigcerdà and its peripheral bluffs coincide with outermost glacier positions recorded by the oldest moraines. Good views towards the west also highlight an occurrence of range-top surface (S), conspicuous on the Campcardós massif (~2900 m; see also digital elevation model in Fig. 13.1). Walking over and down from the col brings you to the small kame terrace of Salit, clearly impounded by the recent and middle lateral moraines—here bunched close together and difficult to distinguish from one another for want of good exposures. Note that they extend ~80 m below the elevation of the older lateral moraine at the top of the col.

Return to Enveitg and continue on the N20 to Latour-de-Carol. As you exit Enveitg, above the station the road skirts around the most recent lateral moraine, which displays large boulders of granite. Here, this younger moraine overlies the middle moraine, attested by an outcrop of weathered till 250 m after the bend in the road to your right (parking is difficult). At Latour-de-Carol, follow the road to Iravals and drive through the resort. The Romanesque chapel of Saint-Fructueux (Sant Fruitós), which hosts a painted gothic retable, stands on a frontal moraine corresponding to an undated recessional stade. After a few hairpin bends, the road rises up the flank of the main moraine. Stop at the ridgetop, spot height 1286 m.

Stop 5. Iravals: frontal moraines of the Querol glacier
The main excursion from this point requires a 1.5 hr walk (7 km), but part of the way can be covered by vehicle from Puigcerdà. A brief walk along the ridgetop track towards the northwest provides wide-angle views of the landscape. The terminal moraine complex of the Querol glacier (Fig. 13.9) is unique because it is the only occurrence in the Pyrenees where three different generations of moraines, each tied to a specific interval of the Pleistocene, are still preserved in the landscape (see **Stop 3** and **Stop 4** for some aspects) (Gourinard 1971a).

The youngest lateral moraine forms a sharp-crested ridge above spot height 1286 m (Fig. 13.13). The LGM age of the moraine was inferred from three boulder exposure ages (25.3 ± 2.8, 24.6 ± 3.5, 23.9 ± 2.9; weighted mean age: 24.6 ± 0.6 ka; Delmas 2009; Pallàs et al. 2010). A younger recessional moraine deposit is exposed in embankments along the access road (Les Devesetes). Further towards the ice front, the lateral moraine flattens out and splits into at least five frontal arcs, here consisting of ribbons of large boulders each separated by a topographic furrow (Fig. 13.13A). The innermost arc is the direct continuation of the LGM-dated lateral unit. The other four ribbons are undated but, given their more advanced states of postdepositional degradation, they likely correspond to earlier glacial stades of the Late Pleistocene: perhaps MIS 3 (29–57 ka), MIS 4

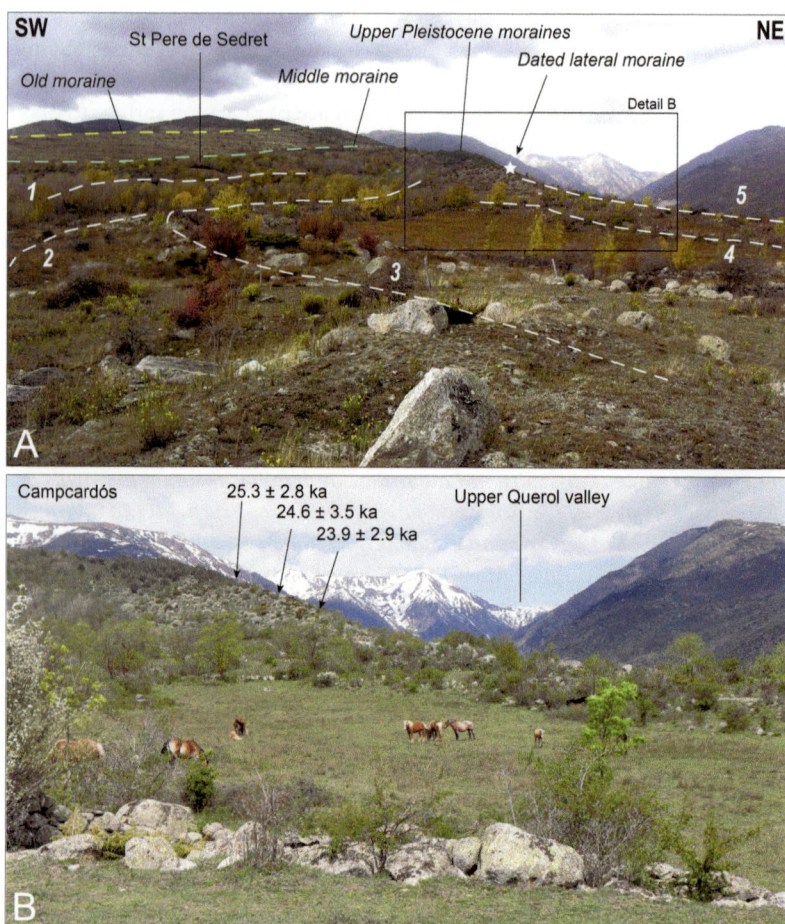

Fig. 13.13   Iravals moraine (Late Pleistocene). A View from the outermost ribbon (international border landmark '461'). Four boulder-rich ribbons of Late Pleistocene moraine can be distinguished, with also a line of large glacial boulders in the wooded furrow to the west (labelled as '1'). To the NW, note the intermediate and older lateral moraines. **B** Close-up of the LGM moraine, near spot elevation 1286 m. This is the innermost moraine, noted 5 in photograph A

(57–74 ka), MIS 5b (84–94 ka) or MIS 5d (104–116 ka). The Querol River cuts through these successive moraines close to La Vinyola farm. The maximum distance separating the innermost (dated) and outermost frontal moraines is 700 m. These ribbons of boulders can be followed as far as landmark no. 462, effectively a large boulder 2.5 m tall, part of the outermost moraine, standing on the international border between France and Spain (for a visit: walk ~800 m to the SE, 25 min return).

From spot height 1286 m, walk down the trail to Senillers. The trail crosses a topographic furrow (containing a few residual boulders from the outermost moraine) between the Late Pleistocene and intermediate (middle) moraines. The Senillers (middle) moraine can be followed up to the ruined buildings of Sant Pere de Sedret, or down to the village of Saneja. Residual boulders are scarce, soils are thick and mature, and the entire landform hosts fields and meadows. Along the trail down to Saneja, a large number of granite boulders display typical spheroidal weathering features, usually with their cores still fresh (Fig. 13.14B); 400 m before reaching the village, a small quarry (now backfilled) previously provided an exposure of the weathered till deposit (Fig. 13.14C). Attempts at exposure dating this moraine have yielded a spread of [10]Be ages ranging from the LGM to the early Late Pleistocene (Calvet et al. 2011). Such age dispersal is probably ascribable to postdepositional erosion of the moraine, entailing boulder exhumation from the surrounding till matrix; or to in-situ boulder weathering and the outer layers getting peeled off (Fig. 13.14D, E). Compared to most Late Pleistocene moraines in the area, the intensely weathered state of the till overall suggests an age compatible with MIS 6 (123–191 ka) rather than with previous speculations of MIS 5b or 5d (Calvet et al. 2011).

At Saneja, continue southward and aim for the summit of Puig de Sant Martí. The hill is an outcrop of schist south of which Miocene sediments extend out into the basin. Here, the entire landscape was covered by an early till deposit, the old moraine, with glacial boulders of granite still preserved on the hilltop. The view from this vantage point embraces the entire Querol glacial system (Figs. 13.9s, 13.15A, B). The older Querol glacier in the basin was a piedmont glacier forming three lobes, with the western lobe skirting around Sant Martí hill, the middle lobe coaxial with the current Querol valley, and the eastern lobe advancing between Llaurà and Puigcerdà hospital. The intermediate and youngest glaciers were classic valley glaciers.

Return to Saneja and follow the road towards Guils de Cerdanya, which reaches the older right-margin moraine (Fig. 13.15B). At the base of the moraine, a small quarry exposes its deeply weathered till material (42°26′40.5″N, 01°53′14.3″E). On the ridgetop itself, on the south side of the hairpin bend in

◀**Fig. 13.14   Glacial landforms of the Querol around Iravals–Latour-de-Carol A** View from Iravals moraine towards the Querol valley. Note postglacial landslide, and stairway formed by the three moraines on the valley side; here the intermediate and youngest generations are poorly distinguishable. **B** Vestige of the intermediate moraine of Saneja, near Senillers, international border landmark '459'. Note scarcity of boulders standing proud of the land surface, and ubiquitous agricultural land use; undesired boulders have been removed by farmers. **C** Section through the intermediate moraine, weathered and rubified (exposure no longer visible). **D, E** Granite boulders of the intermediate moraine (deep weathering of outer layers), here displaced and amassed along a field boundary (border landmark '459')

the road, note the numerous glacial boulders of granite, often weathered to the core. This moraine appears to split into two or three ribbons further down the valley. Follow the moraine ridgetop up-valley and then cut across the fields back to Senillers and to your starting point at spot height 1286 m.

Optional excursion into the Querol glacial trough and to Col du Puymorens (32 km return)

At Latour-de-Carol, take the N20 northbound. Note the landslide scar in the hornfels outcrop above the village, most likely produced in paraglacial conditions soon after Late Pleistocene glacier recession (Fig. 13.14A). The cross-section of the Querol (Carol) valley is reasonably U-shaped, but its floor was not substantially overdeepened and is thus populated by a large number of bedrock 'whalebacks', mostly in granite, rather than by large and sediment-filled rock basins. Three [10]Be exposure-dating results obtained from ice-polished exposures along the valley have yielded a succession of ages (21. 2 ± 2.8 at Latour, 22 ± 4.1 at Quers, and 21.7 ± 3.4 at la Fullatera bridge, just downstream of Porta rock basin), all indicating rapid deglaciation towards the end of the LGM (Pallàs et al. 2010). This chronology conforms exactly to the scenario documented along the Têt glacial stairway (Itinerary 7). The road eventually reaches Porta rock basin, located at the junction with the Campcardós tributary valley. The treads of LGM ice-marginal kame terraces (elevation: 1950 m) suggest a glacier ~450 m thick at the time. A large number of avalanche corridors occur on the valley sides. Those at Montfilla (carriages of a passing train overturned by a powder-snow avalanche in 1935) and Serradal (road and railway line cut off in 1972) are particularly threatening. The corridor at Coma Cervera used to be a threat to the village of Porta (one death in 1826), but forest regrowth indicates a decline in avalanche frequency at this location. By continuing towards Porté, you reach the upper Querol valley and join up with Lake Lanoux (Lanòs). This was the largest glacial lake of the Pyrenees even before it was further enlarged after being closed off by a dam (3 hr walk, round trip). Another option is to drive up to Col du Puymorens (1917 m) to reach **Stop 6**.

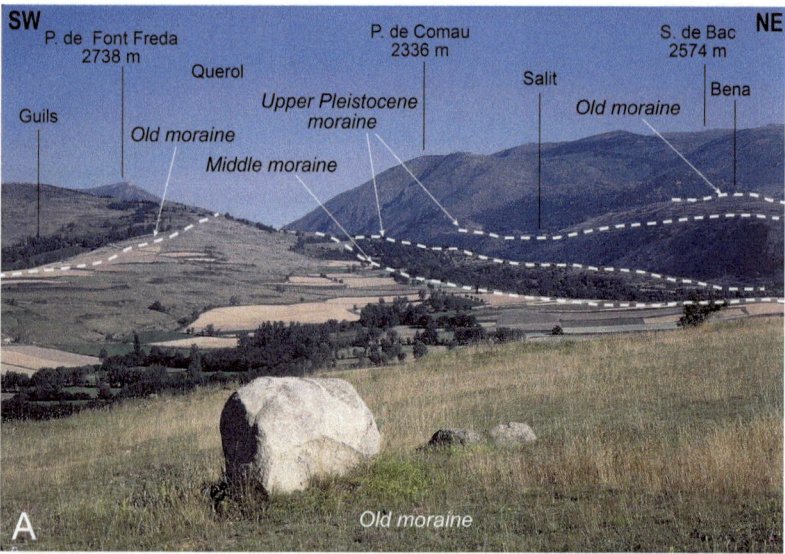

Fig. 13.15 The Querol glacial piedmont, viewed from Puig de Sant Martí de Saneja.
A Looking north up the valley. Note contrast between the rugged and bouldery youngest
moraine and the two older generations, much smoother and covered by crops or pasture.
B Looking NW (older moraine at Guils). Agricultural land use is ubiquitous. In the background: slopes in hornfels and schist, bearing vestiges of pediment P2 (Devesa Canal)

Stop 6. Col du Puymorens (Coll Pimorent)

This mountain pass is in fact a very broad palaeovalley of P2 affinity (Figs. 13.16A, 13.17), which suffered drainage beheading from an agressor stream in the north and now feeds to the Ariège catchment. This drainage capture is explained by the greater steepness of the Ariège stream channel compared to the Segre, which in this area lies comparatively much further from its marine base level and initially flows through the elevated Cerdagne Basin instead of dropping off rapidly to its nearest piedmont. During the Late Pleistocene, the Puymorens operated as a transfluence col through which the Ariège glacier, which was more abundantly supplied by precipitation from the Atlantic and accordingly endowed with a much lower equilibrium line altitude (1800 m vs. 2200 m), spilled southward into the Querol valley (Fig. 12.3). This lasted until the end of the LGM, extending to 18–17 ka as shown by numerous [10]Be exposure ages obtained from lateral moraines in the Ariège and Orri glacial troughs (Fig. 13.16B, C; Reixach et al. 2021).

When at Latour-de-Carol, take the D34 towards the railway station. Before reaching the station, the road crosses the railway line and follows the Querol downstream. At la Vinyola, by the sewage works, the road crosses the outermost Late Pleistocene frontal moraine. Here it forms a low ridge less than 10 m high and contains some large boulders, such as at Els Rocs. On the other side of the moraine, the tread of the outwash fan slopes fairly steeply towards Puigcerdà railway station. This is generation T1, the fan head currently incised 15–20 m by the Querol. The road crosses the Spanish border and enters Puigcerdà; on the south side of the railway station, take the N260 towards La Seu d'Urgell. T1 extends on the left side of the Querol and is recognisable by its poorly weathered alluvial deposits; after the bridge, and extending on the right side of the river, however, Pla de Malamort is a vestige of T2, here barely 2–3 m above T1. The sections along the road show that the alluvium is in an advanced state of weathering, capped by an eluviated soil with a clay-rich B horizon. Farther upstream, this terrace grades to the intermediate moraine previously encountered at Saneja and which displays identical weathering characteristics.

Optional excursion towards Guils de Cerdanya and Feixa plateau (elevated vestiges of erosion surface P1)
The road to Guils passes through Saneja, and thus revisits **Stop 5**. Above Guils, the road follows the top of the older moraine up to 1450 m. You eventually reach the cross-country skiing resort of Guils-Fontanera (1900 m), then the Feixa–Roc Roig plateau up a well-maintained track (2200 m). This low-gradient surface cross-cutting hornfels and granite is a vestige of range-flank pediment P1, i.e., lower than summit surface (S), which extends across the top

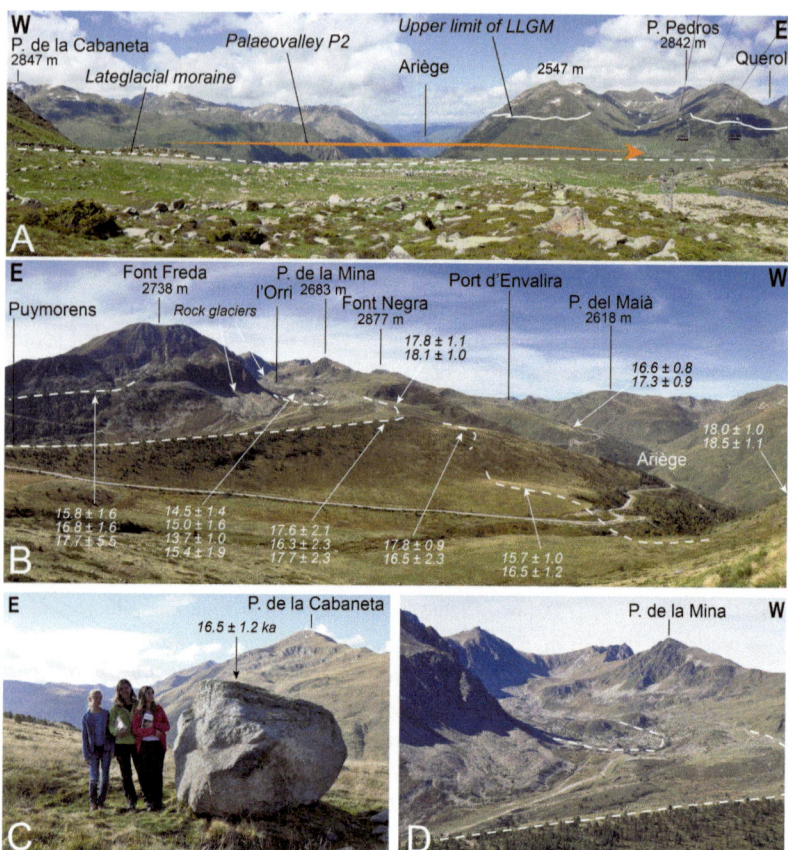

Fig. 13.16 Col du Puymorens (Coll Pimorent). A View from Orri de la Vinyola (elevation: 2180 m). The col is effectively a vast plateau separating the south-facing Querol from the north-facing Ariège catchments, and is the vestige of a broad palaeovalley (generation P2) captured by the Ariège, here the 'agressor' stream. Foreground: Lateglacial moraines of Orri valley. **B** View from the north (elevation: 2020 m, track rising up to Pic de Tossal Mercader); [10]Be exposure ages obtained from the population of moraines at the col, here documenting the early stages of icefield fragmentation and glacier recession after the diffluence which was occurring here during the LGM. **C** Exposure-dated boulder from the Ariège lateral moraine, NW of the col. **D** Lateglacial moraines in Orri valley, here seen from the track rising up to Pic de Tossal Mercader (2200 m) north of the col

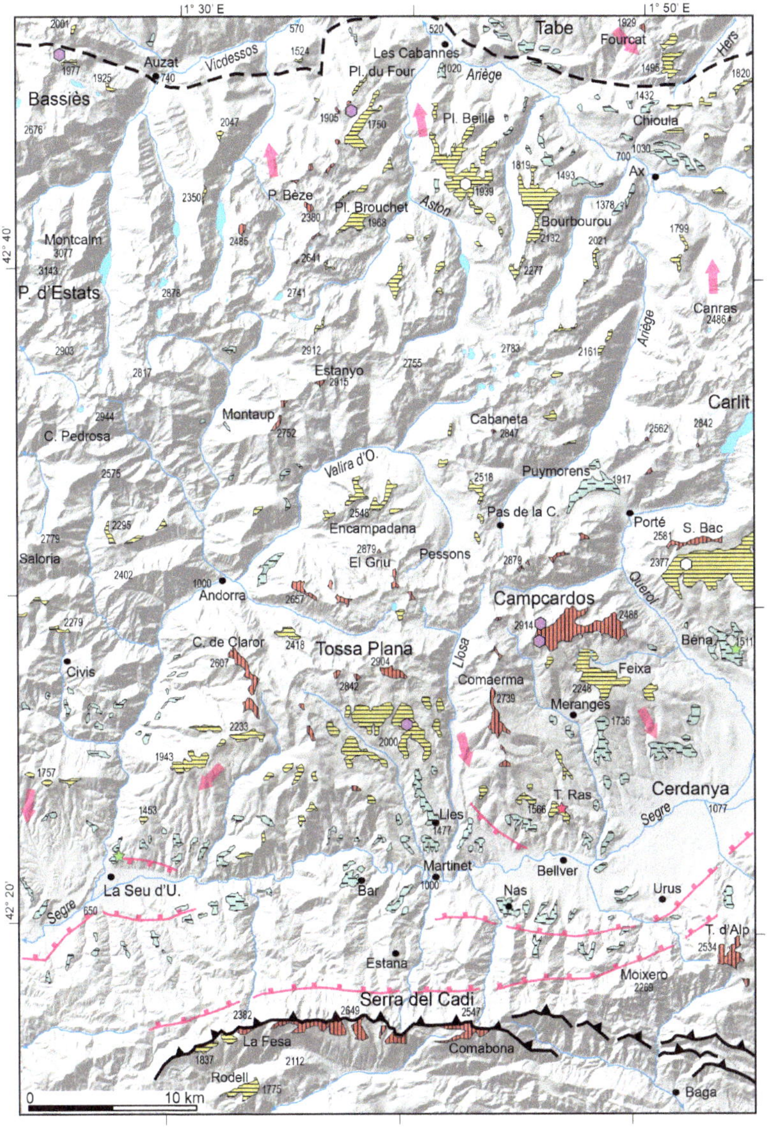

Fig. 13.17 Geomorphological map of the Axial Zone, western part. For symbols and ornaments, see Fig. 11.4, Itinerary 6, and Fig. 10.12, Itinerary 5. Thick dashed line: North-Pyrenean Fault

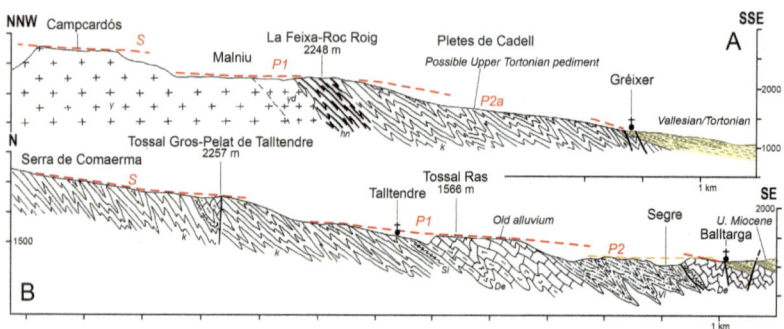

Fig. 13.18 Cross-sections through the Campcardós massif. A Profile through the eastern plateau area. **B** Profile west of Riu Duran. Note southward tilt of the range-top erosion surface, S; range-flank pediment P1 plunges below the Vallesian fill sequence of the Cerdagne Basin; pediment P2 itself forms two generations in this area, but the upper, older vestige (P2a, previously encountered around Bena) is more conspicuous in cross-section A. Key to geological ornaments: γ, Mont-Louis monzogranite; γd, Granodiorite; *hn*, Hornfels; *k*, Schist (Jujols Group); *Or*, Ordovician conglomerate and schist; *Si*, Silurian schist; *De*, Devonian limestone and calcschist; *Vi*, Visean flysch

of the Campcardós massif (clearly visible from here when looking north, and scalloped by glacial cirques; Figs. 13.17, 13.18A, 13.20A). These elevated surfaces have yielded a number of apatite fission-track and (U–Th)/He cooling ages which all indicate—as in the eastern Carlit, Madrès and Canigou massifs, but also among similar surfaces to the north in Ariège—rapid cooling during the Paleogene, followed by protracted residence near or at the surface since the end of the Paleogene (Fig. 13.19) (see also Box 13.1).

The Pleistocene Malniu glacier covered only half of Feixa plateau (Fig. 13.20A), but its conspicuous sequence of moraines has yielded quite a comprehensive ^{10}Be and ^{36}Cl chronology of glacier fluctuation (Pallàs et al. 2010; Palacios et al. 2015; Andrés et al. 2018) (Fig. 13.9). Its eastern frontal moraine was produced at the time of MIS 5b (84 ± 2.2 ka), with readvances during MIS 3 (54.2 ± 1.4, 45.7 ± 1.2, 44.0 ± 1.2 ka) and the LGM (23.9 ± 0.7 and 22.1 ± 0.6 ka). The LGM (23.8 ± 1.6, 27.2 ± 0.7) is also documented by the well-preserved lateral moraine on Feixa plateau (Fig. 13.20A, B); it nonetheless also contains a reworked boulder with MIS 3 signatures (35.8 ± 6.1 ka). Deglaciation here occurred early, and is documented by several recessional moraines (20.5 ± 0.9 ka; and 17.5 ± 0.8 ka for the innermost stadial position). From this plateau, it takes ~5 hr (round trip) to reach the extensive range-top surface of the Campcardós massif—a spectacular plateau well worth the visit (see Part I, Fig. 3.2 therein), and which also hosts impressively large periglacial stone polygons (Fig. 13.21).

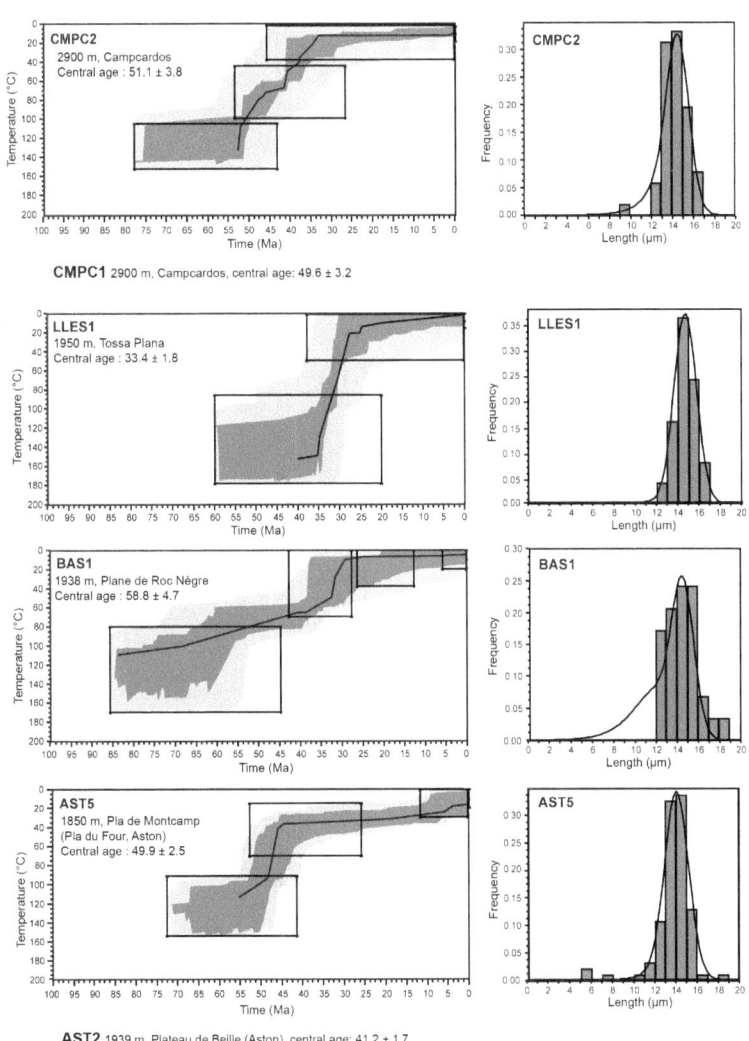

Fig. 13.19 Rock cooling models for bedrock samples collected from S and P1 in the Campcardós, Tossa Plana, and Aston massifs. Note the uniform cooling histories among the samples, with rapid denudation around 35 Ma followed by residence at shallow crustal depths since late Eocene time. In the case of the Aston and Bassiès data, which are situated north of the Mérens Fault (i.e., north of Col du Puymorens and thus not part of the Itinerary but inserted here for contrast), cooling models are rather different: cooling began earlier than among samples south of that fault, as already noted from Fig. 12.5. After Gunnell et al. (2009), modified

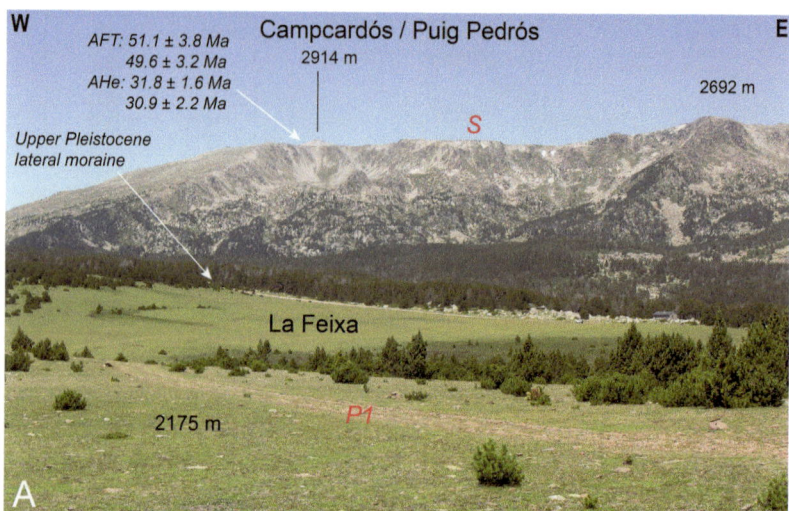

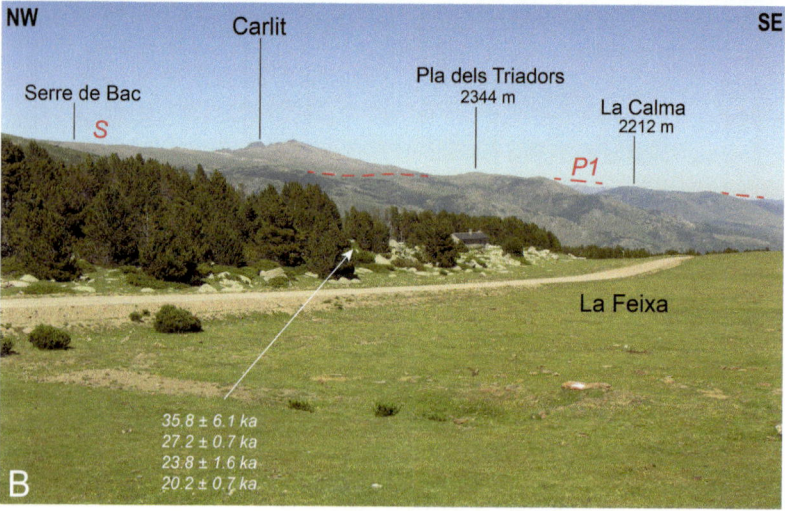

Fig. 13.20 Feixa plateau, Campcardós massif. **A** General view; clear separation between range-top and range-flank surfaces, S and P1; west of the glacial cirques, note extremely regular slope profile of the cyclic erosional scarp connecting the two generations of erosion surface. Malniu moraines in middle ground, with ice-scoured cirque floors and headwalls scalloped into the side of the summit plateau. Apatite fission-track (AFT) and (U–Th)/He ages (i.e., apatite helium, or AHe) after Gunnell et al. (2009). **B** Malniu lateral moraine, Late Pleistocene. Unlike the Querol piedmont lobe lower down, no Middle Pleistocene deposits have been preserved on the Feixa plateau. [10]Be exposure ages after Pallàs et al. (2010)

Fig. 13.21 **Large stone polygons on Campcardós plateau**. Location: near spot elevation 2739 m, Molleres de Puigpedrós

Back on the main itinerary, continue along the N260 towards Bolvir. Before the village, the road cuts across a vestige of upper terrace T3 (+60 m), distinctive by its rubified alluvial material. It could correspond to the outwash fan descending from the oldest moraine of the Querol piedmont glacier (Fig. 13.9; *Torre del Remei: 42°25′03.5″N, 01°53′30.3″E; isolated hill of El Castellot). In a quarry below El Castellot, outcrops of midfan Vallesian beds display well-rolled pebbles of granite, schist and abundant quartz* (Fig. 13.5B) *and dips of* 10–15° *to the SE. At Ger, take the road to Meranges (GIV-4031), where the badlands around the village of All provide good exposures of the Vallesian sequence. The exposures are best approached from the top, i.e., at Serra d'All and Serrat de la Ginebrada. The lower exposures of the sequence can be reached after a short walk from All cemetery (Les Guilleteres, former tile factory).*

Stop 7. Vallesian sequence at All

The gullies expose the stratigraphy of a Miocene fan similar to the fan previously encountered at Llo–Saillagouse (Itinerary 7, **Stop 5**), but smaller. The clastic material consists mostly of schist and hornfels, with a small proportion of granite pebbles and cobbles, often deeply weathered. The entire sequence dips 20° S to SE and displays synsedimentary faults, particularly E–W faults along the boundary with the schist below the village of Gréixer (Fig. 13.22A–C). The road cuts across these faults at several locations, but exposures are currently degraded and overgrown by vegetation. Boulder beds are scarce and gravel clast sizes decrease rapidly upward in the stratigraphy as well as downfan (Fig. 13.22D, E). Ochre, yellow and bright red colours prevail, but the corresponding beds disappear beneath increasingly sand- and silt-rich, greyish to blue-toned distal deposits displaying multicoloured waterlogging features. This palaeofan is the base of the Vallesian sequence; further south, it plunges beneath thick palustrine and deltaic deposits containing lignite beds. These fill much of the basin's depocentre around Sanavastre (south side of the Segre River).

Return to the N260, which after Isòvol enters a small epigenetic gorge in Devonian limestone, cut here by the Segre along the northern edge of the Cerdagne half-graben (Tossal de Balltarga, on the left side of the Segre), and in that respect similar to the epigenetic gorges encountered in the Conflent Basin (Itinerary 6). Epigenetic incision occurred during the Quaternary, probably from a time when the river flowed across the tread of P2, which here bevels the basement (e.g., at Tossal de Balltarga) and the upturned Vallesian beds (Fig. 13.4B). *This gorge separates the main Cerdagne Basin from a satellite extension of it at Bellver and known as Petita Cerdanya, or La Batllia. As you exit the gorge, turn left onto the N-260R, crossing the river, then towards Prats on the LP-4033b, where it will be possible to examine good exposures of Upper Miocene outcrops.*

Stop 8. Coll de Saig, lacustrine facies with plant impressions

The lacustrine outcrops (blue clays and diatomites) only occur in the SW of the basin and are associated with lignite beds and sandy deposits suggesting a deltaic facies (Fig. 12.2). Embankments in the road up to Coll de Saig (42°21′49.7″N, 01°53′30.3″E) provide exposures of the lacustrine beds with plant impressions, known and studied over one century ago at this and other sites in the area (Fig. 13.23D, E) and also studied for their pollen content (Rérolle 1885; Menendez Amor 1955; Bessedik 1985; Martin-Closas et al. 2005; Suc and Fauquette 2012; Barrón et al. 2016). The flora is very diverse, mostly warmth- and moisture-loving, with evergreen megatherms and mesotherms typical of subtropical environments (among the thirty or so taxa, highlights include *Taxodium*, *Cinnamomum*, *Sassafras*, *Persea*, *Ficus*, *Caesalpinia*, *Cassia*, *Sapindus*, *Celastrus*, *Sapotacites*, etc.), associated with deciduous temperate mesotherms (*Acer*, *Quercus*, *Juglans*, *Carpinus*, *Betula*, *Tilia*, *Castanea*, *Platanus*, etc.) as well as meso- to microtherms such as *Fagus* and conifers (*Abies*, *Pinus*).

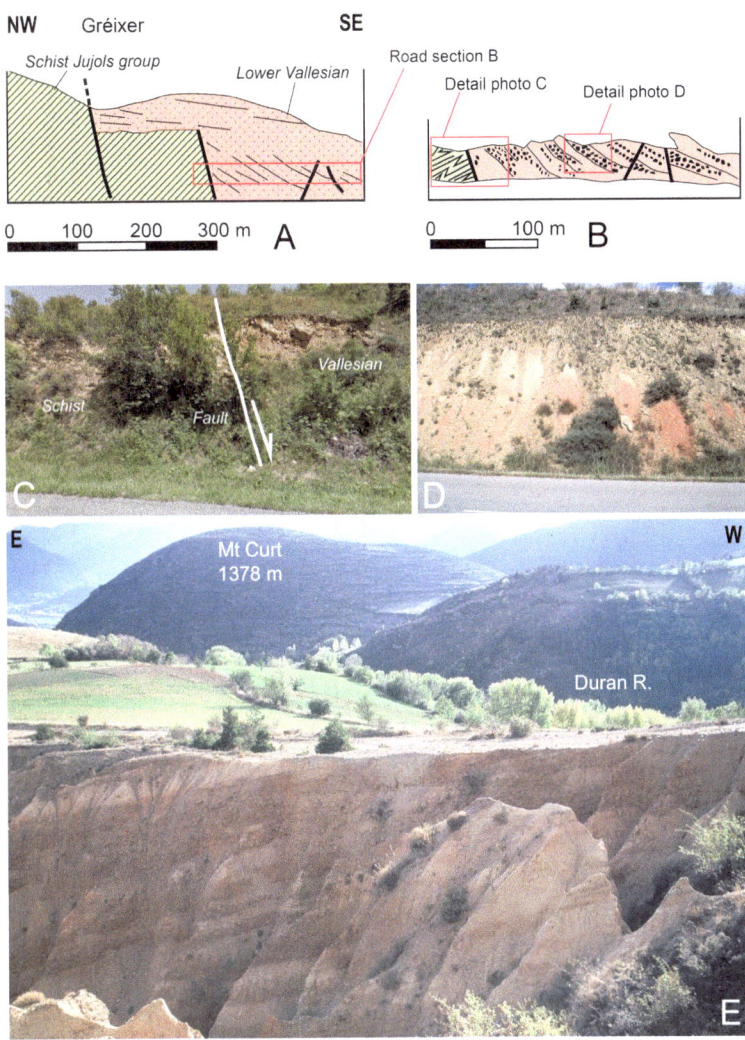

Fig. 13.22 NW border of the Cerdagne Basin at the foot of the Campcardós massif.
A Faulted boundary between the basement and the Vallesian series below Gréixer (road-side section, after 3.5 km on the way to Meranges). **B** Detail of the outcrop. A, B after Roca (1986), view redrawn and modified. Note synsedimentary characteristics of the fault. **C** Exposure of the fault. **D** Close-up of view the Vallesian facies; beige, ochre and red siltstone containing pebble-filled palaeochannels. **E** Vallesian proximal facies, Serrat de la Ginebrada badlands; note steep dips to the south

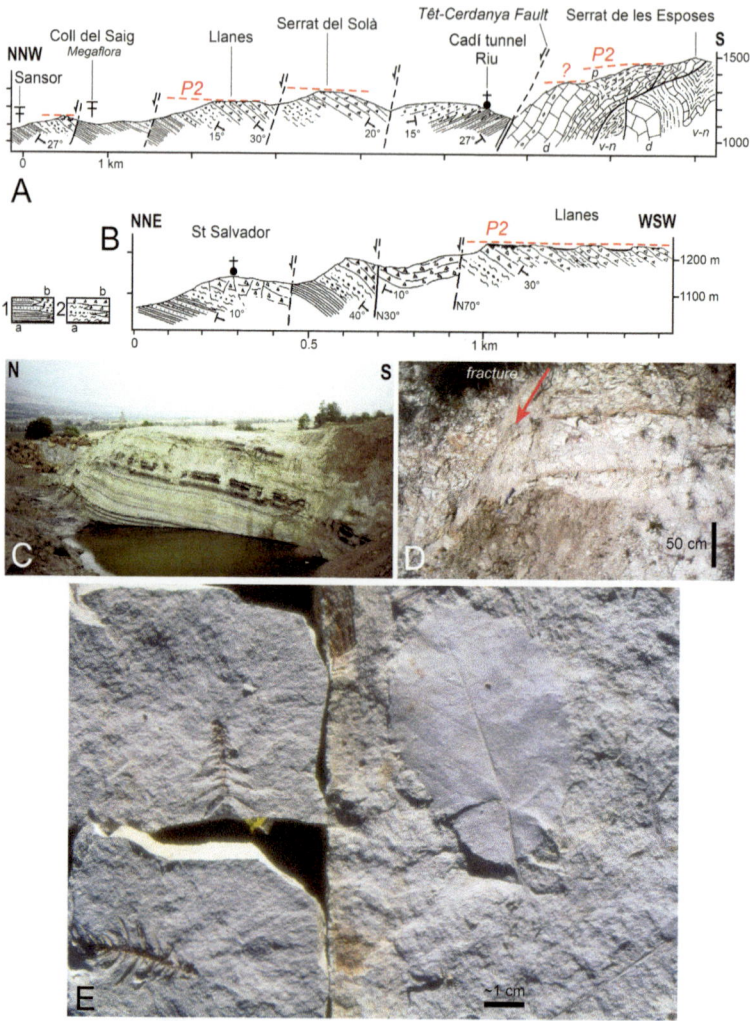

Fig. 13.23   Aspects of the Upper Miocene stratigraphy in the Cerdagne Basin. A General cross-section. **B** Geological section near Prats. Key to geological ornaments—**1**: lower series (Vallesian); **1a**: distal facies (lacustrine clays, sand); **1b**: coarse proximal facies; **2**: upper series (Turolian); **2a**: red detrital facies; **2b**: conglomerate beds (carbonate-rich); basement lithology—d: Devonian limestones; v–n: Visean–Namurian flysch; p: late Carboniferous to Permian volcanics. **C** Lignite quarry at Sansor (Vallesian), here in 1988; note dips to the south. **D** Vallesian clay deposit containing paleofloral remains at Coll del Saig (road embankment, north side); note extensional fractures. **E** Close-up of leaf impressions at Coll del Saig

Immediately north of the col, a normal fault has offset the entire sequence and thrown up the carbonate-rich upper conglomerate series of Turolian age. It overlies lignite-bearing grey clays and sand beds (outcrops at Sansor/Sampsor) that dip south and are interpreted as an association of lacustrine and deltaic facies (Figs. 13.4B, 13.23A, C). The lignite quarries today are disused and overgrown. The boundary between the two series, Vallesian and Turolian, is nowhere clearly exposed in the Cerdagne Basin, but an unconformity has been postulated (Agustí and Roca 1987; Roca 1986, 1996), partly on the basis of the relatively large lacuna inferred from the absence of biozones MN 11 and MN 12 (Fig. 12.2).

Continue as far as Prats, and aim to leave your vehicle and walk over the Upper Miocene outcrops that extend south of the village (5 km, 75 min return).

Stop 9. Prats: cross-section through the Upper Miocene stratigraphy

This excursion showcases the lacustrine lower series and conglomeratic upper series of the basin fill, the latter forming table-top hills south of Prats, between Serrat del Solà and Urús (Figs. 12.2, 13.4B, 13.23A, B). The succession of discrete Turolian conglomerate outcrops from Coll de Saig to Prats has, in part, been interpreted as a population of landslide units (geological sheet of Puigcerdà), but other authors view it as resulting from a succession of normal faults (Gourinard 1971b; Calvet 1996). The breccia and conglomerate of the upper series correspond to debris fans made up of Devonian limestone, calcschist, and Visean schist clasts supplied by outcrops from the southern boundary of the basin (Tossa d'Alp and Moixeró massifs, whose bounding fault scarps display large triangular faceted spurs). Towards the north, where outcrops correspond to the base of the sequence, the gravel-filled palaeochannel units are thin and occur as interlayers between thick beds of bright red siltstone displaying a boxwork pattern of calcium carbonate concretions and nodules. The gravel beds become thicker towards the top of the series and towards the south, and also contain increasingly large (and always rounded) individual clasts (typically 1 m in diameter near Urús). The carbonate cement is pink and compact at the base of the beds, and more vacuolar at the top. The upper series also thickens rapidly towards the south, i.e., towards the boundary fault, suggesting a growth of accommodation space promoted by synsedimentary tectonics along the range front.

A first unit of upper-series conglomerates forms a low-gradient butte on which stands the chapel of Sant Salvador de Predanies, 400 m from Prats village. The conglomerate caps red siltstone beds with interlayers of gravel, and the base of the slope exposes grey arkose and lacustrine clays (the exposures may be obscured by vegetation). From the chapel, walk to the base of the uppermost local scarp. At the col, 150 m to the south of the chapel, the light grey lacustrine clays reappear,

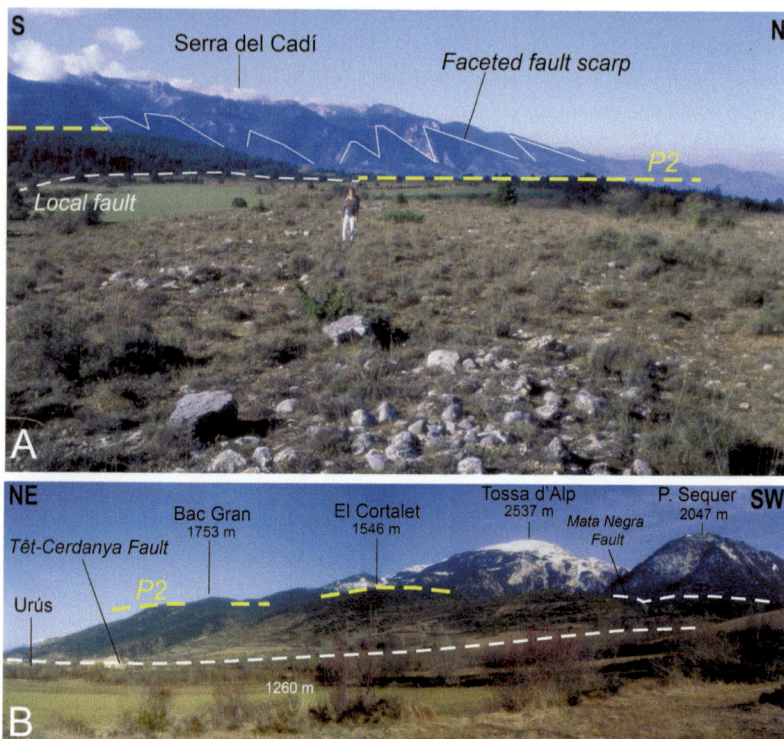

Fig. 13.24 Landscape around Prats–Urus. A Pediment P2 (Llanes plateau), cross-cutting Turolian conglomerate beds (see sections A and B, Fig. 13.23). Background: triangular faceted fault scarp of Serra del Cadí. **B** Triangular faceted spurs along the Cerdagne boundary fault (viewed here from near Urús, and looking towards the Tossa d'Alp and Moixeró massifs)

overlain once again by the red pelites and conglomerate caprock, all dipping 25° to the south. The fact that the stratigraphy below the chapel, and then again south of it, is undisturbed suggests fault-related offsets rather that gravitational disturbance of the stratigraphy. Follow the path leading eastward along the hill flank and joining the small valley upstream of Prats; then walk up that valley along the track (suitable for vehicles). Here, the conglomerate unit appears to undulate and shows fractures. The small, horizontal plateau surface at Llanes (Figs. 13.23B, 13.24A) is an erosional rather than a structural surface: over a width of 500 m, the topography cuts across at least two siltstone outcrops (hosting ploughed fields) and three conglomerate beds, all dipping steadily 30° to the south. This table-top hill is a vestige of post-Turolian pediment P2, also present at the top of Serrat del Solà and Serrat d'Urús (Figs. 13.4B, 13.23A). These pediments lack alluvial cover but display a few pockets of *terra rossa* containing allochtonous rolled quartz pebbles (absent from the underlying conglomerate bedrock). Continue onward as far as the col below Serrat del Solà, from where you get a good view to the south over the depression of Riu de Cerdanya. This depression was eroded out of north-dipping Vallesian lacustrine and gravel outcrops in an area where they adjoin the boundary fault. The triangular faceted spurs also loom large, here displaying two vertically tiered segments separated by an erosional bench at 1500 m. As in the Têt valley (see Itinerary 6), this low-gradient bench could be the hallmark of pediment P2 in the footwall block (Figs. 13.4B, 13.17, 13.23A, 13.24B). The upper belt of faceted spurs, which are very steep, seem linked to the Mata Negra Fault (Fig. 12.2).

From Prats, head back westward and aim for Bellver on the LP-4033a. As you enter Bellver, turn left (road to Talló and Santa Magdalena). Several spurs carved out of Neogene outcrops and hosting conspicuous topographic benches occur along the southern boundary of the basin. Previously mistaken for lakeshore beach berms, these flats actually cross-cut a south-dipping Turolian sequence and correlate instead with pediment P2. The road rises up the spur hosting the hamlet of Santa Magdalena. Stop 2.4 km after Bellver, ca. 1170 m near the motocross scramble. The local mammalian fossil site of Turolian age bears the name of the hamlet situated at the base of the spur.

Stop 10. Can Vilella fossil-rich site

The predominantly midfan and distal facies here represent the base of the Turolian sequence. Palaeochannels several metres thick and containing pebbles >10 cm in diameter have cut and filled thick masses of reddish-brown claystone and finely bedded sandy silt containing abundant carbonaceous debris and gastropod shells. The previously encountered grey lacustrine or palustrine clay beds with plant impressions are also present, but in subsidiary quantities. While situated 20–25 m above the base of the upper Turolian series, those beds host a rich assemblage of small mammalian fossils known as the Can Vilella assemblage and ascribed to continental biozone MN 13. At the top of the sequence, palaeomagnetic results detect a normal polarity, but it has proved impossible to discriminate between chrons C3An.2n (~6.3 Ma) and C3An.1n (~6.7 Ma; Agustí et al. 2006), both eligible. Details of the sedimentary petrology can be examined among the motocross racetracks (Fig. 13.25), or in the wide gully visible to the west, slightly upstream of the stopping point.

Fig. 13.25 Turolian upper series at Can Vilella. Mammalian fossil site (MN 13) in the finer-textured deposits; cut-and-fill channels, pebbles supplied from the south

*Return to Bellver and drive through the old village centre to join up with the road to Pí, then onward to the hamlet of Olià, to the west of Pí. Ignore on your right the road to Santa Eugènia (disused lignite mines in lacustrine Vallesian beds) and follow instead the road to Nas. Unremarkable exposures of Vallesian lacustrine and deltaic facies occur at the base of the hills, but at the turning off to Olià (42°21′20″N, 01°44′25″E) a section reveals a gravel bed rich in small quartz pebbles but also containing granite—and thus indicating a northern provenance (Fig. 13.26C). The Turolian series is recognisable in the second bend in the road between Olià and Nas by the red siltstone outcrops at the base of the scarp, with the stratigraphy evolving upward to increasingly massive conglomerate beds (Fig. 13.26B, D). After reaching the plateau, turn right (i.e., away from Nas) and follow the track until you reach the western edge of the plateau (42°20′44.9″N, 01°44′10.3″E). A 30 min return trailwalk to the south leads you down the slopes of Serrat de l'Avetosa to the site of interest described as **Stop 11**.*

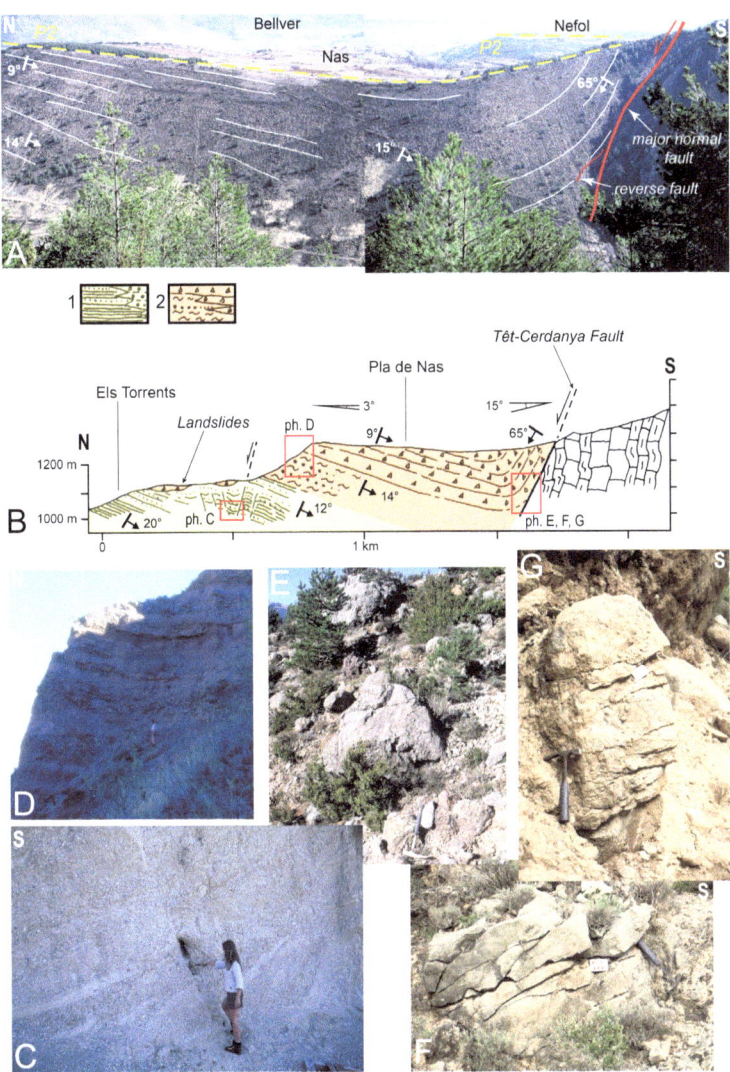

Fig. 13.26   Faulted edge of the Cerdagne Basin south of Bellver. A General view of Nas Plateau, Turolian proximal deposits. **B** Cross-section through Nas plateau. In A and B, note asymmetric deformation of the Turolian sequence, with the beds cross-cut by pediment P2 and the pediment itself warped by tectonic deformation. **C** Olià section; tectonically deformed Vallesian sand and gravel beds. **D** Outcrop of Turolian red beds at Nas plateau (north edge). **E** Very coarse fanhead facies (Turolian limestone boulders). **F, G** Compressional (south-vergent) tectonic shear fractures through large boulders at the boundary fault

Stop 11. Nas: contact between the Turolian sequence and the boundary fault (Fig. 13.26)

As you descend into the ravine, the trailwalk displays exposures of the entire Turolian series (here almost 300 m thick), and particularly the boulder beds of its proximal facies (Fig. 13.26E). The boulders were mostly supplied by Devonian limestone outcrops along the faceted fault scarp immediately upstream, but the more distal red beds also contain clasts of Permian volcanics, red Triassic and yellow Upper Cretaceous sandstone, and nummulitic limestone supplied by the north-facing Serra del Cadí range front. Southerly dips vary from 10 to 15°, with a slight decrease up sequence probably reflecting synsedimentary fault growth (Fig. 13.26A, B). Fault drag has abruptly upturned the bedded sequence, which displays dips of 60°N along the fault itself. These rock deformations are very clearly cross-cut by the tread of pediment P2 (Pla de Nas, Fig. 13.26A, B), but post-P2 tectonic deformation is detectable from a component of southward tilting of P2, and by its slight warp into the shape of a syncline. At the bottom of the trail, walk eastward up along the fault-line gully; on your left you should see a scarp in Turolian breccia coinciding with a reverse fault plane, with south-vergent displacement. A large boulder with several faults slicing through it (Fig. 13.26G) used to occur at the base of the exposed fault, but it was washed away during a recent flash flood. A careful search should nonetheless reveal a number of sheared cobbles and boulders among neighbouring exposures (Fig. 13.26F). These features overall indicate a late episode of tectonic inversion along this E–W segment of the Têt–Cerdagne Fault, probably during the Quaternary (Goula et al. 1999; Calvet 1996, 1999).

Return to Bellver, cross the Segre, and take the small road rising up towards Ordèn–Talltendre. The hairpin bends expose a succession of Hercynian imbricate thrusts, collectively compressed into a syncline and consisting of Upper Paleozoic rocks of the Axial Zone (Silurian schist; Devonian: mostly carbonate; and Visean flysch). Stop at the col (1513 m) and walk to Tossal Ras to observe the view from its west side.

Stop 12. Tossal Ras–Talltendre

The Hercynian structures in Devonian limestone are bevelled here by the vestige of an erosion surface across Tossal Ras (La Reduta on some maps). The surface slopes to the SE (Figs. 13.17, 13.18B, 13.27A) and plunges beneath the Vallesian basin deposits near Sansor (see **Stop 7**). This vestige belongs to generation P1 and extends northward above Talltendre, ca. 1800 m, where a further topographic tier occurs below the residual masses of Pelat de Talltendre–Tossal Gros (2257 m);

Fig. 13.27 Pediment P1 at Tossal Ras–Talltendre. A The erosion surface cutting across deformed Devonian limestones, here viewed from the SE of the plateau. Note other low-gradient components of the landscape on the horizon, and S–P1 erosion surface stairway. **B** Allochtonous pebbles collected from the surface of Tossal Ras: Paleozoic quartz, quartzite and sandstone, often displaying brownish weathering rinds. **C** Grike (solution fissure) in limestone pavement filled with sand, clay, and small rolled pebbles of schist

these ranges themselves bear vestiges of summit surface, S (Fig. 13.18B). The configuration is thus similar to the range-top / range-flank stairway previously described in the Campcardós / Feixa area (visited earlier and clearly visible from here over to the NE) and in the Carlit massif (Itinerary 7); and also to the Pla dels Horts area on the north side of the Conflent Basin (see Itinerary 6). Just before reaching the col, the road embankments show limestone karst fissures (karren) filled with quartz sand, clay, and small siliciclastic gravel (Fig. 13.27C). A scatter of quartz, quartzite and schist pebbles also occurs across the plateau surface (Fig. 13.27B). These indices indicate that pediment P1 functioned as a transport surface for debris being supplied by the residual massifs to the north (see also Box 13.1). From this elevated vantage point, views of the entire Cerdagne

Basin and its faulted southern boundary are excellent (Fig. 13.28). The fault-related faceted spurs are clearly distinct from the Cadí mountain front, which is clearly recessed away from the fault line and at least partly a structural homocline carved out of the Mesozoic and Cenozoic marine cover sequence that rests unconformably on the Axial Zone basement.

Return to Bellver and back to the N260, turning off towards Martinet and La Seu d'Urgell. From Bellver onwards, the Segre channel enters a major knickzone descending to the late Neogene basin of La Seu. The connecting gorge between the Cerdagne and Seu basins displays a series of deeply entrenched meanders. Stop at the first meander (Sant Martí dels Castells; a residual stretch of the old road provides a convenient stopping area to your right).

Stop 13. Evidence of Quaternary neotectonics at Martinet

The section is located at the upper end of the meander, by the old roadside ($42°21'56.2''$N, $01°42'53.2''$E) (Fig. 13.29A). You should observe here a reverse fault where Silurian black schist overrides some poorly weathered large cobbles of Quaternary terrace T1 (Goula et al. 1999; Calvet 1999). The fault itself displays grey fault gouge; the fault plane strikes N120°E, dips 35° to the north, and displays striations documenting pure reverse fault motion (Fig. 13.29B). Broken and striated pebbles are also present. Microtectonic measurements report a transpressional regime. Hydrothermal cementation with geodes of white calcite and greenish-blue metallic mineral veins are also visible just below the fault plane. The tectonic nature of these features has been challenged by Baize et al. (2002, Table II therein), however, because the Silurian schists, which underlie the Devonian limestones (the latter make up the entire northern valley side), may have instead promoted a listric rock slope failure: the alleged fault could thus be the sole of this listric plane of failure. There is, however, no landslide scar further up on the hillside, and the hydrothermal concretions and close geographic proximity with the neotectonic indices observed at Nas (same strike, same stress regime; **Stop 11**) plead in favour of a common Quaternary neotectonic event at Nas and Martinet.

At Martinet, cross the Segre and take the LV-4055 to Montellà, a village situated at the westernmost extremity of the Cerdagne Basin (Upper Miocene outcrops). A poorly maintained forest track from here rises up to Torras refuge, from where a trail continues onward to the limestone ridgetop of Serra del Cadí, and offers a full

Fig. 13.28 Neogene basin at Bellver, Serra del Cadí in background. View from the road to Talltendre. Noted faceted spurs and erosional benches (generation P2) at their base, and truncated spurs at their top (~ 1600 m)

Fig. 13.29   **Late Quaternary reverse fault at Martinet**. See text for debated interpretation. **A** General view: Silurian unit overriding alluvial sheet T1. **B** Close-up of fault plane with striations

geological cross-section of the south side of the Cerdagne Basin and of the Serra del Cadí.

From Montellà, drive back to the bottom of the valley and turn left onto the road to Estana, which follows another valley that hosts the well-known religious sanctuary of Mare de Deu de Bastanist. Near Villec, the road passes over an outcrop of Upper Carboniferous or Permian andesite, resting unconformably on the Hercynian basement and capped to the west by Triassic and Upper Cretaceous marine cover rocks (sandstone and limestone beds, impacted by a small, south-vergent Paleogene thrust). The latter form the Pla de l'Àliga (1599 m)—a possible vestige of erosion surface P1. This small unit of Mesozoic cover is situated in the Cerdagne hanging-wall and backs onto the Cerdagne boundary fault, which here is offset by ~1 km to the south of the main boundary fault and is known as the Estana–Cava Fault. The road crosses the fault upstream of Villec and winds its way up to Estana through Ordovician schists and quartzitic sandstones. Continue through Estana and drive up to Coll de Pallers, 800 m after the village to the south.

Stop 14. Estana: contact zone with the Serra del Cadí

This stop offers insights into the Mesozoic and Cenozoic cover sequence of the Axial Zone, and brings new detail to the tectonic history of the Cerdagne boundary fault. Views from the col area (Fig. 13.30A) show Tossal del Quer to the northwest (1817 m), above the village of El Querforadat. Note that Ker, Kar, sometimes Gar, Ger, indicate pre-Indo-European place names—from the root k^har, often encountered throughout this region and relating to rocks (Querforadat means 'hole in the rock') and rocky areas (typically pebble beds, outwash deposits). Similar to Pla de l'Àliga, and a continuation of it on the opposite side of the valley, Tossal del Quer is a unit of sedimentary cover rocks 700–800 m thick, here augmented by detrital Garumnian (Maastrichtian to Paleogene) and Ilerdian (early Eocene) alveoline limestone beds (see also Itinerary 9, Fig. 14.4A therein). To the south, the stratigraphic package rests unconformably on the Silurian outcrops that occur immediately north of the col, and its dips to the south increase rapidly (to >45°) as the sequence thickens. Overall, the sequence involves 1500 m of Carboniferous to Permian volcanics, Permian to Triassic burgundy-toned sandstones, marine Upper Cretaceous beds, and Garumnian deposits; and, resting over them, an additional 1 km or more of Ilerdian alveoline limestones. These Eocene limestones make up the Serra del Cadí, a huge wall of white rock rising to altitudes of ~2600 m and lightly chiselled by small, north-facing glacial cirques.

This kilometre-high Serra del Cadí clearly does not scale with the more modest landforms generated by the Estana–Cava Fault. This fault effectively has no geomorphological expression in the area between Coll de Pallers and Pla de l'Àliga. The ridge in that area is locally bevelled by low-relief topography, which

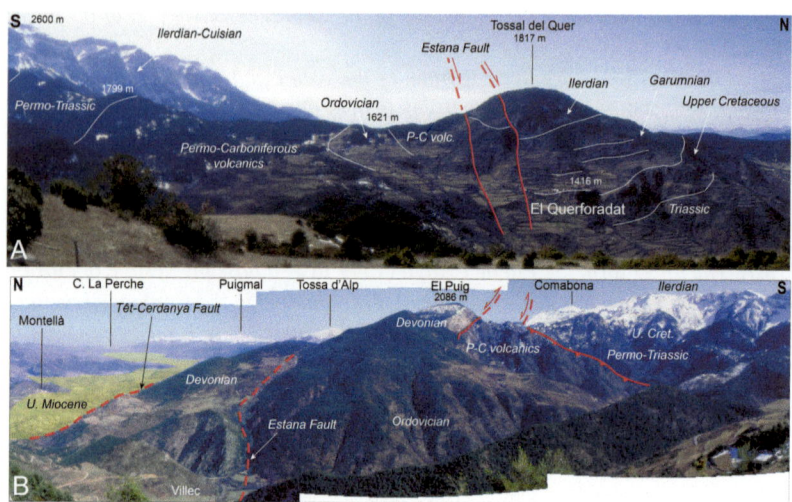

Fig. 13.30 Serra del Cadí viewed from Estana. A Looking west; Estana–Cava Fault and Querforadat hangingwall. **B** Looking east; note limited topographic expression of the Estana–Cava Fault compared to the much bolder Cerdagne boundary fault highlighted at many previous stops. See text for further explanations

cross-cuts indiscriminately (at roughly uniform altitudes) the Hercynian basement (Paleozoic carbonates, softer schists, and Silurian quartzites) and, on the other side of the Estana–Cava Fault, the Campanian–Maastrichtian limestones and their overriding sequence of Muschelkalk limestones, Santonian–Campanian (yellow) sandstones, and grey Maastrichtian limestones already encountered at Pla de l'Àliga. The Cadí scarp rises 4.5 km to the south of the fault line. Here the prominent faceted spurs typical of the footwall uplands east of Montellà (Fig. 13.30B) are absent, suggesting that normal throws generated during the earlier stages of Mediterranean extension in the early Miocene were followed by comparatively low fault activity along the Estana–Cava Fault.

The relative relief of the Serra del Cadí, which like most massifs surrounding the lower Cerdagne Basin displays vestiges of range-top surface, S (Fig. 13.17), is probably not, however, a legacy of this early Miocene faulting episode. Whereas the summit of Pla de l'Àliga, around 1600 m, most likely represents a vestige of P1, nowhere in the region is the vertical offset between S and P1 of such a large magnitude (here: ~1000 m). By inference, the Serra del Cadí, therefore, was more likely thrown up along another (hitherto unmapped) fault cutting through the Permian and Triassic outcrops further south. This fault exists and has been mapped further east, in an area where Mata Negra (1914 m), a block of Upper Cretaceous cover rocks, is downfaulted relative to the Paleozoic outcrops forming the base of the Cadí range front (Fig. 12.2). That fault thus probably extends westward to the area around Prat d'Aguiló–El Puig–Prat de Cadí, an area where several south-vergent Paleocene thrusts slice through the basement and its cover sequence. As in the case of the Mérens Fault (see Itinerary 6), these Alpine thrusts are likely to have undergone tectonic inversion during the late Miocene, generating extensional throws and tectonically enhancing the Cadí scarp face as a north-facing range front at a time when the Cerdagne Basin itself was growing along its own neotectonic boundary fault (Fig. 13.30B).

From the col, a well-marked trail (2.5 hrs return) should lead you to Prat de Cadí and its small frontal moraine produced by a composite Pleistocene glacier descended from the cirques around Canal de Cristall. The trail progresses through the entire Permian and Triassic detrital sequence; the steep mountain front above Prat de Cadí is carved out of a stack of middle Triassic limestones, Upper Cretaceous carbonate sandstones, Garumnian detrital units, and most of all the Ilerdian limestones, i.e., the same lithostratigraphic units as those previously described at the base of the Estana–Cava Fault at Tossal del Quer and Pla de l'Àliga.

Return to Martinet and follow the main road to La Seu d'Urgell for 8 km, until you reach El Pont de Bar. This is the steepest and narrowest part of the Segre gorge, and the channel knickzone coincides with an area of frequent slope instability (large rock slope failures on the valley sides around Toloriu in Devonian schists and carbonates, which rest on a thick mass of highly ductile Silurian schists). The most recent slope failure in 1982 caused the left-bank hamlet to be abandoned and reconstructed further downstream on the right side of the river.

Optional excursion to Lles basin (35 km return, northward from Martinet on the LV-4036) Lles de Cerdanya is a vast etch basin in granite, surrounded by ridges of Devonian limestone and Jujols Group schist (Comaerma–Pelat de Talltendre to the east, and Monturull–Pla de Llet to the west; Fig. 13.17). The north rim of the basin is nonetheless shaped out of the same leucocratic granites and granodiorites as its floor (massifs of Tossa Plana: 2904 m; and Tossal Bovinar: 2842 m). Vestiges of range-top surface, S, occur at the summit of those massifs (Fig. 13.31A–C). The road passes over a succession of low-gradient areas possibly corresponding to generations P3 and P2, e.g., at Travesseras (1170–1200 m, and Lles (1480–1550 m), respectively. The granite is deeply weathered, and chaotic masses of corestones at the weathering front are partly exposed. Above Lles, the road follows the top of a long, low-gradient spur forming a succession of two distinct tiers around 1700–1750 m, then 1950 m (cross-country skiing resort of Cap del Rec; AFT and (U–Th)/He rock-cooling results obtained from this site are illustrated in Figs. 13.19 and 13.31A). P2-related erosional benches become difficult to tell apart from P1, which seems more widely developed as a range-flank pediment in the upper catchment areas ca. 2350 m around the upstanding Tossa

Fig. 13.31   Lles etch basin and Tossa Plana massif. A General view from road to Béixec. Vestiges of pediment P1 can be traced as a series of benches around 2300 m; generations P2 (here, as in Cerdagne around Bena, also displaying two sublevels) and P3 form a stairway above the Segre valley. Apatite fission-track (AFT) and (U–Th)/He ages (i.e., apatite helium, or AHe) after Gunnell et al. (2009). **B** Tossa Plana summit; tilted vestiges of range-top surface, S. **C** Vestiges of summit surface, S, on Tossa Bovinar; on the horizon, L'Orri is a summit in Spain situated between the Segre and Noguera Pallaresa catchments. Geological structures on L'Orri are also bevelled by range-top surface, S

Plana, which was only lightly glaciated on its south side. From Cap del Rec it is possible to walk up to the summit of Tossa Plana (5 to 6 hrs return). From this summit erosion surface, the views are stunning (Fig. 13.31B, C), particularly over the summit surfaces towards the east and west, but also towards the glacial cirques of the Andorran side, which are much more conspicuous than on the Spanish side. In the Lles basin, Late Pleistocene glaciers produced well-preserved lateral moraines (exposure-dated by Palacios et al. 2015). The largest were generated by the Arànser and Llosa glaciers (Fig. 12.3).

Box 13.2 History of the Cerdagne Basin and Its Enclosing Mountain Ranges: An Overview

Evidence provided by itineraries 7 and 8 supports a late rise (ca. 7–6 Ma) of (i) the Cerdagne Basin itself and of (ii) its enclosing mountain ranges relative to the basin. The evidence is based on piecing together a geomorphological puzzle that connects a population of mountain landforms with basin analysis. Most of the relevant landscape evolution history involved a succession of post-P1 base-level changes (Fig. 13.32), with periods of tectonic activity driving range uplift, exacerbating catchment denudation, and increasing accommodation space in the Cerdagne Basin; and periods of relative base-level stability during tectonic lulls, promoting pediment growth around the basin (see Box 13.1). A possible Tortonian pediment, P2a (Fig. 13.32), has been locally identified and associated with the stratigraphic lacuna between the two successive (Vallesian and Turolian) fill sequences of the Cerdagne Basin. Given that pediment P2, the most ubiquitous and extensive, cuts across basement rocks of the Axial Zone as well as upturned sedimentary sequences of the Cerdagne Basin, evidence consistently supports a post-Turolian age for P2, but with a documented resumption of tectonic activity thereafter, including rock deformation during the Quaternary.

The biostratigraphy of the Cerdagne fill sequence itself has also independently documented regional uplift of the Pyrenees in the last 10 Ma, and particularly during the last 6.5 Ma. The subtropical, humidity- and warmth-loving plant assemblages contained in the Vallesian and Turolian sequences could not have thrived at their current elevations (~1500 m: top of the sediment fill in the Cerdagne Basin). Vestiges of similar floras occur in the sedimentary record of the coastal basins at much lower altitudes, e.g., in the form of plant remains in the Empordà and Vallès basins (Sanz de Ciria 1994), and as pollen remains in Languedoc (Suc and Fauquette 2012).

Based on palaeoaltimetric inferences commonly used in palaeobotany, Suc and Fauquette (2012) inferred that the floor of the Cerdagne Basin at the beginning of Vallesian time (11–10 Ma) lay at a maximum palaeoelevation of 200 m; by the beginning of Messinian time (7–6 Ma), it had risen to between 250 and 600 m (wide bracket explained by the age uncertainty over fossil samples and palaeomagnetic data at Can Vilella). These results have been confirmed by a recent study using a Coexistence Approach (CA) method (Barrón et al. 2016). Similar conclusions have arisen from a study of oxygen isotope variations ($\delta^{18}O$) in the dental enamel of rodent fossils from Can Vilella and fom Castelnou 3 (the latter a fossil deposit of similar age: see Itinerary 2); and by an analysis of charophytes and gastropods. The calibrated evidence documents ~500 m of surface uplift since 6.5–6.1 Ma (Huyghe et al. 2020). Palaeoaltimetric methods, however, commonly diverge and those results have since been undermined by another study based on a CLAMP approach (Climate Leaf Analysis Multivariate Program), which concluded that the elevation of the Cerdagne Basin was already 1100–1550 m in Tortonian time (Tosal et al. 2021).

▶**Fig. 13.32 Morphotectonic evolution of the Cerdagne half-graben since mid-Miocene** time (after Calvet and Gunnell 2008, modified). Key to geological ornaments—**1**: Hercynian monzogranite and granodiorite. **2**: Cambrian to Ordovician schist with metamorphic aureole. **3**: Caradocian and Silurian schist. **4**: Devonian limestone series and Carboniferous flysch. **5**: Devonian limestone upper nappe. **6**: Lower Vallesian sequence: high-energy fluvial, lacustrine and palustrine facies. **7**: Upper Turolian series: debris-flow deposits containing large boulders, torrential, fluvial and minor palustrine deposits. **8**: Vertical uplift; extension; compression. **9**: Clastic routing during the late Miocene. Chronology—**1**: pediment (P1) expands during mid Miocene time after tectonic movements have deformed the range-top surface (S) and generated a first series of vertical offsets along the Têt–Cerdagne Fault. **2**: Vallesian series is deposited during a first extensional phase on the southern boundary fault; limited uplift allows pediment P2a to develop. **3**: vigorous surface uplift and faulting of southern block (Tossa d'Alp, Serra del Cadí, Puigmal, Canigou). **4**: Fault throws increase rapidly on the boundary fault during late or post-Turolian times; the Segre River begins to cut a valley into the basin. **5**: pediment P2 expands, cutting across upturned Turolian strata (from Coll de la Perxa to Nas). **6**: Quaternary vertical uplift increases topographic elevation of the Axial Zone and fragments the half-graben into compartments separated by fresh scarps. See Box 13.1 and Itineraries 1 to 6, where landscapes in other Neogene basins and in the Corbières display occurrences of Pliocene to Pleistocene local pediments labelled P2 and P3 (in Cerdagne, the presence of an intermediate level of Late Tortonian age between P1 and P2 explains its name P2a)

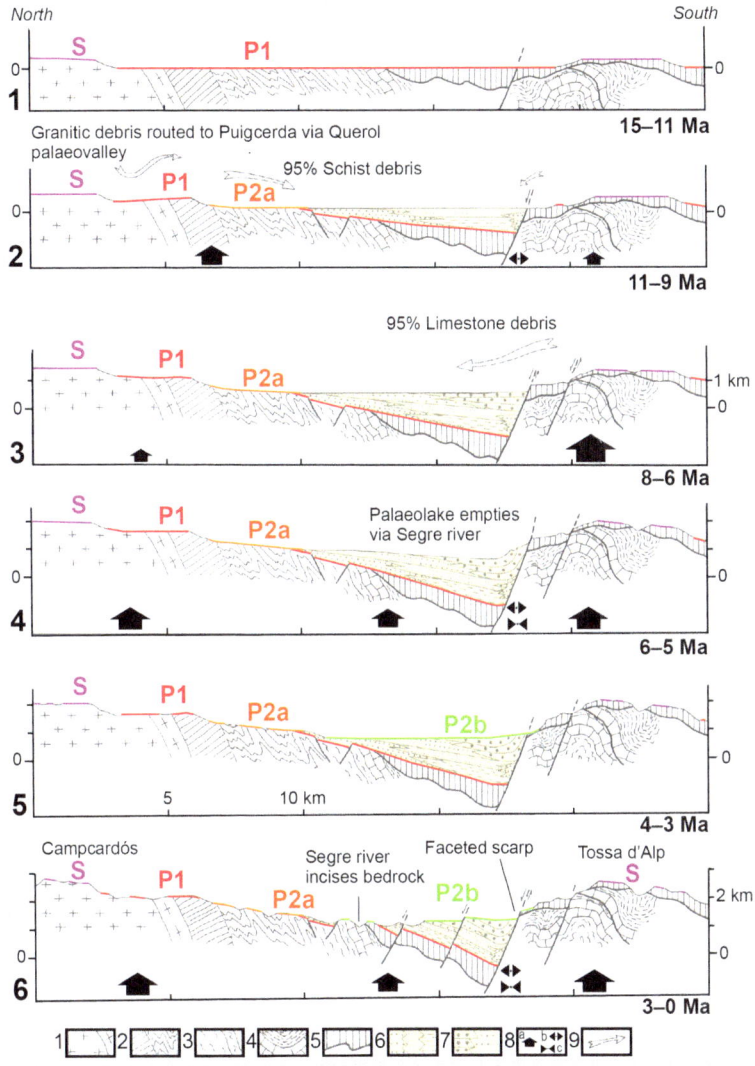

North   South

1 15–11 Ma

Granitic debris routed to Puigcerda via Querol
palaeovalley   95% Schist debris

2 11–9 Ma

95% Limestone debris

3 8–6 Ma

Palaeolake empties
via Segre river

4 6–5 Ma

5 4–3 Ma

5   10 km

Campcardós   Segre river   Faceted scarp   Tossa d'Alp
                        incises bedrock

6 3–0 Ma

A further, debatable, conclusion has also been inferred from the palaeobotanical assemblages contained in the stratigraphy. Following their predecessor Menendez Amor (1955), Suc and Fauquette (2012) concluded that, whereas the Cerdagne Basin had risen substantially since the late Miocene, the relief surrounding the Cerdagne Basin had remained largely unchanged for the last 10 Ma. Implicitly, basin uplift had thus somehow remained decoupled from its regional mountain environment. To support this view, the authors considered that the mix of heat-loving, mesothermic and stenothermic taxa retrieved from the sediments signalled the presence throughout the basin's history of altitudinal vegetation belts on the surrounding mountain sides. The palaeoelevations required among the adjacent massifs to support such a variety of taxa during the late Miocene (10–6 Ma) are postulated to have been 2200–2500 m, i.e., quite similar to present altitudes. In addition to being incompatible with the geomorphological, stratigraphic and tectonic evidence discussed in this guidebook, however, any inference advocating a broad vertical distribution of vegetation belts on local mountain sides during the Miocene relies on an implausible uniformitarian hypothesis which assumes that the ecophysiology of plant taxa has not evolved in the last 10 Ma. The archaeobotanical approach also underplays the impact of Quaternary climatic cycles on the evolution of phenological attributes, and overlooks the significance, as recently as during the Middle Ages (Durand 2003) and even locally in more recent time, of beech and fir stands present, for example, in the coastal plains of Languedoc (relict beech forest of La Valbonne, elevation: 200–300 m) and in the Albères (relict beech forest of La Massane; elevation: 600–800 m).

A more reasonable palaeoaltimetric interpretation based on modern botanical analogues has arisen from a recently rediscovered Vallesian depocentre in Val d'Aran (upper Garonne valley, central Pyrenees), in which the fossil floras suggest a basin-floor palaeoelevation of 700–1000 m in Vallesian time (~12–9 Ma). This implies 900 to 1200 m of uplift since that time, to which must be added the 440 m, fault-controlled upthrow of the Maladeta footwall block—i.e., a total of 1340–1640 m of uplift since ~10 Ma (Ortuño et al. 2013). These independent values for the central Pyrenees, which imply that values of topographic and crustal uplift during this short interval of geological time have been approximately identical, are in good agreement with regional uplift values based on geomorphological criteria obtained for the eastern Pyrenees (Calvet and Gunnell 2008; Gunnell et al. 2008, 2009; Delmas et al. 2018; Calvet et al. 2021; this volume).

References

Agustí J, Oms O, Furió M, Pérez-Vila M-J, Roca E (2006) The Messinian terrestrial record in the Pyrenees: the case of Can Vilella (Cerdanya Basin). Palaeogeogr Palaeoclim Palaeoecol 238:179–189

Andrés N, Gómez-Ortiz A, Fernández-Fernández JM, Tanarro LM, Salvador-Franch F, Oliva M, Palacios D (2018) Timing of deglaciation and rock glacier origin in the southeastern Pyrenees: a review and new data. Boreas 47:1050–1071

Agusti J, Roca E (1987) Síntesis bioestratigráfica de la fosa de la Cerdanya (Pirineos orientales). Estud Geol (Madrid) 43:521–529

Baize S, Cushing M, Lemeille T, Granier B, Grellet B, Carbon D, Combes C, Hibsch C (2002) Inventaire des indices de rupture affectant le Quaternaire, en relation avec les grandes structures connues en France métropolitaine et dans les régions limitrophes. Mém H. S. Soc Géol Fr 175:142 p

Barrón E, Postigo-Mijarra JM, Casas-Gallego M (2016) Late Miocene vegetation and climate of the La Cerdanya Basin (eastern Pyrenees, Spain). Rev Palaeobot Palynol 235:99–119

Bessedik M (1985) Reconstitution des environnements miocènes des régions nord-ouest méditerranéennes à partir de la palynologie. PhD thesis, University of Montpellier 2, 162 p

Bull W (2007) Tectonic geomorphology of mountains: a new approach to paleoseismology. Wiley–Blackwell, 330 p

Calvet M (1996) Morphogenèse d'une montagne méditerranéenne : les Pyrénées orientales. Documents du BRGM 255, Orléans, 1177 p

Calvet M (1999) Régime des contraintes et volumes de relief dans l'Est des Pyrénées. Géomorphol Rel Proc Environ 3:253–278

Calvet M (2004) The Quaternary glaciation of the Pyrenees. In: Ehlers J, Gibbard P (eds), Quaternary Glaciations – Extent and Chronology, part I: Europe. Elsevier, Amsterdam, pp 119–128

Calvet M, Gunnell Y (2008) Planar landforms as markers of denudation chronology: an inversion of East Pyrenean tectonics based on landscape and sedimentary basin analysis. In: Gallagher K, Jones SJ, Wainwright J (eds), Landscape Evolution: Denudation, Climate and Tectonics Over Different Time and Space Scales. Geol Soc Lond Spec Publ 296: 147–166

Calvet M, Delmas M, Gunnell Y, Braucher R, Bourlès D (2011) Recent Advances in Research on Quaternary Glaciations in the Pyrenees. In: Ehlers J, Gibbard PL (eds) Quaternary glaciations, extent and chronology, a closer look, Part IV, Elsevier, 15, pp 127–139

Calvet M, Gunnell Y, Laumonier B (2021) Denudation history and palaeogeography of the Pyrenees and their peripheral basins: an 84-million-year geomorphological perspective. Earth-Sci Rev 215:103436

Delmas M, Calvet M, Gunnell Y, Voinchet P, Manel C, Braucher R, Tissoux H, Bahain JJ, Perrenoud C, Saos T, Aster Team (2018) Terrestrial ^{10}Be and electron spin resonance dating of fluvial terraces quantifies Quaternary tectonic uplift gradients in the eastern Pyrenees. Quat Sci Rev 193:188–211

Delmas M (2009) Chronologie et impact géomorphologique des glaciations quaternaires dans l'est des Pyrénées, PhD thesis (unpubl.), Université de Paris 1, 523 p

Durand A (2003) Les paysages médiévaux du Languedoc, X^e-XIIe siècles. Presses Universitaires du Midi, collection Tempus, Toulouse, 480 p

Gourinard Y (1971a) Les moraines de la basse vallée du Carol entre Latour et Puigcerda (Pyrénées orientales franco-espagnoles). C R Acad Sci Paris D 272:3112–3115

Gourinard Y (1971b) Détermination cartographique et géophysique des failles bordières du fossé néogène de Cerdagne (Pyrénées orientales franco-espagnoles). 96ème Congrès National des Sociétés Savantes, Toulouse 1971, Sciences, t. II, pp 245–263

Goula X, Olivera C, Fleta J, Grellet B, Lindo R, Rivera LA, Cisternas A, Carbon D (1999) Present and recent stress regime in the eastern part of the Pyrenees. Tectonophysics 308:487–502

Gunnell Y, Zeyen H, Calvet M (2008) Geophysical evidence of a missing lithospheric root beneath the Eastern Pyrenees: consequences for post-orogenic uplift and associated geomorphic signatures. Earth Planet Sci Lett 276:302–313

Gunnell Y, Calvet M, Brichau S, Carter A, Aguilar JP, Zeyen H (2009) Low long-term erosion rates in high-energy mountain belts: insights from thermo- and biochronology in the Eastern Pyrenees. Earth Planet Sci Lett 278:208–218

Huyghe D, Mouthereau F, Ségalen L, Furió M (2020) Long-term dynamic topographic support during post-orogenic crustal thinning revealed by stable isotope (δ^{18}O) paleoaltimetry in eastern Pyrenees. Sci Rep. https://doi.org/10.1038/s41598-020-58903-w

Martín-Closas C, Permanyer A, Vila MJ (2005) Palynofacies distribution in a lacustrine basin. Geobios 38:197–2010

Menendez Amor J (1955) La depression ceretana Española y sus vegetales fósiles. Características fitopaleontológicas del Neogeno de la Cerdaña Española. Mem Real Acad Sci Ex Fis Nat Madrid, XVIII, 345 p

Pallàs R, Rodès A, Braucher R, Bourlès D, Delmas M, Calvet M, Gunnell Y (2010) Small, isolated glacial catchment as priority targets for cosmogenic surface exposure dating of Pleistocene climate fluctuations, southeastern Pyrenees. Geology 38:891–894

Palacios D, Gómez-Ortiz A, Andrés N, Vázquez-Selem L, Salvador-Franch F, Oliva M (2015) Maximum extent of Late Pleistocene glaciers and last deglaciation of La Cerdanya mountains, Southeastern Pyrenees. Geomorphology 231:116–129

Reixach T, Delmas M, Braucher R, Gunnell Y, Mahé C, Calvet M (2021) Climatic conditions between 19 and 12 ka in the eastern Pyrenees, and wider implications for atmospheric circulation patterns in Europe. Quat Sci Rev 260. https://doi.org/10.1016/j.quascirev.2021.106923

Rérolle L (1885) Etudes sur les végétaux fossiles de Cerdagne. Rev Sci Nat Montpellier 3, 4, 167–191, 252–298, 368–386

Roca E (1986) Estudi geológic de la fossa de la Cerdanya. PhD thesis, University of Barcelona, 109 p

Roca E (1996) The Neogene of Cerdanya and Seu d'Urgell intramontane basins (eastern Pyrenees). In: Friend PF, Dabrio CJ (eds) Tertiary Basins of Spain. The Stratigraphic Record of Crustal Kinematics. Cambridge University Press, Cambridge, pp 114–118

Sanz de Ciria Catalán A (1994) La evolución de las paleofloras en las cuencas cenozoicas catalanes. Acta Geol Hisp 29:169–189 (publ. 1996)

Suc JP, Fauquette S (2012) The use of pollen floras as a tool to estimate palaeoaltitude of mountains: The eastern Pyrenees in the Late Neogene, a case study. Palaeogeogr Palaeoclim Palaeoecol 321–322:41–54

Tosal A, Verduzco O, Martín-Closas C (2021) CLAMP-based palaeoclimatic analysis of the late Miocene (Tortonian) flora from La Cerdanya Basin of Catalonia, Spain, and an estimation of the palaeoaltitude of the eastern Pyrenees. Palaeogeogr Palaeoclim Palaeoecol 564:110186

Itinerary 9. Tectonic, Palaeogeographic, and Landscape Evolution of the South-Vergent Nappe Structures and the Iberian Clastic Piedmont

14

From Martinet to La Seu d'Urgell (417 km; additional excursions: 56 km; allow for 3 days, or more when including additional excursions). Possible accommodation at Berga, Sant Llorenç de Morunys and Solsona (Fig. 14.1).

This itinerary explores the southern fold-and-thrust ranges of the eastern Pyrenees, also briefly mentioned in the context of Itinerary 4. The focus will be on nappe structures (Pedraforca and Cadí units) and on the synorogenic Paleogene conglomerate sequences that were produced during the emplacement and tectonic deformation of those nappes (Fig. 14.2). Itinerary 9 is also the southern equivalent of Itinerary 5, which provided insights in the history and composition of Paleogene conglomerate deposits in eastern Aquitaine. Greater magnitudes of postorogenic uplift and deep fluvial incision across the Iberian piedmont afford opportunities for probing deeper into the tectonic structures and synorogenic foreland stratigraphy than was possible along the northern range front. Like no other landscape in Europe, the Pyrenean pro-wedge is thus exceptionally well suited to reconstructing the palaeogeography and other aspects of an intracontinental orogen's syn- and post-tectonic evolution. The chronology of collision-related tectonic deformation in the eastern Pyrenees is quite well established, including the timing of its termination. Combining itineraries 5 and 9 (with components of Itinerary 4) overall provides a full cross-section of the orogenic wedge. It does not, however, document the Neogene clastic sequences of the Ebro and Aquitaine basins, which occur more abundantly farther to the west because Neogene rifting in this eastern part of the orogenic wedge has substantially altered the sediment routing systems and preservation potential of the Neogene record (Calvet et al. 2021). A precise and continuous record of clastic output by the orogen during the last 25 Ma is thus provided by the fill sequences preserved in the extensional basins encountered in itineraries 1, 2, 6, 7 and 8. The area covered by Itinerary 9 nonetheless hosts a number of previously unreported or poorly known landforms and deposits that hold important clues about landscape evolution during the Neogene, thus completing the puzzle assembled for the eastern Pyrenees.

© Springer Nature Switzerland AG 2022
M. Calvet et al., *Geology and Landscapes of the Eastern Pyrenees*, Geoguide,
https://doi.org/10.1007/978-3-030-84266-6_14

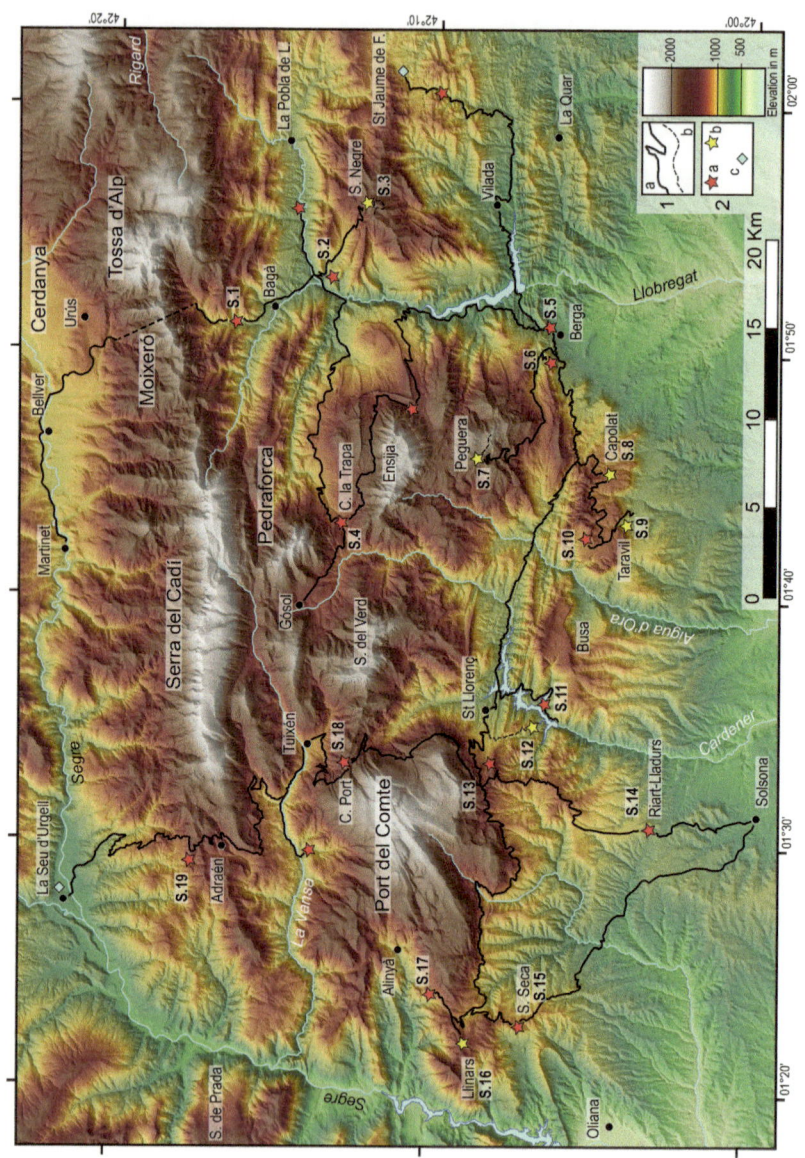

◄**Fig. 14.1 Itinerary 9: route map and overview**. Key to symbols—**1**: Roads and trails. **1a**: Roadways, irrespective of road rank, size, or quality. **1b**: Trailwalks. **2**: Stopping points (outcrops, exposures, landforms, or landscapes). **2a**: Roadside stops. **2b**: Roadside stops involving a short walk, or marking the start of a more major trailwalk. **2c**: Cultural stops. Stops numbered S.1 to S.19 refer to the sequence of roadside stops

*From Martinet, take the N260 towards Bellver, then turn off on the N260R and E09-C16 towards the Cadí tunnel. The route affords good views of the Cerdagne fault scarp, with pediment P2 forming a conspicuous topographic bench above the lowermost band of triangular faceted spurs (see Itinerary 8, **Stop 9**, and* Fig. 13.24B). *After the highway services area the road is sunk in the faulted contact zone between steeply upturned Vallesian conglomerate beds and the Visean–Namurian basement flysch. The fault line striking through Serrat de les Boïgues can be inspected by taking the exit to the services (direction Urús) and then walking along the trail towards the tollbooths. The main highway otherwise drops down into a gorge, then enters the tunnel through outcrops of Devonian limestone. The tunnel is 5 km long, cuts all the way across the limestone mass (Moixeró massif) and comes out into a synclinorium of Visean flysch on the other side. The synclinorium hosts the local stream (Riu de Gréixer), a tributary of the Llobregat River. From here, the road follows the valley and cuts across the entire belt of Mesozoic and Cenozoic sedimentary rocks resting on the southern flank of the Axial Zone. The stratigraphy dips 50°–60° to the south. Given that stopping places are scarce along this road, a brief description of successive outcrops is given here on the basis of the local geological map (sheet La Pobla de Lillet;* Vergés et al. 1994). *Before reaching Gréixer you should notice at first the light pink tones of Permian ignimbrites, resting unconformably on the underlying basement units, followed just after Gréixer by purple Lower Triassic sandstones and mudstones. At the first viaduct (Gréixer viaduct), also note a steep homoclinal ridge in Middle Triassic white limestone. Marine Upper Cretaceous beds cannot be seen from the road, and outcrops of Garumnian deposits are likewise poorly visible apart from the ridge of Danian lacustrine limestone just before reaching another, larger viaduct (Bac de Diví viaduct). Better access to all of these is provided after* **Stop 1**. *Stop at the first rest area after the viaduct.*

Stop 1. Bac de Diví: the Canigou–Cadí nappe
The stopping area is situated just after a trench in upturned beds of Ilerdian alveoline limestone, with easy access for closer inspection. The top of the outcrop also hosts a good viewing point of the stratigraphy. Looking north (Fig. 14.3), steep dips in the Devonian limestone are clearly visible, with the pink ignimbrites

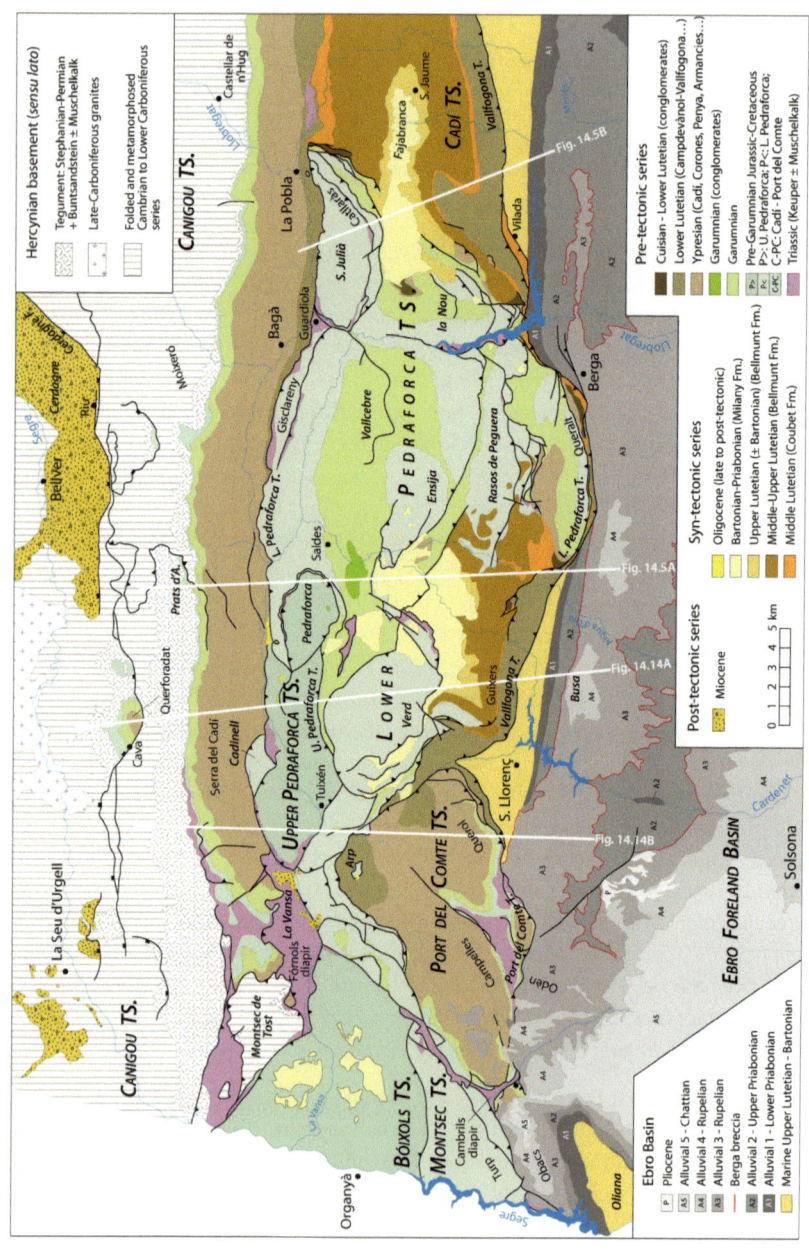

◀**Fig. 14.2** **Map of Alpine sedimentary rock outcrops in the southern Axial Zone (AZ), South-Pyrenean Zone (SPZ) and northern Ebro Basin east of the Segre River.** The map aims to show the complexity of the geological structures, tectonic chronology and depositional palaeogeography in this part of the orogenic wedge. From a structural perspective, the area displays (see Part I, Fig. 1.2 therein, for a more general view):
(i) the South-Pyrenean Zone. It consists of the Pedraforca nappe (its upper and, most extensively, lower units) and of the Bóixols and Montsec units, seen here pinching out in the vicinity of the Arp klippe northeast of Cambrils; they broaden out further west, where they form the South-Pyrenean Central Unit (SPCU).
(ii) the Canigou–Cadí nappe. It consists partly of Hercynian basement rocks from the Axial Zone (Canigou unit) and partly of a younger, overlying sedimentary sequence (Cadí nappe); the Cadí nappe terminates at the Vallfogona Thrust (local name for the South-Pyrenean Frontal Thrust). The Port del Comte unit is part of the Cadí nappe, displaced southeastwards.
(iii) the tectonically deformed part of the outer Iberian foreland zone (northern Ebro Basin).
Map units within the SPZ schematically distinguish the following geological units:
(i) pre-tectonic units, older than Pedraforca nappe emplacement: early to mid-Lutetian at the latest. Tectonic deformation actually also occurred in Ypresian and most of all early Lutetian time within the Pedraforca nappe itself as it advanced southward, as well as within its relative autochton, here the underlying Cadí nappe—which at the time was the southern part of the south-Pyrenean foredeep (Ripoll Basin) and was situated where the SPZ now occurs.
(ii) syn-tectonic units, which record jointly the tectonic deformations of the Cadí nappe and the Pedraforca klippe: mid-Lutetian to Rupelian. After emplacement of the Pedraforca nappe, the Cadí nappe beneath it began to undergo deformation during mid-to-late Lutetian time; tectonic processes continued through the Bartonian and Priabonian (time when the lower units, such as the Freser unit, developed, as well as the flexural fold between the AZ and SPZ generating the so-called Ripoll syncline); nappe advance ended in Rupelian time, when the Vallfogona Thrust came to its final standstill.
The north of the Ebro Basin, i.e., the outermost part of the Iberian foreland basin, underwent tectonic deformation mainly during the Oligocene.
As the region continued its geomorphological transformation in post-tectonic time (i.e., from the end of the Oligocene onward), a major phase of extensional tectonics occurred during the late Miocene, generating sediment-filled intermontane basins such as Cerdagne and La Seu (both in the Axial Zone), and possibly also La Vansa (in the SPZ).
Note the complex tectono-stratigraphy of the Pedraforca nappe, involving:
(i) numerous out-of-sequence thrusts, progressively younger from south to north;
(ii) progressive burial of the corresponding imbricates by clastic deposits, often conglomerate-rich (i.e., ancient range-front alluvial fans), of mid- and mid-to-late Lutetian, Bartonian and Priabonian age.
Sources 1:50,000 scale IGME maps (sheets: La Pobla de Lillet, Berga, Oliana, Gósol, Sant Llorenç de Morunys; see Vergés et al. 1994; Mató et al. 1994; Berastegui et al. 2017; Ardèvol et al. 2020a, b), 1:50,000 scale ICGC maps (sheets: Berguédà, Alt Urgell, Cerdanya), 1:25,000 scale ICGC maps (sheets: La Pobla de Lillet, Gombrèn, Berga; see Martínez et al. 2013; Tallada et al. 2017).

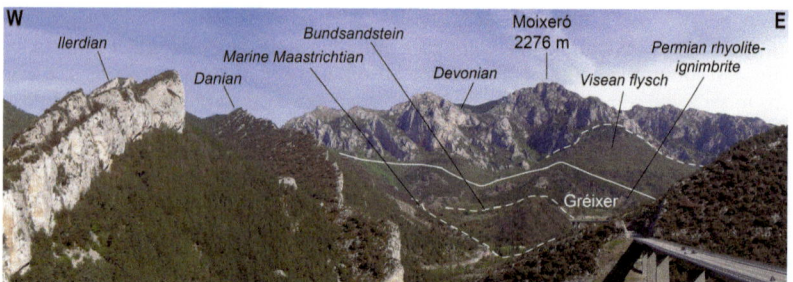

Fig. 14.3   The Cadí nappe unit from Bac de Diví viewing point. Note the tegument of Permian volcaniclastic rocks resting on the Hercynian basement of the Axial Zone, followed in the middle ground by the Mesozoic and Cenozoic cover-rock sequence, which displays large lacunae towards its base

followed by the Danian limestone also discernable. Looking south, a ridge in thinner limestone is an outcrop of the Corones Fm. (early Cuisian). Other outcrops are shrouded in forest. The large topographic mass obstructing the view to the south is the Pedraforca nappe, overriding Cuisian–early Lutetian marine marls (the latter made conspicuous by their gullying features). The stratigraphy of the Canigou–Cadí nappe in this area is strikingly different from what was previously encountered 15 km further west at **Stop 14**, Itinerary 8. East of La Comabona (2547 m) and onward past Coll de Pendís, the Ilerdian to early Lutetian limestone mass forming the Serra del Cadí frays into a series of much thinner limestone beds amid a thick succession of shales and marls. The impact of this lateral facies change on structural landforms is considerable because the massive limestone wall of Serra del Cadí pinches out and gives way to a belt of limestone ridges never exceeding elevations of 1700–1600 m, carved up into a succession of 'flatirons' along the south side of the Tossa d'Alp massif (Fig. 14.4).

From a structural perspective, the Canigou–Cadí nappe (Paleozoic basement and its Mesozoic and Paleogene cover sequence) forms a frontal ramp fold, identical to the one observed at Coustouges further east (Itinerary 4, **Stop 4**) and a direct continuation of it. Unlike the occurrence at Coustouges, where the floor thrust plane of the nappe could be observed through a large window in the Cadí–Canigou unit (Itinerary 4, **Stop 8**, and Figs.9.9 and 9.15), here the thrust is deeply buried (perhaps ~3 km) and with no outcrops (Fig. 14.5A).

The Mesozoic and Paleogene stratigraphy of the Cadí unit can be further inspected by exiting the highway at Bagà, not long after **Stop 1**. By driving 5 km back up the Gréixer valley along road GR7, many stopping points provide

Fig. 14.4   South side of the Pyrenees viewed from Coll de Roc de la Clusa. Foreground: homoclinal ridge of Cap dels Banyadors (Campanian–Maastrichtian limestone; spot elevation: 1432 m). Middle ground: Cadí nappe, which includes Paleozoic outcrops (Devonian limestone) and a younger sedimentary cover sequence (mostly Eocene); note lateral transition from the massive carbonates of the Serra del Cadí in the west to marls with thin limestone intercalations in the east. Background: snow-capped granite massifs of the Axial Zone, also source areas of the Paleogene conglomerates encountered at Serrat Negre (see Figs. 14.5, 14.7, 14.8)

direct access to the Corones Fm. (red deltaic sandstones, late Ypresian), Upper Cretaceous limestones, outcrops of Lower Triassic sandstone (supporting the piers of Gréixer viaduct), and lastly the Permian ignimbrites.

From Bac de Diví, continue your onward journey along the main highway, which cuts through the Corones beds, then through an outcrop of white limestone (spot elevation: 934 m) followed by another of marly limestone (spot elevation: 919 m) before reaching the broad Bagà basin, eroded out of light grey marls (Armàncies Fm., mid- to upper Cuisian, perhaps lowermost Lutetian, 1000–1500 m thick). Several exposures of the marl occur in the road embankment and among deep gullies visible on the right side of the valley (El Bastareny). These marls are submarine continental slope deposits, contemporaneous with the deepening flexural foredeep that would eventually become the Ripoll syncline during the Eocene. Before reaching the Guardiola tunnel, precisely at milestone 116.7 km you will encounter a first outcrop of Campdevànol–Vallfogona Fm. These are early Lutetian turbidites, darker than the Armàncies marls and containing siliciclastic components (mainly fine-textured, ochre sandstone beds). Just before the tunnel entrance, the road embankment exposes a 150 m stretch of Keuper (Upper Triassic evaporites). These form the base of the Pedraforca nappe, which lies structurally above the Cadí nappe (Fig. 14.5B).

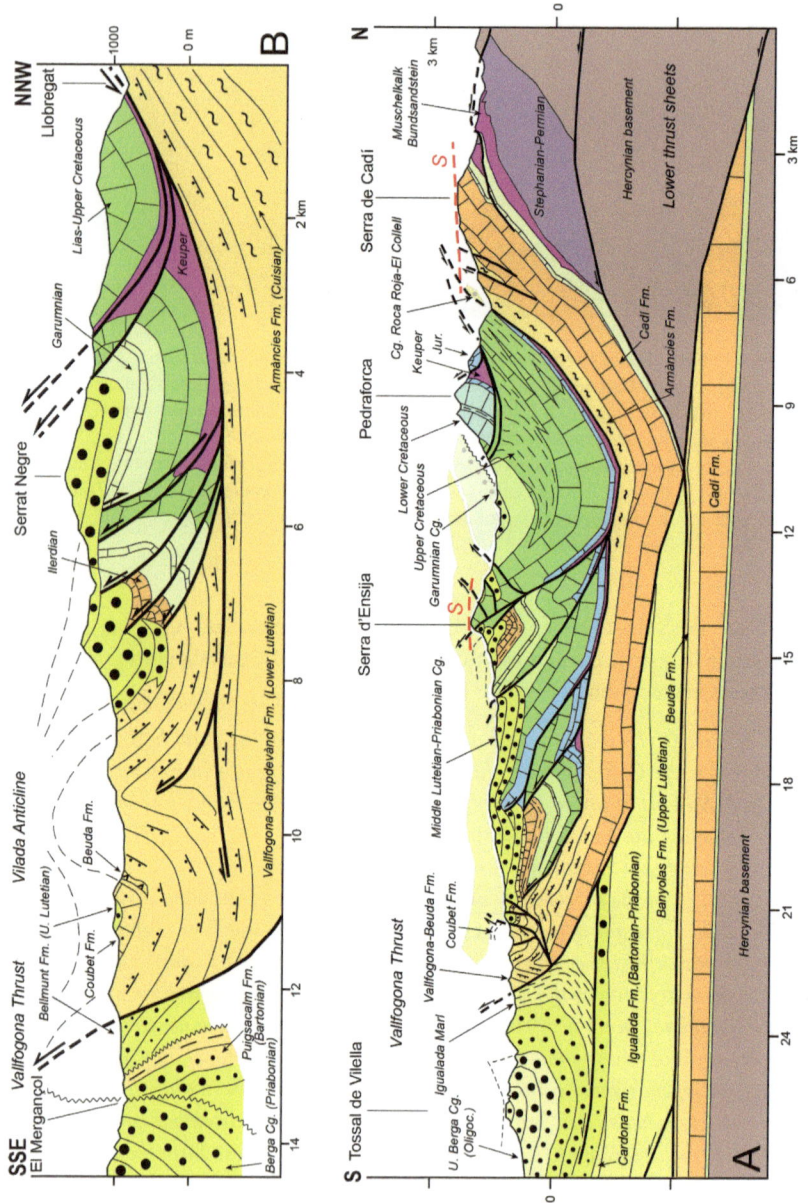

◄Fig. 14.5 Geological cross-sections through the South-Pyrenean fold-and-thrust ranges. A General N–S section through the South-Pyrenean Zone west of the Llobregat valley. The two Pedraforca nappe units override the Canigou–Cadí nappe, which itself overrides the Ebro Basin. Garumnian and Paleogene conglomerates initially buried all the geological structures of the South-Pyrenean Zone. The present-day relief results largely from exhumation of these structures by widespread removal of the clastic overburden when the eastern Pyrenees and their foreland areas underwent regional uplift and tectonic reactivation during the Neogene (see Calvet et al. 2021). **B** Eastern termination of Pedraforca nappe. See Fig. 14.2 for location. The Paleogene, syntectonic conglomerate beds at and around Serrat Negre have buried the nappe front as well as its out-of-sequence imbricates (progressively younger from south to north). After Vergés et al. (1994) and Martínez et al. (1988, 1996, 2001), redrawn and modified

For close-up views of the Campdevànol–Vallfogona Fm., exit the highway after the tunnel and follow road B403 to La Pobla de Lillet (7 km return). Good exposures of the siliciclastic facies can be reached after the access road to Esplet campsite. East of La Pobla, a kilometre-sized olistolith of Keuper (mostly gypsum), lodged in the Campdevànol–Vallfogona Fm., records the advance of the Pedraforca nappe into the foredeep.

Box 14.1 Stratigraphy of the Eocene of Catalonia

In this GeoGuide, focus on lithostratigraphy is mostly restricted to indications relevant to understanding the Pyrenean landscape. Additional details can be obtained from the geological maps and their handbooks, and from abundant journal literature available since the 1960s. The existing body of studies contains a profusion of lithostratigraphic units bearing local names, many of which are used in the legends and taxonomy of IGME and ICGC geological maps. Correlations with the international stratigraphic scale are not always straightforward.

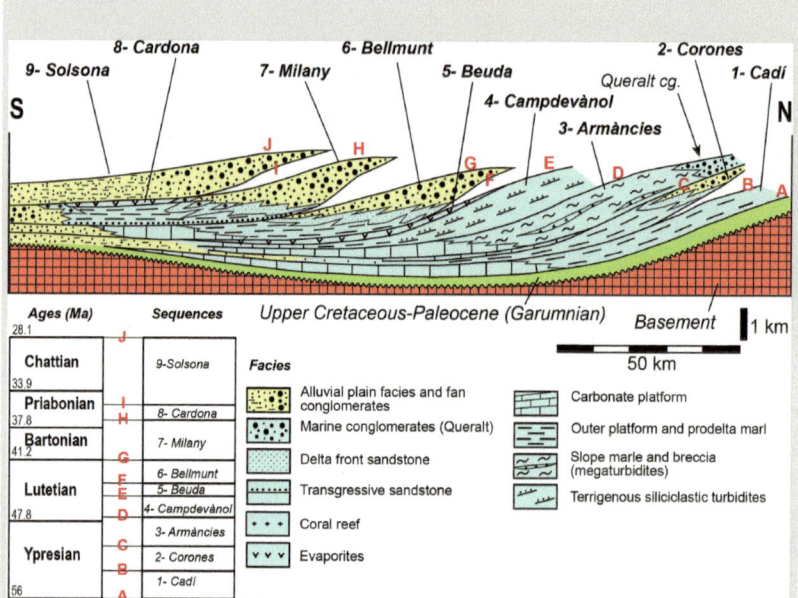

Fig. 14.6 **Paleogene depositional sequences and their broad facies characteristics on the southern piedmont of the eastern Pyrenees.** After Puigdefàbregas and Souquet (1986) and Burbank et al. (1992a), redrawn and modified. Detailed lithostratigraphic formations within each unit are not mentioned. This N–S synthetic cross-section is situated slightly east of the Pedraforca massif and does not document the presence of sub-Garumnian Mesozoic basins in the Axial Zone. For more complete reconstructions and more thorough balanced cross-sections, see Vergés and Martínez (1988), Vergés et al. (1995), and Vergés (1999)

Within that framework, nine depositional sequences of chronostratigraphic significance (Fig. 14.6) have been defined for the Eocene and early Oligocene (e.g., Puigdefàbregas and Souquet 1986; Burbank et al. 1992a; Barnolas and Gil-Peña 2001; Barnolas et al. 2019). These lithostratigraphic units are distributed from north to south as a diachronous succession recording terrestrial, shelf, and continental slope environments. The entire system prograndes southward, following a pattern consistent with the growth of the orogenic wedge, the advance of the nappes, and gradual rollover of the foredeep. The sequence has been dated on the basis of marine fossil assemblages, but magnetostratigraphic evidence (Burbank et al. 1992a) has also shown that some older shelf deposits were locally reworked and redeposited onto the continental slope (Tosquella and Samsó 1996). Given these palaeogeographic complications, some uncertainty remains over the age of the youngest marine deposits in the Ripoll syncline. Magnetostratigraphic data suggest that marine sedimentation could have ceased in Cuisian (late Ypresian) time (Garcés et al. 2020, Fig. 6 therein), but could equally have extended to early Lutetian time (Burbank et al. 1992a; Tosquella and Samsó 1996; Verges et al. 1998; Barnolas and Gil-Peña 2001; Barnolas et al. 2019). The latter option appears supported by the fact that Shallow Benthic Zone 13 (SBZ 13) has been detected at the top of the Armàncies Fm. and likewise in the Campdevànol–Vallfogona Fm. (Tosquella and Samsó 1996). This biozone documents the Lower Lutetian between the upper part of Chron 21r and lower part of Chron 20r, i.e., between 47.8 and 45 Ma (Molina et al. 2011; Vandenberghe et al. 2012). This age bracket provides a minimum age for the emplacement of the Pedraforca nappe, which overrides both of these marine formations.

The top of the marine sequence contains the Beuda gypsum beds, which document a brief episode of seclusion from open marine conditions by the advancing nappes. Littoral facies rocks are still prevalent in the Coubet Fm., while the core of the Ripoll syncline consists of the Bellmunt Fm. (Busquet et al. 1992). Internally, the continental Bellmunt Fm. contains two chronostratigraphic units named Bellmunt (base) and Milany (top); both document a subaerial range-front alluvial environment (> 2 km of conglomerate beds). The Milany unit encompasses the top of the Bartonian and perhaps the base of the Priabonian stage. The Bellmunt Fm. has been dated by three mammalian fossil assemblages at Sant Jaume de Frontanya (one of the richest Paleogene fossil sites in the Pyrenees), with level SJF 3 (750 m above the top of the marine beds) correlating with European Land Mammal biozone MP 15 (41–40 Ma, i.e., Bartonian); and the other two levels (SJF 2, SJF 1), situated at lower elevations, ascribed to MP 14 (~41–43.5, i.e., late Lutetian; Busquet et al. 1992; Badiola et al. 2009; Bonilla-Salomon et al. 2016).

Drive back from La Pobla de Lillet to Guardiola de Berguedà, and take the BV4021 on the left towards Sant Julià de Cerdanyola. The first hairpin bend ascends the lowermost imbricate of the lower Pedraforca nappe. The Lower Jurassic (Lias) marls

and limestones from the base of the nappe are poorly exposed, but the road subse-
quently ascends through thick, well-exposed Santonian and Campanian limestones
and marls. Sant Julià village is nestled in a perched syncline in the Campanian–
Maastrichtian carbonate sequence, which contains marl interlayers. Santa Maria
de les Esposes hermitage is worth a brief stop for the view.

Stop 2. Les Esposes hermitage: views of the Pedraforca and Cadí nappes
From here you will get views down into the Bagà basin, over to the Moixeró massif to the north, and across to other side of the Llobregat valley where the lower
Pedraforca nappe (already beneath your feet) continues westward (Figs. 14.4,
14.5). The core of the nappe hosts the perched syncline of Vallcebre, surrounded
by outward-facing scarps in Danian limestone. The barren summits of the Serra
d'Ensija (2322 m; see **Stop 4**, **Stop 7**) are also within view, but Pic Pedraforca
(2509 m; see **Stop 4**), which occurs in the upper Pedraforca nappe unit, is hidden
from view.

Drive through Sant Julià and take the first left after the bridge. From here you will
be following a large Quaternary debris cone along the left side of the stream. About
1.8 km after the village, turn right onto the forest track that rises to Roc de la
Clusa, from where the summits of Serra del Cadí and Pedraforca will be in full view
(Fig. 14.4). The track (in fairly good condition) cuts through the Keuper evaporites
at the base of the lower Pedraforca imbricate around 1400 m , then rises through
another imbricate of marine Upper Cretaceous limestone with overturned beds
dipping north, itself overriding a third imbricate consisting of Upper Garumnian
red marls and sandstones, visible in the small valley head just before reaching
col. This third unit is overlain by Ilerdian alveoline limestone further west. The
entire sequence has been compressed into a syncline, but its internal architecture
remains mostly concealed by the thick drape of syntectonic conglomerate backfill,
here displaying its weakly deformed uppermost strata (Fig. 14.5B). Outcrops of the
conglomerate occur at Coll de Roc de la Clusa. Park at the col.

Stop 3. Roc de la Clusa, Serrat Negre: conglomerate beds containing granite
boulders
Paleogene conglomerates occur on all the local summits of the area as far as
the top of Serrat Negre (1776 m). The Pedraforca nappe here is entirely buried
beneath them (Birot 1937). The conglomerates belong to the Milany and Bellmunt sequences. On the south flank of the massif, they present themselves as 5
distinct units (Figs. 14.5B, 14.7). The units are deformed with decreasing intensity towards the top and each separated by an unconformity, thereby documenting

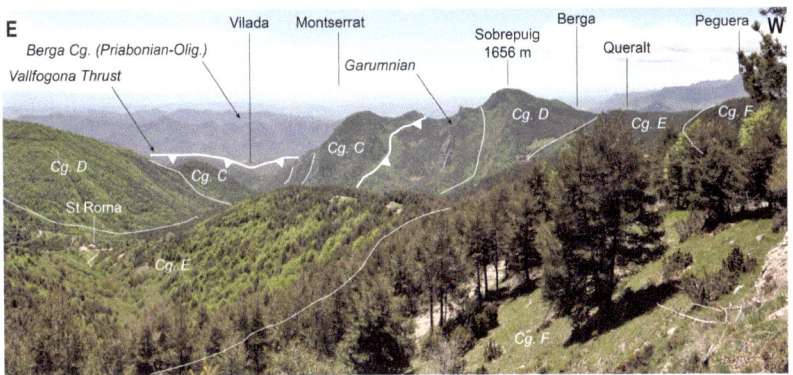

Fig. 14.7   The Paleogene syntectonic conglomerate units, here viewed to the south from Serrat Negre. Letters C to F label the successive range-front alluvial fans listed on the geological sheet of La Pobla de Lillet (Vergés et al. 1994). Fans A and B are outside the frame, but outcrops occur immediately NE of Serrat Negre

successive stages of out-of-sequence thrusting followed by a definitive standstill and burial of the nappe beneath the debris from the eroding mountain range (Martínez et al. 1988). By linear extrapolation of sedimentation rates calculated for the chronostratigraphic levels situated between biozone levels SJF 1-2 (MN 14) and SJF 3 (MN 15; see Box 14.1), the top of the Bellmunt Fm. is potentially early Priabonian (an age also put forward by the IGME map of La Pobla de Lillet; Verges et al. 1994). Other authors (Garcés et al. 2020; and 1:25,000 scale ICGC map of La Pobla de Lillet by Martínez et al. 2013) have suggested that the sequence is exclusively Bartonian, in that case advocating a somewhat earlier termination of tectonic nappe emplacement and deformation (Box 14.1).

First walk west towards Roc de la Clusa (20 min return). The ridge is capped by a highly cemented conglomerate bed, either vertical or slightly overturned (Fig. 14.8B). It contains an abundance of pebbles of Paleozoic source rocks, mixed in with boulders of more local limestone but totally lacking granite clasts. An outcrop containing metre-sized granite boulders, however, occurs at the SW extremity of the ridge. Its geometric link with the rest of the conglomerate deposits remains unclear.

Exposures on the east side of the massif further document the palaeogeography of synorogenic denudation in the Paleogene mountain range. These exposures can be accessed by following the track east of the col, which takes you all the way around Serrat Negre (40 min on foot; note: this is a capercaillie protection zone,

Fig. 14.8 Paleogene conglomerates capping Serrat Negre. **A** Facies comprising almost exclusively of granite, south face of Serrat Negre (fan F); note apparent absence of clast cementation, faint bedding structures (dashed lines), and very gentle dip (to the north). **B** Facies comprising mainly of limestone, west of Coll de Roc de la Clusa; note the presence of very large boulders and tectonically overturned bedding structures (perhaps fan E). **C** Conglomerate outcrop containing cobbles of Paleozoic lithologies, NE face of Serrat Negre (fan E or F); here: Devonian griotte limestone (D) and granite (Gr); note steep dips (to the south). **D, E** In situ granite boulders, several metres in diameter, exhumed from their comparatively finer conglomerate matrix; Serrat Negre ridgetop

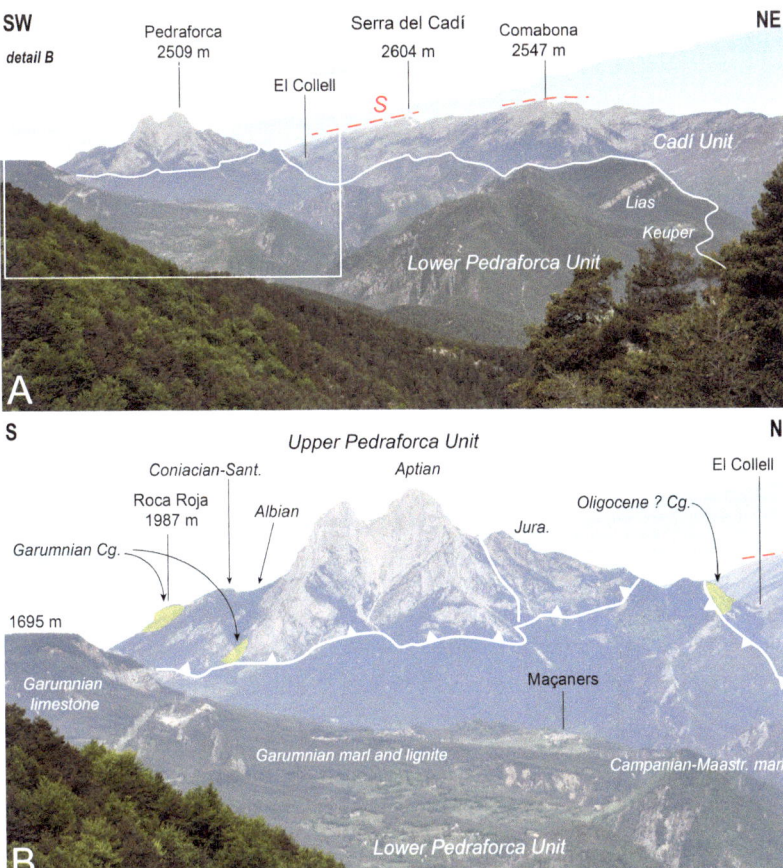

Fig. 14.9 Pedraforca massif, here viewed from the track leading to Coll de Roc de la Clusa. A General view. The two Pedraforca nappe units rest on the Cadí unit; note the low-angle topographic bevel cross-cutting the steeply-dipping limestone beds of Serra del Cadí. **B** Close-up of the Pedraforca twin peaks. Vestiges of various generations of poorly dated conglomerate deposits (Cg.) are distributed across the landscape; note also north edge of the perched syncline of Vallcerbe, in Garumnian limestone

with regulated access at certain times of the year). On the north-east side of Serrat Negre, you should encounter a conglomerate bed consisting exclusively of limestone pebbles and dipping 70°S; the series, however, becomes rapidly enriched in clasts of granite and Paleozoic limestone pebbles (including Devonian griotte limestone; Fig. 14.8C), with bedding likewise displaying steep dips to the south. At the first hillside spur (42°11′56″N, 01°56′05″E), dips decrease and granite boulder sizes exceed 2 m. The south-facing part of Serrat Negre (42°11′51″N, 01°55′55″E) consists almost exclusively of granite debris (boulders up to 3 m). They occur in beds with a gentle dip to the north (Fig. 14.8A) over a minimum cumulative thickness of 50 m. By following the fire break to the ridge summit, you will encounter granite boulders (Fig. 14.8D, E) reaching maximum size of 2.3 × 2 × 3.5 m (42°11′54″N, 01°55′57″E). Lithologies include granodiorite and light-toned and coarse-textured monzogranite (often porphyritic and displaying fluidal structures), alongside items of aplite (dark- and light-coloured), hornfels, and Jujols- or Canaveilles-type schist.

Granite provenance can only be from the Mont-Louis batholith, with a perfect match for all of the aforementioned textural and compositional varieties. The nearest outcrop of granite, however, is situated 30 km to the north, on the other side of the present-day massifs of Moixeró and Tossa d'Alp. This implies a landscape very different to the current scenery, with a mountain front displaying steep, high-energy catchments cutting more deeply into the far-field granite outcrops of the north than into the near-field belt of non-igneous Paleozoic outcrops between Moixeró and Tossa d'Alp (Fig. 14.4). This suggests the latter may have been situated in a comparatively low-energy environment, perhaps even south of the Paleogene mountain front, in which case the Pyrenees mountain range at the time was much narrower than it is today. Based on modern analogues, where gneiss boulders of a similar calibre to some granite boulders encountered here at Serrat Negre occur in the Têt gorge at Villefranche-de-Conflent (Itinerary 6), mean channel gradients of 9–10% would be required to ensure bedload transport from granite outcrops in the Carlit or Campcardós areas. Back in late Bartonian to early Priabonian time (i.e., approximately 38–35 million years ago), we would thus need to imagine (i) massifs in that part of the Axial Zone rising to elevations 2–3 km above the current range-top paleoplain in the Carlit, Campcardós and Tossa Plana massifs; and (ii) the southern pro-wedge forming a relatively steep taper towards the clastic piedmont in the vicinity of Serrat Negre.

The journey onward from the col can be continued along forest tracks to Sant Romà de la Clusa and Vilada (Fig. 14.7), although Vilada is visited later as part of another excursion. This route provides a cross-section through the entire sequence of crumpled conglomerate beds and their underlying marine marls, which are exposed in an erosional window at Vilada. However, the preferred itinerary gets you back to the E09-C16 via Sant Julià and Guardiola. Follow the Llobregat valley southward, and eventually turn right on the B400 towards Saldes. While confined to outcrops of Garumnian, the road follows the north edge of the Vallcerbe perched syncline, with the belt of Danian limestone forming the outward-facing scarps all around it. Hilltops within the synclinal valley display a collection of unconformable conglomerate deposits. These have been correlated with the Milany and Bellmunt units visited at Serrat Negre. After Saldes, the road to Gósol skirts around the southern base of the Pedraforca twin peaks before reaching Coll de la Trapa amid exposures of light red limestone conglomerate.

Box 14.2 The Pedraforca Nappe Controversy

The tectonic origin of the Pedraforca nappe has fuelled a long-lasting controversy, briefly summarised here. The link between the enigmatic Pedraforca massif (see Box 14.3) and a possible tectonic nappe, which at the time was believed to have advanced northward from the south, was first propounded by Astre (1924, 1935) and Jacob et al. (1927) (concerning the work by Astre on this topic, which includes a remarkable map, see Bilotte and Aragonés, 2010; for a more extensive overview of the work by Toulouse geologists on the Pyrenees, see Bilotte 2010). The hypothesis was soon opposed, however, by Aschauer (1934) and Birot (1937), who interpreted the Pedraforca massif instead as a mushroom structure with north- and south-vergent thrusts on its margins. This scenario prevailed until the 1960s (Guérin-Desjardins and Latreille 1961). Séguret (1972) subsequently upset the consensus by advocating the existence of two nappe units resulting from a décollement between the Hercynian basement and its post-Paleozoic cover rocks, and involving transport (mainly by gravitational gliding but also by thrusting) of the cover sequence over a distance of 30–40 km down the southern flank of the Axial Zone. This model has prevailed ever since, corroborated by deep boreholes as well as by seismic data such as the ECORS-Pyrenees profile (see Part I). A dynamic (push) component to nappe displacement, driven by synorogenic compressive stresses within the Hercynian basement, is currently considered to have been the dominant contributing process (Muñoz et al. 1986; Vergés 1999; Vergés and Martínez 1988; Muñoz 1992; Vergés et al. 1995, 2002; Martínez et al. 1996, 2001, 2013, 2020). A few dissenting voices, standing by earlier views while also invoking moderate thrusting from the west (Ullastre and Masriera 2004), appear to have fallen silent.

Stop 4. Coll de la Trapa, Pedraforca massif, and upper Pedraforca nappe
The upper Pedraforca nappe unit (Box 14.2) broadens out extensively to the west as part of the South-Pyrenean Central Unit (SPCU, see Part I, Fig. 1.2 therein), with the Pedraforca massif coinciding here with its easternmost termination and forming an almost entirely isolated klippe (the name of the Pedraforca massif has also been given to the much more extensive tectonic unit; Martínez et al. 2001). In structural terms, the upper Pedraforca nappe (Figs. 14.5A, 14.9A, B) is a hanging-wall ramp. Its basal thrust, highlighted by a level of Campanian and Maastrichtian limestone breccia (usually interpreted as an intermediate unit between the lower and upper Pedraforca nappes), dips to the south and forms a gentle syncline. The thrust cuts through the entire Mesozoic cover sequence. From base to top and north to south, the stratigraphy of the south-dipping sequence includes units of Keuper (evaporites), Lias (limestone), Dogger (massive dolomite), Aptian (thick limestone with a marl interlayer, which explains the dip between the two main peaks—Pedraforca means 'forked rock'), Albian (marl), and lastly Cenomanian–Turonian–Santonian (limestone, forming the southern face of the massif). This stratigraphic sequence is very similar to the sequence encountered in the North-Pyrenean Zone and is, therefore, probably a displaced component of the cover sequence previously situated in the northern half of the Axial Zone (i.e., north of the Hospitalet or Mérens faults; see Part I, Fig. 1.2). The nappe was transported southward from this root area over the lower Pedraforca nappe, which itself probably is a component of the cover sequence originally covering the southern half of the Axial Zone (south of the Hospitalet or Mérens faults), and thus having undergone a comparatively shorter distance of travel. Emplacement of the upper nappe would have occurred in latest Cretaceous time, with its leading edge undergoing burial by Garumnian molasse at the time (Figs. 14.5A, 14.9B). The upper nappe subsequently continued its southward advance, but this time locked to the lower nappe and shearing through the floor of Keuper evaporites.

The steep, rocky peaks of the Pedraforca massif are unusual for the southern Pyrenees, which overall tend to display more rounded, massive landforms. A clue to this paradox is provided by the fact that the massif is surrounded by a girdle of unconformable conglomerate outcrops, their most elevated occurrence recorded at an altitude of 2050 m. It is thus possible that, during the Paleogene, the summit area was entirely buried by these synorogenic foreland conglomerates

(Fig. 14.5A). The unusually fresh appearance of the twin peaks would thus result from exhumation and exposure during the Neogene of an ancient, perhaps late Cretaceous or early Paleogene, mountain landform (Box 14.3).

Box 14.3 Pic Pedraforca and the Conglomerate Conundrum

Much of the geomorphology of the area hinges on establishing different facies and generations of conglomerate deposits, many of which are difficult to distinguish from one another and to date precisely. For example, limestone conglomerates at Coll de la Trapa, exposed on both sides of the road, are apparently a continuation of the fine-textured red and ochre beds that appear to lie stratigraphically beneath the Danian limestone unit 3 km to the ESE. They could thus be late Maastrichtian. The 1:50,000 scale Berga geological sheet produced by the IGCC, and the IGME sheet of Gósol (Martínez et al. 2020), however, both ascribe a Danian / Paleocene age to these red beds, in which case we should envisage a lateral facies change rather than a diachronous relationship between them. From the col, the southern ridge of the Pedraforca massif is clearly visible, with its plateau of unconformable red beds know as Roca Roja (1987 m; Fig. 14.9B). Given that they exclusively contain locally supplied limestone clasts, the formation has been correlated with the deposits at Coll de la Trapa and assigned a Maastrichtian (Vergés and Martínez 1988; Ullastre and Masriera 2004) or Paleocene (Martínez et al. 2020) age.

This Roca Roja outcrop and the Coll de la Trapa red beds had previously been correlated by Astre (1935) and Birot (1937) with the extensive outcrops of thick Eocene conglomerate, the base of which grade to the marine Cuisian–Lutetian beds encountered immediately to the south of Coll de la Trapa (Aygua de Valls catchment). These conglomerates belong to the Bellmunt and Milany continental sequences, even perhaps to the Berga–Solsona sequence in the case of its most elevated occurrences. They reach elevations of 2000 m on the SW flank of Serra d'Ensija (the large, overthrust anticline blocking the horizon to the south of Coll de la Trapa), with isolated residual outcrops of conglomerate even mapped all the way up to its summit, ca. 2250 m. The beds display a succession of increasingly younger unconformities towards the north (Fig. 14.5A; see also **Stop 7** and associated figures), where they rest on the overriding units of the lower Pedraforca nappe (Martínez et al. 1996). Other enigmatic outcrops of conglomerate have been mapped, e.g., on the floor of Gósol basin, where they form an angular unconformity with the conspicuously folded Garumnian series, and beneath the northern base of the lower Pedraforca nappe, north of Pic Pedraforca and at Coll del Colell—another site also named Roca Roja (hereafter named Roca Roja N; elevation: 2041 m).

The most recent ICGC and IGME maps attribute the Gósol conglomerate occurrence to the late Eocene, and the Roca Roja N–El Colell occurrence to the Oligocene (its facies is compatible with other Oligocene deposits in the region). Both occurrences contain local inputs from Eocene and Upper Cretaceous limestone but also cobbles of Paleozoic sedimentary rocks, which suggests that the Serra del Cadí was not yet a topographic barrier to these far-field inputs from the north. Based on its pollen content, which is compatible with bioindicators reported from late Miocene deposits of the Cerdagne Basin (see Itinerary 8), the Roca Roja N deposit has been ascribed a late Neogene age (Durand-Delga et al. 1989). This view is also notionally compatible with the existence of the small Neogene tectonic basin of La Vansa (see **Stop 18**), merely ~20 km to the west (Fig. 14.2). This interpretation was extended to the Gósol conglomerates by Ullastre and Masriera (2004) who, from these indirect clues and based on the fact that the conglomerate deposits contain only few clasts from the Cretaceous Upper Pedraforca Unit, suggested that Pic Pedraforca was a young landform generated in the context of post-Miocene neotectonics rather than an exhumed Paleogene landform. Note that a similar hypothesis has been put forward for Pic de Bugarach, an equally conspicuous, thrust-related landform at the northern front of the Pyrenees (see Itinerary 5). Observations relating to post-Miocene compressional tectonics, whether in Cerdagne (Itinerary 8) or in Roussillon at the base of the Albères massif (Itinerary 2), would accredit this interpretation.

Gósol (11 km return) is an optional destination for examining the west face of the Pedraforca massif and the unconformable outcrops of conglomerate resting on Serrat del Puig (1499 m), SW of the road. Close inspection of the Roca Roja N conglomerates (Fig. 14.9B) requires returning towards Saldes, turning off before the village onto the road leading to the Lluis Estassens mountain refuge, and following the long forest track that rises up to Coll d'El Colell (18 km return).

The main itinerary from Coll de la Trapa otherwise aims for the SE via a small mountain road leading up to Coll de Pradell (1732 m). Good exposures of Garumnian red beds (probably early Garumnian) can be seen to the left of the road. The

Fig. 14.10 The Berga conglomerates (Priabonian to Oligocene). A General view of the depositional series in the Llobregat valley, here vertically upturned (see dashed lines in reservoir section) or even overturned beneath the Vallfogona Thrust. Dips decrease progressively to the south. Note likely occurrence of a former pediment surface (P1 or P2), now substantially dissected, cross-cutting the upturned fan sequences (here lower and mid-level Priabonian alluvial fan deposits). **B** Bed of mid-level conglomerate, containing abundant granite clasts and displaying lenses of laminated sand. **C** Bed of upper-level conglomerate (early Oligocene age) on the Sant Andreu de la Serreta scarp below the road (BV 4241); note the granite pebbles

north limb of the Ensija anticline rises by at least 600 m to the south of the road. Its Campanian–Maastrichtian marine limestone structure has been exhumed from overlying Garumnian beds. East of Coll de Pradell, the road enters an elongated vale overlooked on one side by the previously encountered Danian limestone homocline (southern edge of the Vallcebre perched syncline), and on the other by the Ensija anticline. The floor of the vale contains a number of disused lignite quarries in the grey basal series of the Garumnian sequence, with the younger red series lying above it (Rosell et al. 2001; Marmi et al. 2014; Gómez-Gras et al. 2016). The K/T boundary is situated either near the top of these red deposits or at the base of the Vallcebre limestone rock face. This area hosts dinosaur remains, with exceptionally well-preserved footprint trails observable on steeply dipping slabs at Fumanya quarry (a geosite is open to visitors). Immediately after the quarry, the road joins the BV4025. Follow it as far as Figols, all the way through outcrops of early Garumnian. The road thereafter descends into the Llobregat gorge (cut in Upper Cretaceous marine limestone) and joins up with the E09-C16 highway. Follow signposts to Berga.

At the village of Cercs, the road cuts through a Garumnian-cored syncline, overthrust in the north and east by imbricates from the lower Pedraforca nappe. The village hilltop is capped by a small, unconformable deposit of Bartonian–Priabonian conglomerates. Exit the highway after the tunnel, 2 km after Cercs, and turn right onto the C1411z. This is the old Berga road, where stopping will be easier than along the main highway. The Pedraforca thrust front is exposed in the ravine below the south face of the spur penetrated by the tunnel (presence of associated Keuper evaporites). Outcrops of Lutetian marine marls, which belong to the underlying Cadí nappe, can be seen on the far side of the ravine along the road. At the next rocky spur, 200 m further south, the road reaches the Cadí nappe front; this is effectively the South-Pyrenean Frontal Thrust, also known as Vallfogona Thrust. In map view, the leading edges of the Cadí and Pedraforca nappes are thus closely spaced but still distinguishable. The Cadí nappe front is, furthermore, shielded from view by thick Quaternary slope breccia. The road thereafter continues through outcrops of the 1500-m-thick lower and middle Berga conglomerate sequence, which in this area has been folded over beneath the thrust and exhibits steeply upturned beds as a result. The Bartonian to early Priabonian marine series beneath the conglomerate sequence lacks outcrops in the vicinity. Looking east, you will notice a succession of bold conglomerate ridges bordering the eastern arm of La Baells reservoir (Fig. 14.10A). The lower levels of the Berga conglomerates (part of the Solsona sequence) are mid-Priabonian and consist of alternating sandstone and red mudstone. Exposures can be observed along the road for the next 2 km. The middle levels are late Priabonian and consist of much coarser conglomerate (Fig. 14.10B).

These mid-sequence units become subsequently prevalent all the way to Berga. Stop at the large car park on the left just after having cut through the conglomerate ridge that overlooks the town (C1411z–BV4241 crossroads).

Stop 5. Berga: contact between the orogenic wedge and the Ebro Basin
Several stops are recommended in order to appreciate the full extent and significance of the Berga conglomerates and their Paleogene depositional context. Here we begin with the upper beds of the middle Berga sequence, which can be inspected by walking back north along the same road to the main ridge overlooking the town. From exposures in the road embankment, you will note that clast sizes increase up-sequence (up to 30–40 cm, 50 cm in the case of one granite cobble), with inputs from Paleozoic outcrops attaining 60% (various granites, hornfels, psammitic schist, quartzite, black jasper, Devonian griotte limestone). Consistent with the upper conglomerate beds previously described at Serrat Negre (**Stop 3**), provenance analysis of the clast population thus indicates that drainage catchments at that time were quarrying outcrops from the southern half of the Axial Zone, i.e., around the latitude of the modern Cerdagne Basin. A total absence of debris from L'Hospitalet gneiss, situated further north, also provides clues about the palaeogeography of the Paleogene Pyrenees. Additional inputs include local grey limestone and Upper Cretaceous yellowish sandstone.

The conglomerate strata forming the razorback ridge north of Berga have been sheared into two tectonic units and thrust southward over the upper conglomerate sequence (of estimated Rupelian age). Near the top of the Priabonian middle sequence, a thin but continuous layer of angular limestone debris (1–5 m thick, detected over an area of 75 km^2) containing only rare rolled Paleozoic components (not visible here) has been attributed to a single, large-magnitude sedimentological event reminiscent of a megaflood outburst caused by the rupture of a natural dam in an alpine palaeovalley (Mató et al. 1994; Tallada et al. 2017).

Good views of the Berga basin from the south side of the road reveal a town situated at an elevation of 650–700 m in a narrow syncline filled with upper conglomerate, and closed off to the south by a broad, low-relief anticline (Fig. 14.11). These are foreland folds, in the present case a ramp anticline generated by a blind thrust connecting at depth to the Vallfogona Thrust. Initially rising obliquely from these deeper levels, the thrust plane is then guided at shallower levels by the Cardona evaporites (Priabonian age), which document definitive withdrawal of the Eocene sea and closure of the Ebro basin. The upper conglomerate unit (Oligocene age) is 1000–1500 m thick and consists of at least two upward-coarsening sequences. Over to the west, a thick bed of conglomerate

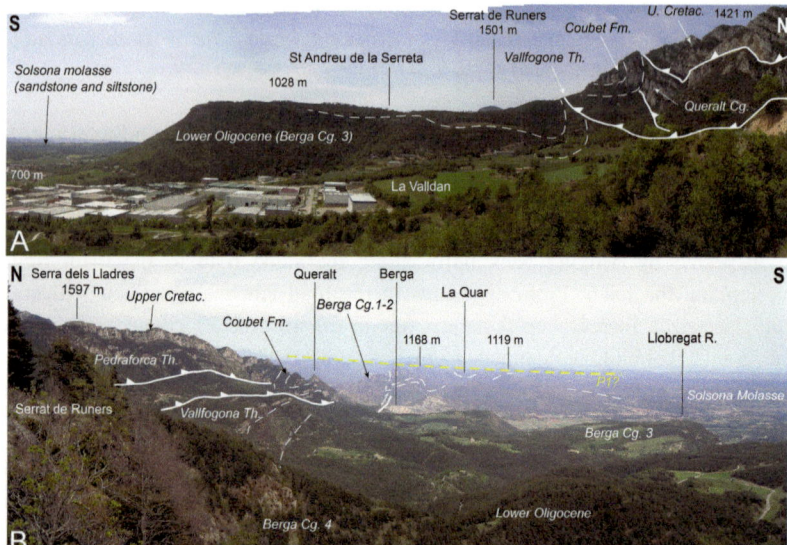

Fig. 14.11   Berga mountain front and contact with Paleogene piedmont. **A** Looking west from the BV4241 to Sant Llorenç, 1 km west of the junction to Peguera–Queralt. Note how the Pedraforca Thrust, just below the Queralt conglomerates, advanced over the Vallfogona marls and lodged itself beneath the Coubet Fm. The Vallfogona Thrust has strongly upturned the upper alluvial series, generating in the process a sharply asymmetric synclinal fold in these Oligocene clastic beds. The lowermost piedmont sequence (Berga Cg. 3) is capped by a slab of conglomerate forming the Sant Andreu de la Serreta plateau; Serrat de Runers, in the background, coincides with the top of the next alluvial sequence in the stratigraphy: Berga Cg. 4. **B** Looking east from the summit of Serrat de Runers. The thrust front was initially buried by the overfilled clastic basin but later exhumed as a result of post-orogenic denudation. In the middle ground, note the Queralt razorback ridge and the frontal folds affecting the conglomerates beneath the Vallfogona Thrust: La Quar syncline, Berga anticline. An hypothetical piedmont erosion surface (generation P1 or P2) may cross-cut at least some of those structures

(Figs. 14.10C, 14.11A) forms the Sant Andreu de la Serreta–Capolat ridge (elevation: ~ 1050 m). Its rock face overlooks a depression hosting softer sandstone and pelitic outcrops (site of La Valldan industrial park). The backslope of Capolat ridge also cradles a perched syncline (Fig. 14.11B), which itself contains another sequence of younger pelitic sandstone and conglomerate deposits forming the topographic mass of Serra dels Tossals–Serrat de Runers (1527 m, partly hidden from view but encountered later). The last recorded tectonic deformation event occurred after emplacement of this upper sequence of Oligocene conglomerate, although the precise age of its youngest beds is unknown. A Chattian or Aquitanian tectonic event is feasible. The modern landscape, however, owes little to this tectonic legacy and is predominantly a response to post-orogenic differential erosion. Denudation has removed thicknesses of conglomerate locally attaining 1000 m (altitude of the Llobregat river channel at Berga: 470 m; top of conglomerate outcrops: 1500 m), and exhumed the previously buried mountain-front structures as a result (Fig. 14.11B). This allows a more detailed forensic analysis of syntectonic structures preserved in the rock record than was possible in the less vigorously uplifted retro-wedge environment encountered in Itinerary 5. The structural geomorphology of the area was described long ago (Birot 1937), its most prominent feature being the asymmetric anticlinal valley that extends from Berga to Solsona. Note the bold double homocline on its north side, lacking opposite-facing counterparts other than a succession of much thinner and more discontinuous scarps reaching elevations of ~800 m south of Berga, and ~1000 m further west. The comparatively subdued morphology on the south flank of the anticlinal valley is explained by a transition along strike from the coarse-textured, strongly cemented proximal facies (Berga conglomerate) to the soft beds of the distal facies (Solsona molasse), which consist of thick interlayers of marl between sandstone beds.

Optional excursion to Sant Jaume de Frontanya (56 km return)

This itinerary from Berga offers an exploration of the Eocene stratigraphy from the Lower Lutetian marine marls to the conglomerate fill sequences of the Ripoll syncline, which rest unconformably on the Pedraforca nappe and whose upper beds were viewed from at **Stop 3**. The mammalian fossil sites of Sant Jaume de Frontanya are also encountered along the way. Drive back north the same was as you came into Berga before **Stop 5** (see Box 14.1) and follow the C26 towards Ripoll. The road crosses La Baells reservoir over a bridge and skirts along the Vallfogona Thrust, cutting across it at several places—either south of it through red beds belonging to the lower Berga units, or north of it through blueish grey marine marls (Vallfogona Fm., Lower Lutetian). After Vilada, the lower and middle Berga units (Priabonian), still conspicuous in the landscape to the south (ridges of grey conglomerate), rest at first on the continental Bellmunt Fm. (angular unconformity), then on Lower Bartonian marine to deltaic beds. This conformation is not immediately apparent because all of these stratigraphic units are steeply overturned beneath the Vallfogona Thrust; thus the dip of the overturned Lutetian–Bartonian beds is to the north, whereas the Priabonian beds dip south (Fig. 14.5B). At Borredà, take the BV4656, which hosts the best exposures of marine and continental Eocene strata in the region, all age-calibrated by magnetostratigraphic and palaeontological data (e.g., Burbank et al. 1992a; Busquet et al. 1992; Ramos et al. 2002). The road cuts through the Vallfogona Thrust after ~1.8 km of travel from the road junction, and offers continuous roadside exposures of thin-bedded, grey and ochre flyschoid marls

(Vallfogona Fm.), also containing thicker sandstone beds—one of which forms the neighbouring ridge (elevation: ~1100 m). The Coubet–Can Bernat Fm. (alternating beds of red sandstone and grey or beige mudstone, compatible with a coastal environment) starts to appear after 4 km. From 5 km onwards, you will encounter the fluvial Bellmunt Formation, displaying its characteristic succession of red sandstone and mudstone beds with interlayers of grey palustrine deposits. Fully continental conglomerate beds begin to appear after ~ 6 km; the SJF mammalian deposits were collected from this section of the traverse (Busquets et al. 1992c; Bonilla-Salomon et al. 2016). The road eventually reaches a farmland landscape on a plateau extending at the base of homoclinal scarps carved out of massive conglomerates. Outcrops of these late Bartonian, and possibly Priabonian, conglomerates (seen at **Stop 3**) occur at elevations of up to 1500 m and are folded into a syncline.

Sant Jaume de Frontanya hosts an eleventh century Romanesque church. It is one of the finest in Catalonia, displaying three apses with typical decorative blind arches (Lombard bands) and a dome above the intersection between transept and nave. Return to Berga following the same route in reverse.

Queralt and Rasos de Peguera (38 km return)

This itinerary from Berga explores the leading edge of the Pedraforca nappe, particularly the razorback ridge hosting Queralt sanctuary (Solé Sugrañes and Clavell 1973). Queralt also affords the best views of the contact between the South-Pyrenean Zone and the foreland basin. The summit to the north (Peguera, 2050 m) provides additional clues about the early stages of landscape evolution after tectonic deformation ceased in late Oligocene time.

Following the CV4241 towards Sant Llorenç de Morunys, a short drive (4 km each way) offers good exposures of the basal nappe imbricates. For the first 1.5 km you begin by driving through continuous exposures of vertically upturned lower- and mid-level Berga conglomerates. Similar-looking beds of fine-textured conglomerate and pink or red mudstone reappear after ~ 3 km, forming a rocky spur at the junction with the road to Queralt (spot elevation: 921 m). These beds rest on an underlying occurrence of vertically-dipping marls and thin-bedded marly limestone, both part of the marine Vallfogona Fm. (early Lutetian) and thus belonging to the Cadí nappe. The abrupt geological boundary between the two stratigraphic units is a tectonic discontinuity, probably the Vallfogona Thrust. The marine marls transition laterally to massive conglomerates (not visible from the roadside). These and the Coubet Fm. drape the leading edge of the Pedraforca nappe, and are thus coeval with its definitive standstill (Figs. 14.2, 14.11A).

Turn back to the initial starting point (road junction to Queralt) and take the BV4242, which cuts through outcrops of red strata (perhaps Priabonian), followed by lighter-toned marly limestone (Vallfogona or Coubet Fm.). The main bedrock ridge is cut

*by a short gorge. The road enters it and the gorge walls reveal 250-m-thick over-
or upturned beds of grey conglomerate known as 'Queralt breccia' (see Part I,
Figs. 1.2 and 2.3). The road subsequently rises through a wooded valley cut in
Cuisian to Ilerdian marine marls and limestones, themselves resting on the upper
Garumnian beds and thereby indicating that the stratigraphy in this area has been
entirely overturned. Leave your vehicle at the sanctuary's parking area and head
for the viewing point 400 m to the west along the trail that follows the southern
edge of the ridgetop.*

Stop 6. Mare de Déu de Queralt

Queralt means 'elevated rock' (for other place names with similar etymological
roots, see Itinerary 8). The conglomerate beds beneath your feet are thick and
nearly vertical. The pebbles are poorly rolled, poorly sorted, and dominated by
alveoline limestone from a local source despite some quartz and clasts of Paleozoic. The beds here dip north, but with fanning dips to the south. The sequence
is marine because clay interlayers sampled in the lowermost 60 m from the base
contain large foraminifera diagnostic of the middle and late Cuisian, and perhaps
earliest Lutetian (Garrido-Megías 1972; Solé Sugrañes and Clavell 1973).

The origin and palaeogeography of the conglomerates, which are possible
equivalents of the lower Palassou sequence (encountered in Itinerary 5) on the
north side of the range (Calvet et al. 2021), have long been controversial. According to Garrido-Megías (1972), the configuration resulted from a syntectonic
unconformity burying the nappe front at an early stage of its southward advance.
Solé Sugrañes and Clavell (1973), however, considered instead that we were
here in the presence of a submarine fan emplaced on the continental slope, conformable with the Cuisian and Ilerdian strata beneath it, and situated on the edge
of the shelf environment (Fig. 14.6) in which the Cadí series was being deposited
at the same time (Solé Sugrañes 1980). The conglomerates, in that case, belong
fully to the sedimentary sequence of the lower Pedraforca unit. They rest over
the thin marine Lower Eocene series of the Pedraforca nappe, which itself originally was part of the cover sequence previously resting on the basement of the
southern Axial Zone. The clastic fan likely documents the early stages of when
Pedraforca nappe advanced for the first time into a marine environment, a palaeogeographic event also heralded a little earlier in the stratigraphy by the deltaic
red beds of the Corones Fm. Modern maps support this interpretation by showing
that these conglomerates are themselves buried by a subsequent, unconformable
clastic sequence, which is either the base of the Bellmunt Fm. or the Coubet Fm.,
resting on the Vallfogona Fm. (the latter previously observed before the drive up
to Queralt; Fig. 14.11A; Mató et al. 1994; Martínez et al. 1996; Tallada et al.

2017). Emplacement of the Pedraforca nappe over the Cadí nappe would overall have occurred within a very short interval of geological time, chiefly during early Lutetian time (~48–44 Ma).

The view from Queralt across the foreland basin also displays the serrated profile of Montserrat (~60 km away). Its conglomerate mass is homologous with the Bellmunt–Milany–Berga sequence but was produced by the Catalan Coastal Range, not by the Pyrenees. The lower-lying area between these two landmarks is an expanse of sandstone and mudstone outcrops (Solsona molasse sourced by the Pyrenees, Artès molasse sourced by the Catalan Ranges), gently folded in the foreground by late-orogenic crumpling over the blind thrust and forming the Quar synform and adjacent Berga antiform. These folds, including the tighter folds carved into successions of massive bedrock ridges closer to the Vallfogona Thrust in the east (Fig. 14.11B), display accordant summits (uniform elevation ~ 1000 m). While it is tempting to speculate that an imaginary drape tangential to these summits once corresponded to erosion surfaces P1 or P2—which must have existed in this region as they do extensively elsewhere in the Pyrenees (Calvet et al. 2021)—these are all structural landforms and lack conspicuous vestiges of planar topography (Fig. 14.10A). The landscape thus predominantly evidences a process of deep dissection by rivers, mostly after ~12 Ma (Calvet et al. 2021).

Looking west, note the previously encountered Capolat homocline, carved out of upper Berga conglomerate. Cradled by the plinth of Capolat beds (compressed into an asymmetric, perched syncline), note also above it the stack of sandstone benches and uppermost conglomerate levels rising to elevations of 1500 m (Serrat de Runers). Limestone homoclines such as Serra dels Lladres (1597 m), carved out of the overturned frontal fold in outcrops of Maastrichtian, Ilerdian, or Queralt breccia, rise discretely above the elevated floor of conglomerate fill (Fig. 14.11B). The topographic boundary between the mountain range and its piedmont thus owes little to the tectonic compression that crumpled the Paleogene conglomerate sequences in late Oligocene time, but rather to subsequent uplift and fluvial dissection into these complex older geological structures, at one time mostly drowned by successions of range-front alluvial fans fed by the eroding Hercynian basement and by the various nappes of post-Paleozoic cover rocks initially encasing the Axial Zone.

Leaving Queralt, join the BV4243 and turn left towards Espinalbet. The itinerary follows the axis of a syncline that soon widens to basin proportions. It is cored by Ilerdian marls and limestones, and is lodged between the frontal and Peguera imbricates of the Pedraforca nappe. Near the 1600 m elevation contour, the road reaches the base of these Peguera imbricates, which rest on a bed of Keuper evaporites and Lower Jurassic marly limestone. Just before reaching the Rasos de Peguera refuge at 1800 m, the road cuts through an outcrop of conglomerate containing locally supplied limestone clasts and reddish-brown mudstone beds (Fig. 14.12D, E). *The outcrop forms a syncline, widening to the west (Cap de la Serra). These conglomerate beds correspond to the Bellmunt Fm. (mid-Lutetian); further west, they overlie the previously encountered Coubet Fm. in an area where both have buried the Pedraforca nappe front* (Fig. 14.13D). *Some thin occurrences of these formations are found sandwiched between the Peguera imbricates, but it is likely that they previously covered and buried the nappe units much more extensively because vestigial outcrops also occur at the summit* (Fig. 14.12C) *as well as on the northern backslope of the massif* (Puig et al. 2020). *Leave your vehicle at the ski resort car park and walk to the pistes ascending towards Rasos de Baix to the ESE (2058 m; 40 min return walk).*

Stop 7. Rasos de Peguera, vestiges of an elevated erosion surface
The out-of-sequence imbricates collectively constitute the Pedraforca lower nappe (frontal unit, Peguera units, Ensija unit). They were emplaced progressively from south to north during middle and late Eocene time (Fig. 14.5A). The chronology is attested by the diachronous succession of syntectonic conglomerates resting unconformably on each of these imbricate units (Martínez et al. 1996, 2020; Puig et al. 2020). The conglomerate sequence is particularly thick in the topographic low situated between Serra de Peguera in the east, Serra d'Ensija to the north, and Serra del Verd to the WNW, and hosting the Aigua de Valls and Aigua de Llinars streams. The basal conglomerates rest on the Lower Lutetian marine beds, the Campdevànol–Vallfogona Fm. (bearing, at the top, the Beuda gypsum, visible from **Stop 13** at Serra de Guixers), and the Coubet Fm. A marine or littoral intercalation consisting of Upper Lutetian marls occurs within the conglomerate mass 600–700 m above the sequence base. A younger conglomerate sequence (upper series), at least 800 m thick and containing > 50% debris from the Axial Zone, rests unconformably over much of the previously mentioned stratigraphy. These upper conglomerates occur all the way to the top of Serra d'Ensija. Their age is considered either Bartonian–Priabonian (geological sheet: Gósol), or Priabonian–Rupelian (sheet: Sant Llorenç de Morunys). It remains difficult to be more precise because these coarse-textured deposits lack index fossils, and merely speculating about links between conglomerate units and tectonic imbricates can lead to circular reasoning.

Massive drowning of the Pedraforca unit by syntectonic alluvial-fan and valley-fill deposits has long been documented (e.g., Birot 1937). Such rapid

Fig. 14.12   Rasos de Peguera and Serra d'Ensija. **A** View from the parking area at Peguera ski resort. Four syntectonic conglomerate units (Lutetian to Priabonian) have buried the out-of-sequence imbricates of the Pedraforca nappe (see also Fig. 14.5A). The Serra d'Ensija was originally entirely buried beneath piedmont conglomerates (estimated age: Priabonian) (vestiges encountered up to an elevation of 2293 m). Note the vestiges of post-Oligocene range-top erosion surfaces on all the massifs in the landscape, probably generated in relation to this regionally raised base level. **B** Erosion-surface vestiges on Rasos de Peguera and Ensija. **C** Residual outcrop of limestone-rich Lutetian conglomerate preserved on Rasos de Baix. **D, E** Limestone-rich Lutetian conglomerate; this series belongs to the Bellmunt sequence and has been lightly compressed into a syncline (Peguera refuge, elevation: 1750 m)

aggradation implies flexural subsidence generating accommodation space in the foreland zone, but probably also a relatively low-relief orogenic structure, with the tectonic units of the evolving South-Pyrenean pro-wedge thereby tending to disappear beneath the mass of debris being pumped out of the southern Axial Zone by steep drainage catchments. The raised local base levels along the overfilled mountain front resulted in attenuating the relative relief of structural landforms in the pro-wedge, thereby facilitating the production of local range-top erosion surfaces on individual massifs rising above the floor of the conglomerate fill (note that such an interpretation of the landscape was initially ruled out by Birot 1937, who claimed that low-gradient topography on range tops in the area was structural rather than erosional; see Box 13.1 on the distinction between structural and erosional land surfaces. Geological maps, however, were nonexistent at the time, road access was limited, and even topographic maps of the area were imprecise).

The summit area of Peguera (Fig. 14.12B) actually consists of three oblate humps, each hosting 4 to 5 ha of erosional topography cross-cutting geological structures and sloping gently towards the south: El Pedro (2077 m), Rasos de Baix (2057), Pla de la Bassa (1990 m). At all of those locations, the land surface cuts across the 25°–50° dips to the north of the entire Mesozoic stratigraphy (Campanian to Maastrichtian, including frontal imbricates of Jurassic strata), and likewise cuts across the unconformable outcrops of conglomerate lying on Rasos de Baix and at Pla de la Bassa (elevation: ~2000 m). Exposures can be inspected at close quarters as you walk up towards to Rasos de Baix (42°08'19"N, 0°46'19"E; Fig. 14.12C). The view to the north (Fig. 14.12A) shows the Sierra d'Ensija, also eroded at the top (Pla d'Ensija, ~2280 m) and also cross-cutting an anticline in Campanian–Maastrichtian and its conglomerate envelope, preserved at the summit. In the distance, note similar summit bevels on the Serra del Cadí to the north, and on Serra del Verd and Port del Comte to the west (Fig. 14.13D). These will benefit from closer investigation at forthcoming stops.

Return to Berga and take the BV4241 towards Sant Llorenç de Morunys. The road ascends the homocline overlooking Valldan industrial park. These massive outcrops of Oligocene conglomerate contain well-rolled pebbles (long axis ~ 10–30 cm), mostly consisting of Mesozoic and Cenozoic limestone, but with a few occurrences of quartz, quartzite, and occasionally granite (Fig. 14.10C). Below this massive bed, the Berga–Valldan basin mostly displays a stratigraphy of sandstone and red or ochre mudstone, the latter predominant.

Onward from Sant Andreu de la Serreta, the road follows the base of the uppermost Oligocene sequence, here mainly consisting of light-pink sandstone and mudstone, with sporadic beds of conglomerate (small pebbles) forming a plateau hosting arable farmland. The sequence dips 10°–20° to the north but you will find it tilted vertically as you approach the Vallfogona Thrust after a travel distance of ~300 m. The road thereafter follows the resulting asymmetric syncline all the way to the tunnel at Coll de Jouet. At the tunnel entrance, turn off onto the road to Capolat on your left. You will soon be reaching topographic benches in sandstone and mudstone (total thickness: ~200 m), hosting a series of hamlets along the base of this uppermost Oligocene clastic sequence. The main highlights of the sequence can be appreciated at Stop 8.

Stop 8. Late Oligocene conglomerates at Capolat

From the hamlet of Sant Martí de Capolat you will get good views of coarse-textured conglomerate strata forming a succession of uniformly thick beds (~10 m on average) below Serrat de Runers (spot elevation: 1501 m; Fig. 14.13A). A footpath up the east side of Serrat de Runers offers close-up views of the outcrops (90 min return). The entire sequence here is ~200 m thick, massively cemented, with pebble sizes increasing up sequence (30–40 cm). Granite clasts are fairly abundant at the base but become sparser at the top. The assemblage documents source areas situated in the Axial Zone, with lithological diversity comparable to clast compositions recorded among the Priabonian conglomerates of the middle units encountered at **Stop 5**. The summit offers wide-ranging views to the north of the previously described massifs of Peguera, Ensija, Pedraforca, Cadí (Fig. 14.13D), and farther to the east the Berga basin (Fig. 14.11B). The upper conglomerate sequence can also be inspected from the roadside, with good exposures in Runers ravine, 2.3 km after Capolat. Here the lowermost bed of the sequence is ~20 m thick, and the full thickness of the coarse-textured conglomerate beds is close to 300 m.

Continue to the hamlets of Trasserras and Taravil. At Taravil–Cal Bertran, turn off onto the narrow road to L'Espunyola and park at the first hairpin bend, 1.3 km after the junction. It is impossible to stop anywhere further along this road. Expect to walk 1 km and back to visit the outcrops.

Stop 9. Lower Oligocene conglomerates at Taravil

The road is entrenched in the bedrock. It cuts through a crag in the lower units of Oligocene conglomerate (Fig. 14.13B) and provides views to the south into the broad anticlinal valley eroded out of sandstone and yellow mudstone outcrops and hosting the Berga–Solsona road. The conglomerates here are up to 200 m thick, forming massive beds and containing well-rolled pebbles > 10 cm

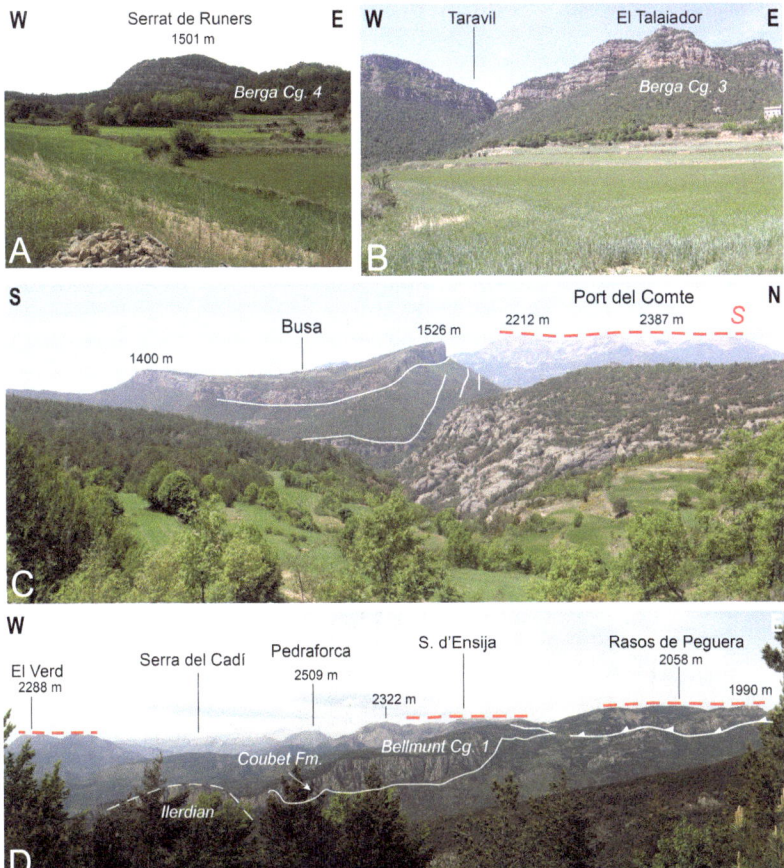

Fig. 14.13 **Oligocene Berga conglomerates: upper alluvial sequence around Capolat.**
A Berga Cg. 4 (uppermost unit) at Serrat de Runers, viewed from the road to Capolat. **B**
Berga Cg. 3, viewed from the foot of the gorge near Taravil. **C** Looking west from the road to
Vilella. Fanning dips in Oligocene conglomerates, here exposed in Aigua d'Ora gorge. Busa
perched syncline displays the youngest Oligocene units of the piedmont sequence. Port del
Comte and its range-top erosion surface in the background. **D** Cadí–Pedraforca massifs from
Serrat de Runers. Note widespread occurrence of low-relief range tops, the main exception
being the twin peaks of Pedraforca

in length. Granite clasts occur in greater proportions here than at sites previously encountered further east.

Return to Taravil and continue towards Vilella. The road skirts round Tossal de Vilella (upper conglomerate sequence, here containing abundant hornfels and granite pebbles as well as quartzite and Permo-Triassic red sandstone). Forest tracks from Taravil provide access to the Tossal. Stop 800 m before Vilella, where a belvedere provides good views to the west.

Stop 10. Vilella viewing point

The Aigua d'Ora gorge cuts across the strike of the Berga conglomerates, and the resulting exposures complement insights gained later from occurrences at Sant Llorenç de Morunys. The main feature here is the (slightly asymmetric) Busa syncline (Figs. 14.13A, 14.14C), folding the upper conglomerate sequence in the same way as at Serra dels Tossals–Serrat de Runers (seen at **Stop 8**), and thus recording the last stage of deformation of the piedmont deposits by the growing orogenic wedge during crustal convergence. This panoramic view provides possible support to the hypothesis that the top of the deformed piedmont sequence originally graded to the range-top erosion surface of Port del Comte massif (further examined at **Stop 15**). The view further suggests that uplift of the South-Pyrenean massifs relative to the clastic piedmont occurred along the strike of this monocline, and that uplift occurred after the youngest Oligocene beds (of imprecise age) were deposited, i.e., during the Neogene.

Return to Coll de Jouet and turn back onto the road to Sant Llorenç de Morunys. After the tunnel, the road follows a topographic furrow in marls situated along the boundary between the Vallfogona Thrust and razorback ridges in upturned Berga conglomerate (Fig. 14.14A). The thrust follows the base of the slope to the north (mountain side in marl and sandstone, locally with limestone strata). A large gypsum quarry (Beuda gypsum) can be accessed from Guixers (the turning to reach it is off to the right). Serra de Guixers (1439 m), to the north, is capped by the Bellmunt Fm. (mid- to late Lutetian), with exposures of Coubet Fm. at its base. The road itself is hosted by the middle to late Eocene marine series over which the Berga fan accumulations were emplaced. These greyish blue marls (Igualada marls) contain sandstone and reef limestone beds (Bartonian–early Priabonian) and are exposed in road embankments, particularly before reaching the bridge over the Cardener River. North of Sant Llorenç, the base of the marine series displays late Lutetian beds overturned and dipping 70°N beneath the Vallfogona Thrust (Puig et al. 2020). After the bridge over the Cardener reservoir, turn left towards Solsona. This itinerary cuts through the entire Berga continental sequence and reveals its

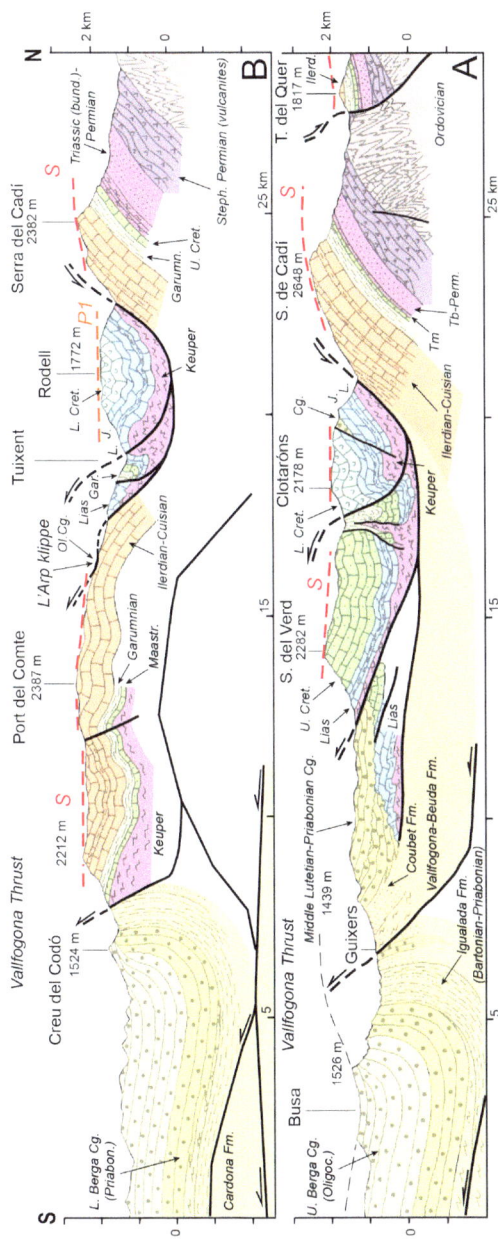

Fig. 14.14 **Two cross-sections through the Cadí and Pedraforca nappe units.** **A** Section through Serra del Verd. **B** Section through Port del Comte. See Fig. 14.2 for location. Stratigraphy and lithology—Steph.: Stephanian; Tb.: Lower Triassic, Buntsandstein; Tm.: Middle Triassic, limestone; L.: Lias (Lower Jurassic), limestone and marl; J.: Middle and Upper Jurassic, dolomitic limestone; L. Cret.: Lower Cretaceous (Urgonian facies), limestone; U. Cret.: Upper Cretaceous, limestone and sandstone; Gar.: Garumnian; Ilerd.: Ilerdian; Ol. Cg.: Oligocene, conglomerates. Note widespread occurrence of range-top erosion surfaces cross-cutting geological structures, possible link between surface S and the topmost Oligocene beds despite their subsequent deformation (in A), and the occurrence of two generations of erosion surface, S and P1 (in B). After Calvet (1996), redrawn and modified

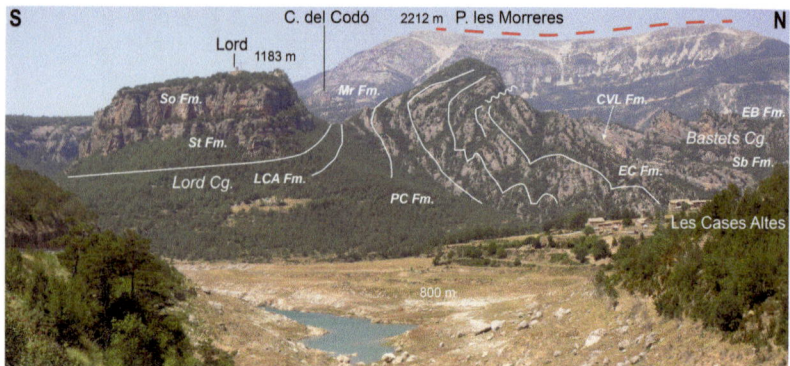

Fig. 14.15   Cardener section in Priabonian–Oligocene syntectonic conglomerates, look-ing west. Background: Port del Comte thrust front and range-top erosion surface; note very small cirques along the range front. Bastets and Lord stratigraphic 'members' sensu Riba (1976). Formations after Ford et al. (1997) and Williams et al. (1998). Abbreviations of geological formations (Fm.)—EB Fm.: El Bastet; Sb Fm.: Sobirana; CVL Fm.: Camps de Vall-Llonga; EC Fm.: El Castell; PC Fm.: Pont de les Cases; LCA Fm.: Les Cases Altes; St Fm.: Santuari; So Fm.: Sobol; Ll Fm.: Llobeta; Mr Fm.: Mirador. Note major erosional unconformity between PC Fm. and EC Fm

well-known succession of internal unconformities (Riba 1973, 1976; Ford et al. 1997; Williams et al. 1998).

Stop 11. Section through Cardener gorge
Follow the road to Llosa del Caval hydroelectric dam (13 km return, with sev-eral stopping opportunities along the way; Figs. 14.14 and 14.15). The top of the marine series crops out 150 m to the south of the LV4241–C462 road junction (grey and yellow sandstone with grey siltstone beds). A thin bed of bioclastic limestone at the top correlates with biozone SBZ 18 (Puig et al. 2020), i.e., late Bartonian / early Priabonian (Vandenberghe et al. 2012). An early Priabonian age was estimated for the marl series (Solé Sugrañes 1972); a magnetostratigraphic sequence places chron 16n.2n at the base of the sandstone and marl directly overlying the marine beds, i.e., an age of exactly 36 Ma (mid-Priabonian; Car-rigan et al. 2016). The overlying continental sequence (2500 m thick) has been subdivided into 3 groups or 'members' (Riba 1976), internally displaying 11 for-mations (Ford et al. 1997; Williams et al. 1998). The road first cuts through the continental Priabonian (pink pelitic beds with increasingly thick interlayers of conglomerate strata; the four thickest beds of Bastets conglomerate can be

observed before the Vallonga tunnel and viaduct). Clast composition is varied but dominated by Mesozoic and Cenozoic limestones, apparently with no granites at this location. The Eocene–Oligocene boundary (inferred from magnetostratigraphic data; Carrigan et al. 2016) is situated just after Vallonga viaduct, but the Sant Llorenç geological sheet (Puig et al. 2020) places the Vallonga beds south of the viaduct in the late Priabonian. Between kilometre marks 19 and 17 (milestone indications as given on Catalonia's 1:25,000 scale topographic sheets), the west side the road provides good views (Fig. 14.15) of the fanning dips and the main angular unconformity at the base of the 'Lord member' (sensu Riba 1973, 1976). These strata (Cases Altes Fm.) appear to be slightly younger than the base of chron C12n, i.e., 31 Ma, but with some uncertainty in the position of the upper sample points (Carrigan et al. 2016). The uppermost sequence (Mirador Fm., formerly 'Busa member') is 300–400 m thick and the unconformity appears to continue northward along its base, but it has not been dated. Sedimentation rates of 278 m/Ma inferred from the full Oligocene sequence (Carrigan et al. 2016) would place the top of the extant stratigraphy around 28–30 Ma. From Cases Altes (kilometre mark 17.3) onwards, dips tilt progressively to the north. After the dam, the road reaches once again the base of the Oligocene, here displaying frequent granite clasts. The core of the anticline hosts outcrops of continental Priabonian.

Drive back towards Sant Llorenç, then take the LV4241 towards Coll de Jou. At the second hairpin bend, 1.5 km after village, turn left onto a small road signposted to Santuari de Lord, and try to park near the tunnel. Plan for a 4 km return walk (~1 hr).

Stop 12. Cross-section and panoramic view from Santuari de Lord
This walk takes you through exposures of the syntectonic unconformities, with opportunities for appreciating conglomerate clast composition and provenance. The tunnel cuts through the first set of beds of the Bastets 'member'; granite pebbles are initially uncommon (tunnel entrance) but become abundant at the tunnel exit and include 40 cm cobbles including schist, Paleozoic sandstone, and Permo-Triassic conglomerate (Fig. 14.16A). In the Lord 'member', granite clasts appear randomly and in small quantities, but inputs from Axial Zone lithologies are overall abundant. The summit at Lord sanctuary provides good views of the syntectonic unconformities and of the Cardener gorge (Fig. 14.17A, B).

Return to the main road and drive up to Coll de Jou, from where the south face of Port del Comte rises abruptly to the north. Approximately 100 m after col, take the small track (left off the road to Solsona) signposted to Tossal de la Creu del Codó.

Fig. 14.16 **Sedimentary petrology of the Priabonian and Oligocene conglomerates**. **A** Bastets 'member'; well-rolled high-energy fluvial facies, rich in granite clasts; exposure is at south exit of tunnel on road to Lord. **B** Busa 'member', Mirador Fm., on west face of Creu del Codó; debris-flow facies containing poorly rolled debris of local limestone, and containing numerous clasts > 0.5 m in diameter

Fig. 14.17 Syntectonic unconformities viewed from Lord hermitage. A Looking northwest; note Creu del Codó belvedere (Stop 13) and the scarp face of Pic de les Morreres displaying a small glacial cirque (GC). **B** Looking east towards the Cardener valley; Vallfogona Thrust in the background, overriding steeply overturned marine marls (70°N) of Bartonian–Priabonian age. For abbreviations of lithostratigraphic units, see Fig. 14.15

Leave your vehicle at the car park and walk 600 m eastward up a track (closed to vehicles) to the viewing point.

Stop 13. Creu del Codó viewing platform

This spectacular geosite displays a number of explanatory boards and panels, and provides a good opportunity for recapitulating the many observations accumulated since leaving the Berga area. Tossal de la Creu del Codó (spot elevation: 1524 m) is carved out of the youngest conglomerate unit (Mirador Fm.), which also forms the Serra de Busa perched syncline (1526 m) on the eastern horizon. Its equivalent to the west is Roca de Canalda, visible from the parking area or from the Tossal summit, but not from the viewing point (see also **Stop 14**). Near-horizontal, rust-coloured slabs of conglomerate occur up to elevations of ∼1700 m, and are wedged or crumpled beneath the frontal thrust where the advancing nappe has ploughed into them. Immediately around you at Creu del Codó, note the heterogeneity in clast sizes, with pebbles abraded but often poorly rounded, and consisting exclusively of Mesozoic and Cenozoic source rocks: abundant nummulitic limestone; grey, beige, and marbled limestone; yellow sandstone; rare schist and Paleozoic sandstone (small pebbles only); no granite (Fig. 14.16B).

The view to the east displays the syntectonic deformation patterns of the conglomerates around Lord and Busa, as well as foreland folds such as the Busa syncline and its successor anticline to the south (Fig. 14.18). To the NE, the Sant Llorenç furrow in marls is overlooked by the Serra de Guixers homocline (Bellmunt Fm.), with beds steeply upturned in the immediate vicinity of the Vallfogona Thrust. The white patch of the gypsum quarry (Beuda Fm.) helps to locate the relative position of the Cadí nappe and its early Lutetian stratigraphy. The Peguera, Ensija and El Verd massifs, with their respective range-top erosion surfaces, are also in full view (Fig. 14.19). This generation of range-top erosion surfaces was produced at a time of late- or post-orogenic relative base-level stability, in the context of an overfilled foreland and prior to deep reincision of the piedmont.

The Vallfogona Thrust passes through Coll de Jou and the mountain front of Port del Comte massif. The Cadí nappe thus ploughed into the conglomerate stratigraphy of the Paleogene apron of alluvial fans, and towers here by ∼700 m above its crumpled and eroded outcrops (see also **Stop 14** and associated illustrations). The frontal scarp of Port del Comte (Figs. 14.14B, 14.15, 14.17A) is carved out of the same Ilerdian and Cuisian limestone series that also forms the Serra del Cadí further north. Half-way up the scarp face, the limestones rest on Garumnian beds, with a thin layer of marine Upper Cretaceous immediately beneath them before reaching the thrust sole in Keuper evaporites. Unlike the

Fig. 14.18 **View from Creu del Codó looking east, highlighting the full sequence of syntectonic conglomerates and their unconformities exposed in the Cardener valley.** U: main erosional unconformity; CB Fm.: Casa Blanca Fm., equivalent to the lower Berga alluvial units (Priabonian). For other abbreviations, see Fig. 14.15

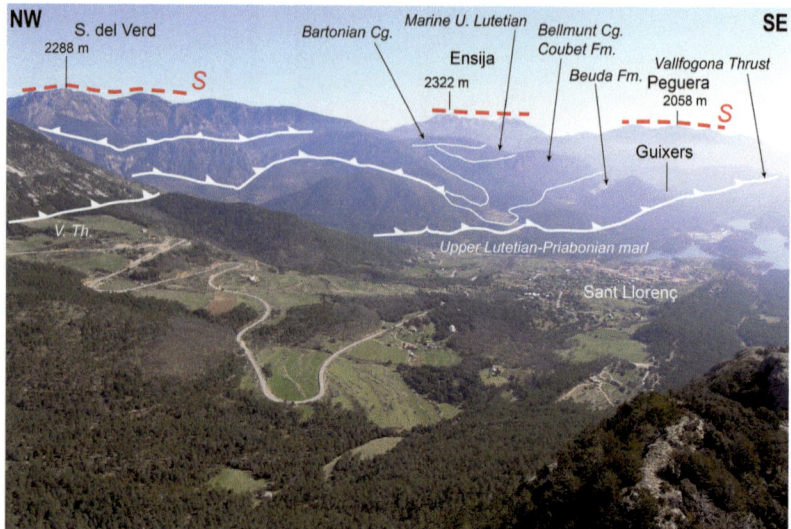

Fig. 14.19 View from Creu del Codó looking NE, displaying the multiple thrusts of the Pyrenean mountain front. Note Lutetian–Bartonian conglomerates overridden by the thrusts, and interlayer of Upper Lutetian marine marl sandwiched between two conglomerate units (not shown in cross-section of Fig. 14.14A). Conspicuous Beuda gypsum quarry, sunk into the top of the Vallfogona Fm., is on far right. Red dashes: range-top erosion surfaces. V. Th.: Vallfogona Thrust

site encountered further east around Berga, where post-tectonic fluvial incision into the syntectonic conglomerates has exhumed the underlying tectonic structures, here in the west the topography appears to be a more direct expression of late or post-Oligocene thrust tectonics. A component of post-tectonic exhumation nonetheless cannot be ruled out because the Mirador Fm. (apparently of Rupelian age) attains elevations of 1700 m at Roca de Canalda, and thick Chattian molasse deposits have been mapped immediately to the south, near Solsona. Former outcrops of these deposits (fine-textured molasse and/or proximal conglomerate) may thus have been thinned by denudation in this area near the Cadí nappe front.

The summit of Port del Comte massif is an extensive plateau (not seen from this viewing point). It is a vestige of range-top erosion surface, S. As in all massifs of the South-Pyrenean fold-and-thrust belt, glacial landforms are scarce or absent. A small, south-facing cirque glacier (Fig. 14.17A) benefiting from wind-blown snow accumulation (the summit plateau acting as a reservoir) under dominant northwesterly winds would explain the conspicuous scallop in the limestone range front beneath Puig de les Moreres (2212 m).

At Coll de Jou, take the LV4241b towards Solsona. The ridge-top itinerary cuts through a syncline filled with massive conglomerate beds belonging to the Lord unit, and reaches the base of the unit at kilometre mark 14. The frontal anticline here is drowned by conglomerate deposits forming extensive plateau topography. The road crosses the plateau in a straight line. Stop at its southern edge, at kilometre mark 10.

Stop 14. Pla de Riart and Neogene alluvial fan at Castell de Lladurs

This Neogene deposit has only been recently reported and partially mapped. Its low-gradient surface forms a plateau sloping away from the landforms in Oligocene conglomerate (~1100 m around Hostal del Cap del Pla to ~990 m around Castell de Lladurs). The fanhead is positioned near the base of the Port del Comte scarp face (Fig. 14.20A, B), where residual fanhead pediments occur around 1200 on the Serra de Canalda, Serra de Sant Martí-Cavallera and Serra d'Odèn, now reduced to dissected mesas and spurs. The fan's distal boundaries are poorly defined because of intense dissection in the claystone- and sandstone-rich Solsona molasse (Fig. 14.20C). The erosion surface, which cross-cuts the folded conglomerate and molasse sequence entirely and also displays in this area components of a well-preserved alluvial cover, is consistent with pediment generation P2—previously encountered around the highland basins of the Axial Zone and in the North-Pyrenean Zone (e.g., itineraries 4, 5, 6, 7, 8). The planar topography here hangs 400 m above the Cardener valley floor (a tributary of the Llobregat in the east), and 420 m above the Ribera Salada, a tributary of the Segre in the west. Landform preservation is partly positional (drainage divide between the Segre and Llobregat watersheds), and partly ensured by lithology (massive and highly resistant Paleogene conglomerates to the north). The alluvial fan deposits covering the pediment mostly consist of reworked pebbles quarried from the syn-tectonic outcrops of Oligocene conglomerate to the north. The alluvial formation can be observed in two quarries, one on Cavaloll mesa east of Castell de Lladurs, the other along the road to Solsona (Fig. 14.21). A stop at this roadside quarry reveals its main sedimentological features, which include heterogeneity in pebble sizes (Fig. 14.21A, C), an absence of clearly-defined bedding or palaeochannel units, with rare interlayers of sandy silt (Fig. 14.21B, D). The base of the formation, a ravinement surface, is exposed at Fonts de Lladurs around 920 m,

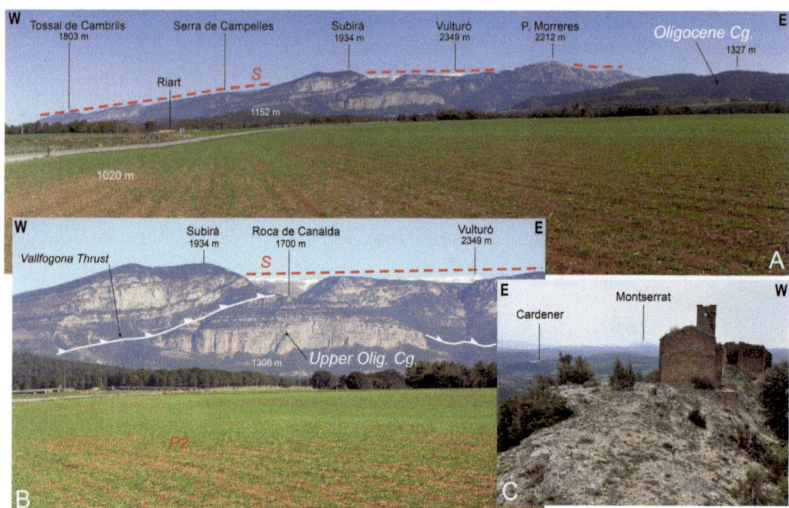

Fig. 14.20 Pla de Riart, a late Neogene pediment (generation P2). A General view of the plateau (foreground), looking north. This range-flank pediment forms a bench below topographic masses in Oligocene conglomerate and connects to the Port del Comte thrust front, far beneath the range-top vestiges of S. **B** Close-up view focused on the Vallfogona Thrust, which here has slightly warped the stack of massive conglomerate. The conglomerates of Roca de Canalada (probable equivalent to the Mirador Fm., Busa 'member') exclusively contain local input from the Port del Comte massif. **C** Southern termination of Pla de Riart. The strongly cemented Neogene conglomerate hosts the chapel and castle of Lladurs (Castell de Lladurs). Background: extensive lowland basins eroded out of the softer Paleogene molasse (see, e.g., Fig. 14.22), with Montserrat in the far distance

indicating the deposit could be up to ~70 m thick. Electron Spin Resonance dating, coupled with $^{26}Al/^{10}Be$ burial results (work in progress), indicate a minimum age of between 1 and 3 Ma for the base of the Lladurs formation.

*Continuing to Solsona through a landscape of clay- and sandstone-rich yellow molasse (Fig. 14.22), with dips to the S and SW, the scarcity of pebble beds highlights the rapid rate of longitudinal bedload grain-size decrease among the synorogenic alluvial fans. The geological map situates these molasse deposits chronostratigraphically in the Chattian, but the two nearest mammalian fossil sites of Tarrega and Calaf, respectively situated 50 and 30 km from this area, correspond to biozones MP 23 and MP 22, both in the Rupelian (*Barbera et al. 2001; Costa et al. 2011; *synthesis in* Calvet et al. 2021). *Onward from Solsona, 16 km*

Fig. 14.21   **Late Neogene alluvial deposit of Pla de Riart–Lladurs**. **A** High-energy, poorly sorted fluvial deposits, here strongly cemented at the top of the sequence beneath Castell de Lladurs. **B** Ochre and red silty-sand levels, 4 m thick, ca. 40 m below the fan surface. **C** Base of the Lladurs deposit, consisting of high-energy, unsorted fluvial debris (LV4241 roadside quarry). **D** General view of the quarry; ochre and red silts in B lie at the top of this uniform, pebble-rich unit; dated section is at the base

Fig. 14.22 Solsona molasse (sandstone and mudstone, estimated Chattian age). Exposure along road LV4241 at Costa del Bou, 3 km north of Solsona. **A** General view. Note small normal faults (downthrow to the NE), possibly a consequence of back-arc extensional tectonics in the western Mediterranean during the Neogene; however, a similar set of faults striking NW–SE, reported in Paleogene beds beneath Pla de Riart (IGME geological sheet of Sant Llorenç), has not been observed to offset the tread of Pla de Riart or its alluvial cover. **B** Detail of sandstone-filled palaeochannels

Fig. 14.23 Views from **Mirador de Serra Seca (north edge of the Oliana anticline)**. **A** The Oliana anticline and its envelope of synorogenic conglomerates (looking west). AS 1 to AS 5 (see also Fig. 14.2): 'alluvial systems' based on IGME geological sheet of Oliana; **1**: mid-Priabonian; **2**: late Priabonian; **3**: early Rupelian; **4**: late Rupelian; **5**: early Chattian. **B** Conglomerates of units AS 4 and AS 5, partly burying the thrust front (looking north). Note the ridges in white Ilerdian limestone drowned by the conglomerate and barely emerging from it

to the east, salt tectonics have produced a diapir at Cardona. The evaporites were generated ca. 36 Ma, at a time when the sea definitively retreated and the Ebro Basin became internally drained. Cardona is also a worthwhile cultural stop, with an eleventh century collegiate church and associated Romanesque buildings in the fortified citadel above the town.

Back at Solsona, take the C26 westward, then turn right after 3 km on the LV4011 towards Cirera and Cambrils. The ridge-and-ravine landscape in syntectonic Oligocene molasse is eventually succeeded by the coeval conglomerate belt at Montpol, here draping the Oliana frontal anticline (a large, breached anticline with the town of Oliana situated in the anticlinal valley). After Montpol, the road follows the ridgetop in Oligocene conglomerates. Stop after 6 km at the Serra Seca viewing tower (1234 m), which can be reached after a short walk.

Stop 15. Mirador de Serra Seca, and the Oliana foreland anticline

The viewing tower stands on massive syntectonic conglomerate beds, preserved here from post-orogenic erosion. The alluvial fan sequence contains, as elsewhere, internal unconformities. Around the outer flanks of the Oliana anticline, the aggradational sequence has been subdivided into six 'alluvial systems' resting on the Eocene marine marls (Igualada Fm.), only five of which are encountered east of the Segre River (Figs. 14.2, 14.23). The Oliana marine marls were originally age-bracketed as Bartonian (chron 18 to base of chron 17; Burbank et al. 1992b) but have been more recently revised to the Priabonian (top of chron 16n, i.e., 36 Ma; Alluvial System 4 belongs to chron 12r, i.e., 33–31 Ma; Costa et al. 2010, 2013). The conglomerate sequence at Serra Seca extends up the stratigraphy for a further 600 m above the level of chron 12r. By linear extrapolation of sedimentation rates calculated from the lower portion of the sequence (180 m/Ma; Burbank et al. 1992b), it is thereby possible that the top of Alluvial System 5 is of early Chattian age, as otherwise indicated by the Oliana geological map (Berástegui et al. 2017). System 5 contains a large number of pebbles sourced by Paleozoic outcrops of the Axial Zone, such as dark psammitic schist, quartzite, and hornfels. Granitoid abundance is extremely variable depending on the palaeochannel unit considered, with only rare occurrences around the viewing tower, but populating up to 80% of clast assemblages further upstream, including boulders 0.5–1 m in diameter, for example by the roadside near Cambrils (Fig. 14.24).

The Oliana homocline, carved out of Alluvial System 1, is conspicuous in the SW (Fig. 14.23A). Outcrops of systems 2 and 3 occur in the Móra depression and its eastern hillside, system 4 underpins the Serra dels Obacs escarpment, and system 5 caps all of the ridgetops to the north (e.g., Pla de las Guàrdies, 1402 m).

Fig. 14.24 Sedimentary petrology of upper conglomerate units (base of Alluvial System 5). Exposure situated between Serra Seca and Cambrils. Palaeochannels are rich in well-rolled granite clasts. **A** General view. **B** Close-up view

Note the white razorback ridges of Ilerdian limestone exposed in ravines at the base of Pla de las Guàrdies (Fig. 14.23B). They form an extension of the Cambrils syncline and were clearly drowned by the syntectonic conglomerate sequence. Scanning the horizon from north to west, note the scarp faces of Turb (1620 m) and Aubenç (1612 m) in Upper Cretaceous limestone, barely rising above the floor of conglomerate. They correspond to the thrust front of the South-Pyrenean Central Unit (SPCU, westward continuation of the lower Pedraforca nappe unit); as these outcrops continue westward, they become a lot more prominent in the landscape, such as at Serra de Montsec thrust-front scarp. The Aubenç ridgetop lacks planar land surfaces, but the fact that the ridge lines are accordant and cross-cut massive Campanian limestone as well as outcrops of Aren sandstone (Maastrichtian) could suggest a poorly preserved erosion surface in this area, rising above the overfilled piedmont base level and reincised more recently.

The view to the east (Fig. 14.25) displays the erosional surface crowning the summits of Port del Comte massif (Calvet et al. 2021). Tight folds in the Ilerdian to early Lutetian limestones, with 30°–60° dips, are clearly cross-cut by the planar land surface (generation S), particularly around Tossal de Cambrils and Prat Major (~1800 m). The plateau rises steadily towards the ENE, cross-cutting bedding patterns around the 2300 m summit area in the east. Synorogenic crustal denudation recorded by the conglomerate sequences was largely responsible for denudation of the nappe stack and for the sculpture of structural landforms. This scenario is neatly illustrated by the small tegument of conglomerates burying the Arp klippe (Lower Jurassic; Fig. 14.14B) on the north flank of Port del Comte massif (Martínez et al. 2020), and by the unconformable Alluvial System 3 draping the Tossal de Cambrils (Berástegui et al. 2017; Fig. 14.26A). However, the decisive erosional bevel in the landscape (range-top generation 'S') cross-cuts the unconformable and folded or overthrust conglomerate units as well as their underlying structures. Surface 'S' is thus clearly post-Oligocene. The conglomerates were initially very thick, as can be appreciated from the vestiges of Alluvial System 4 still resting on the southern flank of Port del Comte and abruptly folded over the Cambrils massif (Fig. 14.26A, B).

Continue on the road as far as Cambrils, which provides good views of the southern flank of Port del Comte and access to exposures of the granite-rich base of Alluvial System 5 (Fig. 14.25). After Cambrils you eventually join up with the L401. Turn left towards Llinars, continue through the village, and drive up to Coll de Boix (1250 m). At the col, turn left onto a small road that doubles back westward above Llinars and takes you the site of Cal Mestre (4208′50″N, 01°21′58″E).

Fig. 14.25 Views from Mirador de Serra Seca towards the east and Port del Comte. Unconformable conglomerate sequences have been stripped from the range front and exhumed the Vallfogona Thrust as a result. The dips and folded structures of Port del Comte are clearly bevelled by the range-top erosion surface. The late Neogene Pla de Riart (pediment P2) is also conspicuous on the skyline. See Fig. 14.23 for the ages of AS 1 to AS 5

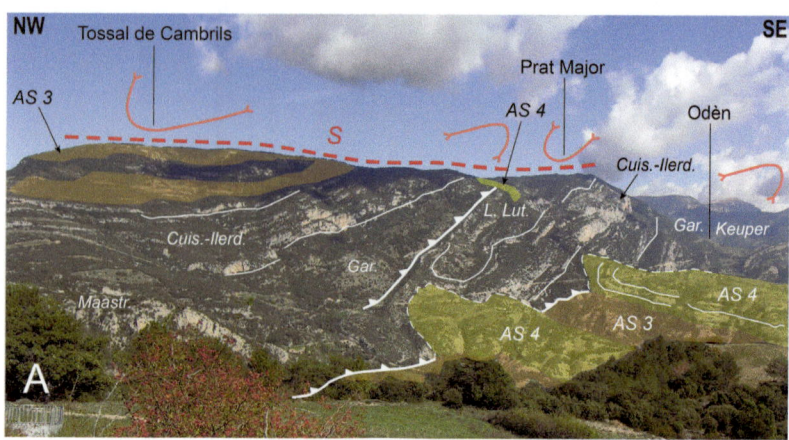

Fig. 14.26   Burial of the Port del Comte thrust front by Oligocene piedmont conglomerates. Age of alluvial systems AS 1 to AS 5: see caption for Fig. 14.23. **A** View from road linking Serra Seca to Cambrils. The conglomerates from unit AS 3 rest unconformably on the folds and thrusts in the massif, but are nonetheless locally sheared by the frontal thrust. Unit AS 4 forms a simple monocline, with evidence of onlap onto the folded structures. Range-top surface, S, cross-cuts the entire stack and is thus younger than the geological structures as well as the conglomerates. **B** Detail of the monocline from road L401 (Cambrils to Odén)

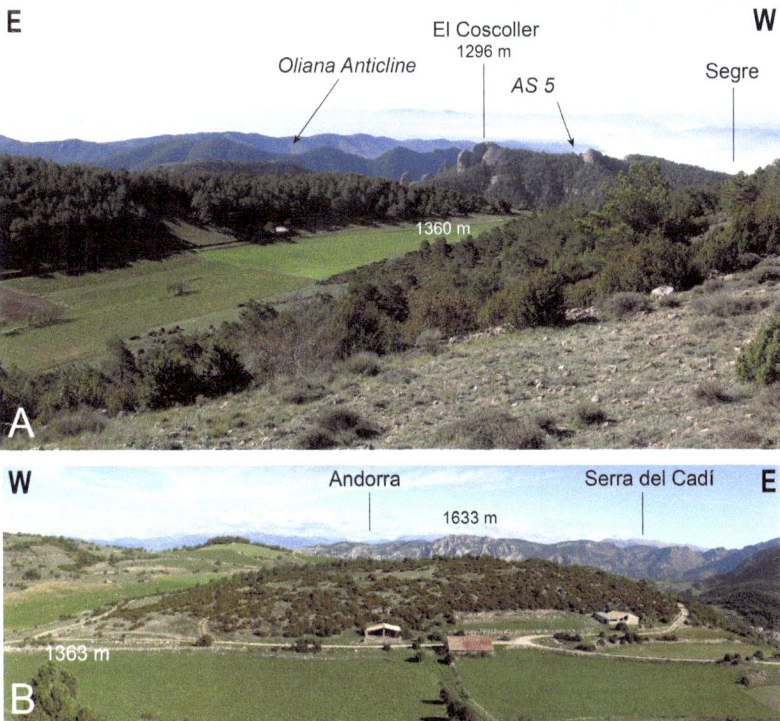

Fig. 14.27   Llinars: a suspended dry valley. A View of its lower segment; note bold erosional landforms in 'Alluvial System 5' conglomerates (see also Fig. 14.23B). **B** View of its upper segment (two tributary streams). On the horizon, to the north: high range around Andorra (Axial Zone) and Serra del Cadí

Stop 16. Hanging dry valley at Llinars

This 200-m-wide palaeovalley floor, also displaying a left-bank tributary, lies at an elevation of 1361 m, i.e., +870 m above the current Segre River channel (Fig. 14.27). Other dry-valley occurrences at lower elevations also occur in the region, e.g., through the eastern part of the Montsec thrust-front homocline (Calvet et al. 2021). Here, the ancient river flowed in from the NE and cut a shallow valley into the Campanian limestone, the massive nature of the rock perhaps explaining good landform preservation. The valley floor contains a bedload lag deposit consisting of well-rolled pebbles of local limestone but also including a

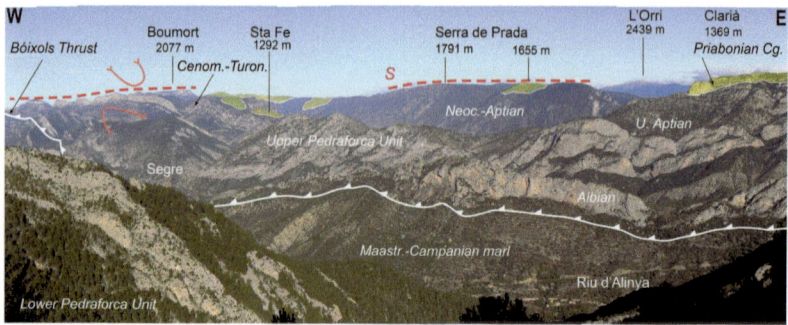

Fig. 14.28   View of the South-Pyrenean Central Unit from the road to Alinya. Note widespread distribution of range-top erosion surfaces cross-cutting the fold and thrust structures prominent on the west side of the Segre gorge (Boumort and Serra de Prada massifs). Numerous unconformable vestiges of Paleogene conglomerate beds are likewise distributed widely up to high elevations. L'Orri massif, which also displays a low-gradient range-top plateau, is part of the Axial Zone

large number of resistant, siliciclastic lithologies from the Axial Zone (dark sandstone, quartzite, micro-conglomerate, quartz). Most are probably reworked from the locally available Paleogene conglomerates capping the ridgetop marked on maps as Lo Pedró. Although there is currently little further evidence to support the hypothesis, this palaeovalley could perhaps record the first regional stages of post-orogenic incision into pediment generation P1, thus perhaps coinciding with generation P2 following similar landscape patterns previously observed in the Corbières massif (Itinerary 2), the Pays de Sault (Itinerary 5), around the Cerdagne Basin (itineraries 7 and 8), and at Pla de Riart (**Stop 14**).

Return to Coll de Boix and take the L401 towards Alinya as far as kilometre mark 15.5 (42°09′53.5″N, 01°23′56.5″E). The area offers good views of the Segre gorge and the South Pyrenean Central Unit–Pedraforca nappe.

Stop 17. Panoramic view of the Segre gorge
Over to the NW, note the succession of folds affecting the nappe units of the central Pyrenees, which extend westward from the Pedraforca area (Fig. 14.28): in the foreground is the lower Pedraforca unit, or Montsec nappe (Upper Cretaceous over Lower Jurassic), followed in the background by the upper Pedraforca unit, or Bóixols nappe (thick Jurassic and Lower Cretaceous sequence). Their tightly

folded stratigraphy has given rise to a population of accordant ridgelines extending ~1500–1600 m on either side of the Segre valley, rising to 1800–2000 m in the Boumort massif and among the massive limestones of the Serra de Prada, where planar summit landforms are extensive. As in the case of Port del Comte, these range-top plateaus are tentatively correlated with generation S (as in the Axial Zone), but only by default and with no corroborating evidence because limestone does not lend itself to thermochronological dating (compare with results presented in the context of itineraries 6, 7 and 8). Residual patches of Paleogene conglomerate are widespread in the landscape on both sides of the Segre. The IGCC regional map correlates them with the conglomerates of La Pobla de Segur, situated further west, and considers them to be of Bartonian and Priabonian age. Range-top surface, S, cross-cuts these syntectonic conglomerate beds as well as the underlying geological structures (Calvet et al. 2021). As a result, denudation has removed from this local depocentre most of the unconformable conglomerate deposits, assumed to have been initially continuous with other outcrops in the region. The N–S strike of the Segre valley is probably guided by a saddle between the comparatively more uplifted Boumort and Cadí–Pedraforca massifs, with the river perhaps having superimposed itself on the low-gradient erosion surface.

From here you can reach the Segre valley via Coll de Nargó, and then drive onward to La Seu d'Urgell up the Segre gorge. However, an alternative mountain route aiming east provides access to further conglomerate outcrops and underlying geological structures previously only seen from a distance. Return to Coll de Boix and take the L401, which follows the southern edge of Port del Comte along the Vallfogona Thrust. Between Cal Xinquet and Odèn (kilometre marks 22–24), the road cuts through several outcrops of 'system 4', here draped over the frontal thrust but forming an abrupt flexure and resting unconformably over the fault-bend fold with 30° dips to the south (Fig. 14.26B). Near Odèn, the road enters a large breach in the frontal anticline, with deep exposures down to the Keuper beds. The road then once again cuts across the thrust south of Subira limestone massif (kilometre mark 31). Roadside exposures, particularly around Sant Martí (kilometre mark 34) and below the topographic bench at La Borda (1450 m), show nearly horizontal or slightly north-dipping conglomerate beds containing small pebbles rich in lithologies from the Axial Zone (at places 100% from Paleozoic source rocks), such as psammitic schist, hornfels, and granite. Correlating these deposits with any specific alluvial unit is speculative (perhaps Alluvial System 3, maybe 4; Fig. 14.25), but the provenance analysis indicates quite clearly that the Port del Comte massif did not yet exist at the time. From the rocky spur at kilometres marks 35–36, you should get good views of Roca de Canalda and the rust-coloured breccia capping the Paleogene sequence between 1400 and 1700 m (Fig. 14.20B). The strata can be inspected close up above the road; exposures will show that clast composition is local limestone exclusively—particularly Eocene limestone, a strong indication that this deposit records the rise of Port del Comte as a prominent massif in the growing orogenic wedge. At kilometre mark 41, you finally reach Coll de Jou (i.e., previous **Stop 13***).*

At Coll de Jou, follow the signs to Port del Comte ski resort. The road follows the mountain side and offers great views of Sant Llorenç de Morunys and Serra del Verd. The top of El Verd (2282 m) consists of a plateau (~20 ha, another vestige of Pyrenean range-top surface, S), with the land surface cross-cutting the 20°N to NE dips of the Santonian to Maastrichtian limestone sequence. The road eventually reaches the C462. Follow it as far as Coll de Port (1660 m), where you should get fine views from a platform on the right side of the road.

Stop 18. Coll de Port, panoramic views of Tuixén basin

The col lies at the base of the lower Pedraforca nappe unit. From here you can also drive to the Arp klippe (Fig. 14.14B), which is an outlier of the lower nappe on the north flank of Port del Comte, by driving up the small road to the ski resort of Tuixén–La Vansa. The tectonic boundary between the lower and upper Pedraforca nappe units lies at the bottom of the valley, and the massifs rising above Tuixén to the north and east are carved out of the upper unit. The stratigraphy consists of thick reef limestone of Neocomian–Barremian age (Prada Fm.) overlying reef limestone of Upper Jurassic age and dark Dogger dolomites. The entire sequence has been compressed into steep or overturned synclines. The Serra del Cadí also looms large, closing off the view to the north and concealing the massifs of the Axial Zone. Its thick Ilerdian–Cuisian–early Lutetian limestone sequence, with 40°–50° dips to the south, is structurally homologous to Port del Comte and, at elevations of 2500–2600 m, towers over the Pedraforca nappe units.

These complex tectonic structures have been bevelled by a late- or post-orogenic erosion surface, and this area is one of the few in the South-Pyrenean Zone (another exception is the Montsec massif, to the southwest of here; see Calvet et al. 2021) where both generations S and P1, previously recognised in the Axial Zone and the North-Pyrenean Zone (see all previous itineraries), are distinguishable. Here the summit level, S, clearly bevels the steeply-dipping limestone stratigraphy on Serra del Cadí (Fig. 14.29B), as previously documented by Calvet (1996) and Ullastre and Masriera (2004, Fig. 7 therein). This range-top surface, also recognised at Sierra del Verd and Port del Comte, is itself warped

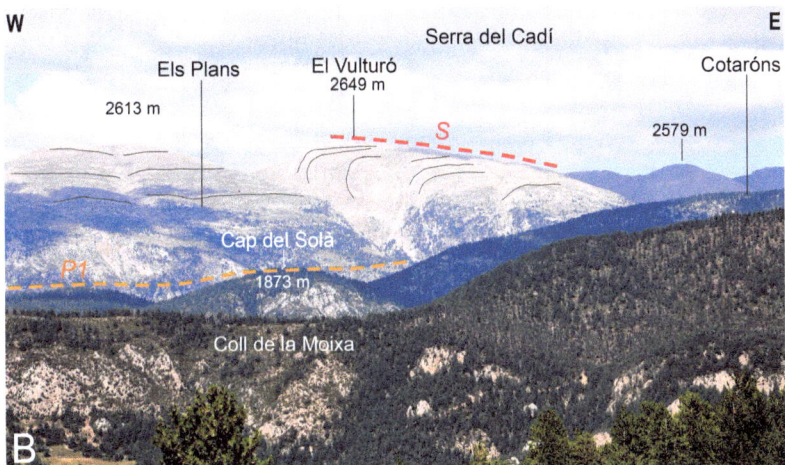

Fig. 14.29   Views to the north from Port del Comte massif. A Panorama at Coll de Port. Vestiges of pediment P1 are well preserved compared to other areas farther east. The pediment surface cross-cuts structures belonging to the upper Pedraforca nappe unit (Urgonian limestones at Pla de Rodell–Roca de Santaló). Note offset in the Serra del Cadí summit ridge generated by the Cornellana Fault. As a result, depending on the age of fault movements, El Goleró plateau can be interpreted either as an offset component of S, or as a vestige of P1. **B** Looking towards Serra del Cadí from Port del Comte ski resort. The topographic gradient of the mountain side is clearly more gentle than the steep dips in the bedrock, indicating that the south face of Serra del Cadí corresponds to a tilted vestige of surface S rather than to the dipslope of the Ilerdian–Cuisian limestone structure

into a synclinal structure by late-orogenic tectonic deformation, with dips of 10°–15° to the south on the Serra del Cadí (Fig. 14.14). A younger erosion surface, probably linked to generation P1 given its occurrence as a population of partial pediments, cross-cuts very tight folds in the upper Pedraforca nappe. Vestiges of it occur slightly above Coll de Port (1750–1850 m), at Pla de Prat to the east of Tuixén, and likewise at Pla de Rodell and Roca de Santaló (north of Tuixén, noted Tuixent on some maps and roadsigns; Fig. 14.29A, B). This generation P1 also displays conspicuous pyramidal peaks (residual inselbergs or monadnocks) such as Cadinell (2112 m) and Cotarons (2179 m). Another vestige of range-flank surface P1 occurs at El Goleró (2044 m), at the westernmost extremity of the Serra del Cadí, although this could also be a downfaulted vestige of S on the west side of the Cornellana Fault (Fig. 14.29A).

The abrupt western termination of the Serra del Cadí is enigmatic. One reason is probably structural because the Cadí nappe has been disrupted in this area by the Fórnols salt diapir, which strike N–S at the location where the Cenozoic limestone sequence breaks off. The entire system appears to plunge westward beneath Montsec de Tost, an imbricate of Paleozoic rocks considered to represent the eastern extremity of the Nogueres Zone. Neotectonic causes, however, cannot be ruled out. The impact of post-orogenic tectonics is tentatively supported by the N–S saddle followed by the Segre valley noted earlier at **Stop 17**, and by the N–S Cornellana Fault (Fig. 14.29A), which has vertically dropped the entire mass of the Cadí range, as well as the small Neogene Fórnols Basin, down to the west (Guérin-Desjardins and Latreille 1961; see also Gósol geological sheet: Martínez et al. 2020).

The road winds down into Tuixén basin, with views of Serra del Verd in the east and of the rock faces of La Roca de Santaló in Urgonian limestone. Tuixén basin is lodged in Lower Jurassic marls or highly fractured limestone and Keuper evaporites between the upper and lower Pedraforca nappe units. After Tuixén, follow the C462 towards La Seu d'Urgell.

At kilometre mark 53, you should reach a junction with the small road descending into the Vansa valley. If you follow it as far as Sorribes (12 km return), you should reach the outcrops of a Neogene fill sequence. Exposures start to appear after Molí de Fórnols, and the most valuable can be found along the road between Sorribes and Sant Pere de la Vansa (Fig. 14.30A). The entire sequence dips 25°–30° to the SE and displays a stratigraphy of alternating coarse clastic beds (poorly rolled, locally sourced pebbles and breccia in a red matrix) and beige lacustrine mudstone. Like its larger neighbouring analogues of Cerdagne and La Seu d'Urgell, this basin (named here La Vansa, Fig. 14.2) is probably of late Miocene age. If correct, a Miocene age

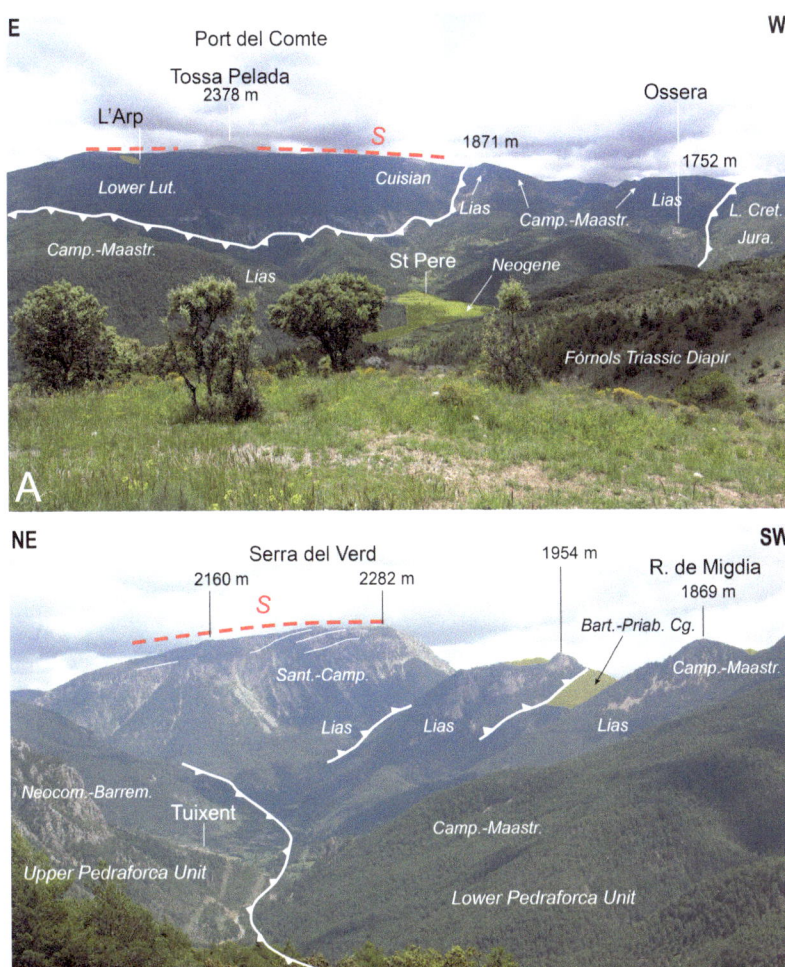

Fig. 14.30   Views to the south from the road to Fórnols. A Port del Comte. This massif corresponds to a window in the Pedraforca nappe (see also Figs. 14.2, 14.14B), exposing rocks from the Cadí nappe and cross-cut by erosion surface, S. Note position of the Arp klippe and its unconformable vestige of Paleogene conglomerate; note also the small Neogene basin of Sant Pere de la Vansa (see Fig. 14.2). **B** Serra del Verd; this massif is part of the lower Pedraforca nappe (see also Figs. 14.2, 14.14A), and is also cross-cut by range-top surface, S (not angle between S and the steeper dips of Upper Cretaceous limestones). Tuixent basin occurs at the boundary between the upper and lower Pedraforca nappe units

*for the Roca Roja conglomerates previously encountered north of Pic de Pedraforca (see **Stop 4** and Box 9.3;* Durand-Delga et al. 1989) *is likewise plausible.*

Drive back to the main road to La Seu, which rises up the south flank of Serra del Cadí via Cornellana and Fórnols. Between Fórnols and Coll de Bancs, views of the Serra del Cadí to the east show the acute angle between the erosional bevel of the range-top surface and dips in the Eocene limestones, as well as the range-flank vestige of P1 at Pla de Rodell. Views over to Serra del Verd (Fig. 14.30B) *and Port del Comte are also very clear. A small inselberg on the Port del Comte plateau hosts the highest elevations* (Fig. 14.30A).

The road follows the edge of the Fórnols diapir. At Adraén, it cuts through the Upper Cretaceous marine limestone of the Serra del Cadí, then enters an outcrop of Triassic strata and, at its base, the quartz-rich conglomerates and reddish purple mudstones of Permian to Triassic age. Stop at Mirador de la Trava, an excellent viewing point from a bend in the road (kilometre mark 72; 42°17′25″N, 01°29′17″E).

Stop 19. Panoramic views of La Seu Basin

The exceptionally broad view embraces 60 km of the southern Pyrenean mountain range from the Boumort massif in the WSW to the Campcardós massif in the ENE. In structural terms, most of the mountain range visible from here is in the Axial Zone, but it also shows part of the Nogueres Zone in the west (see Part I, Fig. 1.2 therein) and a portion of the South-Pyrenean Central Unit (SPCU, effectively a large klippe; Part I, Fig. 1.2), with the Serra de Prada and Serra de Boumort being part of it. In this area, the Permo-Triassic basin fill resting unconformably on the Hercynian basement is at its thickest, with excellent exposures along the road and in the local landscape (Fig. 14.31A). Outcrops of the Cadí nappe basal thrust occur only in a small window at Sort, in the valley of the Noguera Pallaresa, 30 km from here to the WNW.

Whereas vestiges of range-top surface, S, in the Axial Zone are widespread east of the Valira valley and La Seu Basin (Camp de Claror–Monturull, Tossa Plana, Campcardós), glacial landforms are particularly scarce (Fig. 14.31B). The erosion surfaces extend to the west on L'Orri massif (Lower Paleozoic schist, 2400 m), but mostly give way to serrated alpine peaks NW of Andorra (Coma Pedrosa: 2945 m; Saloria: 2789 m; Fig. 14.32). No evidence currently suggests that erosion surfaces ever had a chance to develop among the most elevated massifs of the Axial Zone in the central Pyrenees (Calvet et al. 2021). In the South-Pyrenean Zone, however, for example around L'Orri, summit surfaces appear to have entirely cross-cut the structures of the Nogueres nappe unit, forming accordant, low-relief summits around 1900 m a.s.l.—whether in Triassic cover

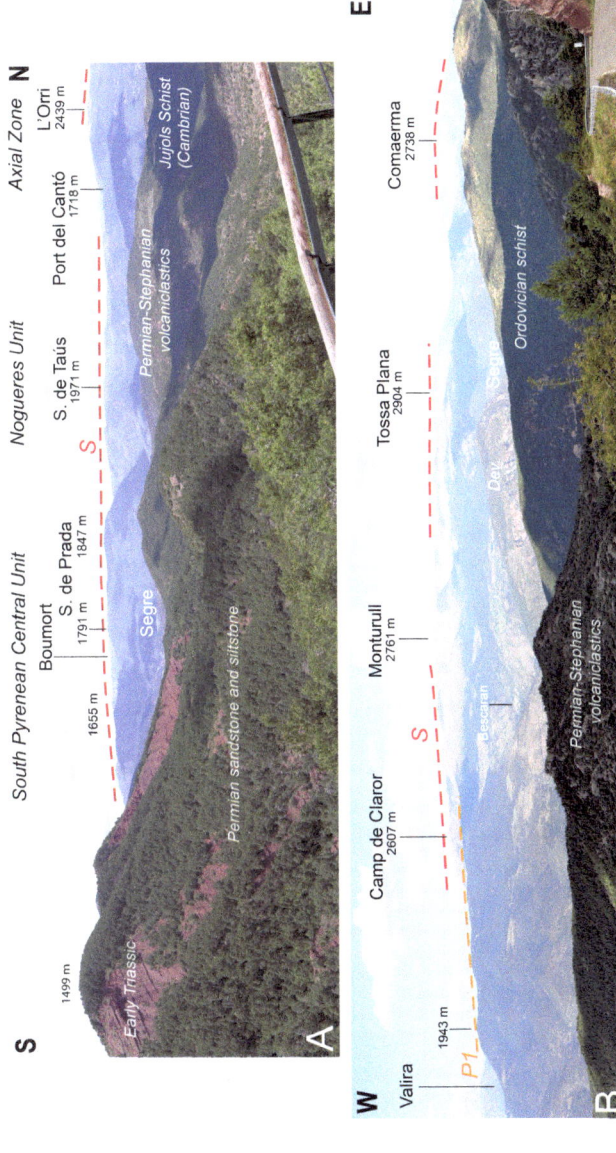

Fig. 14.31 Views from the col at Adraén–Mirador de la Trava. A Looking west (Axial Zone to the South-Pyrenean Central Unit). Foreground: red Permian to Triassic series. Further back: uniform skyline documents a widespread erosion surface truncating most of the underlying alpine geological structures. **B** Looking NE to the eastern Axial Zone, with its widely distributed vestiges of range-top surface, S. The Campcardós massif is hidden by the Comaerma ridge

Fig. 14.32   View from the col at Adraén–Mirador de la Trava. Scenery looking north and down into the Seu Basin

rocks or Devonian limestones and softer outcrops of Visean flysch. This sur-face extends over the Boumort massif as well as the reef limestones of Serra de Prada, thus cutting across these tectonic units of the south-central Pyrenees (Fig. 14.31A).

The city of La Seu d'Urgell (Fig. 14.32) locates a late Neogene half-graben coeval with the Cerdagne Basin further east. Unlike the Cerdagne, however, this basin was largely cleared of its deposits by post-Miocene fluvial erosion. Thus, although its remaining Vallesian deposits have been dated (Chevalier 1909; Hartevelt 1970; Agustí et al. 1981) and are similar to those of Cerdagne, they have been mostly cleared by the Segre and its tributaries. Accordingly, the basin's boundaries are deeply dissected and display no faceted fault scarps of a comparable stature to those encountered in itineraries 7 and 8 (Calvet 1996).

The southern edge of La Seu Basin presents itself as a series of plunging spurs (the road follows one of them), some of which widen out locally into planar landforms at a range of elevation, some as low as 1200–1000 m. These topographic benches could represent pediment generation P2 but, compared to similar landforms around the Cerdagne Basin their morphology (here in early Paleozoic schist) is less clear-cut

(Fig. 14.32). *They certainly extend well above the highest Quaternary alluvial terrace of the Segre, which has been mapped at +190 m above the modern valley floor north of the city. The southernmost boundary fault, which is a westward extension of the Estana-Cava Fault encountered in Itinerary 8, appears locally to lose its scarp relief where pediments occur, particularly along the road to Serrat del Solà. The road crosses the fault line three times between kilometre mark 79 and the Artedó crossroads, and can be located wherever Stephanian andesites come into abrupt contact with the Cambrian to Ordovician schists. The inner boundary fault, which displays exposures of late Miocene sedimentary deposits in the hangingwall, strikes past Cerc. It has been dissected into poorly preserved triangular faceted spurs, and scarp relief nowhere exceeds 80–100 m. You soon enter the city after briefly driving over the late Pleistocene alluvial terraces flooring the basin. The Romanesque Saint Mary's cathedral and cloister, with sculpted granite capitals (twelfth century), is the main cultural highlight of La Seu d'Urgell.*

With a choice of three major destinations out of La Seu, you can travel eastward to Perpignan via itineraries 8, 7 and 6 (150 km); northward to Toulouse via Andorra and the Ariège valley (205 km); or southward to the Ebro Basin, e.g., to Barcelona (172 km) via Solsona, or to Zaragoza (279 km) via Lerida (Lleida).

References

Agustí J, Gibert J, Moya S, Cabrera L (1981) Roedores insectivores del Mioceno superior de La Seu d'Urgell. Acta Geol Hisp 14:362–369

Ardèvol L, Solé X, Escuer J, Martínez A, Puig C, Samsó JM (2020) Mapa geológico España 1:50,000, Memoria explicativa, hoja Sant Llorenç de Morunys (292). IGME, Madrid, 100 p

Ardèvol L, Solé X, Losantos M, Puig C, Escuer J, Casanovas J (2020) Mapa geológico España 1:50,000, Memoria explicativa, sheet Gósol (254). IGME, Madrid, 132 p

Ashauer H (1934) Die östliche Endigung der Pyrenäen. Abh. Ges. Wiss. Gött Math Phys Kl 10:2–115

Astre G (1924) Sur les unités tectoniques des Sierras del Cadi, de Port del Compte et de quelques massifs voisins. C R Acad Sci Paris 178:2106–2109

Astre G (1935) Sur l'âge des orbitolines de Pedra Forca et sur la structure de ce pic. Bull Soc Hist Nat Toulouse 67:115–122

Badiola A, Checa L, Cuesta MA, Quer R, Hooker JJ, Astibia H (2009) The role of new Iberian finds in understanding European Eocene mammalian paleobiogeography. Geol Acta 7:243–258

Barbera X, Cabrera L, Marzo M, Parés JM, Agustí J (2001) A complete terrestrial Oligocene magnetobiostratigraphy from the Ebro Basin, Spain. Earth Planet Sci Lett 187:1–16

Barnolas A, Gil-Peña I (2001) Ejemplos de relleno sedimentario multiepisódico en una cuenca de antepaís fragmentada: la Cuenca Surpirenaica. Bol Geol Min Esp 112:17–38

Barnolas A, Pujalte V, Schmitz B (2019) South Pyrenean Foreland and Basque–Cantabrian Paleogene Basins. In: Quesada C, Oliveira JT (eds) The geology of Iberia: a geodynamic approach, Springer, pp 7–40

Berástegui X, Pí ME, Escuer J, Casanovas J, Samsó JM, Arbués, P., Martínez, A., Vilella, L. (2017) Mapa geológico de la Hoja n° 291 (Oliana). Mapa Geológico de España, 1:50,000 scale, IGME. Memoria explicativa 111 p

Bilotte M (2010) 1912–1928 : L'aventure pyrénéenne de Charles Jacob. Bull Soc Hist Nat Toulouse 146:63–70

Bilotte M, Aragones E (2010) Gaston Astre et la géologie de la région Cadí-Pedraforca. Histoire d'une cartographie inédite réalisée de 1922 à 1927. Treb. Mus. Geol. Barcelona, 17:27–43

Birot P (1937) Recherches sur la morphologie des Pyrénées orientales franco-espagnoles. Baillière Édit., 318 p

Bonilla-Salomon I, Minwer-Barakat R, Vianey-Liaud M, Moya-Sola S (2016) Middle Eocene rodents from Sant Jaume de Frontanya (eastern Pyrenees, northern Spain) and biochronological implications. J Vertebr Paleontol 36(4):e1121149–e1121152

Burbank DW, Puigdefabregas C, Muñoz JA (1992a) The chronology of the Eocene tectonic and stratigraphic development of the eastern Pyrenean foreland basin, northeast Spain. Geol Soc Am Bull 104:1101–1120

Burbank DW, Verges J, Muñoz JA, Bentham P (1992b) Coeval hindward- and forward-imbricating thrusting in the south-central Pyrenees, Spain: timing and rates of shortening and deposition. Geol Soc America Bull 104:3–17

Busquets P, Ramos-Guerrero E, Moyà-Solà S, Agustí J, Colombo F, Checa L, Khöler M (1992) La Formación de Bellmunt (Unidad del Cadí, Pirineo oriental): aportaciones bioestratigraficas de los sistemas lacustres y palustres asociados. Acta Geol Hisp 25:109–116

Calvet M (1996) Morphogenèse d'une montagne méditerranéenne: les Pyrénées orientales. Documents du BRGM, Orléans, 255, 1177 p

Calvet M, Gunnell Y, Laumonier B (2021) Denudation history and palaeogeography of the Pyrenees and their peripheral basins: an 84-million-year geomorphological perspective. Earth-Sci Rev 215:103436

Carrigan JH, Anastasio DJ, Kodama KP, Par JM (2016) Fault-related fold kinematics recorded by terrestrial growth strata, Sant Llorenç de Morunys, Pyrenees Mountains, NE Spain. J Struct Geol 91:161–176

Chevalier M (1909) Note sur la « cuencita » de la Seo de Urgell (Province de Lerida, Espagne). Bull Soc Géol Fr 158–178

Costa E, Garcés M, López-Blanco M, Beamud E, Gómez-Paccard M, Larrasoaña JC (2010) Closing and continentalization of the South Pyrenean foreland Basin (NE Spain): magnetochronological constraints. Basin Res 22:904–917

Costa E, Garcés M, Sáez A, Cabrera L, López-Blanco M (2011) The age of the "Grande Coupure" mammal turnover: New constraints from the Eocene–Oligocene record of the Eastern Ebro Basin (NE Spain). Palaeogeogr Palaeoclim Palaeoecol 301:97–107

Costa E, Garcés M, López-Blanco M, Serra-Kiel J, Bernaola G, Cabrera L, Beamud E (2013) The Bartonian–Priabonian marine record of the eastern South Pyrenean foreland basin (NE Spain): a new calibration of the larger foraminifers and calcareous nannofossil biozonation. Geol Acta, 11, 17–193

Durand-Delga M, Méon H, Masriera A, Ullastre J (1989) Effets d'une phase tectonique compressive, affectant du Miocène supérieur, daté palynologiquement, dans la zone de la Pedraforca (Pyrénées catalanes, Espagne). C R Acad Sci Paris 308(II):1091–1098

Ford M, Williams EA, Artoni A, Vergés J, Hardy S (1997) Progressive evolution of a fault-related fold pair from growth strata geometries, Sant Llorenç de Morunys, SE Pyrenees. J Struct Geol 19:413–441

Garcés M, López-Blanco M, Valero L, Beamud E, Muñoz JA, Oliva-Urcia B, Vinyoles A, Arbués P, Cabello P, Cabrera L (2020) Paleogeographic and sedimentary evolution of the south-Pyrenean foreland basin. Marine Pet Geol 113:104105

Garrido-Megías A (1972) Sobre la colocación del manto de Pedraforca y sus consecuencias: una nueva unidad tectónica independiente, "el manto del Montsec". (Vertiente surpirenáica, provincias de Huesca y Lérida). Bol Geol Min Esp LXXXIII-Ill:242–248

Gómez-Gras D, Roigé M, Fondevilla V, Oms O, Boya S, Remacha E (2016) Provenance constraints on the Tremp Formation paleogeography (southern Pyrenees): Ebro Massif vs Pyrenees Sources. Cretaceous Res 57:414–427

Guérin-Desjardins B, Latreille M (1961) Etude géologique dans les Pyrénées espagnoles entre les Ríos Segre et Llobregat (Provinces de Lérida et Barcelone). Revue de l'Institut Français du Pétrole, 16, 922–940

Hartevelt JJA (1970) Geology of the Upper Segre and Valira valleys, central Pyrenees, Andorra-Spain. Leidse Geol Meded 45:167–236

Jacob C, Fallot P, Astre G, Ciry R (1927) Observations tectoniques sur le versant méridional des Pyrénées centrales et orientales. 14e Congr. Géol. Int., Madrid, 2:335–411 (1926)

Marmi J, Oms O, Vila B, Galobart A, Estrada R, Dinarès-Turell J (eds) (2014) Field trip guide: the Maastrichtian paleoenvironmental record of the Tremp formation (southeastern Pyrenees, Iberian peninsula). In: Reconstructing the terrestrial End-Cretaceous paleoenvironments in Europe (Tremp, September 16–20th, 2014), field trip guide and abstracts book. Paleontologia I Evolució, Memòria Especial 7, pp 1–42

Martínez Rius A, Berástegui l, Losantos M (1996) Corte geológico N-S en el Pirineo oriental: emplazamiento en una secuencia de bloque superior ("fuera de secuencia") de las laminas cabalgantes que forman el manto inferior del Pedraforca. Geogaceta 20:450-453

Martínez A, Vergés J, Muñoz JA (1988) Secuencias de propagación del sistema de cabalgamientos de la terminación oriental del manto del Pedraforca y relación con los conglomerados sinorogénicos. Acta Geol Hisp 23:119–127

Martínez A, Berástegui X, Losantos M, Schöllhorn E (2001) Estructura de los mantos superior e inferior del Pedraforca (Pirineos orientales). Geogaceta 30:183–186

Martínez Rius A, Tudela M, Tallada A, Copons R (2013) Mapa geologic de Catalunya, escala 1:25,000, La Pobla de Lillet, 255–1–2 (71–22). Institut Geologic de Catalunya-Institut Cartografic de Catalunya

Martínez A, Losantos M, Domingo F, Samsó JM, Saula E, Soriano C, Schöllhorn E, Gisbert J, Casas JM, Caus E (2020) Mapa geológico de la Hoja nº 254 (Gósol). Mapa Geológico de España 1:50,000 scale, IGME. Memoir by: Ardèvol L, Solé X, Losantos M, Martínez A, Puig C, Escuer, J Casanovas J, 132 p

Mató E, Saula E, Martínez A, Muñoz JA, Vergés J, Escuer J (1994) Mapa geológico de España 1:50.000 scale, sheet Berga (293), IGME. Accompanying memoire, 66 p

Molina E, Alegret L, Apellaniz E, Bernaola G, Caballero F, Dinarès-Turell J, Hardenbol J, Heilmann-Clausen C, Larrasoaña JC, Luterbacher H, Monechi S, Ortiz S, Etxebarria

XO, Payros A, Pujalte V, Rodríguez-Tovar FJ, Tori F, Tosquella J, Uchman A (2011) The Global Stratotype Section and Point (GSSP) for the base of the Lutetian Stage at the Gorrondatxe section, Spain. Episodes 34:86–108

Muñoz JA (1992) Evolution of a continental collision belt: ECORS-Pyrenees crustal balanced cross-section. In: McClay KR (ed) Thrust tectonics, Chapman and Hall, New York, pp 235–246

Muñoz JA, Martínez A, Vergés J (1986) Thrust sequences in the Eastern Spanish Pyrenees. J Struct Geol 8:399–405

Puig C, Pí E, Samsó JM, Saula E, Martínez A, Vilella L, Caus E (2020) Mapa geológico de la Hoja n° 292 (Sant Llorenç de Morunys). Mapa Geológico de España, 1:50,000 scale, IGME. Memoir by: Ardèvol, Solé, Escuer J, Martínez, Puig C, Samsó JM, 100 p

Puigdefàbregas C, Souquet P (1986) Tecto-sedimentary cycles and depositional sequences of the Mesozoic and Tertiary from the Pyrenees. Tectonophysics 129:173–203

Ramos E, Busquets P, Vergés J (2002) Interplay between longitudinal fluvial and transverse alluvial fan systems and growing thrusts in a piggyback basin (SE Pyrenees). Sed Geol 146:105–131

Riba O (1973) Las discordancias syntectónicas del Alto Cardener (Prepirineo catalán), ensayo de interpretación. Acta Geol Hisp 1–6

Riba O (1976) Syntectonic unconformities of the Alto Cardener, Spanish Pyrenees: a genetic interpretation. Sed Geol 15:213–233

Rosell J, Linares R, Llompart C (2001) El 'Garumniense' prepirenaico. Rev Soc Geol Esp 14:47–56

Séguret M (1972) Étude tectonique des nappes et séries décollées de la partie centrale du versant sud des Pyrénées. PhD thesis, Public. USTELA, Série Géol Struct n°2, Université de Montpellier, 155 p

Solé Sugrañes L (1972) Nota sobre una discordancia en el Eoceno medio del Prepirineo oriental. Acta Geol Hisp VII:1–6

Solé Sugrañes L (1980) Nappe du Pedraforca et autochtone de la Sierra de Cadí. In: 26th International Geological Congres, Paris, Excursion vol 42, G 18–3, pp 168–171

Solé Sugrañes L, Clavell E (1973) Nota sobre la edad y posición tectónica de los conglomerados eocenos de Queralt (Prepirineo oriental, Prov. de Barcelona). Acta Geol Hisp VIII:1–6

Tallada Masquef A, Copons Llorenç R, Casanova i Petanàs J, Martínez Rius A (2017) Mapa geologic de Catalunya, escala 1:25,000, Berga, 293–1–1 (71–23). Institut Geologic de Catalunya-Institut Cartografic de Catalunya

Tosquella J, Samsó JM (1996) Bioestratigrafía y litoestratigrafía del Paleoceno superior-Eoceno inferior del sector oriental de la cuenca surpirenaica. Acta Geol Hisp 31:3–21

Ullastre J, Masriera A (2004) Pedraforca: estratigrafía y estructura (Pirineo Catalán, España). Treb Mus Geol Barcelona 12:11–52

Vandenberghe N, Hilgen FJ, Speijer RP, with contributions by Ogg JG, Gradstein FM, Hammer O (2012) The Paleogene period. In: Gradstein et al. (eds) The Geologic Time Scale, Elsevier, chap. 28:855–921

Vergés J, Martínez A (1988) Corte compensado del Pirineo oriental: Geometría de las cuencas de antepaís y edades de emplazamiento de los mantos de corrimiento. Acta Geol Hisp 23:95–105

Vergés J, Millán H, Roca E, Muñoz JA, Marzo M, Cirés J, Den Bezemer T, Zoetemeijer R, Cloetingh S (1995) Eastern Pyrenees and related foreland basins: pre-, syn- and post-collisional crustal scale cross-sections. Mar Pet Geol 12:893–915

Vergés J, Martinez-Rius A, Domingo F, Muñoz JA, Losantos M, Gisbert J, Fleta J (1994) Mapa geológico de España, 1:50 000 scale, sheet 225 La Pobla de Lillet, IGME, Madrid (Memoir by: Vergés J, Martinez-Rius A, Fleta J, Pujadas J, Tosquella J, Samsó JM, Sanz J, Barberà M, Berástegui X)

Vergés J, Marzo M, Santaeularia T, Serra-Kiel J, Burbank DW, Muñoz JA, Giménez-Montsant J (1998) Quantified vertical motions and tectonic evolution of the SE Pyrenean foreland basin. In: Mascle A, Puigdefabregas C, Luterbacher HR, Fernandez M (eds), Cenozoic Foreland Basins of Western Europe. Geol Soc Spec Publ 134:107–134

Vergés J (1999) Estudi geològic del vessant sud del Pirineu oriental i central. Evolució cinemàtica en 3D, Servei Geològic, Monografia Tècnica, n. 7, 192 p

Vergés J, Fernàndez M, Martínez A (2002) The Pyrenean orogen: pre-, syn-, and post-collisional evolution. In: Rosenbaum, G, Lister, GS (eds), Reconstruction of the evolution of the Alpine-Himalayan Orogen. J Virtual Expl 8:55–74

Williams EA, Ford M, Vergés J, Artoni A (1998) Alluvial gravel sedimentation in a contractional growth fold setting, Sant Llorenç de Morunys, southeastern Pyrenees. Geol Soc Lond Spec Publ 134:69–106